Dedication:
To the memory of my beloved sister, Susan

Statistics Editor: Curt Hinrichs
Senior Editorial Assistant: Jennifer Burger
Production: Greg Hubit Bookworks
Print Buyer: Randy Hurst
Designer: John Edeen
Copy Editor: Carol Reitz
Technical Illustrator: Lotus Art
Cover: Stuart Paterson
Compositor: Beacon Graphics
Printer: Maple-Vail Book Manufacturing Group

This book is printed on
acid-free recycled paper.

Duxbury Press
An Imprint of Wadsworth Publishing Company
A division of Wadsworth, Inc.

1 2 3 4 5 6 7 8 9 10 — 98 97 96 95 94

Library of Congress Cataloging-in-Publication Data

Farnum, Nicholas R.
   Modern statistical quality control and improvement / Nicholas R. Farnum
      p.   cm.
   Includes bibliographical references and index.
   ISBN 0-534-20304-3
      1. Process control—Statistical methods.   2. Quality control–
Statistical methods.   I. Title.
TS156.8.F37   1993
658.5'62'015195—dc20

                                                      93-9510
                                                      CIP

PREFACE

Modern Statistical Quality Control and Improvement is designed to be used as a classroom text and as a reference for practitioners of statistical process control (SPC). The text emphasizes modern quality philosophies and techniques, including implementation/management issues, current SPC software, a decreased emphasis on acceptance sampling, the study of measurement systems, calibration, and an overview of advanced topics such as experimental design, time series methods, analysis of means, and short-run production.

In the academic setting, the text is designed as a one-semester course for business school students at the upper division or MBA level. Students in such courses are normally expected to have a prerequisite of at least one semester of statistics. The text can also be used as a supplementary text for engineering courses, which tend to offer a more mathematical treatment and a slightly different array of topics (e.g., reliability, accelerated life tests, and burn-in). To introduce notation and terminology, we include a review chapter on statistics and probability. In academic courses, such review material can be covered at a quicker pace, but not skipped, since these chapters contain quality-specific subjects that are not usually found in general statistics courses (e.g., confidence intervals for p when *no* nonconforming items are found in a sample, bias-correcting factors for s and R, etc.).

Every effort is made to assure that the material is mathematically correct, but mathematical proofs and derivations are not emphasized. In their place, we substitute in-depth descriptions of the concepts and rationale underlying the various methods. Examples focus on the mechanics of the calculations and the interpretation of the results. For the most part, the data used in the examples either are "real" or refer to a real situation or process.

Control chart calculations should be done "by hand" at least once so that students fully understand how the charts work. After that, when it comes to using these methods in practice, statistical methods are more efficiently handled by computer. To introduce the use of SPC computer software, we use printouts from several packages throughout the text. In particular, *SPC-PCII, SPC1+, STATMAN,* and *NWA Quality Analyst* printouts are used. Beyond this, printouts from the general-purpose package *Minitab* (time series, SPC, and experimental design) are used. The text is not tied to any specific software, but attempts are made to provide samples of the many formats and styles that might be encountered on the job. The computer printouts in no way supplant the development of the necessary formulas. Printouts are used only in examples and serve to add "realism" to what is, after all, a very real field of study.

In the past, most books on statistical quality control have been divided into two parts: half on charting theory/examples and half on acceptance sampling. Many modern quality programs now de-emphasize sampling inspection because it tends to be an "after the fact" procedure, capable of catching gross problems but not otherwise offering the real-time control associated with SPC. Nonetheless, sampling inspection does have valid applications. Also, regardless of whether one views it as good or bad, MIL-STD-105D (newly revised to version 105E) has become a pervasive and sometimes misused sampling system in many industrial settings, especially in defense-oriented businesses where 105D is often required by contract. (Chapter 11 highlights the changes that were made in revising 105D to version 105E.) For these reasons, acceptance sampling occupies a smaller role in this text than it does in many others. Our approach, apart from describing sampling plan concepts, emphasizes the use and misuse of sampling plans. Important points are made (such as the sample size being more of a driving force than the lot size, and the need to use switching rules in MIL-STD-105E).

As it applies to quality assurance, experimental design has recently become an exciting topic. The Taguchi approach, though somewhat controversial, is in large part responsible for the renewed interest in design. *Modern Statistical Quality Control and Improvement* is not a text on experimental design, but the topic is so important that it is discussed at several points throughout the book. The point is made early in Chapter 1 that design techniques have been very successfully used to understand processes and to control sources of variation. In Chapter 10, basic analysis of variance techniques are used to estimate the components of measurement error variation. As a general process improvement technique, the terminology and mechanics of experimental design are discussed in Chapter 13. Each of these topics could easily fill an entire text, so the presentation is necessarily brief but still of great benefit in rounding out the reader's knowledge of statistical methods.

An entire chapter is devoted to measurement systems (see Chapter 10) and related concepts. Important practical concepts such as measurement error, calibration, repeatability, and the role of the National Institute of Standards and Technology (NIST) are discussed. Some instructors may wish to skip this chapter, and doing so will not disrupt the flow of the text. Practitioners, however, eventually need to be familiar with this material, for it is in the production environment that questions about data quality immediately come to the surface. Making real decisions about real production processes requires confidence not only in the statistical methods used but also in the measurements upon which they depend.

Advanced methods and applications of SPC in special situations are presented in Chapter 12. For example, this chapter discusses "low-volume" techniques, deviations-from-nominal charts, and applications in automated manufacturing, especially the use of time series methods for autocorrelated data.

References to current publications dealing with SPC and productivity issues appear throughout the text. Regarding topics mentioned, but not covered in the text, the phrase "the interested reader may consult the text by" occurs many times. Our goal in doing this is to make the text useful beyond its own covers, as a guide to further studies in the field. Also, to increase its usefulness to students and practitioners, we make extensive use of cross-referencing. Knowledgeable readers can ignore these references, whereas others may find them to be welcome time-savers when searching for related material on a given subject.

The effectiveness of any tool depends on *how* it is used, *where* it is used, and even *if* it is used. Presenting the statistical methods and concepts is one thing, but implementing these methods in an ongoing production process is another matter altogether. Certain recommendations concerning management's role, computer software, control of upstream processes, interactions with vendors and customers, implementation plans, and the new role of the QA department are now common to many successful quality programs. These topics are discussed throughout, and Chapter 14 is devoted to them.

The text tries to maintain a professional tone that avoids the needless use of acronyms and that does not embrace one quality philosophy to the exclusion of others. For the most part, acronym use is limited to those that have stood the test of time. Similarly, we attempt to give a fair representation of the differing views espoused by the major contributors to the subject.

A reasonable number of exercises are included at the end of each chapter. The exercises are of two types: some that are purely mechanical (to reinforce the basic calculation/techniques) and some that depict small scenarios/problems and ask the reader to provide the appropriate analysis. A complete solutions manual is available from the publisher to instructors who adopt this book.

ACKNOWLEDGMENTS

Several people have helped this text proceed from proposal to manuscript to printed page. Their collective input has greatly improved the final product. In particular, I would like to thank Victor Lowe of the Statistical Methods Office at Ford Motor Company's Corporate Headquarters for his detailed comments on various versions of the manuscript and for our frequent conversations on statistics and quality. I am also grateful to Eric Ziegel of AMOCO Corporation for his page-by-page comments and contributions to the exercises, and to Richard Counts of the Oak Ridge National Laboratories for his numerous contributions over many years. I also acknowledge my indebtedness to John Schulz, Dave Braun, and many others at LORAL Aerospace. In addition, the helpful comments of the following reviewers are gratefully acknowledged: John W. Adams, Lehigh University; Frank B. Alt, University of Maryland; William F. Fulkerson, Deere and Company; Mark Gershon, Temple University; C. P. Kartha, University of Michigan, Flint; Farzad Mahmoodi, Clarkson University; Victor R. Prybutok, University of North Texas; Dale O. Richards, Brigham Young University; Larry J. Stephens, University of Nebraska at Omaha; David C. Trindade, Advanced Micro Devices; Chamont Wang, Trenton State College; Wayne L. Winston, Indiana University; Steven A. Yourstone, University of New Mexico.

I would also like to thank Alex Kugushev of Wadsworth Publishing Company for his valuable suggestions concerning the organization of the chapters, as well as my editor, Curt Hinrichs, and the staff of Wadsworth for their support and enthusiasm for this project. Greg Hubit deserves special thanks for his truly professional coordination of the copyediting, artwork, typesetting, and myriad production tasks required to translate a manuscript into a book. Finally, I thank my wife, Laura, for her support throughout this and all my projects.

Nicholas R. Farnum

BRIEF CONTENTS

CONTENTS

1

Basic Concepts and Terminology

A study of the interface between statistics and quality improvement requires an understanding of the basic terminology of both fields. This chapter introduces the fundamental terms, definitions, and concepts of quality control along with an overview of the specific statistical techniques used in the pursuit of quality. Subsequent chapters present detailed developments of the statistical methods described in this overview.

CHAPTER OUTLINE

1.1   INTRODUCTION

Whenever one group (producers) strives to create products, services, or works of art, its efforts are evaluated by another group, those who use these creations (consumers). This process of evaluation, of seeing whether something exceeds, meets, or falls short of requirements, is basic to most things we do. 'Quality' is the term often ascribed to the results of these evaluations.

Producers are deeply involved in this evaluation process. In many ways, their interest in quality exceeds that of their customers; beyond meeting consumers' expectations, producers must also be concerned with the economic aspects of quality. Making inferior products can waste labor and material costs and eventually drive down customer loyalty and market share. Conversely, providing superior products or services can reduce internal costs and increase market share.

This text is written for producers and consumers alike. It describes the many statistical techniques that have been developed to help evaluate quality and to use quality as a competitive tool. It also incorporates new attitudes and ideas about quality that have evolved in recent years. In particular, this text addresses issues of process problem solving, customer satisfaction, designing for quality, and process capability, as well as the traditional topics of measurement and process control.

In the study of statistical quality control, one commonly encounters a great deal of jargon, some developed by quality practitioners and some arising from the many statistical techniques used. This, of course, is to be expected whenever the methods of one discipline are applied to those of another. Consequently, readers of this text are likely to fall into one of two groups: those with experience in the quality professions who want to more fully understand the statistical procedures involved, and those with a statistical background who want to see concrete applications of the methods they have studied. Chapter 1 has been written with both groups in mind.

Chapter 1 contains a broad overview of the entire text. Our purpose is to provide the reader with a working vocabulary of the quality control and statistical terms that appear throughout subsequent chapters. Statisticians, for example, will learn that quality professionals usually refer to measured data as either 'variables' or 'attributes' data (not 'continuous' and 'discrete'), while quality professionals will learn that not all data sets are equal and that it is the *method* of collecting data (the sampling design) that is paramount in solving quality problems. This chapter, then, attempts to describe the various methods of statistical quality control without becoming involved in the mechanics of using these methods.

1.2   DEFINITIONS OF QUALITY

The term 'quality' admits many definitions. Nonetheless, one can attempt to list various characteristics that quality products ought to possess. In this section, we present such a list and then describe two *operational* definitions of quality in currrent use: the 'conformance to specifications' and 'Taguchi' definitions. Both approaches make use of terminology that is required later.

DIMENSIONS OF QUALITY

There are several dimensions to quality. Products can be judged by how well they perform, how durable they are, how many options they have, whether they are easy to service, and how aesthetically pleasing they appear. Less tangible quality characteristics include how solidly a car door closes, how comfortable a hand-held camera feels, and how "user-friendly" an appliance or software package is. Creating a product that consumers will eventually judge to be of high quality requires that producers first determine consumers' requirements in these diverse dimensions—from very specific performance requirements to the less specific, but equally important, subjective requirements.

In recent times, attempts have been made to differentiate between these different aspects of quality (Garvin 1987). For example, what follows is one possible list of the components of quality:

- **Performance** (Will the product do the job?) Customers evaluate a product based on whether it will perform specific functions and on how well it performs them. *Example:* Two software packages may both be able to sort through databases, but one may significantly outperform the other with respect to the speed of execution or the size of the files it can handle.

- **Added features** (Does it have features beyond the basic performance characteristics?) Quality is often associated with products that do *more* than expected. *Example:* In addition to an airline's primary function of transporting passengers safely and on time, it may emphasize secondary features such as free beverages or friendly service.

- **Reliability/durability** (Will it last a long time?) We usually want products that not only work but also continue to perform well over a reasonably long period of time. *Example:* In the automobile industry, customers' opinions about quality are heavily influenced by the frequency with which things go wrong on a new car.

- **Conformance** (Was the product made exactly as its design specified?) Products are judged to be of higher quality when they, and each of their constituent parts, meet the requirements placed on them. *Example:* Manufactured parts that are slightly too large or small can cause quality or performance problems when they are used as components in more complex assemblies.

- **Serviceability** (Is it easy to fix?) A product's quality is often linked to how easily and quickly it can be repaired. *Example:* Part of the quality provided by service-sector businesses depends on the speed and courtesy with which mistakes in billing or order placement are corrected.

- **Design** (Does it look better than the competition?) Subjective criteria such as styling and aesthetic appeal can affect one's view of quality. *Example:* A food product may be highly nutritious, but manufacturers have also learned to add artificial colorings to enhance the fresh "look" that consumers expect.

- **Reputation** (Does the company have a history of attention to quality?) Without knowing all the details about a product, customers often rely instead on the company's reputation for quality. *Example:* Customer loyalty varies directly with the number of successful encounters with a company's products.

It is clear from this list that defining "quality" is not as simple as it might first seem. The next section discusses the traditional definition of quality, which addresses only the aspect of conformance. Modern definitions, on the other hand, try to encompass more, if not all, of the aspects of quality mentioned above.

THE TRADITIONAL DEFINITION OF QUALITY

Our concern with quality is certainly not new. Mankind has always had an eye for quality and has taken steps to attain it, but the organized study of quality concepts and methods did not begin until the early 1900s. Prior to that time, quality concerns were primarily confined to craftsmen, guilds, and artists.

The working definitions of quality used in the early 1900s focused almost entirely on the aspect of **conformance**. Quality was then thought of as the ability of a manufactured part to meet (i.e., conform to) the specifications placed on it. If the length of a part was supposed to be 1 inch, to within an allowed tolerance of 0.005 inch, then quality parts were very simply those with lengths between 0.995 inch and 1.005 inches.

TRADITIONAL
DEFINITION OF
QUALITY

> **Quality** is the conformance to specifications.

This definition is certainly clear and easy to apply. With this definition, a company's goal is primarily to stay within product specification limits.

Over the years, some modifications were made to the traditional definition. Juran (1974) expanded the definition to *"quality is fitness for use,"* and the American Society for Quality Control (ASQC 1983) gave the definition *"Quality is the totality of features and characteristics of a product or service that bear on its ability to satisfy given needs."* Although these two definitions sound much less specific than the traditional one, they do encompass a correspondingly wider range of quality characteristics.

TAGUCHI'S DEFINITION OF QUALITY

Until around 1980 the traditional definition of quality as conformance to specifications was the one most commonly used in the United States and Europe. Even expanded versions such as "quality is fitness for use" were not, in practice, interpreted much differently from the traditional view. In 1980, however, Japanese engineer Dr. Genichi Taguchi introduced another definition that offers a different view of quality and the statistical methods we use to attain quality.

Dr. Taguchi's definition emphasizes the total loss that a product imparts to society.

TAGUCHI'S DEFINITION OF QUALITY	**Quality** is the loss imparted to society from the time the product is shipped.

This definition represents a significant change in the way of thinking about quality. Instead of trying to define all the good things that a quality product should be, Taguchi emphasizes instead the 'losses' that a product can create. That is, when a product fails to perform correctly, when it breaks down, or when some of its parts do not conform to specifications, it creates undesirable costs to society ('society' includes the manufacturer as well as the customer). These costs can be measured in the form of energy and time (to fix the problem), in monetary form (e.g., when paying for replacement parts), as loss of goodwill by the customer, or as lost market share.

With Taguchi's definition, a company's quality goal becomes to try to minimize any possible loss to society. Although it may sound a little magnanimous, the practical import of focusing on 'loss to society' is that it draws attention to all the various problem areas that may eventually drive customers away. Such a goal also incorporates the aspects of quality outlined in the previous section. For example, it means that one should try to make products reliable, since unreliable products will impart repair costs to both customers and manufacturers (e.g., warranty repairs).

Although there can be little confusion when applying the traditional definition of quality, applying Taguchi's definition requires additional explanation. In particular, Taguchi provides a simple mechanism, the **loss function,** for translating the somewhat abstract concept of 'loss to society' into an operational concept that is as easy to apply as the traditional definition. Loss functions assign measurable penalties that are proportional to the distance a quality characteristic is away from its desired target value. (Loss functions are described in detail in the next section.) The farther a characteristic is from its target, the larger the numerical 'loss' incurred. Using the loss function results in completely new quality goals. For example, when compared to the traditional definition in regard to the quality of conformance, the loss function approach implies that merely meeting specifications is not sufficient. No longer are parts that are specified to have a length of 1 inch with an allowed tolerance of 0.005 inch considered to be good as long as their lengths are between 0.995 and 1.005 inches. Instead, the loss function indicates that the parts with lengths closer to the target of 1 inch are better (i.e, they impart less loss to society) than those whose lengths are farther from 1 inch, *even though the latter parts may have lengths between the specification limits of 0.995 and 1.005 inches.*

Taken to its natural conclusion, Taguchi's definition implies that the goal is to make all the parts exactly the same, each with a length equal to the target value and with no variation in lengths from part to part. At first sight, this goal may seem to be impossible, unrealistic, and very costly to achieve. In practice, however, not only is such a goal realistic, but the costs associated with achieving it can be relatively small. (More is said about this in Section 2.6.)

1.3 BASIC QUALITY TERMINOLOGY

As the field of quality control has grown, so has its associated collection of terminology, jargon, and definitions. This section gives a brief, and admittedly nonexhaustive, introduction to terminology that is basic to quality practitioners and to the study of quality control methods.

SPECIFICATION LIMITS

Products and their subcomponents are designed to perform certain tasks. When designs are translated into actual products (i.e., creating the products), it becomes necessary for someone to precisely define the various **performance characteristics** of the product and each of its subcomponents. For manufactured products this usually means specifying the desired physical measurements that the components should have and the desired quality characteristics that the final product should have. A car door, for example, can be neither too wide (or it may not close properly) nor too narrow (or it may fail to latch solidly). Specifications for service industry products, on the other hand, are often in the form of maximum acceptable waiting times for service, strict rules for processing bills and orders, or guidelines for handling transaction errors. If, in addition, there is a single measured value that corresponds to the most desired quality level, we refer to this value as the **nominal**, or **target**, value of the characteristic.

Who decides what the specifications will be? Traditionally, in manufacturing, engineering departments have almost exclusively set specifications and requirements, while the production department as well as the customer have been left out of the design process. The usual explanation for doing this is that customers could not possibly understand the many technical requirements and measurements involved. The customer's input traditionally came in the marketplace when a purchase decision was made. This long-standing practice is now changing. Companies are beginning to realize that it makes sense to involve not only the manufacturing department but also the final purchaser when setting product specifications.

Section 2.5 discusses new methods used to incorporate customer requirements and manufacturing knowledge into the design process.

In the service sector, on the other hand, it is hard *not* to have the customer involved in setting the specifications. Unlike in manufacturing, where customers usually have little contact with engineering and manufacturing staff, purchasers of service products routinely call the various departments of a service organization to ask questions, file complaints, or select different service options. In some cases, it is thought that the specifications concerning the *way* a service is delivered are at least as important as the product specifications themselves (King 1987).

Having established a desired level of performance, one must provide for some flexibility in achieving that level. This is done by setting **specification limits** on the performance characteristic. These limits delineate the range of measurements that one will accept as "close enough" to the desired level and that one believes will not cause serious deterioration in performance.

SPECIFICATION LIMITS	The largest allowable value that a quality characteristic can have is called its **upper specification limit** (USL); the lowest allowable value is called the **lower specification limit** (LSL).

In discussing the traditional definition of quality, we used the example of parts that were to be 1 inch long with a tolerance of 0.005 inch. In standard quality jargon, these parts are said to have a nominal length of 1 inch, an upper specification limit of 1.005 inches, and a lower specification limit of 0.995 inch. Quite often, the long phrase "specification limits" is abbreviated to simply "the specs," a practice we sometimes use in this text. For instance, in our example, part lengths could be said to have specs of 1″ ±0.005″.

Some characteristics (such as the breaking strength of a material) have only lower specification limits, others have only upper limits (like the time required to process a transaction). In these cases, the characteristic is said to have a **one-sided tolerance.** When both specification limits exist, the characteristic is said to have a **two-sided tolerance.**

One final bit of terminology related to specification limits concerns what happens when these limits are used in creating the product. In manufacturing, for example, if the process used to make certain parts never (or, at most, rarely) results in parts that fall outside of their specification limits, the process is said to be capable of meeting the specs and is then called a **capable process.** This concept is explored more fully in Chapter 8.

CONFORMANCE VERSUS NONCONFORMANCE

For many years, the quality of finished products and parts was classified as simply "good" or "bad." The terminology most often used was that some parts were **defective** whereas others were **nondefective.**

Recently, more careful distinctions have been made for classification schemes. These distinctions arise from the simple truth that not all problems have the same degree of seriousness. Some problems are catastrophic and prevent the part from ever being used, whereas others may be more cosmetic or perhaps do not cause serious functional difficulties. These distinctions have given rise to alternative terminology for discriminating between the various defect levels.

In quality control, products that do not meet their specifications are called **nonconforming** (Freund 1985), and the problems or flaws in nonconforming items are then called **nonconformities**. This does *not* necessarily mean that the product is unfit for any use. For example, the measured concentration of a certain chemical cleaning solution may be slightly lower than its lower specification limit dictates, but it may still perform many of its intended cleaning tasks (except that more of the solution might be needed). On the other hand, a metal bolt made of inferior material and intended for use in an airplane safety system is clearly not in the same class.

NONCONFORMING PRODUCT

A **nonconforming product** is one that has one or more nonconformities that cause it, or an associated product or service, not to meet a specification requirement.

Defective products, by comparison, are considered to be more serious. **Defectives**, which can contain one or more **defects**, represent a departure from a quality standard that is serious enough to interfere with intended usage requirements.

DEFECTIVE PRODUCT

A **defective product** is one that has one or more defects that cause it, or an associated product or service, not to satisfy intended usage requirements.

Defects can then be further classified as to their degree of seriousness. This practice is especially predominant in many acceptance sampling plans (see Chapter 11). For example, in the MIL-STD-105E acceptance sampling system, defects are classified as follows:

Critical Defect	A defect that is likely to result in hazardous or unsafe conditions, or that may prevent the tactical function of a military system or unit of which it is a part.
Major Defect	A defect, other than critical, that is likely to result in failure, or that may reduce materially the usability of a unit for an intended purpose.
Minor Defect	A defect that is not likely to reduce the usability of a unit for its intended purpose.

Deciding what constitutes a nonconforming or a defective product is not as simple as it first appears. In fact, W. Edwards Deming has pointed out that without **operational definitions,** these decisions are either impossible or meaningless (Gitlow and Hertz 1983). To clarify, Deming observes that one must have a working (i.e., operational) definition of the quality characteristics being examined; otherwise, two different people may easily come up with two different methods for measuring and classifying the defective items. Merely specifying that a solution should contain 10% of a certain chemical, for example, is not sufficient. It is unclear whether this means 10% by weight or 10% by volume. To one person it may mean that checking the percentage in each bottle is required, whereas another may check only the percentage in each large batch of the solution; there may even be more than one analytical procedure available for measuring the percentages.

The following list contains the steps necessary for creating operational definitions:

CREATING AN
OPERATIONAL
DEFINITION

1. Specify the exact criterion to be applied to the quality characteristic of some product.
2. Specify how the product or products will be selected.
3. Specify the operation to be performed (e.g., how a measurement or observation is to be performed).
4. Record the results and decide whether or not the condition in step 1 is met.

The article by Gitlow and Hertz (1983) contains several examples of how operational definitions are created.

In the past, little, if any, distinction was made between the terms 'defective' and 'nonconforming,' with 'nonconforming' being the more recently introduced. The tendency still exists in practice and in the literature to use these terms interchangeably, or to simply use the word 'defective' alone. The reader should therefore be cautioned to carefully consider the intended meaning of any occurrence of the word 'defective,' especially in older publications.

THE LOSS FUNCTION

The traditional definition of quality described in Section 1.2 leads to the simple interpretation that characteristics whose measurements exceed the specs (i.e., that are either smaller than the LSL or larger than the USL) are nonconforming. Otherwise they are classified as conforming. Figure 1.1 shows a graphical depiction of this situation. Since the objective seems to be simply getting the measurements into the acceptable region, some authors (Ross 1988) have called this the "goalpost" philosophy. With the LSL and USL viewed as football goalposts, only the measurements that pass between the goalposts are counted as successes, with no further distinction made between successful measurements. Values close to the center of the goalposts (the nominal value) are just as good as ones near, but not outside, one of the specification limits.

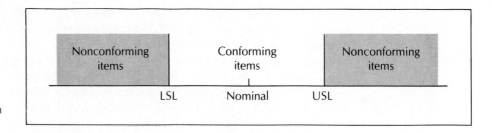

FIGURE 1.1
Traditional Definition
of Quality

As previously mentioned, Taguchi's definition of quality emphasizes that there are, in fact, differing *degrees* of goodness or badness that the goalpost philosophy does not consider.

EXAMPLE 1.1   In a manufacturing process that produces threaded metal bolts, one important quality characteristic is the shaft diameter of the bolts. Since the bolts must fit into corresponding holes on another assembly, if a bolt's shaft diameter is too small, the threads will fail to engage and the bolt will not hold firmly. Shaft diameters that are too large, on the other hand, result in bolts that will not even fit into the holes. Suppose that the specifications for the shaft diameters (in inches) are 0.5 ±0.005. Under the traditional definition, a bolt with a diameter of 0.504 inch would be classified as conforming, whereas one with a diameter of 0.506 inch would be nonconforming. Taguchi's definition, however, could indicate that these two bolts are fairly similar, with neither being especially desirable. The 0.506-inch bolt may indeed not fit at all, but the 0.504-inch bolt may just barely fit into the hole and could possibly bind or stick. More important, if some of the hole sizes tend to be close to their lower spec, then even the 0.504-inch bolt may not fit.

Taguchi quantifies these differences in quality by using a simple quadratic function that compares the measured value to the nominal value. With an upper-case Y used to denote the quality characteristic of interest (e.g., Y could be the bolt diameter from the previous example) and a lowercase y used to represent a particular measured value of Y (e.g., y could be the shaft diameter, in inches, of one particular bolt), the loss function can be written as

$$l(y) = \text{loss associated with the particular value } y = k(y - T)^2 \qquad (1.1)$$

where T is the target, or nominal, value of that characteristic. In essence, equation (1.1) says that the loss is proportional to the square of the deviation of the measured value from the target value. The loss, $l(y)$, is measured in monetary units, and the proportionality constant, k, is determined as in the next example. (See Appendix 12 for an explanation of how the form of the loss function is derived.)

EXAMPLE 1.2   In Example 1.1, suppose that the bolts that are too large can be reworked to reduce their diameters at an average cost of $0.20 per bolt. Substituting the value y = USL (the value of y at which bolts become "too large") into equation (1.1) yields

$$\$0.20 = l(\text{USL}) = k(\text{USL} - T)^2 = k(0.505 - 0.500)^2$$

which can be solved for k:

$$k = \frac{0.20}{(0.505 - 0.500)^2} = 8,000$$

[To be dimensionally consistent with the rest of equation (1.1), k must be measured in units of dollars/inch^2.] Thus, the loss function is $l(y) = 8,000 \cdot (y - T)^2$. This equation can then be used to calculate the loss associated with any other measured value of Y. For example, a bolt with diameter y = 0.502 inch would induce a loss of $l(0.502) = 8,000 \cdot (0.502 - 0.500)^2 = \$0.032$, or about 3 cents. Similarly, bolt diameters close to one of the specification limits, such as diameters of 0.504 and 0.506, would have loss functions of $l(0.504) = \$0.128$ and $l(0.506) = \$0.288$.

In Figure 1.2, the loss function of Example 1.2 has been plotted over the specification limits of Figure 1.1. It is apparent from the figure and from equation (1.1) that the loss function is symmetric; that is, the loss is the same regardless of the direction that y deviates from the target, T. In many situations, this may not be appropriate and a nonsymmetric loss function would describe the situation better.

EXAMPLE 1.3

In the previous example, let us further suppose that parts smaller than the lower spec of LSL = 0.495 inch cannot be reworked and must be scrapped at a cost of $2.50 per bolt (which represents the material and labor costs involved in producing such bolts). Then the following loss function could be used to describe the situation:

$$l(y) = \begin{cases} k_1(y - T)^2 & \text{when } y \leq T \\ k_2(y - T)^2 & \text{when } y \geq T \end{cases}$$

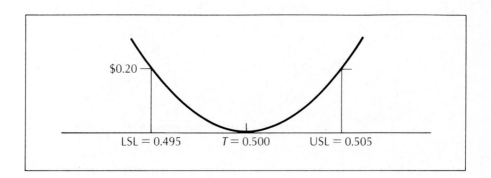

FIGURE 1.2
Loss Function for
Example 1.2

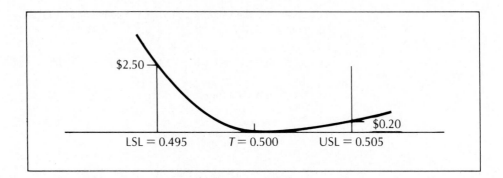

FIGURE 1.3
Loss Function for
Example 1.3

The constants k_1 and k_2 are calculated exactly as in Example 1.2 by substituting $y =$ USL and $y =$ LSL into $l(y)$ and solving:

$$k_1 = \frac{l(\text{LSL})}{(\text{LSL} - T)^2} = \frac{2.50}{(0.005)^2} = 100{,}000 \ (\$/\text{in.}^2)$$

$$k_2 = \frac{l(\text{USL})}{(\text{USL} - T)^2} = \frac{0.20}{(0.005)^2} = 8{,}000 \ (\$/\text{in.}^2)$$

and the resulting loss function would appear as in Figure 1.3.

It is also possible to have one-sided loss functions, which correspond to characteristics with only one specification limit (i.e., a unilateral specification). A good example of such a loss function is given by Barker (1986) concerning the measurement of the strength of a plastic valve used in a lawn mower carburetor. Obviously, there must be a lower specification limit on a characteristic such as strength, but there need be no upper spec. The loss function for the strength of the plastic part would then appear as in Figure 1.4.

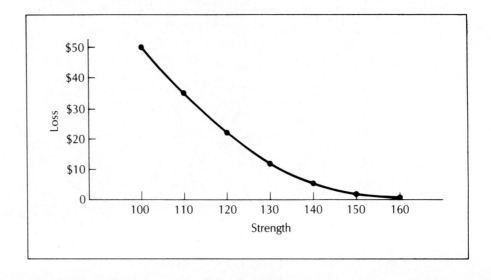

FIGURE 1.4
Loss Function
for Strength of
Plastic Valve

The loss function has more than one use. As seen in the examples, it allows managers to quantify the concept of "loss to society" and apply it directly to their particular products to give a measure of how well they are doing. Later, in a discussion of experimental design (see Chapter 13), it is seen that the loss function can also be used to define performance measures that allow one to optimize a product or process.

TESTING AND INSPECTION

Inspection and testing have always been part of quality control, long before quality methods were organized into a separate field of study. When the first attempts to study quality issues were made at Bell Telephone Laboratories in 1907, one of the first quality-related departments created was dedicated to inspection and testing.

Inspection and testing are done for the purpose of evaluating materials, quality characteristics, parts, products, and processes for conformance to requirements. They are also done to classify items, diagnose problems, and make adjustments. Some authors use the terms 'inspection' and 'testing' interchangeably, but **inspection** is most often used to describe the evaluation process, whereas **testing** refers to specific methods or instruments used for achieving these evaluations.

Inspection can occur just about anywhere in a production process, from the point at which raw materials are received to the inspection of finished goods. Prior to receiving raw materials, some companies go one step further by sending inspectors to evaluate and approve the product designs and production processes used by their suppliers. Table 1.1 summarizes the most commonly used types of inspection according to the point in the production process at which they are used.

There are other ways to classify inspection. **Visual inspection,** as its name implies, refers to the use of human inspectors to evaluate conformance or nonconformance. With recent developments in camera technology, visual inspection also includes some forms of automated inspection using a television. Until very recently, for example, the inspection of the quality of soldered connections on printed circuit boards was based solely on the judgment of inspectors trained in solder technology. Some companies are now beginning to use instruments based on vision (camera) technology or X-ray technology to inspect solder connections. Relying on instruments for inspection is often referred to as **automatic inspection.** In addition, inspection is sometimes done in order to detect and fix problems before a product is delivered. In this case, **rectifying inspection,** or **screening**, is the term used.

TABLE 1.1 Inspection at Different Points in the Production Process

First Article Inspection	The evaluation of product prototypes or designs prior to approval of full-scale production
Receiving Inspection	The examination of incoming parts and other raw materials
In-Process Inspection	Inspection that occurs at any point in the production process prior to final inspection and after receiving inspection
Final Inspection	Evaluation of finished products or materials prior to shipment to customers or distribution channels

---

The amount of inspection done can vary. Consider a group of 1,000 finished parts that are to be inspected prior to shipment to the customer. Inspecting each of the parts once is called **100% inspection.** If the inspection process is unreliable, each of the parts may also be subjected to a second inspection (perhaps by a different inspector). In that case, one is doing 200% inspection. If five people examine the same part, then the part is said to have undergone a 500% inspection, and so forth. Examining every item can be time-consuming or unnecessary, so, instead, companies often base their evaluations on a small sample of items. Examining a representative sample of 50 parts from the entire batch of 1,000 parts is called **sampling inspection** (see Chapter 11 on acceptance sampling).

To most practitioners, **testing** refers to the use of instruments and other techniques for measurement and, ultimately, evaluation. Testing is done for the following reasons:

- To determine the conformance or nonconformance of a part, product, or characteristic

- To find and fix potential problems

- To gain more information about the behavior of a new part or process

- To make adjustments to settings or characteristics

- To sort products into different categories of usage

As with inspection, there are various types of tests and test procedures. Testing the breaking strength of metal, for example, necessarily requires destroying the item being tested. Such testing is called **destructive**, whereas testing that leaves a product intact is called **nondestructive testing.** Determining whether an appliance meets certain federal safety standards is called **safety** or **compliance testing.** Finally, much of today's testing is accomplished with minimal human intervention by means of **automatic test equipment** (ATE).

1.4   STATISTICAL TERMINOLOGY

The terminology introduced in previous sections covers many areas of quality practice. With but a few exceptions, the terms introduced are commonly known and used by quality practitioners everywhere. This is not necessarily the case for the *statistical* terminology connected with quality assurance, although there has been a marked increase in the use of statistical methods during recent years.

This section introduces the reader to statistical quality control *without actually using any statistics*. It provides a cursory look at most of the methods covered in this text. In keeping with the theme of Chapter 1, this section is intended to convey a sense of the *purpose* and philosophy underlying these methods, not the mechanics. The detailed calculations and procedures are reserved for later chapters (see also Section 2.5).

PROCESS AVERAGE AND PROCESS VARIATION

As is emphasized again in Section 2.4, in the modern approach to quality, each step in a product's manufacture or each step in a service procedure is viewed as a separate **process** to be performed. Quality efforts can then be directed toward controlling each process step so that the overall procedure performs as well as possible. Efforts can also be directed toward making improvements in the production system itself. For the moment, let us consider only the first goal.

It seems to be a basic property of the physical world that things rarely happen exactly the same way twice. When one is trying to produce threaded bolts with a 0.500-inch diameter, even though one will generally make a lot of bolts with diameters *close* to 0.500 inch, their diameters are, nonetheless, measurably different from one another. It is easy to understand how this happens: different batches of raw metal stock may be involved, the machine operators may change one or more times during the production run, there may be more than one machine used to produce the bolts, the machine(s) may be experiencing some wear as more and more bolts are made, and so on. Indeed, it would be hard to imagine a process that did not experience some variation in its output. Burr (1984) has summarized this concept by saying, "If anyone says that he does not have variation, then he is simply not measuring closely enough."

The natural variation inherent in any repetitive process is called **process variation,** and statistical methods are the tools used to describe this variation and to help make decisions that either reduce or take account of such variability. If it turns out that the successive process results tend to stay in the same general region and don't wander uncontrollably, then we call the center of such a region the **process average.** For a production process that is stable enough to have a process average, it is further desired that this average be close to the target value of the characteristic.

ATTRIBUTES AND VARIABLES DATA

Statistical methods are designed to simplify, describe, and interpret *data*. In applying statistical techniques to quality, a good deal of attention must be paid to how the requisite data are gathered and how they are measured, topics covered at length in Chapter 3. Beyond these important considerations, there is also the question of how one classifies the types of data collected.

This distinction is fairly easy to make since, in practice, there are only two ways that one measures things: either by counting or by reference to a graduated continuous scale. In the statistical community, measurements that arise by counting are called **discrete**, and those that result from using a reference scale *other* than the counting numbers are called **continuous** measurements. An easy way to decide whether a particular measurement is discrete or continuous is to use the following criterion: if it is *conceivably* possible to improve the accuracy of a measurement by using successively more sensitive measuring instruments, then the characteristic being measured is a continuous quantity; otherwise, it is discrete (Snedecor and Cochran 1980). With this operational definition, one can easily classify characterristics such as time, weight, volume, length, pressure, and concentration as continuous, whereas such things as number of defects and number of nonconforming items are discrete.

Admittedly, it is true that measurements of continuous quantities are usually rounded off (because of the inherent limitations of measuring instruments), which gives rise to data that can appear to be discrete. In such instances, however, we still consider the underlying *characteristic* to be continuous and think of the data as continuous.

Quality practitioners use different terms to describe discrete and continuous measurements. In the field of quality control, the term **attributes data** is used to refer to discrete measurements (especially *counted* data), whereas **variables data** is reserved for continuous measurements. These two terms are extensively used to distinguish the various types of control charts and sampling plans of quality control. For example, $\bar{x}$ and R charts (see Chapter 7) are examples of variables control charts, the u chart (Chapter 9) is an attributes control chart, MIL-STD-105E (Chapter 11) is an attributes sampling plan, and MIL-STD-414 is a variables sampling plan. The phrases sound a little strange at first (especially the plural form of the terms), but seem to become more normal with use.

GRAPHICAL METHODS

When recounting the development of statistical quality control methods in Japan, Dr. Kaoru Ishikawa (1989) pointed out that, initially, there was a period of overemphasis in applying sampling inspection and Shewhart's control chart methodology. As Ishikawa said, "We overeducated people by giving them sophisticated methods where, at that stage, simple methods would have sufficed." To address these more basic needs, attention was then shifted toward using some simple descriptive graphs and problem-solving methods.

The Japanese experience reaffirms what statisticians have maintained for a long time, that well-constructed graphs based on reliable data can be powerful tools for decision making. Consequently, graphical techniques are emphasized more in modern training programs than in traditional programs based exclusively on control charts and acceptance sampling.

Creating a visual display of data serves a number of purposes. For one thing, graphing data can produce a clear summary of a situation so that problems can be quickly pinpointed without having to sort through pages of records. Where language and other differences exist, graphs also are a common medium for communicating ideas. Some graphs are useful for sorting out which process characteristics are the critical ones, while other graphs help to focus problem-solving efforts. Last, graphs tend to be impartial, allowing people to make decisions more on the basis of factual evidence and less on the basis of their opinions or hunches.

Of the many graphical techniques available today, five are widely used in quality control: **histograms** (and to some extent stem-and-leaf displays), **Pareto charts, fishbone diagrams** (also called 'cause-and-effect' diagrams), **location diagrams,** and **scatter plots.** These simple graphs are usually used in the early stages of problem solving, before control charting or more complicated procedures are used. Examples of these and other graphical methods are given in Chapter 4.

CONTROL CHARTS

The recognition that variation is unavoidable in every repetitive process is not new. It was well understood by the pioneers of statistical quality control. To handle process variation statistically, W. A. Shewhart introduced the control chart method in 1924 for the purpose of understanding and separating the *sources* of variation.

Shewhart envisioned two types of variation that, when combined, would account for all variation in a process. The first type, **common cause variation,** is the result of the myriad imperceptible changes that occur in the everyday operation of a process. This includes such things as small voltage surges when a machine is operated, the effect of slight changes in temperature or humidity on the shape of a plastic part, and an unnoticed keypunch error. Together, these small disturbances create natural, uncontrollable variation in a process. **Controllable variation,** on the other hand, is variation for which one can find definite causes. Changes in the brand of raw material used, turnover in the workforce, and installation of new machines or the wearing down of old machines are all possible causes of additional variation in a system. Thus, Shewhart thought of any process as consisting of systems of **common causes** and **special causes,** and it was the special causes that control charts were designed to detect.

It is easy to argue philosophically about whether a cause is special or common. One could probably make plausible arguments to show that any cause, under the right circumstances, could be considered either special or common. If we installed an instrument to monitor voltage surges, then voltage surges could viably be considered a source of special causes and steps could be taken to eliminate this problem (e.g., by installing a surge protector). Operationally, the distinction between special and common causes comes down to a question of data and **traceability**. To identify special causes, one needs to have the ability to trace back through the data to pinpoint when changes occurred that might have altered the process.

Control charts are constructed by regularly taking small samples from the output of a process, making some measurements on the sampled items, and then plotting *summary measures* of the results. Figure 1.5 shows a typical control chart. The sample measurements are used to calculate and plot the two **control limit** lines shown on the chart. The region between the control limits is associated with the natural (common cause) variation in the process. If a summary measures falls outside one of the control limits, it is taken as a signal of a special cause. To make effective use of this chart, one must be able to quickly trace back through other process records to determine the likely *cause* of the signal.

Many practitioners and authors have made the observation that a control chart resembles a device that monitors one's heartbeat or pulse (Nelson 1988). The analogy is a good one. The chart is intended to be an ongoing monitor of signals that come from a process, and it offers a method for deciding when a signal may

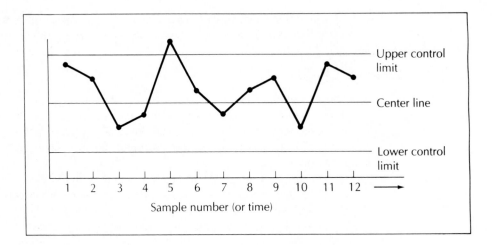

FIGURE 1.5
A Typical
Control Chart

be evidence of problems. Solving these problems, however, requires further exploration of the process.

EXPERIMENTAL DESIGN

Experimental design is a statistical technique for efficiently and effectively investigating research hypotheses. Called **design of experiments** (DOE) by quality practitioners, this technique consists of a set of specific requirements for data acquisition and special methods for analyzing the resulting data. Experimental design is used in quality control for the purpose of accurately sorting out the effects of the many different factors that influence a process or product. It can be used in both the product design phase and the production phase itself.

The origins of experimental design date back to 1922–1923 with Fisher's analyses of agricultural data (Kirk 1982). The subsequent development of this topic has made it a cornerstone of modern statistical practice.

To illustrate the DOE approach, consider the problem of trying to reduce the numbers of defective items produced in a certain manufacturing process. Suppose, for purposes of simplicity, that only two factors could influence the defect rate: the type of raw material used and the speed setting of the machine being used. Then, raw material and speed are **factors** to be investigated, and the defect rate is the **response** variable that these factors may affect. Most DOE techniques require the user to select some **levels** of these factors that are of particular interest; that is, one must choose some factor settings. Suppose, again for simplicity, that one is interested in comparing two particular materials (material A and material B) and two running speeds for the machine (speed 1 and speed 2). The experiment would then involve examining the effects of these two factors, each at two levels, on the response variable, as summarized in Figure 1.6. A critical feature of the experimental design is that responses are measured at *each combination* of the factor levels.

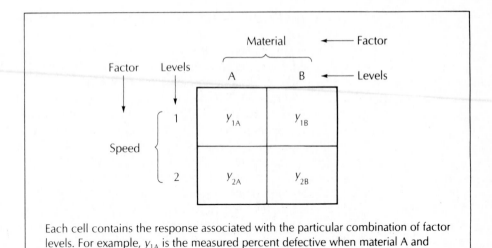

Each cell contains the response associated with the particular combination of factor levels. For example, y_{1A} is the measured percent defective when material A and machine speed 1 are used.

FIGURE 1.6
Components of an
Experimental Design

For more examples of experimental design concepts, we refer the interested reader to Box's (1987) excellent overview of the subject. The mechanics of experimental design are introduced in Chapter 13.

One of the most important features of an experimental design is the attention paid to how the data are to be collected. The realization that not all data carry the same amount of information can come as a surprise to those unfamiliar with statistical tools. In fact, the way the data are gathered is often more critical than the subsequent statistical analyses of the data. As Hahn (1984) points out, the important results to be derived from a well-designed experiment are usually easily obtained from simple graphical analyses of the data. The opposite, however, is also true: data from a badly designed study can be almost useless. Appreciating what can and cannot be concluded from different data collection schemes is fundamental to all experimental design programs.

Another important aspect of DOE techniques is that they represent a very *active* approach to gathering data. The experimenter must make special efforts, as in the example above, to systematically change key process or product variables when taking measurements. These types of data are not normally available from daily log sheets, inspection reports, or other company records. Using company records or any other easily obtainable data represents a *passive* approach to gathering data, and such data cannot usually be expected to yield information required for DOE analyses.

Much work has been done in the field of experimental design since its inception in the early part of this century. Experimental methods for quality control, especially for industrial applications, have developed at the same time. The renewed interest in this subject among quality professionals has been spurred by the notable successes of the Japanese in applying DOE, especially in the area of product design.

In the United States, Box and Wilson (1951) introduced the **response surface** technique for selecting process parameters that produce optimal process improvements. Later, to facilitate the application of response surface methods, Box and Draper (1969) introduced the **EVOP** (evolutionary operation) technique, which allows workers to conduct designed experiments without stopping the production process and without causing the manufacture of defectives. As its name implies, EVOP allows one to slowly change process parameters in an effort to improve a manufacturing process. In England, Davies (1954, 1957) compiled reference texts on the application of DOE methods in manufacturing.

In about 1980, the Taguchi method of experimental design was introduced in the United States. According to Taguchi (1988), development of these methods began in Japan in around 1948 and they were refined over the next three decades. While his approach was based on well-known DOE constructs such as factorial designs and orthogonal arrays, Taguchi had added some completely new ideas and methods, such as the concepts of loss function, signal-to-noise ratio, parameter and tolerance design, and robust design. Although the theoretical basis for the Taguchi methods is still being evaluated, in practice these methods have already been adopted by many U.S. companies. Regardless of the direction that theoretical questions about method may take, there can be no question that Taguchi has been responsible for creating a much greater interest in experimental design throughout U.S. industry than has ever existed before. Chapter 13 describes the elements of the Taguchi method of experimental design.

ACCEPTANCE SAMPLING

In any process there is a need to evaluate the acceptability of the products created or the workmanship involved. This can be accomplished in several ways: by using some type of inspection or testing, by allowing responsible parties to 'stamp' or 'sign off' (i.e., accept responsibility for) finished items, or by using control chart information to ascertain quality levels. This section describes a particular type of inspection called **sampling inspection** or **acceptance sampling.**

Quite often, products are grouped together into batches (called **lots**) before they are shipped to customers, and lot numbers are assigned to each such group of items to facilitate tracing. Then, if any questions or problems arise concerning the product, the lot number can be used to trace these problems back to the particular factors involved when the lot was made. The use of lot numbering is a very old practice, extending back at least as far as fabric marking in medieval Europe (Lerner 1970).

Lots (or batches) are not confined to manufacturing. In service industries, lots take the form of accumulated batches of documents, such as paperwork concerning weekly orders, invoices, checks, or sales slips. For a discussion of acceptance sampling in the field of accounting, the reader is referred to the book by Vance and Neter (1956).

In acceptance sampling, lots are evaluated based on relatively small samples drawn in a representative fashion from the lots. This inspection most frequently occurs in two places. If the producer inspects the lots before sending them, it is called **outgoing inspection.** If the customer inspects the lots before accepting them, it is called **incoming inspection,** or receiving inspection. Figure 1.7 depicts how lots are screened in the two types of inspection.

In either case, the inspection proceeds by taking a sample from the lot and either measuring a particular characteristic of the product or simply counting the number of sampled items that do not meet specifications. When measured (i.e., variables) data are used, a lot is rejected whenever the average of the measurements exceeds some specified limit. For attributes data, lots are rejected when the number (or percent) of defectives in the sample is too large. Lots that pass the inspection are then shipped (if the manufacturer does the inspection) or put into stock (if the customer does the inspection). Because of their relative simplicity, acceptance sampling plans for attributes data have been used more frequently than those for variables data.

Depending on the industry, producers and consumers are very specific about how well a sampling plan is expected to perform. Sometimes they agree on a particular fraction defective, called the **acceptable quality level** (AQL), such that any lot whose fraction defective doesn't exceed the AQL is considered acceptable by the customer. Such sampling plans are designed to have a high probability of accepting lots whose fraction defective is less than or equal to the AQL and a high probability of rejecting the rest of the lots. A simple graph (called the **operating characteristic curve,** or OC curve) of the probability of accepting a lot, versus the possible quality levels of the lot, is used as an aid in evaluating how well the sampling plan will perform in practice. Each sampling plan has its own distinct OC curve, and some popular plans are accompanied by several pages of published OC curves for the user's convenience in selecting the appropriate plan.

One of the most widely used collections of plans for attributes data is Military Standard 105E, which is usually abbreviated MIL-STD-105E. This collection has

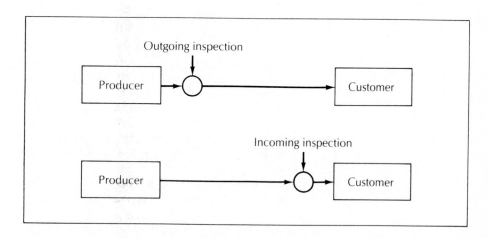

FIGURE 1.7
Sampling Inspection

only recently been updated from the previous, long-standing version MIL-STD-105D. It has become a standard in many industries, especially those involved with government contract work. The 105E system gives tables of AQLs, sample sizes, and associated OC curves so that the user can select a plan that meets both the consumer's and the producer's requirements.

The most familiar acceptance sampling system for variables data is MIL-STD-414, whose application is patterned after MIL-STD-105E. MIL-STD-414 assumes that the process measurements follow a normal distribution. Its tables of sample sizes and OC curves are indexed according to whether one knows the process variation or one is estimating it from the data. As a general rule, acceptance sampling plans for variables data involve much more computational effort than attributes plans. For an overview of the types of acceptance sampling plans available and their intended uses, the reader should see Schilling (1984).

The use of acceptance sampling plans has subsided somewhat in recent years due to the increased emphasis being placed on controlling processes rather than inspecting finished products. It has been argued, for example, that inspecting finished products does not do much to improve product quality; it merely sorts out some of the bad lots. However, acceptance sampling still has a role to play. Although it does not have the sensitivity of control charts and should therefore not be used in place of control charts, there are still many situations where it is economically advisable to use acceptance sampling. New processes, for example, replete with unknowns and uncertainties, are not always immediately controllable with charting techniques. Until such processes *are* brought under statistical control, some amount of inspection is necessary to protect against serious production errors. An excellent compilation of acceptance sampling plans can be found in the text *Acceptance Sampling and Quality Control* by E.G. Schilling (1982).

EXERCISES FOR CHAPTER 1

1.1 A standard letter-sized envelope is 4 inches wide and 9.5 inches long. Normally, 8.5-inch by 11-inch pages are folded in thirds before they are inserted into such envelopes for mailing. Consider page-folding as a process whose measurable output is the width of the folded page.

(a) What specification limits does the envelope size place on the page-folding process?

(b) What is the penalty for exceeding the upper specification?

1.2 If driving a car down a single-lane road is thought of as a process, would the painted lane markings more likely be considered specification limits or control limits?

1.3 (a) In the manufacture of threaded bolts, bolts with thread sizes smaller than the LSL or larger than the USL are not satisfactory and must be scrapped, although scrapped bolts can sometimes be reworked or sold for salvage. Suppose that the LSL and USL for 0.25-inch-diameter bolts are 0.24 inch and 0.26 inch, respectively. Furthermore, suppose that the loss associated with bolts with diameters equal to either the LSL or the USL is $0.02 per bolt. (This takes into account lost labor and materials as well as any salvage value.) Determine the constant k in a symmetric loss function $l(y) = k(y - T)^2$ for this process.

(b) Using the loss function from part (a), calculate the loss associated with bolts with thread sizes of 0.255 inch and 0.265 inch.

(c) Suppose that in a large batch of 100,000 bolts, 70% of the bolts have a diameter of exactly 0.25 inch, 15% have diameters of 0.24 inch, 6% have diameters of 0.23 inch, 2% have diameters of 0.27 inch, and 7% have diameters of 0.26 inch. Calculate the total loss associated with this batch of bolts. What is the average loss per bolt for this batch?

1.4 In service industries, 'waiting time' is often an important quality characteristic. Does waiting time have a one- or two-sided tolerance?

1.5 When affixed to some object, each piece of paper in a pad of adhesive notes must stay in place but must also be easily removable. The strength of the adhesive used is critical in the manufacture of this type of product. In this application, does adhesive strength have one- or two-sided specification limits?

1.6 Of the many shirts made by a clothing manufacturer on a particular day, some were fabricated using cloth from a different dye lot than the others. As a result, a quality control inspector decided that the color of the sleeves did not adequately match the rest of the shirt for this batch. Would these shirts more likely be classified as defective or nonconforming?

1.7 For a general loss function $l(y) = k(y - T)^2$, explain what happens to the average loss per part as the process average shifts farther and farther away from the target value T.

1.8 Give two distinctly different operational definitions for measuring the fuel efficiency of an automobile (fuel efficiency = average miles per gallon).

1.9 Why must a process be considered stable before it makes sense to talk about the process variation?

1.10 Many word processing software packages now include a spell-checking feature; that is, it is possible to check every word in a document for spelling correctness against a dictionary of words included in the software. Considering a document as a 'product' of your typing process, using the spell-checker amounts to 100% inspection of the product.

(a) Give several reasons why this form of 100% inspection will not catch all the spelling errors in a document.

(b) After a 100% inspection with the spell-checker is done, which would be of more value: performing another 100% inspection using the spell-checker, or performing a 100% visual inspection yourself?

1.11 Measurements are to be taken of each of the following characteristics. In each case, indicate whether the resulting measurements would be classified as variables data or as attributes data.
(a) The number of flaws per square foot in a large sheet of metal
(b) The concentration of a certain chemical in a solution
(c) The thread diameter of a bolt
(d) The number of oversized bolts in a batch
(e) The proportion of oversized bolts in a batch
(f) The number of errors on an invoice
(g) The number of invoices in a batch that contain errors
(h) The number of transaction errors in 1,000 banking transactions
(i) The number of errors per 1,000 lines of computer code
(j) The time between successive breakdowns of a machine
(k) The breaking strength of a metal bar

1.12 One method for increasing the likelihood of meeting specifications at final inspection is to tighten the specification limits in all previous (i.e., *upstream*) processes. Although this generally has the desired effect of reducing nonconformances at final inspection, the method has several drawbacks. List some of these drawbacks.

1.13 To reduce costs, a manufacturer uses an inexpensive subcomponent in each of its products. Using this subcomponent lowers the cost of the finished product, but the subcomponent is also known to have a higher failure rate than that of a slightly more expensive brand. List some of the 'losses to society' incurred after these products are shipped to customers.

1.14 Suppose that the total weight of three objects is to be determined by using a standard scale. Use A, B, and C to denote these objects, and consider the following two methods for estimating their total weight:

Method I Each object is weighed separately on the scale. Let x_1 denote the reading for object A, let x_2 be the reading for B, and let x_3 be the reading for C. Their total weight is estimated by $x_1 + x_2 + x_3$.

Method II All three objects are placed on the scale and weighed, giving a reading of y_1. The objects are removed from the scale, then replaced and weighed again. This time the reading is y_2. A third repetition gives a weight of y_3. Finally, the total weight is estimated by averaging the three results to give $(y_1 + y_2 + y_3)/3$.

Both methods use three weighings. Is there any reason why one method may give more accurate results than the other?[1]

1.15 Give several reasons why inspection and test results at a manufacturer's final inspection may not agree with inspection and test results at a customer's receiving inspection.

1.16 If a measuring instrument is out of calibration, explain the effect this instrument has on the production of nonconforming and conforming products.

---

[1]This topic is explored further in Banerjee (1950).

REFERENCES FOR CHAPTER 1

ASQC. 1983. "Glossary and Tables for Statistical Quality Control." American Society for Quality Control, Milwaukee, WI 53203

Banerjee, K. S. 1950. "Weighing Designs." *Calcutta Statistical Association Bulletin* 3:64–76.

Barker, T. B. 1986. "Quality Engineering by Design: Taguchi's Philosophy", *Quality Progress* 19 (no. 12).

Box, G. E. P. 1987. "The Scientific Context of Quality Improvement." *Quality Progress* 20, (no. 6):54–61.

Box, G. E. P and N. R. Draper. 1969. *Evolutionary Operation: A Statistical Method for Process Improvement.* New York: Wiley

Box, G. E. P. and K. B. Wilson. 1951. "On the Experimental Attainment of Optimum Conditions." *Journal of the Royal Statistical Society,* Series B, 13:1-45.

Burr, I. W. 1984. "Management Needs to Know Statistics." *Quality Progress,* July, pp. 26–30.

Davies, O. L. 1954. *The Design and Analysis of Industrial Experiments.* London: Oliver and Boyd.

Davies, O. L. 1957. *Statistical Methods in Research and Production with Special Reference to the Chemical Industry.* London: Oliver and Boyd.

Freund, R. A. 1985. "Definitions of Basic Quality Concepts." *Journal of Quality Technology* 17 (no. 1):50–56.

Garvin, D. A. 1987. "Competing on the Eight Dimensions of Quality." *Harvard Business Review,* Nov.–Dec., pp. 101–109.

Gitlow, H. S., and P. T. Hertz. 1983. "Product Defects and Productivity." *Harvard Business Review,* Sept.–Oct., pp. 131–141.

Hahn, G. J. 1984. "Experimental Design in the Complex World." *Technometrics* 26 (no. 1):19–31.

Ishikawa, K. 1989. *What is Total Quality Control: The Japanese Way.* New York: Prentice Hall.

Juran, J. M. 1974. *Quality Control Handbook,* 3rd ed. New York: McGraw-Hill.

King, C. A. 1987. "A Framework for a Service Quality Assurance System." *Quality Progress,* September, pp. 27–32.

Kirk, R. E. 1982. *Experimental Design,* 2nd ed. Belmont, CA: Brooks/Cole.

Lerner, F. S. F. 1970. "Quality Control in Pre-Industrial Times." *Quality Progress* 3 (no. 6):22–25.

Nelson, L. S. 1988. "Control Charts: Rational Subgroups and Effective Applications." *Journal of Quality Technology* 20 (no. 1):73–75.

Ross, P. J. 1988. *Taguchi Techniques for Quality Engineering,* p. 4. New York: McGraw-Hill.

Schilling, E. G. 1982. *Acceptance Sampling and Quality Control.* New York: Marcel Dekker.

Schilling, E. G. 1984. "An Overview of Acceptance Control." *Quality Progress* 17 (no. 4):22–25.

Snedecor, G. W. and W. G. Cochran. 1980. *Statistical Methods,* 7th ed., p. 17. Iowa State University Press.

Taguchi, G. 1988. "The Development of Quality Engineering." *The ASI Journal* 1 (no. 1). Dearborn, MI: American Supplier Institute.

Vance, L. L., and J. Neter. 1956. *Statistical Sampling for Auditors and Accountants.* New York: Wiley.

2

THE MODERN APPROACH TO QUALITY—
An Overview

This chapter presents an overview of quality systems, philosophy, history, and practice. It provides a working vocabulary for the framework within which modern quality techniques are applied.

CHAPTER OUTLINE

2.1   INTRODUCTION

Quite often, developments in a given field are guided by underlying paradigms that go unchallenged for long periods of time. In much of the history of quality assurance, the goal of meeting specifications has been the operative paradigm, and it has only recently been challenged. As a result, many of the early statistical methods used in quality were developed with the goal of *controlling* manufacturing processes to operate within specifications.

In the modern application of quality methods, the paradigm has shifted to one of *continuous improvement.* Continuous improvement is not at odds with meeting specifications; it simply goes further in its demands. Statistically controlled processes are still desired but are now considered to be only a part of the overall approach to quality.

In this chapter, we begin with a brief account of important events in the history of statistical methods for quality improvement (see Section 2.2). Because of their great impact on the field, the names of Deming, Juran, Taguchi, and others appear throughout this development. A brief summary of their contributions is given in Section 2.3. The remaining sections of the chapter present the framework of the modern approach to quality. In particular, systems diagrams (see Section 2.4) serve as a convenient device for relating most of the newly developed vocabulary of quality improvement, from 'continuous improvement,' 'prevention/detection approaches,' and 'customer/supplier relationships' to 'design quality' and 'off-line methods.'

2.2   HISTORY OF QUALITY CONTROL

In quality technology, as in most technical fields, the acceptance and usage of new methods follow a process of evolution rather than simple accumulation. For example, some techniques that were popular 40 years ago have fallen into disfavor; on the other hand, the computational complexity that slowed the widespread use of some early methods no longer poses a problem in today's computer-intensive environment. Finally, methods used for years in other countries have caused great hope and controversy when recently introduced in the United States. Even the philosophical approach to the field as a whole has changed over the years.

Following the quality mandate that data should be presented in a simple, understandable form, our approach to the history of quality control consists of brief chronological listings, or timelines, of major developments in the United States, Japan, and Great Britain. Obviously not every historical event is (or can be) included, but the list ought to provide a good foundation for further reading.

It should also be mentioned that there are some very good narrative histories of quality developments. Many of these have been included in the references. In particular, the interested reader will find much of interest in Morrison (1987), Ishikawa (1985), and Duncan (1986).

Finally, while studying the following chronologies, the reader should keep in mind that there have definitely been periods of widespread usage and periods of little usage of quality control methods. The periods of highest usage seem to have been during World War II and again starting in the late 1970s and continuing into the present. During the war, controlling the costs of labor and limiting the waste of precious raw materials motivated the concern with quality methods. Such concerns diminished after the war, most notably in Great Britain because of the small number of practicing industrial statisticians at that time (Morrison 1987). Today, however, besides providing the benefits of cost and scrap control, quality methods are viewed worldwide as a strategic marketing and productivity tool (Garvin 1987).

HISTORY OF QUALITY METHODS IN THE UNITED STATES		
1700–1800	Craftsmen immigrating from Europe bring quality workmanship standards; Eli Whitney introduces the concept of standardized, interchangeable parts (Golomski 1976).	
1907–1908	AT&T company begins inspection and testing of manufactured and installed products and materials (Wadsworth, Stephens, and Godfrey 1986)	
1920s	Bell Laboratories forms a quality department with emphasis on quality, reliability, testing, and inspection.	
1924	W. A. Shewhart introduces the concept of control charts in a memorandum at Bell Laboratories (Duncan 1986).	
1928	Acceptance sampling is refined and developed by H. F. Dodge and H. G. Romig (Duncan 1986).	
1931	Shewhart publishes *Economic Control of Manufactured Product,* outlining the control chart method.	
1938	W. Edwards Deming invites Shewhart to present control chart seminars to U.S. Department of Agriculture.	
1940	The War Department publishes a military guide for using control charts to analyze data (Duncan 1986).	
1940–1943	Bell Lab engineers develop sampling inspection standards for the U.S. Army.	
1942–1945	A. Wald develops a sequential sampling technique, which is considered classified information until after World War II.	
1942–1946	Training courses in statistical quality control are given to industries and armed services; over 15 societies for quality control are formed (Duncan 1986).	
1944	The journal *Industrial Quality Control* begins publication.	
1946	Various quality control societies merge to form the American Society for Quality Control (ASQC).	
1946–1949	Deming is invited to give statistical quality control seminars to Japanese universities and businesses.	

HISTORY OF QUALITY METHODS IN THE UNITED STATES (continued)		

	1951	A. V. Feigenbaum publishes *Total Quality Control*.
	1951–1960s	Feigenbaum introduces the concept of total quality control.
	1954	J. Juran is invited by the Japanese to give seminars on quality management (Deming 1982).
	1954	British statistician E. S. Page introduces the concept of cumulative-sum control charts.
	1957	J. Juran and F. M. Gryna's classic *Quality Control Handbook* is published
	1969	*Industrial Quality Control* ceases publication and is replaced by the management-oriented *Quality Progress* and the technically oriented *Journal of Quality Technology.*
	1980	G. Taguchi's experimental design methods are introduced to U.S. companies.
	1984	The American Statistical Association establishes the Ad Hoc Committee on Quality and Productivity, emphasizing the role of statisticians in the quality field.
	1988	The U.S. Congress establishes the Malcolm Baldrige National Quality Award (similar to the Deming Prize established in Japan in 1951).

RECENT HISTORY OF QUALITY METHODS IN JAPAN		

	1941	British Standard BS600 is translated into Japanese.
	1946	W. Edwards Deming travels to Japan with the Economic and Scientific Security Section of the U.S. Department of War to help occupation forces rebuild Japan (Mann 1988).
	1946	K. Koyanagi forms the Japanese Union of Scientists and Engineers (JUSE).
	1948	The U.S. Department of Defense invites Deming to continue his work in Japan.
	1948	G. Taguchi begins work on experimental designs for industrial applications (Taguchi 1988).
	1949	JUSE invites Deming to teach statistical quality methods for industry.
	1950	Deming begins seminars for Japanese managers.
	1950 onward	Statistical quality methods are taught in Japan.
	1950	K. Ishikawa introduces cause-and-effect diagrams.
	1951	JUSE establishes the Deming Prize for significant achievements in quality and quality methods (Bush and Dooley 1989).
	1960–1962	QC circle concept is introduced by Ishikawa; the first QC circle is registered in 1962.

1976–1977	Taguchi publishes volumes 1 and 2 of *Experimental Designs.*
1979	G. Taguchi and Y. Wu publish *Introduction to Off-Line Quality Control.*
1986	G. E. P. Box and others visit Japan and note widespread use of experimental design methods (Box 1988).

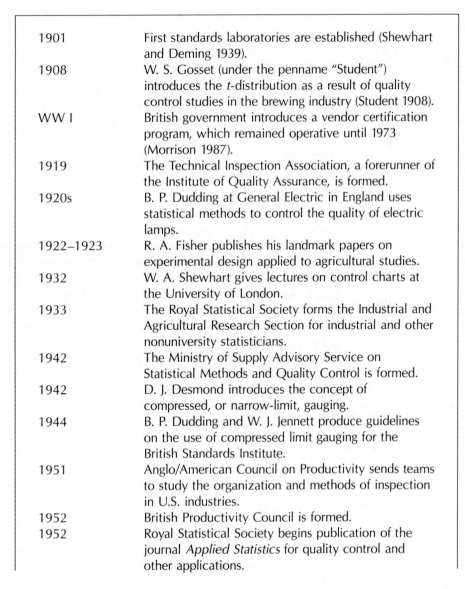

HISTORY OF
QUALITY METHODS
IN GREAT BRITAIN

1901	First standards laboratories are established (Shewhart and Deming 1939).
1908	W. S. Gosset (under the penname "Student") introduces the *t*-distribution as a result of quality control studies in the brewing industry (Student 1908).
WW I	British government introduces a vendor certification program, which remained operative until 1973 (Morrison 1987).
1919	The Technical Inspection Association, a forerunner of the Institute of Quality Assurance, is formed.
1920s	B. P. Dudding at General Electric in England uses statistical methods to control the quality of electric lamps.
1922–1923	R. A. Fisher publishes his landmark papers on experimental design applied to agricultural studies.
1932	W. A. Shewhart gives lectures on control charts at the University of London.
1933	The Royal Statistical Society forms the Industrial and Agricultural Research Section for industrial and other nonuniversity statisticians.
1942	The Ministry of Supply Advisory Service on Statistical Methods and Quality Control is formed.
1942	D. J. Desmond introduces the concept of compressed, or narrow-limit, gauging.
1944	B. P. Dudding and W. J. Jennett produce guidelines on the use of compressed limit gauging for the British Standards Institute.
1951	Anglo/American Council on Productivity sends teams to study the organization and methods of inspection in U.S. industries.
1952	British Productivity Council is formed.
1952	Royal Statistical Society begins publication of the journal *Applied Statistics* for quality control and other applications.

HISTORY OF QUALITY METHODS IN GREAT BRITAIN (continued)	1954	E. S. Page introduces the concept of cumulative-sum control charts (Page 1954).
	1957	British Productivity Council publishes books on control charts for many specific industries.
	1961	National Council for Quality and Reliability (NCQR) is formed as part of the British Productivity Council.
	1970s	British government eliminates funding for the NCQR.
	1970s	NCQR and Institute of Quality Assurance merge to form the British Quality Association.

2.3   DIFFERENT QUALITY PHILOSOPHIES

Many people have contributed to the statistical literature on quality control, but only a handful have had a lasting influence on *implementing* quality improvements. In the latter group, the names Deming, Juran, Feigenbaum, and Taguchi should be familiar to all. This section briefly delineates the key contributions and approaches of these individuals.

THE DEMING PHILOSOPHY

W. Edwards Deming is undoubtedly the most well known of the group mentioned above. It is he who conducted the first quality control seminars for Japanese managers (from 1946 onward) and who actively promoted Shewhart's methods in the United States and Japan. He is credited with helping Japan become an undisputed quality leader, and in his honor the Japanese created the Deming Prize (1951), a prestigious award given to Japanese companies that are able to demonstrate outstanding improvement in quality control (see Section 14.6).

Over the years, Deming evolved a series of recommendations outlining his approach to quality. It is interesting to note that Deming, a professional statistician by training, emphasizes simple, people-oriented tenets. This is in line with his fundamental belief that the primary responsibility for quality (or lack thereof) lies with management. His oft-quoted dictum that over 85% of quality problems are solvable only by management, not by the workforce, is key to the Deming philosophy. His recommendations to management have become known as Deming's 14 points:

| DEMING'S 14 POINTS FOR MANAGEMENT[1] | 1. | Create constancy of purpose for the improvement of products or services. |
| | 2. | Adopt the new philosophy. |

---

[1] I am indebted to Victor Lowe of the Ford Motor Company for this version of Deming's 14 points.

3. Cease dependence on inspection to achieve quality.
4. End the practice of awarding business on the basis of price tag alone. Instead, minimize total cost by working with a single supplier.
5. Improve constantly and forever, every process for planning, production, and service.
6. Institute training on the job.
7. Adopt and institute leadership.
8. Drive out fear.
9. Break down barriers between staff areas.
10. Eliminate slogans, exhortations, and targets for the workforce.
11. Eliminate numerical quotas for the workforce and numerical goals for management.
12. Remove barriers that rob people of pride of workmanship. Eliminate the annual rating or merit system.
13. Institute a vigorous program of self-improvement for everyone.
14. Put everybody in the company to work to accomplish the transformation.

In his book *Quality, Productivity, and Competitive Position* (1982), Deming gives additional explanation of the 14 points:

1. 'Constancy of purpose' means that a company should seek continual quality improvements and that it should devote the requisite resources and efforts toward this end.

2. 'Adopt the new philosophy' means that a company should reject old beliefs about quality (that some defectives, poor workmanship, and service are to be tolerated) and sincerely strive to put quality as a prime objective.

3. Wherever possible, seek to replace inspection (the 'detection' approach) with process control (a 'prevention' approach).

4. Vendors should be chosen on the basis of their ability to supply quality products, not solely their ability to supply low-cost products. Eliminate the practice of giving contracts to the lowest bidder (without regard to quality). Take advantage of building long-term relationships with suppliers.

5. Seek continuous improvement everywhere, not just in manufacturing; that is, improve the quality of all the company's activities, from marketing, accounting, and manufacturing through payroll, engineering, and procurement.

6. Train the workforce in the tools, both statistical and nonstatistical, needed to improve quality.

7.   Improve supervision, empower workers to make decisions about a process, and require management to make necessary process changes and exhibit leadership in quality improvement.

8.   Eliminate employees' fear of asking questions about processes and procedures.

9.   Different departments should become familiar with one another's operations and take advantage of this knowledge to improve products and processes.

10. Numerical goals, productivity targets, quality targets, and so on should be eliminated. Instead, emphasize continuous improvement. Stressing targets will not increase quality, but stressing quality will achieve such goals.

11. Work standards and quotas are another form of numerical target and should be eliminated.

12. Give workers the power to control their particular process or operation. Eliminate performance reviews and merit systems for promotions and salary.

13. Everyone should be trained in quality techniques. In particular, the quality department is no longer solely in charge of determining quality or using statistical methods.

14. Only top management can create the atmosphere and structure that will make these points work.

Deming's philosophy requires a great change in management style for some companies. Under his system, managers are required to be leaders and facilitators, not merely authority figures. They must play an active role in changing systems to improve quality, not the passive role of creating targets that the workforce is required to meet.

JURAN—ORGANIZING FOR IMPROVEMENT

Joseph Juran is a quality pioneer whose span of activity parallels that of Deming. Juran was also invited by the Japanese to conduct quality control seminars (from 1954 onward), and he has consistently been at the forefront of quality control. From the late 1950s through the 1970s, Juran's name was probably even more familiar in the United States than Deming's because of his book *Quality Control Handbook* (coauthored with F. M. Gryna in 1957), which became a standard reference for quality organization and practice over the years.

Whereas Deming emphasizes statistics and the role of management, Juran's strength is in implementation and organization for change. Juran describes his steps for solving quality problems as a 'breakthrough sequence,' which can be summarized as follows:

JURAN'S
BREAKTHROUGH
SEQUENCE[2]

1. Establish the existence of a problem; that is, present data and arguments that *prove* to others, especially management, that a problem exists and is worthy of study.
2. Pare down the list of possible projects to the 'vital few' whose solution will provide the greatest impact. Pareto charts are used at this stage.
3. Organize for change: obtain agreement on the project's aim, obtain necessary permissions and authority, obtain necessary diagnostic skills, establish time horizons, and plan for implementing the solution.
4. Collect and analyze the data. Statistical tools are used at this stage.
5. Ascertain the effect (on people as well as on processes) of proposed changes and try to limit resistance to these changes.
6. Make the changes.
7. Put controls in place to maintain the improvements (sometimes called 'foolproofing' by Juran).

Note that Juran's breakthrough sequence is more a problem-solving technique than it is a list of management objectives. From a management standpoint, Juran, like Deming, subscribes to the view that the majority (about 80%) of quality problems can be fixed only by management, not by the workforce (Juran and Gryna 1980, p. 139). He also stresses companywide quality control. In fact, his introduction of total quality control (TQC) to Japan in 1954 eventually led some to believe that Juran's influence on Japan was even greater than Deming's (Port 1991). To show their appreciation of Juran's work in Japan, the Japanese Union of Scientists and Engineers (JUSE) suggested that a Juran Medal be established for companies that demonstrate continuous quality improvement for a five-year period. Juran, however, politely declined this honor.

Juran was also a critic of the 'zero defects' movement when it arose in the 1960s, but not for the same reasons as those provided by Taguchi and the 'continuous improvement' philosophy. According to Juran, the weakness in the zero defects approach was the assumptions that the workforce was the source of the majority of quality problems and that, with proper motivation, they could be motivated to improve their performance. Such assumptions are at odds with both his and Deming's belief that the majority of quality improvements reside in management's hands.

Along more statistical lines, Juran is known as the inventor of the Pareto chart (see Section 4.6). This tool was developed to support step 2 in his breakthrough sequence but has since become a very generally applicable problem-solving tool in quality improvement.

---

[2]See Juran (1964).

FEIGENBAUM—TOTAL QUALITY CONTROL

Armand V. Feigenbaum originated the concept of companywide, or total, quality control in 1951. **TQC** stresses that quality should be the concern of all facets of the organization, not just the quality department. Modern versions are usually called **TQM** (total quality management) programs, since this acronym further eliminates any implied emphasis on both quality control (rather than quality improvement) and the quality department's role.

It might be more accurate to describe modern TQM programs as Japanese versions of Feigenbaum's original definition of TQC. Ishikawa (1985, p. 90) notes that Feigenbaum originally suggested that TQC be supported by a well-trained group of quality specialists, rather than placing responsibility for quality more directly within the departments of an organization. To emphasize this distinction, Ishikawa and others refer to their approach as companywide quality control, or **CWQC**. However, common to both Feigenbaum and the Japanese is the realization that companywide quality improvements do not spring forth automatically but require an organized approach.

TAGUCHI—THE LOSS FUNCTION AND ROBUST DESIGN

Taguchi's concept of **loss function** echoes throughout modern quality control. At the most fundamental level, the loss function idea results in new attitudes about quality. *Any* deviation from target will result in some loss to society. This directly challenges the more traditional view that 'fitness for use' or 'conformance to specifications' is sufficient, since products that function and are within specifications may still impart some loss to society. Thus, continuous improvement must always be the goal. Furthermore, the philosophy of zero defects is at odds with the loss function, since once the zero defect level is reached, there is little incentive to improve.

Besides giving a theoretical basis for the goal of continuous improvement, Taguchi has redirected the focus of statistical techniques by concentrating on the **design** phase, instead of on production. In addition, he has provided a statistical approach to **robust design,** the development of products that can function well under a variety of different operating conditions. His approach has been to simplify standard experimental design methods and to give a simple procedure for their application. This simplified approach has met with great enthusiasm by practitioners and has generated a renewed interest in DOE techniques in general. A detailed discussion of Taguchi's DOE techniques is given in Chapter 13.

Practitioners should be aware that these methods have been the subject of some criticisms on theoretical grounds and that clear published descriptions of the details of his methodology have appeared only quite recently. As the debate subsides, Taguchi's methods will undoubtedly evolve and take their proper place in statistical practice.

2.4	THE SYSTEMS APPROACH

PROCESS DIAGRAMS

Borrowing from the field of systems analysis, we can view each step in a manufacturing line or a service procedure as a separate **process** to be performed. In addition, every process must have its specific inputs and outputs, as depicted in Figure 2.1. Take, for example, the procedures involved with diagnostic medical tests. Many steps are involved from the time a physician requests that a particular test be done until the test results arrive from the laboratory. Test samples must be transported by courier to the testing lab, followed by analyzing of the samples, recording the results, obtaining signatures or stamps, returning the results, and entering them into the patient's chart. The 'transportation' step can be thought of as the input to the 'samples analyzed' step, whose output leads to the 'return results' step, and so forth.

Since each process step either precedes or follows another step, the entire sequence can be linked as in Figure 2.2 to give a visual summary or flowchart of the complete procedure. Notice that it is even possible to have some subprocesses occurring at the same time. Flowcharting a production process gives a simple model of an operation, which, in turn, facilitates communication about the various steps

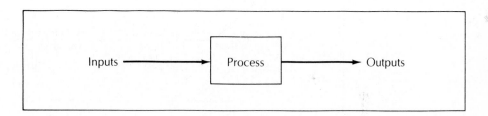

FIGURE 2.1
Process Diagram

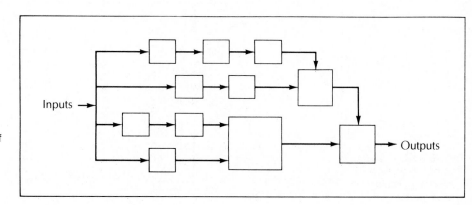

FIGURE 2.2
Typical Breakdown of
a Good or Service
into a Series of
Interrelated
Subprocesses

in the operation. Many companies create such diagrams to describe **work flow**— that is, the order in which work flows through the various departments of an organization. Figure 2.3, for example, shows part of a work flow diagram for the manufacture of printed circuit boards.

The process diagrams used in quality control applications also include those at the within-department level, not just diagrams of work flow between departments. Developing diagrams of the detailed steps in a product's creation is called **process flow analysis.** Creating such diagrams can sometimes be an extremely simple task. More frequently it turns out to be an eye-opening experience because process steps often undergo many small changes as time passes and a manager's idea of what is going on in production can turn out to be very different from what is actually happening on the shop floor. As an example, Figure 2.4 shows two process flow diagrams that describe the insertion of components onto printed circuit boards. Figure 2.4(a) was drawn by the manager of the department, whereas Figure 2.4(b) was created by actually walking through and recording each process step.

The language of systems analysis is familiar to most companies today, and process diagrams in some form are commonplace. In quality control, the shift toward

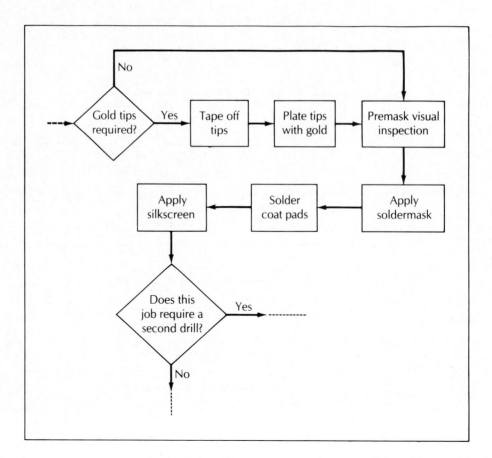

FIGURE 2.3
Part of a Work Flow
Diagram for Printed
Circuit Board
Manufacture

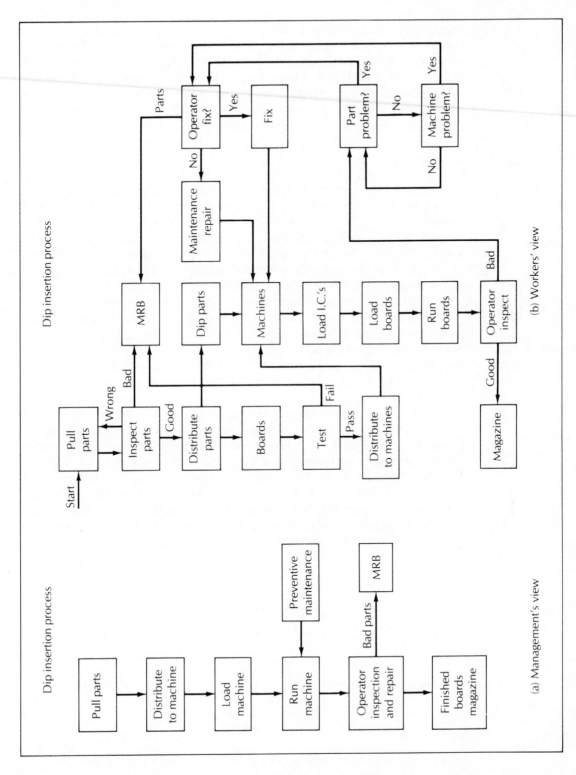

FIGURE 2.4 Printed Circuit Board Process

using this language is fairly recent. Besides the obvious benefits gained from understanding a particular process, systems language allows one to speak of **statistical process control** (SPC) rather than statistical *quality* control. As simple as this slight shift in terminology may be, it tends to lift the sole responsibility for learning and applying statistical methods from the quality department of the company. It implies that all departments are candidates for using these methods, a view espoused by modern total quality control programs.

THE QUALITY SYSTEM

There are a large number of techniques, both statistical and nonstatistical, for improving the quality of a good or service. Names and acronyms for these methods abound. To list but a few, there are SPC (statistical process control) techniques, DOE (design of experiments) methods, TQM (total quality management) programs, QIS (quality information systems), QFD (quality function deployment) techniques, JIT (just-in-time) inventory management methods, vendor certification programs, and cost of quality analyses. Some of these techniques are nonstatistical, and some attempt to blend both management concepts and statistical methods.

When discussing these methods, it helps to have in mind a diagram of a basic quality system. Such a diagram appears in Figure 2.5. One could expand this diagram by including successively more and more detailed subprocess diagrams, but such detail is not needed here. Most of the methods and programs mentioned in the previous paragraph were developed to address one or more of the basic process steps in Figure 2.5.

Vendors, also known as **suppliers**, are the various companies that supply needed raw materials, services, parts, components, subassemblies, and so on to a producer. When, in turn, the finished items from the producer are sold to another company, the roles are reversed and the *producer* becomes a vendor. Some authors take this idea one step further by saying that *every* step in a production process can be thought of as both a supplier (to the following process step) and a customer (of the preceding step). Although we also use this interpretation at times, it should be pointed out that Figure 2.5 depicts only the more traditional case, which considers vendors as the external suppliers of raw materials.

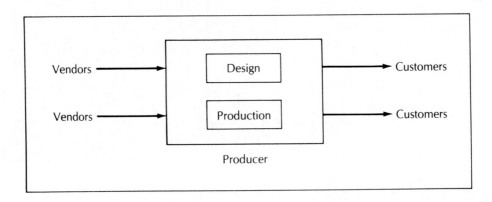

FIGURE 2.5
Quality System
Diagram

Within a company, the quality system can be divided into two phases: the design phase and the production phase. The **design phase** is concerned with the plans for how the product is to be constructed. This usually includes the basic **system design** decisions about the type of technology and materials to be used, the **parameter design** decisions concerned with setting the best target values for the various quality characteristics, and the **tolerance design** phase for prescribing the specification limits for each characteristic of the product.

The **production phase** consists of the actual creation of the product or service. During this phase, raw materials from the vendors are used in conjunction with product designs to make the product. Most traditional quality control methods were originally developed to improve quality during the production phase.

In Section 2.5, a few of the large number of quality-related programs and methods in use today are introduced. Figure 2.5 is a convenient device for keeping track of where these techniques fit into a company's quality system.

2.5 WHERE QUALITY IMPROVEMENT METHODS ARE USED

No single tool can be expected to achieve all of the wide-ranging goals associated with the improvement of quality. Instead, a variety of methods are used, each addressing a particular step in the general quality system (see Figure 2.5). This section takes a closer look at various parts of the quality system and the tools, new and old, that each part uses.

DESIGN VERSUS CONFORMANCE

The quality of a company's product is the result of both the design and production stages (see Figure 2.5). During the design stage, quality efforts are directed toward the quality of the design itself. During the production phase, after the basic design has been established, attention shifts to one's ability to faithfully convert the design into a product.

Quality of design refers to characteristics such as aesthetics, performance, and product parameters and tolerances (Marquardt 1984). A good design is critical to the overall quality of an item, since problems caused by a bad design are often not correctable later in the production stage. Also, design is not limited only to products. Manufacturing *processes,* for example, can also be redesigned in order to improve their performance (Kacker and Shoemaker 1989).

The statistical technique of experimental design, or DOE (see Section 1.4), finds its primary use in the design phase. When used to optimize a product or process, DOE and other optimization techniques are called **off-line quality control** methods. They are 'off-line' because they usually consist of special projects or statistical studies conducted off, or apart from, the normal day-to-day production line. Taguchi (1979) is generally credited with introducing the term 'off-line.'

In the production stage, **quality of conformance** refers to the ability to make a product according to its specifications. It is also concerned with measuring and reducing process variation in order to achieve uniformity or consistency in the resulting products. Quality improvement techniques used at this stage are called **on-line quality control** methods, since they are applied in real time—that is, during the actual running of the production process. On-line methods can also be applied to other parts of the quality system, especially in monitoring the quality of incoming materials and finished products.

Many of the early statistical methods for quality control were on-line methods. From the first Shewhart control charts and subsequent acceptance sampling plans, through the more recent cumulative-sum charts (see Chapter 7), statisticians first focused on methods for monitoring production, particularly in manufacturing industries. To some extent this was inevitable, since early economies tended to be dominated by the manufacturing sector. Although the importance of design quality was also recognized, off-line methods were initially used much less frequently, if at all. This situation is changing rapidly today. Spurred by the notable success of Japanese off-line methods (Box et al. 1988), statistical DOE methods are finding greatly increased usage. This includes a resurgence of interest in factorial designs (see Chapter 13), designs that were developed long before the Taguchi approach and serve as part of the foundation for his methodology.

USING ON-LINE METHODS

On-line methods such as control charts, sampling inspection, and testing can be used at just about any stage in production. When production is viewed as a series of subprocesses (see Figure 2.2), any input or output stage can be a possible place to apply on-line methods. In practice, however, it is neither necessary nor beneficial to use on-line methods everywhere possible. Instead, maximal benefit occurs by following certain guidelines for choosing which process steps to monitor.

One important guideline is to emphasize on-line methods near the *beginning* of the overall production process rather than at the end (Nelson 1988). Practitioners refer to these steps as being **upstream**, whereas later steps are said to be **downstream** processes. The rationale is simple: it is better to catch problems early, especially in complex processes; defective products and processes will otherwise continue to affect the remaining process steps, accumulating needless labor and material costs, only to be pronounced as defective during a final inspection. This approach to improving product quality has been called the **prevention approach** (Figure 2.6). With this approach, a company tracks key process variables *while* the process is running and *before* the final product is made. In this way, potential problems can be spotted quickly and the process can be adjusted immediately before defective products are made.

Some companies, however, use on-line methods only at the end of the production phase so that the resulting charts or data can be used to demonstrate final product quality to their customers. In practice, this turns out to be more an exercise in presenting data than in truly analyzing and improving the production process because, at this late stage, it is usually impossible to pinpoint which upstream processes are responsible after defectives are found. For example, the key factors that affect a process, such as the workers involved, the work shift, materials used,

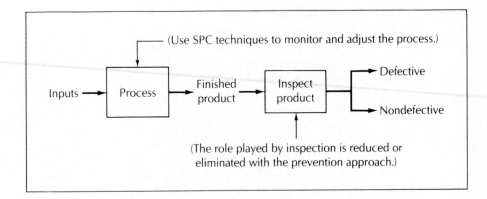

FIGURE 2.6
The Prevention
Approach to
Quality Control

and equipment used, can easily change from day to day. A company that uses control charts or some form of inspection only at the *end* of production is said to be using the **detection** approach (Figure 2.7). Such companies essentially wait until the product is made; then they sort out the good items from the bad and try to use the flaws found in the bad items as indicators of how or where to fix the process.

Prevention is by far the preferable approach and lies at the foundation of modern quality control programs. With this approach, efforts are made to identify the important variables that affect a process and then to examine these variables on a regular basis.

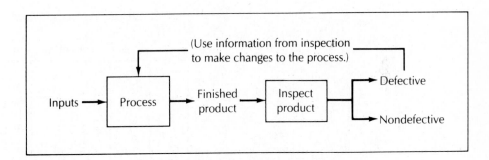

FIGURE 2.7
The Detection
Approach to
Quality Control

CUSTOMER INPUT

Interacting with the customer has traditionally been the job of a company's sales and marketing departments, which create customer enthusiasm for a product by attempting to influence buying behavior (e.g., by using intensive ad campaigns, testimonials, personal relationships) and which conduct marketing research studies of the customer's, as yet, unvoiced wants and needs. The latter method is discussed in this section.

In some ways, ascertaining and reporting consumer opinions and desires in service industries can be much easier than in the manufacturing sector (see

Section 1.3). A great many service products, after all, are created while the customer is present (e.g., in hotels, restaurants, hospitals, travel), so customer input is readily available. Making effective use of such input depends, however, on whether or not a company has created a formal method for summarizing and reacting to such data.

As King (1987) points out, the service customer's input comes in three forms: unsolicited complaints and compliments, solicited feedback via comment cards, and customer surveys. Attention to data quality is an important consideration, since these inputs can suffer from self-selection biases (e.g., some customers with serious complaints will choose not to respond, deciding instead simply not to return). Analyzing such data is frequently done using simple graphical devices such as Pareto charts (see Chapter 4).

In manufacturing, translating customer requirements into product features is a more complicated task. Often-used methods for accomplishing this involve market research studies (e.g., using focus groups, surveys, interviews), sales records (e.g., industry trends, warranty claims), and independent evaluation and testing labs. One tool that has recently been added to this list is **quality function deployment** (QFD).

QFD is a step-by-step method for translating the customer's subjective wants and needs into specific engineering design requirements. A simple matrix diagram is used as a template for performing this translation. Figure 2.8, for example, shows a simplified QFD matrix for an electric hand-held kitchen appliance.

The first step in creating the matrix is to use direct experience with customers, marketing information, or customer surveys to form a list of customer requirements

Customer requirements	Rank	Tighter bearing tolerances	Stronger (heavier) bearing material	Shorter shaft guide	Lighter casing material	Smaller casing dimensions	...
⋮	⋮						
Motor runs more quietly	10	(+)	(+)	(−)	(−)		
⋮	⋮						
Reduced weight	6		(−)	(+)	(+)	(+)	
⋮	⋮						

FIGURE 2.8
Quality Function
Deployment Diagam

or concerns. There can be hundreds of these. The concerns are listed on the left side of the matrix. Next, from frequency of occurrence or some other criterion, the requirements are ranked in order of importance (to the customer). After that, the engineering department is asked to identify the specific engineering characteristics related to each customer requirement. Engineering characteristics may be related to more than one customer requirement, so some type of symbol is used to show the existence, as well as the direction, of the relationship.

For simplicity, Figure 2.8 highlights just two customer concerns: that the appliance be light and that its motor run quietly. In the figure, changing to a lighter casing material has a positive effect (denoted by the + sign in the matrix) on reducing the appliance's weight and a negative effect (denoted by the − sign) on quietness. Other details needed to complete the matrix, such as cost estimates and downstream processes that may be affected, are omitted here to simplify the discussion.

By looking at a completed matrix, one can evaluate customer rankings, estimate costs of making changes, list required process changes, and then choose a strategy that balances these factors. The formal structure of QFD, with team meetings and relationship matrices, tends to coordinate everyone's efforts in a way that is not possible with an informal approach in which each department independently tries to achieve the customer's objectives. More comprehensive references on QFD are Hauser and Clausing (1988) and Sullivan (1986).

VENDOR RELATIONS

Extending quality improvement methods to include one's suppliers follows naturally from the recommendation that process control efforts be concentrated at the upstream stages of production. The reasons for doing this are not new, of course, as evidenced by ancient proverbs involving silk purses and sows ears or, in modern times, by the software programmer's law of "garbage in—garbage out."

One aspect of vendor relations that has received much attention is the question of how many suppliers to use. Using more than one supplier is called **multiple sourcing,** whereas relying on only a single supplier is called **sole sourcing.** In an effort to increase the quality of purchased materials, many companies now try to approach sole sourcing for each individual item purchased. Where there were once ten suppliers of a particular part, the list is sometimes trimmed to one or two suppliers. The objective is not simply to eliminate suppliers, but to increase the feeling of mutual trust and dependence between company and supplier. Strengthening this relationship is one way of improving vendor quality (Broeker 1989).

There are various ways to examine the quality of a vendor's products. In the initial stages, the customer must be sure that the vendor is indeed capable of delivering products that meet specifications. When a vendor is contracted to make a new part or component, the customer must first approve the vendor's design. After design approval, a company inspector is sent to approve the first few items produced by the vendor. This step is referred to as **first-article inspection.** At the same time, inspectors must also evaluate and approve the vendor's manufacturing process to assure that it is capable of making the component. Afterward, during production, company inspectors periodically visit the vendor to monitor progress, or sometimes companies hire an inspector to stay on site at the vendor facility.

Another approach to ascertaining vendor quality is to thoroughly evaluate and approve the vendor's quality system and then give the vendor the right to *approve its own products for shipment,* the understanding being that the vendor's judgment will be accepted and little or no incoming inspection will be performed on these shipments. Obviously, trust is an important factor in **vendor certification** programs such as these. One variant on this approach requires suppliers to include control chart, inspection, or test data with their shipments. In this way, incoming inspection is reduced to examining and approving the shipments based on data from the vendor's quality system.

2.6	THE MANAGEMENT OF QUALITY

From the producer's point of view, attaining quality can appear complicated. In addition to resources devoted to materials and labor, there are costs associated with any quality program. How much time and money should be devoted to statistical quality techniques? What types of new equipment and software will be needed and where should they be placed in the organization for maximal impact? Besides costs, there are decisions about how much employee training is needed and, more generally, how to smoothly effect a transition to a functioning SPC system. Finally, one must consider the possible effects these changes may have on productivity, profit, product development (cycle) time, and long-term competitiveness.

Given all these factors, the approach to managing quality may be formal or informal, but experience suggests that some sort of formal **implementation plan** is best. In fact, many of the formal training and implementation programs used today emphasize how to avoid the pitfalls encountered by earlier, less formal, approaches. This section summarizes some of the questions about quality costs and discusses their relationship to managerial indicators such as productivity, profit, and competitiveness. These topics are taken up again in Chapter 14, where the various steps in an implementation plan are described.

COST OF QUALITY

To simplify the task of deciding how resources are to be allocated to quality-related tasks, quality cost is divided into four categories: internal and external failure costs, appraisal cost, and prevention cost (Juran and Gryna 1980). **Internal failure costs** are those incurred during the production phase because of defective products. Examples include the cost of wasted material and labor used on defective products, the cost of reworking those defective products that are capable of being salvaged, and the extra material and labor devoted to any remedial actions (e.g., retesting and rework). Many companies use a **material review board** (MRB) to temporarily house defective parts and then determine the appropriate action to take on each part. Costs associated with the MRB are included in internal failure costs.

External failure costs are those incurred after a product is delivered to the customer and defectives (undetected by the producer) are discovered. Any cost attributed to the return, replacement, or repair of a product is an external failure cost. The lowered price of 'seconds' (flawed items that may still be acceptable by some customers) is also considered an external failure cost.

In order to detect possible defective items, a certain amount of in-house testing and inspection are necessary. The equipment, its maintenance, and the labor required for such testing are included in the category of **appraisal costs.**

Appraisal and failure costs are generally considered to be the price paid for using the detection approach (see Section 2.5). On the other hand, costs attributed to using the prevention approach are called **prevention costs.** It is this cost category that applies to the methods in this text, since it includes resources devoted to SPC programs, planning, and employee training related to quality improvement.

It is readily apparent that the four cost categories are not independent but, instead, interact with one another. For example, increasing the resources spent on prevention should decrease both types of failure costs and reduce the required appraisal cost. However, since total quality cost is simply the sum of these four categories, increasing prevention cost without bounds will eventually increase the total. This suggests that there might exist an optimal amount to spend on prevention and appraisal in order to counteract the failure costs. Searching for such an optimum requires that a company first perform detailed 'cost of quality' analyses to assess where it stands in relation to the four cost categories. For those interested in this approach, software has been developed to help identify, record, and monitor changes in the cost categories over time.

While some attention to quality cost is certainly needed, problems can occur if one attempts to use cost alone to govern the approach to quality. Managers, for example, sometimes refer to the simple relationship profit = (sales − costs) when deciding what resources to allocate to prevention and appraisal programs. Using this equation, they can easily and mistakenly conclude that any extra money spent on quality (i.e., increasing the cost factor) will thereby reduce profits. However, from the relationships among the four category costs, it is easy to see how increases in prevention costs can lead to decreases in failure costs and eventually total quality cost. Furthermore, sales and quality costs are not independent: reducing prevention and appraisal costs can also lead to higher failure costs and eventually to a decrease in sales. In short, the equation profit = (sales − costs) may provide a good *summary* of what a company has done during a given period, but it is of little use as a tool for *predicting* the effects of changes in cost categories. To effectively use the results of cost of quality analyses requires careful consideration of the true relationships. For an excellent discussion of these relationships, the reader is referred to Kume (1985).

One operational approach to quality costs is to emphasize a program of **continuous improvement** (referred in Japan as '**Kaizen**'). Continuous improvement, now considered fundamental to most quality programs, focuses quality efforts on making innumerable small process and product improvements. From the point of view of costs, modest changes tend to incur modest costs. But more important, the *cumulative* effect of many years of such improvements can often produce better results than the more costly approach of relying on new technologies or more modern equipment (Schneiderman 1986).

PRODUCTIVITY AND COMPETITVENESS

Productivity and competitiveness are close to the top of the list of any company's goals. In this section, the factors that affect them are listed, and the role of quality in attaining them is illustrated.

Labor productivity is defined as the total value of all goods and services produced divided by the total number of labor hours that went into the production process (Baumol 1989). From the point of view of management, increasing productivity is obviously desirable because of the reduced cost per item that results. It is not immediately clear, however, how quality and productivity are related.

Skinner (1986) suggests that attacking productivity problems directly may in fact *reduce* quality. This can happen, for instance, when a company emphasizes cost-cutting programs in order to boost productivity. In this case, quality appraisal and prevention costs are often included in the cuts. More important, though, cost reductions tend to focus on only labor and equipment efficiencies while discouraging new projects and experimentation. When the primary goal is shifted to increasing quality, experience shows that costs usually decrease (and productivity thereby increases). Thus, the effective management of quality can lead to increased productivity, whereas a direct attack on productivity rarely increases quality.

Next, consider the relationship between quality and competitiveness. In manufacturing and services industries alike, quality is now viewed as a strategic tool for increasing competitiveness. Along these lines, the notable success of foreign business competition has now been attributed to an emphasis on quality, the increased communication between research and development, and the speed with which new products are introduced—that is, a reduced product cycle time (Young 1988). Thus, possessing high technology is not enough to guarantee product leadership. A company must have the ability to channel such technology into quality products. Furthermore, the time it takes to introduce new products can often *substitute* for possessing technology, since short product cycle times can give the appearance of technical innovation (Gomory and Schmitt 1988).

REFERENCES FOR CHAPTER 2

Baumol, W. J. 1989. "Is There a U.S. Productivity Crisis?" *Science* 243:611–615.

Box, G. E. P., et al. 1988. "Quality Practices in Japan." *Quality Progress* 21 (no. 3):37–41.

Broeker, E. J. 1989. "Build a Better Supplier-Customer Relationship." *Quality Progress,* September, pp. 67–68.

Bush, D., and K. Dooley. 1989. "The Deming Prize and Baldridge Award: How They Compare." *Quality Progress* 22 (no. 1).

Deming, W. E. 1982. *Quality, Productivity, and Competitive Position,* p. 104. Cambridge, MA: MIT Press.

Duncan, A. J. 1986. *Quality Control and Industrial Statistics,* 5th ed. Homewood, IL: Irwin.

Feigenbaum, A. V. 1983. *Total Quality Control—Engineering and Management,* 8th ed. New York: McGraw-Hill.

Garvin, D. A. 1987. "Competing on the Eight Dimensions of Quality." *Harvard Business Review,* Nov.–Dec., pp. 101–109.

Golomski, W. A. 1976. "Quality Control: History in the Making." *Quality Progress* 9 (no. 7):16–18.

Gomory, R. E., and R. W. Schmitt. 1988. "Science and Product." *Science* 240:1131.

Hauser, J. R., and D. Clausing. 1988. "The House of Quality." *Harvard Business Review,* May–June, pp. 63–73.

Ishikawa, K. 1985. *What Is Total Quality Control? The Japanese Way.* Englewood Cliffs, NJ: Prentice-Hall.

Juran, J. M. 1964. *Managerial Breakthrough.* New York: McGraw-Hill.

Juran, J. M. and F. M. Gryna. 1980. *Quality Control Handbook,* 2nd ed., New York: McGraw-Hill.

Juran, J. M. and F. M. Gryna. 1980. *Quality Planning and Analysis,* 2nd ed., New York: McGraw-Hill.

Kackar, R. N., and A. C. Shoemaker. 1989. "Robust Design: A Cost-Effective Method for Improving Manufacturing Processes." In *Quality Control, Robust Design, and the Taguchi Method,* edited by K. Dehnad. Pacific Grove, CA: Wadsworth.

King, C. A. 1987. "A Framework for a Service Quality Assurance System." *Quality Progress,* September, pp. 27–32.

Kume, H. 1985. "Business Management and Quality Cost: The Japanese View." *Quality Progress,* May, pp. 13–18.

Mann, N. R. 1988. "Why It Happened in Japan and Not in the U.S." *Chance* 1 (no. 3):8–15.

Marquardt, D. W. 1984. "New Technical and Educational Directions for Managing Product Quality." *The American Statistician* 38 (no. 1):8–14.

Morrison, S. J. 1987. "SQC Is Not Enough." *The Statistician.* 36:439–464.

Nelson, L. S. 1988. "Control Charts: Rational Subgroups and Effective Applications." *Journal of Quality Technology* 20 (no. 1):73–75.

Page, E. S. 1954. "Continuous Inspection Schemes." *Biometrika* 40:100–115.

Port, O. 1991. "Dueling Pioneers." *Business Week, Special Issue:*17.

Schneiderman, A. M. 1986. "Optimum Quality Costs and Zero Defects: Are They Contradictory Concepts?" *Quality Progress,* November, pp. 28–31.

Shewhart, W. A., and W. E. Deming. 1939. *Statistical Methods from the Viewpoint of Quality Control,* pp. 4–5. Washington, DC: Department of Agriculture.

Skinner, W. 1986. "The Productivity Paradox." *Harvard Business Review,* July–Aug., pp. 55–59.

"Student." 1908. "The Probable Error of the Mean." *Biometrika* 6:1–25.

Sullivan, L. P. 1986. "Quality Function Deployment." *Quality Progress* 19 (no. 6):39–50.

Taguchi, G. 1988. "The Development of Quality Engineering." *The ASI Journal* 1 (no. 1). Dearborn, MI: American Supplier Institute.

Taguchi, G., and Y. Wu. 1979. *Introduction to Off-Line Quality Control.* Nagoya, Japan: Central Japan Quality Control Association.

Wadsworth, H. M., K. S. Stephens, and A. B. Godfrey. 1986. *Modern Methods for Quality Control and Improvement.* New York: Wiley.

Young, J. A. 1988. "Technology and Competitiveness: A Key to the Economic Future of the United States." *Science* 241 (July 1988):313–341.

3

GRAPHING AND SUMMARIZING DATA

Organizing and summarizing the information in a set of measurements provide a better understanding of the process generating the data. This chapter describes methods for graphically and numerically summarizing measurements on a quality characteristic. It also introduces some of the basic concepts of metrology, the science of measurement.

CHAPTER OUTLINE

| 3.1 | USING STATISTICAL METHODS |

DESCRIPTIVE AND INFERENTIAL STATISTICS

Statistical methods help decision makers draw conclusions from data. Some methods involve mathematical calculations, while others use only simple graphs to summarize the data. In both cases, however, the eventual *purpose* of statistical techniques is inference (Kruskal and Tanur 1978).

It is convenient to distinguish between methods designed for inference and those that primarily describe data (from which inferences are eventually drawn). **Inferential statistics** are those methods whose results can be extrapolated beyond the data to a more general setting. For example, inferential statistics are used when one is estimating an entire day's process variation by examining a small sample from the daily output of a process. Methods such as hypothesis testing, confidence interval estimation (see Chapter 5), and experimental design (see Chapter 13) are considered inferential methods.

Descriptive statistics, on the other hand, are methods for summarizing data. They take the form of either visual displays of the data or numerical summaries. Histograms, Pareto charts, boxplots, stem-and-leaf displays, scatter plots, and location diagrams are typical examples of descriptive graphs (see Chapter 4). Numerical descriptions include such summary statistics as means, medians, ranges, percentiles, standard deviations, and variances (see Section 3.3).

ANALYTIC AND ENUMERATIVE STUDIES

In addition to descriptive and inferential statistics, Deming (1975) has proposed another way to classify statistical studies. He distinguishes between studies done to describe a well-defined population of items and studies undertaken for the purpose of improving or changing the underlying nature of the thing being studied. The former, termed **enumerative studies,** includes virtually all descriptive and inferential statistics. The latter, called **analytic studies,** require inference techniques *beyond* those provided by statistical inference alone.

An enumerative study is conducted, for example, when one takes a sample from the daily output of some process and measures the percentage of defective items in the sample. Even when these results are used to make inferences about the entire day's production, one is said to be performing an enumerative study. As Deming states, enumerative studies tend to answer the question How many?, whereas analytic studies try to answer the question Why?

Analytic studies, which Deming says could alternatively be called comparative studies, are also done in order to choose between various actions that will affect (it is hoped for the better) the process under consideration. For example, a study to compare and choose between two machines for a production process is considered an analytic study. Although this type of study is commonly done by using statistical experimental design methods (see Section 1.4), Deming argues that this is not sufficient, since one can never guarantee that all the conditions present during the study will be present in the future. Because of this, he concludes that analytic studies can never be complete and that the conclusions from such a study must rely on expert, nonstatistical knowledge of the subject studied.

Other disciplines have also found it necessary to distinguish between the two types of studies. For example, in the social sciences, the distinction is made between **external** and **internal validity.** External validity concerns whether the results of an experiment can be generalized *beyond* the subjects of the experiment to the larger population. In words similar to Deming's admonitions about analytic methods, Edgington (1980) states that "inferences about treatment effects for other subjects must be *nonstatistical* [emphasis his] inferences."

What operational use can be made of the distinction between analytic and enumerative methods? In this text, we take Deming's concern about the nonrepeatability of experimental conditions as a *warning* to pay close attention to such things when attempting to extrapolate conclusions beyond the data. Conditions can change over time, and one should accordingly take care to make inferences that are likely to be robust against such eventualities. With that said, however, the history of scientific advancement suggests that our ever-changing world can, nonetheless, be effectively studied with quantitative methods and that statistical methods *will* provide good predictability and useful conclusions that, together with other relevant information, can be used to improve a process.

3.2 VISUALLY SUMMARIZING DATA: THE HISTOGRAM

In quality control and in general statistical practice, graphical methods have become increasingly popular during the last decade. This is partly because of computers, since creating good graphics is now easier than ever before. More important, though, the increased use of visual data displays is a result of the realization that properly chosen graphs can be powerful problem-solving tools. Chapter 4 collects and describes the graphical tools currently used in quality improvement programs. However, to better understand the statistical concepts in this chapter, one of these graphs, the histogram, is introduced now.

The reader will recall from Section 1.4 that process data can be classified as either attributes or variables data (or, in statistical terminology, as discrete or continuous measurements). It simplifies the ensuing discussions to use the following notation when referring to measurements: uppercase letters are used to denote a quality characteristic, while lowercase letters denote *measured values* of that characteristic. Thus, if X represents the characteristic "bolt diameter," then x_1, x_2, x_3, ..., x_n denote the actual measured diameters of a sample of n bolts.

To construct a visual display of the measurements x_1, x_2, x_3, ..., x_n, first imagine a graduated horizontal axis along which these n data values are plotted. Some information, but not a lot, is conveyed by such a plot. It gives a rough picture of the variability in the data and where they are centered, but otherwise it is not much more informative than just a look at the unplotted raw data. Instead, a more informative plot results by first breaking the horizontal axis into a collection of adjacent intervals (of equal lengths) and counting the frequencies with which the measurements fall into these intervals. When bars are drawn above the intervals at heights equal to (or proportional to) their associated frequencies, a graph emerges that gives a better picture of how the data behave. Bar diagrams constructed in this manner are called **histograms**. The primary difference between the histogram and any other bar diagram is that the horizontal scale of a histogram represents a *numerical* scale. Figure 3.1 shows the graph of a typical histogram. Notice that, unlike other bar diagrams, adjacent bars in a histogram share common boundaries to account for the fact that, generally, the values in a particular interval could have occurred anywhere within that interval.

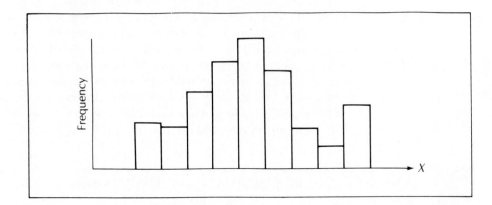

FIGURE 3.1
An example of a
histogram

There is some subjectivity involved in creating a histogram, since one can freely choose the number of intervals, or classes. Guidelines have been proposed for choosing the number of classes, one of which recommends using between 5 and 20 classes. The tradeoff is that too few classes do not give fine enough resolution to the picture, whereas with too many classes the frequencies become small and the resulting histogram conveys no more information than if the data were simply plotted on the horizontal axis. The goal, of course, is to produce a clear picture of the data, and this should be the final guide when selecting the classes.

When computer software is used to generate histograms, there are two approaches to choosing the number of classes. Most programs, as their default choice, use a formula based on the number of measurements, n, to calculate the number of classes to use. Such formulas are designed to give reasonable looking histograms without any guidance from the user. After examining the default histogram, the user may then decide to override the default setting and choose a different number of classes (see Example 3.3).

EXAMPLE 3.1   Rosander (1985, pp 226–227) notes that waiting time is an important factor in judging quality in service industries. The longer a customer waits for service, the lower the perceived quality of the service. Evaluating and minimizing

TABLE 3.1 Elapsed Time (in Hours) for Completing and Delivering Medical Lab Tests
(0.1 hr = 6 min)

6.1	8.7	1.1	4.0	1.7	4.4	2.5	16.2
2.1	3.9	2.2	5.0	17.8	2.9	4.0	6.7
2.1	7.1	4.3	8.8	5.3	8.3	2.8	5.2
3.5	1.2	3.2	1.3	17.5	1.1	3.0	8.3
1.3	9.3	4.2	7.3	1.2	1.1	4.5	4.4
5.7	6.5	4.4	16.2	5.0	2.6	12.7	5.7
1.3	1.3	3.0	2.7	4.7	5.1	2.6	1.6
15.7	4.9	2.0	5.2	3.4	8.1	2.4	16.7
3.9	13.9	1.8	2.2	4.8	1.7	1.9	12.1
8.4	5.2	11.9	3.0	9.1	5.6	13.0	6.4
24.0	24.5	24.8	24.0				

waiting times for medical test results in hospitals provide a good application for a histogram. Shaw (1987) considers elapsed times from the completion of lab tests until the results are recorded on the patients' charts. Table 3.1 shows the recorded times (in hours) it took the in-house courier service to deliver the lab results for 84 lab tests completed during a one week period. For ease of presentation, suppose that a histogram with class widths of 2 hr is used to describe these data. With X denoting the elapsed times, the resulting histogram appears as in Figure 3.2. From the histogram, one can immediately evaluate the timeliness of the courier service. The majority of deliveries are made quickly, most within an 8-hr period. The graph also shows a small number of deliveries with unusually long elapsed times, some as long as two days. Investigating these extreme cases more closely could uncover reasons for the delays and, consequently, lead to suggestions for improving the delivery times.

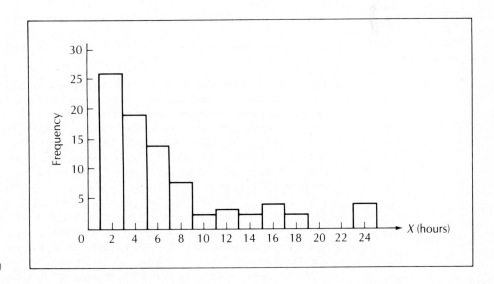

FIGURE 3.2
Histogram of the data in Table 3.1 (class width = 2 hr)

EXAMPLE 3.2   Inspection based on attributes sampling plans (see Section 1.4 and Chapter 11) uses data obtained by counting either the number of nonconformities or nonconforming items in a sample. Suppose, for example, a government contractor uses a sampling plan such as MIL-STD-105E and records the number of nonconforming items in samples of 32 drawn from incoming shipments of parts (required sample sizes are found in tables provided with the MIL-STD-105E plan). With X denoting the number of nonconforming parts in a sample of 32, suppose the most recent ten lots inspected produced these results: $x_1 = 0$, $x_2 = 0$, $x_3 = 1$, $x_4 = 0$, $x_5 = 0$, $x_6 = 3$, $x_7 = 2$, $x_8 = 0$, $x_9 = 0$, $x_{10} = 1$ (i.e., there were no nonconforming items in the first two samples, the sample of 32 from lot #6 had three nonconforming items, etc.). Histograms for integer data are most easily created by using the integers as class *midpoints*. For these data, such a histogram appears as in Figure 3.3.

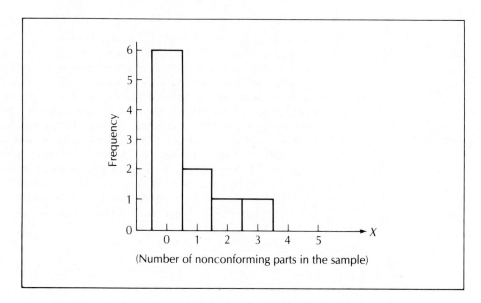

FIGURE 3.3
Histogram of data in
Example 3.2

EXAMPLE 3.3   Printed circuit (PC) boards are used in almost all electronic equipment such as computers, TVs, radios, and appliances. PC boards are the (usually green) cards upon which various electronic components are mounted, with each board performing some specific function within the product. One of the steps in manufacturing PC boards involves putting the boards into a copper solution and using electrolysis to build up the thin layers of copper that eventually form the circuits on the boards. Lab measurements on 50 PC boards recently put through the plating process are shown in Table 3.2. From the *Minitab* computer package, a histogram of these data is given in Figure 3.4. Notice that the *Minitab* package plots histogram frequencies horizontally instead of vertically and that it locates the bars by means of the class *midpoints* instead of the endpoints of each class. With that in mind, it is easy to convert a *Minitab* histogram to the 'standard' format of Figure 3.1. *Minitab* also has the capability of changing the number of classes and printing high-resolution graphics. To illustrate, Figure 3.5 shows four

TABLE 3.2   Copper Plating Thicknesses (in Inches) of 50 Printed Circuit Boards in Example 3.3

Board No.	Thickness	Board No.	Thickness	Board No.	Thickness
1	0.00125	18	0.00170	35	0.00245
2	0.00110	19	0.00255	36	0.00335
3	0.00170	20	0.00155	37	0.00195
4	0.00235	21	0.00295	38	0.00205
5	0.00100	22	0.00240	39	0.00295
6	0.00110	23	0.00255	40	0.00320
7	0.00025	24	0.00225	41	0.00415
8	0.00135	25	0.00205	42	0.00350
9	0.00205	26	0.00155	43	0.00395
10	0.00300	27	0.00230	44	0.00400
11	0.00115	28	0.00315	45	0.00320
12	0.00275	29	0.00280	46	0.00345
13	0.00255	30	0.00185	47	0.00395
14	0.00210	31	0.00300	48	0.00475
15	0.00210	32	0.00270	49	0.00360
16	0.00260	33	0.00310	50	0.00350
17	0.00245	34	0.00290		

high-resolution histograms of the data in Table 3.2 for class widths of 0.0003, 0.0004, 0.0008, and 0.0010 in., respectively. The histograms based on widths of 0.0003 and 0.0004 in. appear to give the best pictures of these data. By comparison, computer packages designed specifically for quality control applications use somewhat more detailed histograms, such as the *NWA Quality Analyst* printout in Figure 3.6 of the Table 3.2 data. The bell-shaped curve (see Section 3.3) drawn over the histogram is intended to approximate what *all* the

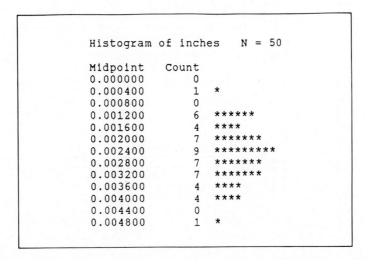

```
        Histogram of inches    N = 50

        Midpoint   Count
        0.000000     0
        0.000400     1   *
        0.000800     0
        0.001200     6   ******
        0.001600     4   ****
        0.002000     7   *******
        0.002400     9   *********
        0.002800     7   *******
        0.003200     7   *******
        0.003600     4   ****
        0.004000     4   ****
        0.004400     0
        0.004800     1   *
```

FIGURE 3.4
Minitab histogram of data in Table 3.2

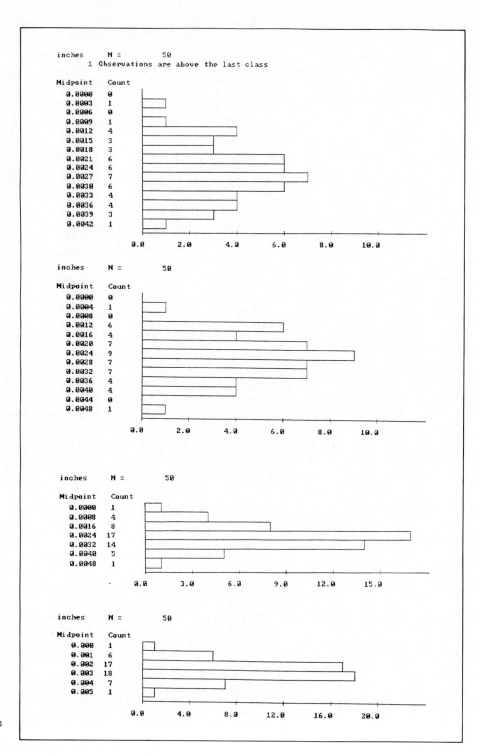

FIGURE 3.5
Minitab
high-resolution
histograms (various
class widths) for data
of Example 3.3

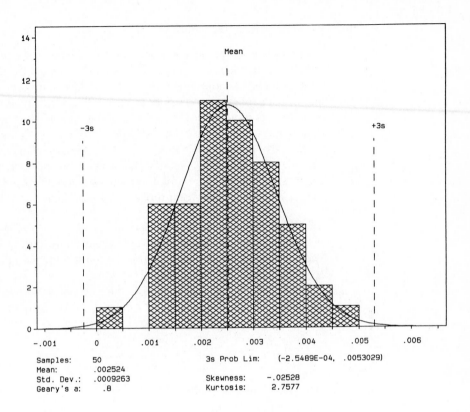

FIGURE 3.6
NWA Quality Analyst
histogram of data in
Table 3.2

Samples:	50		3s Prob Lim:	(-2.5489E-04, .0053029)
Mean:	.002524			
Std. Dev.:	.0009263		Skewness:	-.02528
Geary's a:	.8		Kurtosis:	2.7577

process data would look like if plotted, and numerical measures estimating the process average (Mean) and process variation (Std. Dev.) are recorded also. Histograms for variables data (such as in Table 3.2) can be very helpful for evaluating the capability of a process to meet its specifications (see Chapters 4 and 8).

3.3   NUMERICALLY SUMMARIZING DATA

In addition to visual summaries of data it is convenient to have numerical measures for the same purpose; that is, given a set of measurements $x_1, x_2, x_3, \ldots, x_n$, it is desirable to generate one or two numbers (calculated from the measurements themselves) that summarize the important features of the data. Any number, or summary measure, calculated from a set of data is called a **statistic**.

This section presents formulas for calculating the statistics that are most commonly used in quality control work. For the purposes of this chapter, the discussion focuses on how statistics are used for *describing* data—in particular, for describing central tendency (process average) and dispersion (process variation). In subsequent chapters, the focus shifts to using statistics to make inferences and, consequently, to improve processes or products.

MEASURES OF CENTRAL TENDENCY

If a process is operating smoothly and measurements are made on samples of its output, the stability of the process should be evident in the data. In particular, one often expects the sample measurements to exhibit some sort of **central tendency**— that is, a tendency for the values to stay within some interval, with perhaps more of them occurring toward the middle of the interval than near the endpoints. For stable processes, the center of this region is referred to as the **process average** (see Section 1.4). For nonstable processes, with measurements that wander uncontrollably, there is no central tendency and, instead, one refers to a process average that changes with time. This section introduces commonly used statistics for estimating the central tendency of a process or set of data.

THE MEAN.   The **mean** of a set of data $x_1, x_2, x_3, \ldots, x_n$ is simply the average of all n measurements. Statisticians have long used the term 'mean' as a substitute for 'average' and, following suit, we use the terms 'process mean' and 'process average' interchangeably throughout this text.

Notationally, $\bar{x}$ is used to denote the mean of n measurements on some characteristic, X. In the field of quality control, $\bar{x}$ is most often read as "x bar." The mean, like any average, is calculated by using the formula:

$$mean = \bar{x} = \frac{x_1 + x_2 + x_3 + \cdots + x_n}{n}$$

or, in summation notation:

$$\bar{x} = \frac{1}{n} \sum_{i=1}^{n} x_i \tag{3.1}$$

The mean is the most popular estimate of process average. It also has the interesting property of being the *center of gravity* of the data in the sense that, when the data are plotted along the horizontal axis, the mean falls exactly at the balancing point of the data. For data in histogram form, the mean can be *approximated* as the weighted sum of the class midpoints (weighted by their respective class frequencies). Just as the mean of the raw data lies at the center of the data, so the approximate mean lies at the center of gravity of the histogram. Thus, given a histogram of the data, one can immediately get a good idea of where the mean is by just imagining where a fulcrum would be located that would just balance the histogram. This comes in handy when quick 'ballpark' estimates are needed, since no calculations are required.

In Figure 3.6, the mean is plotted on a histogram of the data from Table 3.2. From equation (3.1), the mean for these data is

$$\bar{x} = \frac{0.00125 + 0.00110 + \cdots + 0.00350}{50} = 0.00252$$

As a measure of central tendency, $\bar{x} = 0.00252$ indeed appears near what one would call the center of these data, but before saying that the process average is about 0.00252 in., one would have to make sure that the process is stable. From the histograms in Figures 3.4, 3.5, and 3.6, there is no way to tell whether the process is indeed stable or not. (In fact, it is *not* stable, which is made evident by a different type of chart in Section 4.5.) For the moment, we simply conclude that the mean of this set of 50 measurements is 0.00252 in.

THE MEDIAN.   In addition to the mean, there are other ways to estimate the center of a set of data, each with its own advantages and disadvantages. Although the advantages of the mean are that it is familiar to all, intuitive, easy to compute, and easy to approximate from a histogram, a possible disadvantage is that it is somewhat sensitive to extreme values (called outliers) in the data. Consider, for example, what would happen to the mean of the data in Table 3.2 if the last observation had been 0.03500 (instead of its actual value, 0.00350). Instead of $\bar{x} = 0.00252$, the mean would then be $\bar{x} = 0.00315$. A change in *one* of the 50 measurements could cause a noticeable shift in our estimate of central tendency.

In quality control work, the sensitivity of the mean is desirable at times, especially when one *wants* to get early warnings of possible problems. In practice, though, a balance must be struck between sensitivity, on the one hand, and the risk of false signals, on the other. In the preceding example, suppose the altered reading of 0.03500 (which is an order of magnitude larger than any of the other measurements) had arisen because someone simply mistyped the true value, 0.00350, when entering the data in a computer. In that case, it might be nice to have a measure that did not react strongly to occasional errors of this sort, since one would not want to shut down an otherwise stable production process because of a keypunch error.

One statistic that satisfies this requirement is the median. Roughly defined, the **median** is the middle observation in the data. To find the median of a set of data $x_1, x_2, x_3, \ldots, x_n$, one must first sort the data from smallest to largest and then count halfway through the sorted list to find the median. In the previous example, changing 0.00350 to 0.03500 in Table 3.2 would cause *no change* in the median. Notation for the median is not as standardized as it is for the mean, so the reader will find an assortment of symbols used by other authors. In this text, $\tilde{x}$ is used to denote the median.

When *n* is an odd number, the median must coincide with one of the x_i's, but when *n* is even, the convention of averaging the *two* middle values is used. In Table 3.1, for example, there are $n = 84$ observations, so the median is taken to be the average of the forty-second and forty-third largest measurements, or $\tilde{x} = 4.6$ hr. From the definition of the median, it appears that about half of all delivery times in Example 3.1 are less than 4.6 hr (and, therefore, half exceed 4.6 hr). For comparison, note that the mean of these data is $\bar{x} = 6.52$ hr.

It is convenient to introduce some additional notation for ordered data. When sample measurements $x_1, x_2, x_3, \ldots, x_n$ are sorted from smallest to largest, the sorted values are denoted by $x_{(1)}, x_{(2)}, x_{(3)}, \ldots, x_{(n)}$ and are called the **order statistics** of the sample. Thus, $x_{(1)}$ is the smallest measurement, $x_{(2)}$ is the next smallest, and so forth; $x_{(n)}$ is then the largest value in the list. With this notation, the median can be written as

$$\tilde{x} = x_{(n+1)/2} \tag{3.2}$$

that is, $\tilde{x}$ is the $[(n + 1)/2]$th observation in the data. When n is even, this notation is understood to represent the *average* of the $(n/2)$th and the $[(n + 2)/2]$th observations.

THE MIDRANGE.   A very quick way of estimating central tendency is to average the smallest and largest values in the data. The resulting estimate is called the sample **midrange** and can be written as

$$midrange = \frac{smallest + largest}{2} = \frac{x_{(1)} + x_{(n)}}{2} \qquad (3.3)$$

The midrange is not very reliable for large data sets, but in quality control, where sample sizes are frequently small, the midrange can be almost as efficient as the mean or median for describing central tendency.

THE TRIMMED MEAN.   Finally, we mention the **trimmed mean,** another option for measuring central tendency. The trimmed mean is a statistic whose sensitivity to extreme values lies somewhere between that of the mean and the median. The trimmed mean has become popular in recent statistical practice because, like the median, it is not overly sensitive to outliers, but it still retains the ability to react to other changes (such as trends) in the data.

The trimmed mean is calculated by 'trimming' off (i.e., discarding) the largest and smallest $\alpha\%$ of the data (where α is called the trimming constant) and then calculating the mean of the remaining observations. It is very common to use a trimming constant of 5%, which means that the smallest 5% and the largest 5% of the data are discarded before the remaining 90% of the data are averaged.

To illustrate the properties of the trimmed mean, consider Example 3.3 once again. With a trimming constant of 5%, the trimmed mean turns out to be 0.00252, which happens to coincide with the mean. As in the previous example, when the last observation in Table 3.2 is changed from 0.00350 to 0.03500, the trimmed mean changes very little, becoming 0.00253.

MEASURES OF VARIATION

Another important characteristic of data is the variation or dispersion they exhibit. Since all process and product measurements vary to some degree (see Section 1.4), the amount of variation is an important factor affecting quality. Excessive variation, for example, is often responsible for the production of nonconforming items.

Because of this, a major component of modern quality improvement programs is the goal of 'continuous improvement,' which, to a large extent, stresses the importance of continually reducing process variation. The effect of reducing variation is that product dimensions become more consistent, processes become more predictable, and fewer nonconforming items result. Reducing variation, of course, requires first having some way to *measure* or estimate process variation from sample measurements.

THE RANGE.   There is one measure almost everyone uses to describe dispersion, whether it be variation in the stock market, food prices, or process measurements. That measure, called the **range**, is simply the difference between the largest and smallest values in the data. With the range denoted by the letter R and with the usual notation for order statistics, the range of a collection of observations x_1, x_2, x_3, ..., x_n can be written as

$$R = x_{(n)} - x_{(1)} \qquad (3.4)$$

In Table 3.2, for example, the largest and smallest values are $x_{(50)} = 0.00475$ and $x_{(1)} = 0.00025$, so the range of these data is $R = 0.00475 - 0.00025 = 0.00450$ (in.).

The range is easy to calculate and interpret but, unfortunately, it is an unreliable measure of variation for large data sets. Consider the delivery time data in Table 3.1. If only the first 80 observations had been available for analysis, the range would have been $R = 17.8 - 1.1 = 16.7$. By comparison, the full set of 84 observations has a range of $R = 24.8 - 1.1 = 23.7$ (about a 42% increase from 16.7). The problem, of course, is that the range depends so heavily on the extreme values in the data. For this reason, one finds only passing references to the range in statistics texts, which generally assume larger sample sizes than those used in quality control applications.

However, the range *is* extensively used in quality control, where samples are small because of time and money constraints. It finds the majority of its application in variables control charts (see Chapters 6 and 7) that usually use successive samples of five or so items. For sample sizes of about ten or fewer, the range is very nearly as good an estimator of variation as the more sophisticated measure, the standard deviation, introduced in the next section (Dixon and Massey 1983). As a result of its efficiency for small samples, and because it is so easy to calculate, the range continues to be used in control chart applications.

VARIANCE AND STANDARD DEVIATION.   Since the usefulness of the range is limited to small samples, other measures of variation are needed for large samples. The two measures most used for this purpose are the sample variance and standard deviation. Each of these statistics makes use of *all* the data, not just the largest and smallest values. This has the effect of reducing the volatility of these measures as sample sizes change and outliers arise.

The **sample variance** of a set of measurements x_1, x_2, x_3, ..., x_n is denoted by s^2 and calculated from the formula

$$s^2 = \frac{1}{n-1} \sum_{i=1}^{n} (x_i - \bar{x})^2 \qquad (3.5)$$

The sample variance measures dispersion by averaging the squared deviations of each value from their center, $\bar{x}$. A divisor of $n - 1$ is used instead of n in the average in order to make s^2 a more accurate estimate of the 'true' process variation. Given that there is a 'true' process variance (denoted by σ^2) underlying any system of measurements, s^2 is considered to be an *estimate* of σ^2, and it can be shown mathematically that using a divisor of $n - 1$ makes s^2 a better estimate of σ^2 than if a divisor of n was used. In the language of statistics, the divisor of $n - 1$ makes s^2 an *unbiased* estimator of σ^2 (see, for example, Meyers 1970, p. 274).

Although equation (3.5) is used to *define* s^2, it is usually easier to *calculate* it from another, equivalent formula:

$$s^2 = \frac{1}{n-1}\left[\sum x_i^2 - \frac{1}{n}\left(\sum x_i\right)^2\right]$$ (3.6)

(The range of summation $i = 1$ to $i = n$ is suppressed to simplify the formula.) Equation (3.6) is often called the short-cut or 'one-pass' method (since it requires passing through the data only once to accumulate the necessary sums) for finding s^2.

Applying either equation (3.5) or (3.6) to the copper thickness data of Table 3.2, one finds that $s^2 = 8.836 \times 10^{-7}$. How is this result to be interpreted? First, notice that the units of measurement of s^2 are always the *squares* of the units in which the x_i's are measured. Thus, in Table 3.2, the sample variance is 8.836×10^{-7} *square* inches! This obvious disadvantage is easily eliminated by taking the square root of s^2, giving rise to a related statistic called the sample standard deviation. The **sample standard deviation,** denoted by s (which explains why s^2 is used to denote the variance), is calculated from

$$s = \sqrt{\frac{1}{n-1}\sum_{i=1}^{n}(x_i - \bar{x})^2} \quad \text{or} \quad s = \sqrt{\frac{1}{n-1}\left[\sum_{i=1}^{n}x_i^2 - \frac{1}{n}\left(\sum_{i=1}^{n}x\right)^2\right]}$$ (3.7)

Since s is not as easily understood as the range, it is convenient to have a way of checking whether one's calculation of s is close to being correct. One way of doing this is to use the inequality (Gutermann 1962)

$$s \leq \frac{R}{2}\sqrt{\frac{n}{n-1}}$$ (3.8)

For most samples, especially larger ones, s should be less than about half the range. If it isn't, a calculation error has probably occurred.

The sample standard deviation serves as an estimate of the 'true' process standard deviation, σ, in the same way that s^2 is considered an estimator of σ^2. Since the Greek letter σ (pronounced 'sigma') is traditionally used to represent the standard deviation of a process or population, practitioners have grown accustomed to referring to either s or σ by the term 'sigma.' Consequently, process variation can be described by how many sigmas (i.e., how many multiples of σ or s) fit within the specification limits.

As given above, $s^2 = 8.836 \times 10^{-7}$ (in.^{2}) for the copper thickness data. Taking square roots yields a standard deviation of $s = 0.00094$ in. Note that this value passes the rough test of equation (3.8) (recall that $R = 0.00450$). The interpretation of s (or σ) is more complex than interpreting a measure of central tendency, so we return to this topic later.

OTHER ESTIMATES OF σ

The underlying process variation, σ, can be estimated in several ways. Using the sample standard deviation, s, suffices in many cases, especially for large sample sizes. In the realm of quality control, though, where small sample sizes tend to be the norm, additional formulas are needed.

It turns out, for instance, that even though s^2 is always a good (i.e., unbiased) estimate of σ^2, the sample standard deviation, s, is somewhat *biased* as an estimate of σ. Although the amount of bias is negligible for large samples and can be ignored, for small samples it cannot be ignored. Fortunately, correcting for this bias is a fairly simple matter: one divides s by an appropriate constant (denoted c_4 in the quality control literature) that takes the bias and sample size into account. Following the statistical practice of putting a ^ over a quantity to denote an *estimate* of that quantity, the revised estimate of σ is

$$\hat{\sigma} = \frac{s}{c_4} \tag{3.9}$$

Equation (3.9) plays an important role in the construction of s charts in Chapter 7. Values of c_4 are found in Appendix 1. Note that the values of c_4 approach 1 as the sample size increases (since the bias then becomes negligible). A useful approximation to c_4, accurate to three decimal places, is given by Sweet and Ong (1983):

$$c_4 \approx e^{-1/2}\left(\frac{n}{n-1}\right)^{(n-3)/2}\left(\frac{6n+1}{6n-5}\right) \tag{3.10}$$

where n is the number of observations used in computing s.

The range, R, can also be used to generate an estimate of σ. In this case, the approximation is given by

$$\hat{\sigma} = \frac{R}{d_2} \tag{3.11}$$

where d_2 is a constant, depending on n, which is also found in Appendix 1. Equation (3.11) is used in the construction of R charts (see Chapter 7) and in estimating process capability (see Chapter 8).

The final estimate we introduce requires no auxiliary tables. The approximation is

$$\hat{\sigma} = 0.39(D_9 - D_1) \tag{3.12}$$

where D_1 and D_9 denote the first and ninth *deciles* of the sample data (Sachs 1982, p. 100). Like equations (3.9) and (3.11), it assumes that the process data are approximately bell-shaped. Defined in a manner analogous to the median, D_1 is the measurement that exceeds the lower 10% of the observations and, correspondingly, D_9 is the number that exceeds the lower 90% of the data. If D_1 (or D_9) happens not to coincide with one of the observations in the data, then (in a manner similar to calculating the median when n is even) one uses the convention of *averaging* the two values closest to D_1 (or D_9).

THE EMPIRICAL RULE

Measurements from a stable process usually exhibit central tendency by bunching up around a mean as depicted in Figure 3.7. Numerous empirical studies have shown that, quite often, measurements stack up in this fashion, which has variously been described as the bell-shaped curve, the normal distribution, and the Gaussian distribution (see Chapter 5). The bell-shaped or (as it is called from now on) **normal curve** has a defining formula (given in Chapter 5) that can be used to plot it, and it has become common practice to superimpose a normal curve over a histogram (as in Figure 3.6). Such curves are meant to depict how one expects a histogram of *all* the process measurements (not just the particular sample) to appear. At times, other curves are used to describe the underlying process, but in many applications, the normal curve is a good approximation.

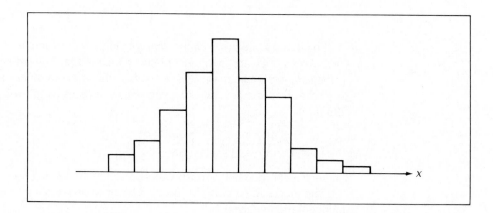

FIGURE 3.7
Measurements
exhibiting central
tendency

Using the normal curve as a model of an underlying process imposes some very specific relationships on the process data, the mean, and the standard deviation. One short list of these relationships, which is referred to as the empirical rule, reads as follows:

THE EMPIRICAL RULE

> - About 68% of all process measurements should lie within 1 standard deviation of the mean.
> - About 95% of all process measurements should lie within 2 standard deviations of the mean.
> - Almost all (about 99.73%) process measurements should lie within 3 standard deviations of the mean.

More detailed calculations involving the normal curve are given in Chapter 5, but the empirical rule is an easily remembered, useful tool for interpreting the standard deviation.

Since the majority of process measurements are expected to fall within 3 sigmas (on either side) of the mean, the total spread of 6 sigmas is sometimes called the **spread** of the process. Process performance and process requirements are then analyzed by comparing the process spread to the specification limits. (See Chapter 8 for further discussion of process capability.) As an example, Figure 3.8 shows a process whose performance (6-sigma spread) is well within its required specification limits.

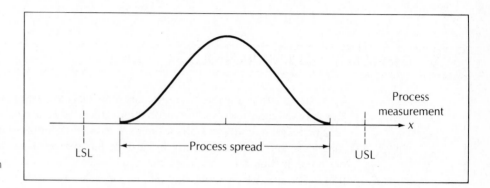

FIGURE 3.8
Process spread
(6 sigmas) lies within
specification limits

EXAMPLE 3.4

In 1988, the Motorola Corporation was one of the first recipients of the Malcolm Baldrige National Quality Award (see Henkoff 1989). Motorola bases much of its quality effort on what it calls its "6-Sigma" program (described in the proceedings of the Quest for Excellence Conference, Washington, DC, April 6–7, 1989). The goal of this program is to reduce the variation in every process to such an extent that a spread of 12 sigmas (i.e., 6 sigmas on either side of the mean) fits within the process's specifications. Figure 3.9 shows what this looks like and, for comparison, also shows a process in which the process

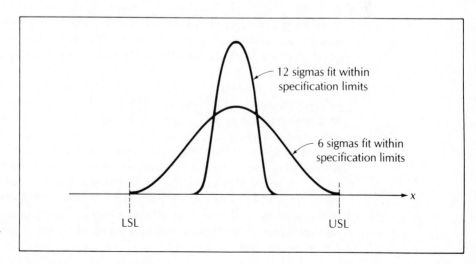

FIGURE 3.9
Example of
Motorola's "6-Sigma"
program

spread of 6 sigmas just fits within the specifications. In the latter case, one must be extremely careful to assure that the process average never slips off target; otherwise, the curve will shift and nonconforming items will become more likely. With Motorola's 6-sigma requirement, however, the process mean can shift by as much as 1.5 sigmas before the likelihood of nonconforming items is significantly increased. Even if the process mean does shift off center by as much as 1.5 sigmas, the Motorola plan assures that only about 3.4 nonconforming items per million parts should result. (See Example 5.15 for the calculation.)

3.4 MEASUREMENTS AND METROLOGY

Up to this point histograms and summary statistics have been generated from process measurements without consideration of how the raw data themselves were obtained. The majority of statistics texts give little attention to the quality of measurements, the tacit assumption being that measurements, especially continuous ones (see Section 1.4), are always made with sufficient accuracy for statistical purposes. The situation is somewhat different in quality control work, where the quality of measurements and the measurement system are of vital importance to the evaluation of product quality. Measurements, after all, determine which items are nonconforming or defective, so measurement inaccuracies can result in mistakenly classifying good products as bad, or bad ones as good.

The study of measurement is called **metrology**. Very broadly speaking, metrology is concerned with two basic questions about process and product measurements. The first deals with the ability to consistently produce measurements of sufficient accuracy and precision (see Section 3.4). If, for example, the specifications for the diameters of metal bolts are 0.500 ±0.005 in., it would not suffice to use a measuring instrument that is accurate to the nearest five-thousandth of an inch, since a bolt whose *true* length is 0.504 in. could then yield a *measured* length anywhere between 0.499 and 0.509 in. As a result, if the measured length was 0.499 in., the bolt would be classified as conforming, while a measured length of 0.509 in. would classify it as nonconforming (and even, possibly, as defective).

The other concern of metrologists is calibration. Even if an instrument is capable of measuring to the required degree of accuracy, it may still give misleading measurements if it is out of adjustment. As an example, consider what happens when you change time zones during an airplane trip. Your watch, which may be capable of great accuracy (perhaps even to within 1 second per month), is now out of calibration, since it is consistently off by a certain number of hours. In this case, it is possible to physically (or mentally) adjust its readings to get the true times, but with most measuring instruments the correct adjustment factor is not so readily available and there is usually no advance warning of when adjustments are needed. When to recalibrate and how much of an adjustment to make become important questions.

MEASUREMENT UNITS AND RESOLUTION

Before making measurements on some quality characteristic, one needs to establish what units of measurement will be used. For example, the characteristic 'length' can be measured either in feet (English system) or in meters (metric system). After the measurement unit is chosen, the next step is to determine how many digits of precision, sometimes called **resolution**, are needed. As noted in previous paragraphs, the resolution is important for correctly classifying measured items as conforming or nonconforming.

There are primarily two methods used to describe the resolution of a measuring instrument. One method is simply to state how many digits of precision the instrument can be expected to deliver; the other defines the resolution as a percentage of the quantity being measured. The following examples illustrate these two approaches.

EXAMPLE 3.5   A digital multimeter is an instrument that measures several electronic characteristics such as voltage, resistance, and amperage and reports its measurements on a digital display (instead of, say, an analog format that has a needle and dial readout). From the product information catalog for one brand of multimeter, it is found that the instrument is capable of measuring DC voltage with resolutions between four and eight digits, depending on the user's preference. At the eight-digit resolution, voltages in the 1-volt range, such as 1.2 volts, can be read as 1.2000000 volts on the meter, while in the 100-volt range, a voltage of 120 volts is read as 120.00000 volts.

The other method for reporting resolution is to specify the **coefficient of variation** (CV) of the measurements. For a set of data $x_1, x_2, x_3, \ldots, x_n$ with mean $\bar{x}$ and standard deviation s, the coefficient of variation is defined by

$$\text{coefficient of variation} = CV = \frac{s}{\bar{x}}$$

For the underlying (or 'true') process mean and standard deviation, denoted by μ and σ, respectively, the CV is defined by

$$CV = \frac{\sigma}{\mu}$$

With measurement systems, reported CVs refer to μ as the true measurement and σ as the variation to be expected in the measurements of μ given by the instrument.

EXAMPLE 3.6   An automated hematology analyzer measures various blood components (e.g., white blood count, red blood count, hemoglobin) in medical testing laboratories. White blood count, for example, is measured in units of thousands per microliter, and the instrument has a coefficient of variation of CV = 0.3% for measuring the white blood count in a sample. Whatever the true white blood count (μ) is, the variation possible in the measured values of this quantity is about 0.3% of μ, or $\sigma = 0.003\,\mu$. If, in addition, one assumes that the measurements follow the empirical rule (page 66), then the range of possible measurements should be 6 sigmas or, since $\sigma = 0.003\,\mu$, about $6(0.003)\mu = 0.018\,\mu$. Thus, the measured value can be expected to be within about 1.8% of the 'true' white blood count. The precision depends on the size of the quantity measured, so small blood counts are measured slightly more precisely than larger ones (over the operating range of the instrument).

Given a set of specification limits for a characteristic, one question that re-mains to be answered is how to choose an appropriate level of resolution for the measurement system. In the metrology community, it has become almost standard to use the '10-to-1' rule, which states that the variation in the measurements should be no more than one-tenth of the allowed variation in the product (i.e., the specification limits). For example, if a specification called for the length of a di-ameter to be within 0.1 in. of the nominal (target) value, then the 10-to-1 rule dic-tates that measurements should be made to the nearest 0.01 in. or less. Although this rule is easy to apply, it is viewed by some as overly conservative (Juran and Gryna 1980, p. 398).

Another frequently raised question concerns the averaging of multiple mea-surements of the same item. It has sometimes been suggested that averaging will reduce the measurement error, which will thereby increase the precision (resolu-tion) of the measurements. Although averaging does tend to eliminate much of the noise that can be present in measurements (due to environmental influences, op-erator errors, etc.), it does *not* have the capability of increasing the resolution of the measuring instrument. Logic alone indicates that this must be the case; other-wise, there would be no need for the sophisticated (and expensive) measuring in-struments that exist today, since one could theoretically achieve any degree of precision by just averaging enough measurements on the same part. The fact that people *do* use these sophisticated devices is itself empirical proof that averaging does not increase resolution. However, in a more simple vein, consider a bath-room scale with a digital display that reads out weights to the nearest pound. After weighing yourself many times in a 5-minute period, it is apparent that the readout consistently reports the same number (i.e., your current weight to the nearest pound) and, therefore, averaging will not change or improve the resolution at all.

MEASUREMENT BY GAGING

Rather than reading measurements from an instrument display, it is sometimes convenient to use gages. **Gages** are devices of known dimensions that can be compared to product dimensions to determine whether the product meets or ex-ceeds specifications. Gages have been used in quality control work from the very beginning and continue to play an important role.

Gages usually come in standard sets, with sizes graduated in specified incre-ments over a particular range. A set of gage blocks, for example, consists of several small metal blocks, each having a specified length slightly different from the rest.

To measure a particular length dimension on a product, one successively places each block next to the dimension to be measured, until eventually finding the gage block that most closely matches. With practice, only a few gages need be tried before the closest match is found. If the measurement marked on this block is within the allowable tolerance for the product (i.e., within the specification limits), the particular part being measured is deemed to be conforming. One virtue of gaging is that gaged measurements are usually much easier to obtain than measurements made with instruments. Compare, for example, how you would use two gages, with lengths equal to the specification limits, to decide whether an object's length is conforming or nonconforming, as opposed to using a micrometer to measure its length. Gaging is almost instantaneous, whereas the micrometer measurements are more time-consuming.

Gages come in a variety of forms. Thin metal strips of different thicknesses are used to set sparkplug gaps in automobiles. For checking hole diameters, pin gages are used; for air pressure, air gages are used; and so on. Gages are also built to test the functionality of a component. These **functional gages** are used by operators to see whether, at some intermediate stage of production, key features of the product function correctly before the product is passed to the next stage of production. As a quick check on whether parts meet their specification limits, **go/no-go gages** are used. Go/no-go gaging makes use of two gages: one set at the lower specification limit and one set at the upper limit. Products are then routinely compared to these gages, with those smaller than the lower gage or larger than the upper gage deemed nonconforming. Finally, go/no-go gages can also be set *inside* the specification limits; that is, the lower gage can be set slightly *higher* than the lower spec limit, while the upper gage is set slightly *lower* than the upper spec. In this sense, the gage limits are narrower than the specification limits. Gaging in this manner is called **compressed limit** or **narrow-limit gaging.**

Some proponents of SPC programs have either ignored or discounted gaging as a process control tool. Instead, they insist that measurement instruments be used wherever possible. This may indeed be appropriate for companies whose budgets accommodate the necessary instrumentation and training costs. However, the fact that it is possible to achieve very nearly the same sensitivity of control using gaged measurements as it is with measured items makes gaging a viable alterative (Stevens 1948). Control chart schemes based on narrow-limit gaging are presented in Chapter 7.

MEASUREMENT ERRORS

ACCURACY AND PRECISION.   Measurements, even those taken repeatedly on the same item, must exhibit some variation. To paraphrase Burr once again, if you don't see any variation in your measurements, you are simply not using an instrument with fine enough resolution. Factors that cause measurements to vary include variation due to operators (different operators use slightly different procedures in their measurements), environmental conditions (temperature differences can greatly affect certain measurements), systematic errors (instruments out of calibration can cause readings to be consistently biased in one direction), as well as instrument resolution. Whatever the cause(s) may be, the resulting variation is called **measurement error,** and it is necessary to understand and estimate the magnitude of this error. Only when the measurement error is small compared to the quantity being measured can one reliably make decisions regarding quality.

There are various ways to estimate measurement error. First, though, let us note that it takes *at least two* numbers to summarize measurement error (Eisenhart 1962). These two measurements are called accuracy and precision. **Accuracy** of measurements refers to their long-run *average*; **precision** refers to their long-run *variation.* In statistical terms, accuracy is estimated by the mean of the measurements, precision by the standard deviation. Accuracy is also defined as the closeness of agreement between observed values and a known reference standard, whereas precision is a measure of the closeness between several individual readings (ASQC 1983).

To show the relationship between these two concepts, several authors have repeated (as we do) the concise example given by Murphy (1961), which uses the analogy of a rifleman firing at a target. With the center of the target taken to be the *true* value of the characteristic being measured and with the rifle shots representing the *measured* values, Figure 3.10 shows the various combinations of precision and accuracy possible in practice. The worst case occurs in Figure 3.10(a), where the shots vary quite a bit (imprecision) and are also systematically off target (inaccurate). The best-case scenario in Figure 3.10(d) shows all the shots packed tightly around the target, illustrating high precision as well as a high degree of accuracy.

Estimation of accuracy and precision is most easily explained by using the following formula, in which the *measured* value, x_m, of a characteristic is composed of its 'true' value, x, plus the error, ϵ, made by the measuring instrument:

$$x_m = x + \epsilon \qquad\qquad (3.13)$$

Equation (3.13) is the basis of much research on measurement error and is used to explain related concepts such as repeatability and reproducibility.

REPEATABILITY AND REPRODUCIBILITY.   **Repeatability** of a measuring instrument refers to how well it is able to repeatedly measure the *same* item under the exact

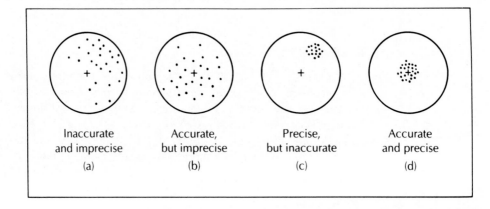

FIGURE 3.10
Accuracy and
precision of
measurements

same conditions. The usual situation in a repeatability study involves *one* inspector using the *same* instrument to repeatedly measure *one* item all during the *same* short period of time. In other words, repeatability is concerned with finding the variation inherent in the instrument itself, and it tries to achieve this by holding every other possible source of variation constant. Conducting a repeatability study often requires the repeated measurement of a known **reference standard**—that is, an item whose 'true' measured value is already known (e.g., precision gages are often used as reference standards). Then, according to equation (3.13), since the 'true' value, x, remains constant during the repeated measurements, the accuracy and precision in the measured values must be caused by that of the measuring instrument itself. Using σ_m^2 to represent the variance of the measurements and σ_ϵ^2 to denote the variance of the error terms, one expects that $\sigma_m^2 = \sigma_\epsilon^2$. Also, since x (the reference standard) remains unchanged, the long-run average of the measurements should equal x (plus any systematic bias in the measurements). Given n repeated measurements $x_{m1}, x_{m2}, x_{m3}, \ldots, x_{mn}$ on some reference standard, it follows that their mean and standard deviation, $\bar{x}_m$ and s_m, provide estimates of the accuracy and precision of the instrument (Burchett 1984).

EXAMPLE 3.7

To estimate the repeatability of an instrument that measures lengths, a 0.300-in. standard gage block was measured five times in succession by an inspector. The resulting measurements were 0.301, 0.303, 0.299, 0.305, and 0.304 in. For these five measurements, the mean and standard deviation were $\bar{x}_m = 0.3024$ and $s_m = 0.00241$, respectively. Since the true length of the gage block is known to be 0.300 in., the accuracy of the instrument must be:

$$\text{accuracy} = \text{average measurement} - \text{reference standard}$$

$$= \bar{x}_m - x$$

$$= 0.3024 - 0.300 = +0.0024$$

That is, the instrument tends on average to produce readings that are slightly (0.0024 in.) higher than the true value. To estimate the precision, assuming the empirical rule applies, the instrument's 6-sigma precision is about $6s_m = 6(0.00241)$, or 0.0145 in. Almost all repeated measurements (using this instrument) should fall within about 0.0145 in. of one another.

Since repeatability measures the best precision possible under ideal circumstances, one should never use an instrument whose repeatability is less than the resolution required by the specifications (on the characteristic to be measured). In practice, though, conditions are rarely ideal, and additional methods are needed to evaluate the measuring capability of an instrument. One way of doing this is to use *expanded* definitions of repeatability, ones that take into account differences between inspectors, environmental conditions, setups, times, and so on. Such definitions give a more realistic picture of how an instrument will perform in the production environment. Of course, the uncertainty associated with including additional factors causes the estimated measurement variation to increase and, therefore, the precision to decrease.

Expanded repeatability studies take the form of experimental designs that efficiently estimate the contribution of various factors to the total measurement error (see Section 1.4 and Chapter 10). Chapter 10 introduces some of these designs. The interested reader may also refer to the excellent survey article by Grubbs (1973) as well as articles by Thompson (1963), Hicks (1980), Hahn and Nelson (1970), and Mandel and Lashof (1987). Also, the *GAGE*STAT* computer software manual offers a good overview of various sources of measurement errors and the experimental designs used to estimate their effects.

Reproducibility is a concept closely related to repeatability. It estimates an instrument's measurement error by giving an *upper bound* on the difference that any *two* independent measurements can be expected to have. For further discussion, see Section 10.7.

CALIBRATION

Physical characteristics are measured by choosing a measurement system (English, metric, or International Standard) and using its particular collection of measurement units (recall that, for example, if electromotive force is the *characteristic* of interest, then volts is the *unit* of measurement). This means that it is necessary to have exact definitions of each unit, and a *physical* example of each unit must be maintained as the ultimate reference standard. In the United States, these reference standards are kept at the **National Institute of Standards and Technology** (NIST).

For example, the NIST maintains a standard 1-volt cell that is taken to be the defining value of the volt. In practice, of course, voltage measurements must be made by instruments (voltmeters) whose measurements must be in agreement (i.e., up to their resolution capability) with the national standard. Using a reference standard to adjust a measuring instrument so that it produces accurate measurements is called **calibration**.

The problem of calibrating the thousands of instruments in everyday use is solved by making several *copies* of the reference standard. The first step, transferring the NIST measurement standard to other calibration laboratories throughout the country, is done by means of copies called **transfer standards.** In turn, calibration labs are responsible for calibrating additional copies called **working standards** that are used by companies and other labs. Eventually, any instrument can trace its accuracy back through this succession of copies and calibrations to the NIST itself. The ability to do this is called the **traceability** of the measurement.

At the company level, where instruments are used for measuring process and product characteristics, the management of calibration is also important. Some instruments, for example, are very sensitive and easily go out of adjustment, whereas others stay in good adjustment for long periods of time. A **calibration program** consists of managing the systematic and periodic recalibration of instruments. Each instrument, depending on its performance, is assigned its own **calibration interval,** which is the interval of time that the instrument can be used in the field before it is returned to a calibration lab for checking and readjustment.

Although the process of calibration sounds straightforward, there are many complex and unresolved issues involved and, to date, there is no complete summary of statistical techniques for calibration. An appreciation for the scope of the problem can be gained by considering the calibration of instruments used for chemical or biological tests, where reference standards consist of chemical or biological samples that do not have the same stability as physical standards tend to have. In addition, the results of chemical analyses or chemical calibrations can often depend heavily on the lab skills of the person performing the tests. The interested reader may wish to consult the discussion by Rosenblatt and Spiegelman (1981) for further reading along these lines.

EXERCISES FOR CHAPTER 3

3.1 The following are measurements (in inches) on a characteristic with specification limits of 2.50 ±0.05 in.:

2.54   2.52   2.50   2.52   2.50   2.50   2.47   2.48   2.51   2.53   2.53
2.51   2.50   2.47   2.49   2.50   2.50   2.50   2.46   2.48   2.48   2.50
2.51   2.53   2.51   2.53   2.53   2.52   2.47   2.51

(a) Create a histogram of these data.
(b) Assuming that the data come from a stable process, calculate the process average and standard deviation.
(c) What percentage of these measurements fall outside of the specification limits?
(d) According to the empirical rule (see Section 3.3), between what two values do you expect to find about 99% of the process measurements?
(e) Explain the difference between your answers to parts (c) and (d).

3.2 Samples of 20 and 50 measurements from the same process were taken on two successive days. On the first day, the average in the sample of 20 was $\bar{x}_1 = 6.03$. The average on the second day was $\bar{x}_2 = 6.45$.
(a) What is the average of all 70 measurements?
(b) Suppose that the medians of these two samples are 6.00 and 6.50, respectively. What can you say about the median of the combined group of 70 observations?
(c) Suppose in part (b) that only the medians of 6.00 and 6.50 are available, the sample sizes having been inadvertently not recorded. What can you say about the median of the combined samples?

3.3 Canning processes involve the high-volume filling of cans with specified amounts of a product. Generally, there is a minimum weight or volume specification that each can must meet or exceed. Underfilled cans are either refilled or, in some cases, scrapped.[1]
(a) In a certain filling process, suppose that the specified fill amount is 10 oz. Furthermore, suppose that the standard deviation of the filling operation is estimated to be $\sigma = 0.8$ oz. Assuming that the empirical rule adequately describes the distribution of weights of filled cans, approximately where should the mean fill rate μ be set so that only 2.5% of the cans are underfilled?
(b) Where should μ be set if it is desired that only 1% of the cans be underfilled?

---

[1]See Boucher and Moshen (1991).

3.4   For a set of 100 process measurements, suppose you calculate the sample standard deviation s to be 0.90. A partial listing of these measurements (the smallest and largest ten readings, in ascending order) is:

{0.20, 0.21, 0.22, 0.22, 0.23, 0.24, 0.24, 0.24, 0.25, 0.25, ..., 1.3, 1.3, 1.4, 1.5, 1.5, 1.6, 1.6, 1.6, 1.7, 1.8}

(a) Explain why the value $s = 0.90$ cannot possibly be correct.
(b) Using only the portion of the data shown here, produce at least two reasonable estimates of the standard deviation.

3.5   The following data are obtained from a process set to make parts with a nominal dimension of 0.400:

```
0.391   0.396   0.404   0.381   0.394   0.411   0.406   0.375   0.399
0.400   0.404   0.394   0.408   0.387   0.398   0.386   0.413   0.393
0.408   0.408   0.396   0.411   0.411   0.380   0.382   0.400   0.393
0.413   0.377   0.396
```

(a) Calculate the sample mean, median, and standard deviation of these data.
(b) Subtract the nominal value of 0.400 from each reading in the data. Calculate the mean, median, and standard deviation of the transformed data. How can these results be used to estimate the mean, median, and standard deviation of the original data?
(c) Break the data into six groups of size 5 each. Find the sample range of each group and find the average of the six results. Then estimate the standard deviation of the process. Does it matter *how* you select the six groups (e.g., horizontally, vertically, or simply randomly)?

3.6   Create a histogram of the data in Exercise 3.5. Using $\bar{x}$ and s from part (a), what percentages of the data fall within the intervals $[\bar{x} - s, \bar{x} + s]$, $[\bar{x} - 2s, \bar{x} + 2s]$, and $[\bar{x} - 3s, \bar{x} + 3s]$?

3.7   An instrument is rated to give length measurements that are accurate to within 1% of the true value being measured. Two parts are measured, one whose true length is 0.500 in. and the other with a length of 2.000 in. For each part, find the range within which you expect the measured lengths will lie.

3.8   A 0.500-in. standard gage block is measured five times using instrument A and five times using instrument B. The resulting measurements are:

Instrument A	Instrument B
0.512	0.520
0.533	0.517
0.481	0.495
0.489	0.489
0.512	0.487

Estimate the accuracy and precision of each instrument.

3.9   Suppose a ruler is graduated in inches, but no finer graduations are given. Explain why measurements made with such a ruler are accurate to the nearest *half* an inch.

3.10   Given the five measurements 1.51, 1.48, 1.56, 1.50, and 1.55, estimate the standard deviation of the process from which these measurements were obtained. [*Careful: The answer is not simply the sample standard deviation, s.*]

3.11   For any constant *T*, elementary algebra shows that the equation

$$\sum_{i=1}^{n} (x_i - T)^2 = \sum_{i=1}^{n} (x_i - \bar{x})^2 + n(\bar{x} - T)^2$$

always holds, where $x_1, x_2, x_3, \ldots, x_n$ represent sample measurements of a quality characteristic with nominal dimension *T*. Use this equation to justify the assertion that, for any symmetric loss function $l(y) = k(y - T)^2$, the average loss associated with the *n* measurements increases as the process average shifts farther from its target value, *T*.

3.12   A measuring instrument is rated to give measurements to within 2% of the true value being measured. More specifically, if an item with a true value of *M* was measured many times, about 99% of those measurements would lie within 2% of *M*. Assuming that the empirical rule (see Section 3.3) applies to these measurements, for any particular value *M*, what is the coefficient of variation of the measurements from this instrument (assume that the instrument's readings have no bias)?

3.13   When measuring surface flatness, deviations above or below a nominal level are often measured as positive numbers. In other words, such measurements record how far 'off target' the surface flatness is, regardless of the direction of the deviations (Pyzdek 1992). When put in histogram form, such data form a folded normal distribution. To obtain an example of such data, use a computer package to generate a sample of 100 observations from a normal distribution with a mean of zero and a standard deviation of one. Change the signs of all the negative readings in the data and create a histogram of the resulting data.
(a) Calculate the mean and standard deviation of the data.
(b) What proportion of the data fall within 1-, 2-, or 3-standard deviations of the mean?

3.14   In a bottling process, a beam of light is passed through the necks of bottles passing by the beam on a conveyor system. Under-filled bottles allow the beam to pass through the bottle and trip a sensing device. Such bottles are automatically removed from the conveyor system. Describe the *shape* of the distribution of fill volumes that you would expect to find in bottles that pass inspection. In particular, do you expect this distribution to be skewed (if so, in what direction) or truncated (if so, on which side)?

REFERENCES FOR CHAPTER 3

ASQC. 1983. *Glossary and Tables for Statistical Quality Control.* American Society for Quality Control, Milwaukee, WI.

Boucher, T. O., and A. J. Moshen. 1991. "The Optimum Target Value for Single Filling Operations with Quality Sampling Plans." *Journal of Quality Technology* 23 (no. 1):44–47.

Burchett, G. L. 1984. "Testing Test Equipment." *Quality Progress* (no. 4):34–36.

Deming, W. E. 1975. "On Probability As a Basis for Action." *The American Statistician* 29 (no. 4):146–152.

Dixon, W. J., and F. J. Massey. 1983. *Introduction to Statistical Analysis,* 4th ed., Table A-8b(1). New York: McGraw-Hill.

Edgington, E. S. 1980. *Randomization Tests,* p. 5. New York: Marcel Dekker.

Eisenhart, C. 1962. "Realistic Evaluation of the Precision and Accuracy of Instrument Calibration Systems." *Journal of Research of the National Bureau of Standards—C. Engineering and Instrumentation* 67C (no. 2).

Grubbs, F. E. 1973. "Errors of Measurement, Precision, Accuracy, and the Statistical Comparison of Measuring Instruments." *Technometrics* 15 (no. 1):53–66.

Gutermann, H. E. 1962. "An Upper Bound for the Sample Standard Deviation." *Technometrics* 4:134–135.

Hahn, G. J., and W. Nelson. 1970. "A Problem in the Statistical Comparison of Measuring Devices." *Technometrics* 12 (no. 1):95–102.

Henkoff, K. 1989. "What Motorola Learns from Japan." *Fortune,* April 24, pp. 157–168.

Hicks, C. R. 1980. "Checking Instrumentation by Statistical Methods." *Journal of Quality Technology* (no. 4):181–186.

Juran, J. M., and F. M. Gryna. 1980. *Quality Planning and Analysis,* 2nd ed. New York: McGraw-Hill.

Kruskal, W. H., and J. M. Tanur. 1978. *International Encyclopedia of Statistics* 2: 1072. New York: Macmillan and Free Press.

Mandel, J., and T. W. Lashof. 1987. "The Nature of Repeatability and Reproducibility," *Journal of Quality Technology* 19 (no. 1):29–36.

Meyers, P. L. 1970. *Introduction to Probability and Statistical Applications,* 2nd ed. Reading, MA: Addison Wesley.

Murphy, R. B. 1961. "On the Meaning of Precision and Accuracy." *Materials Research and Standards* 4:264–267.

Pyzdek, T. 1992. "Process Capability Analysis Using Personal Computers," *Quality Engineering* 4 (no. 3):419–440.

Rosander, A. C. 1985. *Applications of Quality Control in the Service Industries.* New York: Marcel Dekker.

Rosenblatt, J. R., and C. H. Spiegelman. 1981. "Discussion." *Technometrics* 23 (no. 4):329–333.

Sachs, L. 1982. *Applied Statistics.* New York: Springer Verlag.

Shaw, R. A. 1987. "A Quality Cost Model for Hospitals." *Quality Progress* 20 (no. 5):41–45.

Stevens, W. L. 1948. "Control By Gauging." *Journal of the Royal Statistical Society,* Series B, 10 (no. 1):54–108.

Sweet, A. L., and K. L. Ong. 1983. "On a Simple and Accurate Approximation of Control Chart Constants." *IEE Transactions* 15 (no. 4):367–370.

Thompson, W. A. 1963. "Precision of Simultaneous Measurement Procedures." *Journal of the American Statistical Association* 58:474–479.

4

GRAPHICAL TECHNIQUES

Many quality control graphs are used for problem solving, not just data description. This chapter presents the most commonly used graphical methods for detecting and solving quality problems.

CHAPTER OUTLINE

4.1   INTRODUCTION

The use of graphical methods for summarizing and analyzing data has a surprisingly short history. As Tufte (1983) points out, statistical graphics were invented in around 1750–1800, quite a long time after other major mathematical achievements such as the calculus, analytic geometry, and logarithms. Perhaps even more surprising is that most efforts to develop analytically powerful graphical methods have occurred relatively recently, within the last two or three decades. In the literature of statistics and quality control, it is now widely held that well-constructed graphs can more clearly represent data than numerical summaries can (e.g., see Section 1.4 as well as Gunter 1988, Tufte 1983, and Schmid 1986).

With the exception of cause-and-effect diagrams, location charts, and perhaps Pareto charts, the graphical techniques described in this chapter are standardly used statistical graphs that have been applied to quality control situations. This chapter emphasizes their use as problem-solving tools, not just data summarization techniques.

4.2   HISTOGRAMS

The construction of histograms from variables and attributes data was discussed in Section 3.2. We turn now to the applications of histograms in *problem solving* and process capability analysis. For this purpose, the histogram is modified slightly by the inclusion of specification limits on the graph. In this way, the current behavior of a process (represented by the histogram) can easily be compared to the requirements (specification limits) placed on that process.

Figure 4.1 shows some typical histograms that arise when process data are examined, along with a list of the most likely reasons (causes) for their occurrence. *The interpretations given in Figure 4.1 are best applied when the process is known to be fairly stable* (i.e., the process average does not vary markedly over time). Otherwise, they should be used as rough indicators of where problems may lie, since (from Deming's warning about inferences from analytic studies in Section 3.1) changes in underlying conditions could equally well have been responsible for some of these histograms.

Concerning process stability, it is important to recognize that stability *cannot* be judged by looking at the histogram. Other tools, such as runs charts and control charts, are needed to determine stability. For example, a centered histogram contained within the specification limits, like Figure 4.1(h), does *not* necessarily indicate process stability. Such a histogram may, in fact, arise from an unstable process, as is demonstrated in Example 4.2 of Section 4.3.

EXAMPLE 4.1   Various histograms of the copper thickness data of Example 3.3 appear in Figures 3.4–3.6. Suppose that the specification limits for the copper plating

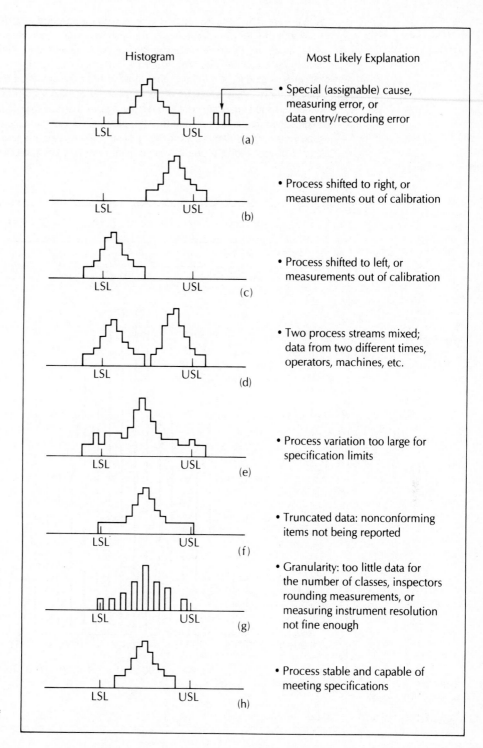

FIGURE 4.1
Typical histograms of
process data

process are LSL = 0.001 in. and USL = 0.005 in. around a nominal (target) thickness of 0.003 in. These limits are included on another histogram of the data (produced this time using the *SPC1+* package) in Figure 4.2. From this histogram it can be seen that the process average (0.0025 in.) is fairly close to the target value (0.003 in.) and that most of the measurements are within the specifications. Only a few measurements are at or below the lower specification, so the situation is not too different from that in Figure 4.1(h), and one might be tempted to conclude that the plating process is stable and capable of meeting its specifications. Unfortunately, as will be seen in Example 4.2, another graphical tool shows (using the same data) that the plating process is not at all

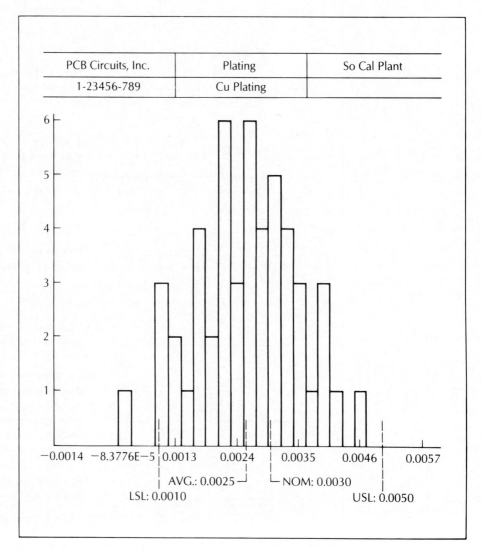

FIGURE 4.2
Histogram of copper plating thickness data of Table 3.2

stable. Thus, although histograms can provide evidence of certain kinds of problems, they cannot guarantee that others do not exist.

Besides detecting possible process problems, histograms are used to evaluate process capability—that is, the ability of a process to stay within its specifications limits. As the preceding example cautions, however, process capability can be assessed only after one feels that the process is indeed stable. For stable processes, capability is then judged by how well **centered** the process is (i.e., how close the center of the data lies to the target value) and by the magnitude of the variation in the measurements. Centered processes with very small variation (compared to the specification limits) are often called **capable** processes. The smaller the process spread (see Section 3.3), the more capable the process is said to be. Figure 4.1(h) shows what a typical histogram from a capable process should look like. Conversely, processes with large numbers of measurements outside the specification limits are said to be **not capable.** A process that is not capable could result because of shifts in the process average, process variability, or both. Beyond these rough definitions, process capability can be quantified more exactly. This topic is taken up in Chapter 8.

4.3   RUN CHARTS

The stability of a process cannot be judged by a histogram alone. *Time* is the essential element left out of a histogram analysis. Keeping track of the *order* in which the measurements are produced is critical for monitoring, adjusting, and controlling processes and, for this reason, control chart techniques are frequently based on sequential or time-based samples (see Chapters 6, 7, and 9).

One of the easiest ways to investigate changes over time is to plot the data versus the order in which the measurements are obtained. Usually this means plotting the measurements versus time or versus the order of production. The resulting graph is called a **run chart,** and it is used to spot possible trends or to assess the effect of known process changes (e.g., changes in workshifts, materials, workforce). The run chart can be used as an indicator of process stability, since, for most stable processes, the chart should resemble a random (i.e., patternless) horizontal scatter as in Figure 4.3(a). Instability or changes in a process are evidenced by noticeable trends or shifts, as shown in the other plots in Figure 4.3. Trends, for example, can be caused by tool wear, the effects of process improvements, slow changes in chemical concentrations, and so on.

EXAMPLE 4.2   Returning to the copper plating thickness data in Table 3.2, we recall that the histogram in Figure 4.2 indicated that most of the thicknesses fall within the product specification limits. If it is assumed that these thicknesses were measured on successive printed circuit boards from the plating process, the run

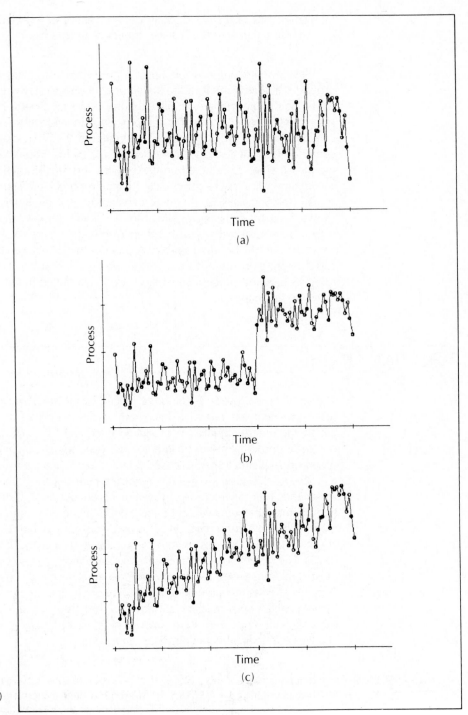

FIGURE 4.3
Run charts: (a) a
stable process,
(b) a process shift, (c)
a trend.

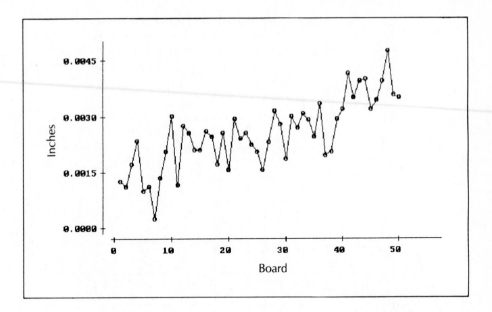

FIGURE 4.4
Run chart of the
copper thickness
data in Table 3.2

chart for these data in Figure 4.4 shows that the ability to meet specifications is probably only a temporary situation, since there is a pronounced upward trend in the data over time. The run chart indicates that the plating process is not stable and that immediate action is necessary to keep it from exceeding the upper specification limit. By comparison, the histogram in Figure 4.2 gave no such indication of impending problems.

Run charts are usually viewed as precursors to more sensitive control chart schemes for detecting process changes. However, what run charts lack in sensitivity for detecting problems, they make up for in ease of application. Unlike control charts that require many calculations and, therefore, present opportunities for errors, the run chart requires only the ability to draw a simple graph. In cases such as complex service operations where large amounts of data exist and training budgets and time are limited, run charts can provide valuable information at minimal cost.

EXAMPLE 4.3

Some service organizations, such as the California Department of Motor Vehicles (DMV), process massive amounts of paperwork and computer transactions. A major problem in such environments is the amount of rework (in the form of correspondence or visits to field offices) that is sometimes necessary to complete required forms. Complicating the situation is the fact that the DMV workforce is not centrally located, but scattered throughout the state in about 160 field offices. In 1987, new policies and procedures were introduced for reducing the percentage of RDFs (Report of Deposit of Fees), a DMV measure of unresolved problems regarding fee payments. As an inexpensive, easily implemented method for evaluating the effects of the new program, run charts were used at the various offices to show the RDF rate before and after the new policies were instituted (see Figure 4.5). The downward trend evident in Figure 4.5 is an indication that the procedures did improve the RDF problem.

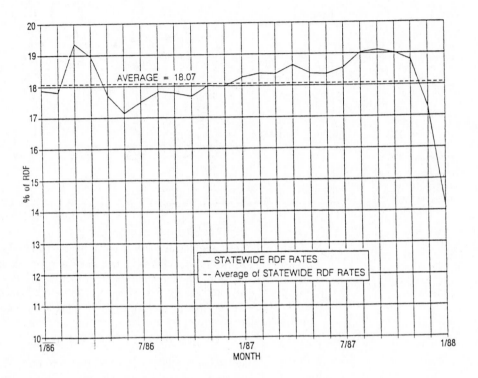

FIGURE 4.5
Run chart of DMV
data in Example 4.3

4.4   STEM-AND-LEAF DISPLAYS

Much of the current interest in graphical tools can be attributed to work done at Bell Laboratories by Princeton statistician John Tukey and others (Tukey 1977). Tukey is the originator of the stem-and-leaf display discussed in this section and the boxplot introduced in Section 4.5. At the same time, quality practitioners have developed new graphs (e.g., the cause-and-effect diagram) and modified others (Pareto diagrams) to become general problem-solving tools (see Sections 4.6 and 4.7).

Recognizing the power of visual displays, the designers of these new tools have emphasized simple graphs that retain much of the information in the original data and that are resistant to outliers (i.e., unusually small or large values in the data). Stem-and-leaf displays, for example, do much the same job as histograms by showing the range and concentration of the data, but they also retain more of the individual readings than do histograms, they are easier to construct (e.g., it is easier to decide what classes to use), and they allow quick estimates of percentiles (especially the median).

Stem-and-leaf displays are usually constructed for variables data—that is, measured data recorded to a certain number of digits of accuracy. The following steps are followed in the construction of a stem-and-leaf display:

CONSTRUCTING A
STEM-AND-LEAF
DISPLAY

1. Examine the data and decide where (between which two adjacent digits) to split each of the measurements.
2. The digits to the left of the split are called the **stems**, which are then sorted from smallest to largest. (Missing stems in the data are filled in where necessary.)
3. The **leafs** are the *single* digits to the right of the split, and the **leaf unit** is the decimal place occupied by the leaf digits. (*No rounding* is used when discarding the remaining digits.) For example, with a leaf unit of 1.0, a leaf of 6 represents the number 6; with a leaf unit of 0.10, a leaf of 6 stands for 0.6; and so on. Stacking all the leafs to the right of their corresponding stems forms the display.

EXAMPLE 4.4

A histogram of the delivery time data of Table 3.1 appears in Figure 3.2. For comparison, two stem-and-leaf displays of these data were constructed using the *Minitab* software package, as shown in Figure 4.6. Note that the leftmost columns in both graphs are neither stems nor leafs, but rather cumulative cell counts (explained later in this example). To generate Figure 4.6(a), the digits are split between the 10's and 1's place so, for example, the number 16.7 splits into a stem of 1 and a leaf of 6 (remember, digits trailing the leafs are truncated, not rounded). Similarly, a stem of 2 and leaf of 4, which occurs four times in the last row of Figure 4.6(a), corresponds to the data values 24.0, 24.5, 24.8, and 24.0. Thus, the leaf unit here is 1, since leafs occupy the 1's place. In Figure 4.6(b), the data are split at the decimal point, making the leaf unit equal to 0.1. In this diagram, a stem of 2 and leaf of 4 corresponds to the data value 2.4.

Notice that deciding on how many classes to use is not the problem with stem-and-leaf displays that it is with histograms. The main decision with stem-and-leaf graphs concerns where to split the digits, which immediately determines the classes (stems) that will be needed. This may require filling in a few stems for which there are no data, such as the stems 10, 14, and 18–23 in Figure 4.6(b), which is analogous to including an empty bar in a histogram when no data fall in a class.

The column to the left of the stem in the *Minitab* printout represents the *cumulative number of observations* from that cell out *to the nearest tail*. In Figure 4.6(a), the entry 9 in the left column means that there are nine data values from that stem to the nearest end (upper end in this case) of the data. The one cell whose count is enclosed in parentheses is associated with the stem that contains the median of the data. This cell count is *not* cumulative, but

```
                (a) Leaf Unit = 1.0

                N  = 84

         14     0 111111111111111
         34     0 2222222222233333333
        (21)    0 4444444444445555555555
         29     0 666677
         23     0 88888899
         15     1 1
         14     1 2233
         10     1 5
          9     1 66677
          4     1
          4     2
          4     2                      (b) Leaf Unit = .10
          4     2 4444
                                        14     1 11122333367789
                                        26     2 011224566789
                                        34     3 00024599
                                       (11)    4 00234445789
                                        39     5 0012223677
                                        29     6 1457
                                        25     7 13
                                        23     8 133478
                                        17     9 13
                                        15    10
                                        15    11 9
                                        14    12 17
                                        12    13 09
                                        10    14
                                        10    15 7
                                         9    16 227
                                         6    17 58
                                         4    18
                                         4    19
                                         4    20
                                         4    21
                                         4    22
                                         4    23
                                         4    24 0058
```

FIGURE 4.6
Stem-and-leaf
displays of the
delivery time data in
Table 3.1

simply represents the number of data values or leafs on that particular stem. Thus, the stem of 0 next to the parenthetical count of (21) must have 21 leafs. The column of cumulative counts and the column containing the median can be used to quickly estimate the median of the data. For example, in Figure 4.6(b), we know that the median must be the (84 + 1)/2th, or 42.5th, observation in the data (using the conventions of Section 3.3)—that is, the average of the 42nd and 43rd observations. Since the first 34 values are accounted for by stems 1, 2, and 3, one only has to count up the remaining distance of 8 and 9 leafs on stem 4 to find observations 42 and 43. These leafs are 5 and 7, so the corresponding data values are 4.5 and 4.7, yielding a median of 4.6 (which agrees with the median calculated in Section 3.3).

In statistical practice, the leftmost digits are usually considered the most important ones, but in quality control work, the rightmost ones are usually of primary interest. The reason for this is that the *leading* digits of process data usually remain fairly constant, so that the information on variability is to be found in the trailing digits. To eliminate unnecessary repetition of the leading digits in the data, the nominal (target) dimension may be subtracted from the raw data and the resulting deviations from the target graphed instead.

EXAMPLE 4.5   Table 4.1 lists the lengths of certain parts whose nominal dimension is 1.5 in. and whose specification limits are ±0.005 in. A stem-and-leaf display of these data appears in Figure 4.7(a). Since only the rightmost digits are changing in these data, the nominal value of 1.5 can be subtracted from each reading and the deviations can also be plotted in a stem-and-leaf fashion as in Figure 4.7(b). Both graphs tell the same story, but the second one may be preferred by practitioners who are already very familiar with the nominal dimension for a part and whose primary concern is with the last few digits.

TABLE 4.1   Part Lengths (in inches) for Example 4.5

1.488	1.492	1.492	1.492	1.492	1.493	1.494	1.495	1.496	1.496
1.497	1.497	1.499	1.499	1.499	1.501	1.501	1.501	1.501	1.503
1.503	1.503	1.504	1.504	1.505	1.505	1.506	1.506	1.508	1.509

FIGURE 4.7
Stem-and-leaf
displays of the data
in Table 4.1

```
        (a) Stem-and-Leaf of Raw Data

Stem-and-leaf of data          N  = 30
Leaf Unit = 0.0010


     1  148 8
     1  149
     6  149 22223
     8  149 45
    12  149 6677             (b) Stem-and-Leaf of deviations from 1.5 inches
    15  149 999
    15  150 1111                Stem-and-leaf of deviations        N  = 30
    11  150 333                 Leaf Unit = 0.0010
     8  150 4455
     4  150 66                      1   -1 1
     2  150 89                      1   -0
                                    6   -0 77776
                                    8   -0 54
                                   12   -0 3322
                                   15   -0 000
                                   15    0 1111
                                   11    0 333
                                    8    0 4455
                                    4    0 66
                                    2    0 89
```

4.5   BOXPLOTS

Boxplots, also called **box and whisker plots,** are graphs that describe the distribution of process data by using Tukey's 'five-number summary' scheme (Tukey 1977). These five numbers are the minimum, the first quartile of the data (denoted by

Q_1), the median, the third quartile (denoted Q_3), and the maximum of the particular data set. The first quartile, Q_1, is defined to be the value that exceeds the lower 25% (i.e., first quarter) of the data after the data are sorted from smallest to largest. Similarly, the third quartile, Q_3, is the number that just exceeds the first 75% of the data. The median, maximum, and minimum were defined in Section 3.3. These five numbers usually provide a good picture of where the process is centered, the process spread, and potential outliers. Also, instead of using the mean, boxplots use the somewhat more stable median to estimate the central tendency of the data.

There is a quick method for approximating the quartiles: notice that, since the median divides the data into upper and lower halves, Q_1 and Q_3 will be very close to (*but not necessarily equal to*) the medians of these two halves. Technically, these two medians are referred to as **hinges**, not quartiles, but for most practical purposes, the slight difference between the exact quartiles and the hinges is not important (Velleman and Hoaglin 1981).

Figure 4.8 shows a typical boxplot depicting the five-number summary of a set of data. The area inside the box represents the middle half of the data, while the line drawn through the box shows where the median falls. The two lines that extend from the sides of the box out to the extreme values are called the 'whiskers.' The following steps outline the construction of a boxplot:

CONSTRUCTING A
BOXPLOT

1. Calculate the median of the data.
2. The median divides the data into lower and upper halves; find the hinges by calculating the medians of these two halves.
3. Draw a rectangle (i.e., the 'box') connecting the two hinges, and locate the median inside.
4. Extend lines out from the hinges to the smallest and largest values in the data.

In order to highlight possible outliers in the data, another version of this graph is used in which the whiskers do not necessarily extend all the way out to the extreme values, but instead are drawn out to a distance of $1.5 \cdot (Q_3 - Q_1)$ from

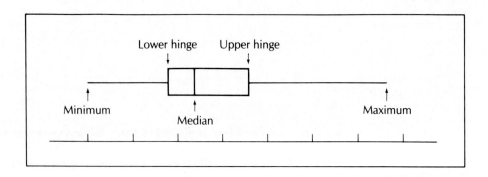

FIGURE 4.8
The boxplot

each side of the box. The spread $Q_3 - Q_1$ is called the **interquartile range,** and boxplots with whiskers one and a half times this length are referred to as **modified boxplots** (Moore and McCabe 1989). Observations that fall beyond the whiskers are highlighted as potential outliers in the data. In this text, the term 'boxplot' is used to refer to either boxplots or modified boxplots.

EXAMPLE 4.6   The delivery time data of Table 3.1 has been examined using both the histogram and stem-and-leaf approaches (see Figures 3.2 and 4.6). For comparison, a boxplot of these data from the *Minitab* package is shown in Figure 4.9. Since *Minitab* uses *modified* boxplots, the delivery times in the 24-hr region clearly stand out from the rest of the data. Notice that *Minitab* uses asterisks to denote potential outliers and that repeated values are plotted only once (i.e., the outliers 24.0, 24.5, 24.8, and 24.0 give rise to three, not four, asterisks). These values could be the result of data entry mistakes, or they could be true outliers. As pointed out in Example 3.1, further investigation of these observations could provide valuable information on the delivery time process. The skewness of the data (i.e., the fact that the preponderance of the times are less than 8 hr while some run as high as 24 hr) is also clearly shown by the leftward orientation of the box and the longer whisker on the right side.

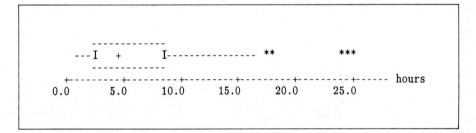

FIGURE 4.9
Modified boxplot of
the delivery time
data of Table 3.1

Stem-and-leaf displays tend to retain more of the information in the data than do boxplots. One way to see this is to notice that from the stem-and-leaf display, one can easily calculate the five numbers required for the boxplot, but, conversely, a stem-and-leaf cannot be generated from only the five-number summary in the boxplot. As a result, boxplots can sometimes hide important features in the data that might otherwise be more easily seen in the stem-and-leaf display. Boxplots are therefore recommended for comparing *several* sets of data rather than for analyzing a single data set.

EXAMPLE 4.7   To compare boxplot and stem-and-leaf displays, consider the two boxplots shown in Figure 4.10. Part (a) was calculated from all 50 plating thickness measurements in Table 3.2. Part (b) was made after eliminating the measurements that correspond to circuit boards 16–34. Both boxplots look much the same, and one might be tempted to conclude that both sets of data are probably very similar. However, from the corresponding stem-and-leaf displays in

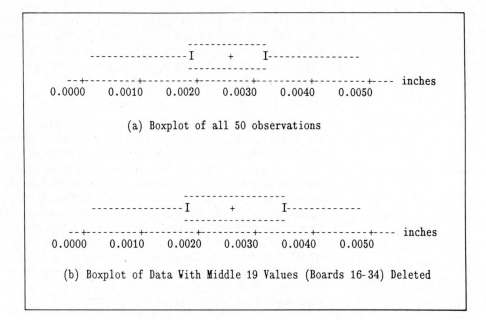

FIGURE 4.10
Boxplots of the
plating thickness
data in Table 3.2

FIGURE 4.11
Stem-and-leaf
displays
corresponding to
the boxplots of
Figure 4.10

Figure 4.11, it can readily be seen that the data sets are different. The first set (i.e., all 50 observations) appears concentrated around the median, whereas the second collection seems to exhibit more dispersion.

From the recommendation of using boxplots when comparing several sets of data, it should also be noted that these plots can be very helpful for comparing successive samples from a process, or for comparing the effects of a process change by looking at boxplots before and after the change. In the first situation, boxplots of the samples can be plotted sequentially (as the successive samples arise) to monitor changes in the process average (in this case, the median) and variation (Iglewicz and Hoaglin 1987).

4.6 PARETO CHARTS

Bar charts that depict data on a single *measured* characteristic are called histograms (see Sections 3.2 and 4.2). The bars are formed by dividing up the horizontal scale into a collection of *classes* and then counting the frequencies with which the measurements fall into these classes. Another type of bar chart, called a **Pareto chart,** is also constructed by counting the class frequencies, but, unlike histograms, the horizontal axis of a Pareto chart represents various *categories* of interest; it is not a continuous graduated scale.

Pareto charts were introduced as a quality control tool by Juran (1964) and are based on attributes (counted) data. After a 'concern' or 'area of interest' is determined to be used as the horizontal axis of the chart, the concern is *subdivided* into categories. Then the number of times that each particular category occurs in the data is counted, and the resulting frequencies are plotted as the heights of the corresponding bars on the chart. These charts focus attention on important factors or problems that affect a process when the bars (and associated categories) are *rearranged* so that the tallest bars are on the left side of the chart. This cannot be done with histograms, since measured data place a very specific ordering on the classes. Categories, on the other hand, can be listed in any order desired, and it has become traditional to put the tallest bars on the left. By drawing attention to the tallest bars, then, Pareto charts help sort out the few serious problems from the many minor ones.

Pareto charts are so named because they usually show a "Pareto effect," first noticed by the Italian economist Vilfredo Pareto (1848–1923) in his studies of the distribution of incomes. The Pareto effect refers to the fact that, in most cases, only a small group of bars in the chart tend to account for a majority of the data. The Pareto effect also occurs in inventory theory, where it is usually called the "80–20" rule, which states that about 80% of a company's sales can be attributed to only 20% of the inventoried items.

EXAMPLE 4.8   In the manufacture of printed circuit boards (see Example 3.3), finished boards are subjected to a final inspection before they are shipped to customers. During this inspection, defective boards are found and the types of defects are recorded. For a particular month, Table 4.2 shows the numbers of boards rejected at final inspection by a circuit board company's inspection department. The Pareto chart for these data appears in Figure 4.12. Note that the chart includes a 'miscellaneous' category, which is conventionally listed last. The reason for this is that the miscellaneous category usually consists of a small

TABLE 4.2   Rejected Printed Circuit Boards Classified by Type of Defect

Type of Defect	Number of Rejected Boards
Poor electroless coverage	35
Lamination problems	10
Low copper plating	112
Plating separation	8
Etching problems	5
Miscellaneous	12

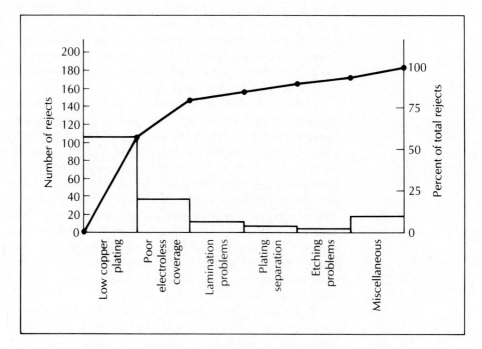

FIGURE 4.12
Pareto chart of data
in Table 4.2

number of rejects that are not worth the effort of being put into separate categories. So, since this category actually represents a group of infrequently occurring smaller categories, it properly belongs on the right side of the chart.

Because it is also desirable to quickly convert category frequencies into their relative percentages compared to *all* the data, Pareto charts normally use two vertical scales in their construction. The left-hand scale represents the original units of measurement, and the right-hand scale shows the corresponding range of percentages. As a visual aid, a cumulative frequency/percentage line is drawn through the corners of the bars so that one can easily judge the additional contribution of successive categories.

In Figure 4.13, the data of Table 4.2 have been replotted using *STATMAN* and

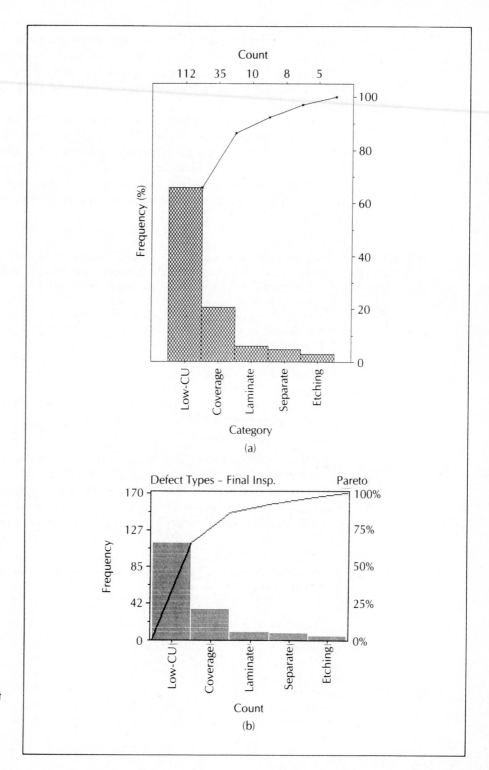

FIGURE 4.13
(a) *NWA Quality Analyst* Pareto chart of data in Table 4.2.
(b) *STATMAN* Pareto chart of data in Table 4.2.

the *NWA Quality Analyst* software packages. (*Note:* Some software packages do not allow complete user control in ordering the categories, so, for purposes of comparison, the 'miscellaneous' category is left out in both graphs.) From either Figure 4.12 or 4.13 one can see in the percentage scale that the first two bars (low copper plating and poor electroless coverage) account for more than 80% of the rejects.

Besides representing frequencies or percentages, the vertical scale on the Pareto chart can also denote *money.* Managers, for example, are often interested in seeing a Pareto chart based on frequencies of rejects converted into a chart in which the vertical scale is in dollars. *This may completely change the ordering of the categories* in the chart, since the categories that contribute most to the dollar volume may turn out to be quite different from those with the largest numbers of rejects.

EXAMPLE 4.9   Industries such as hospitals, banks, data processing, and public utilities handle large numbers of transactions daily. Correctly transcribing information from conversations, computer screens, and application forms is a major concern. Table 4.3 shows a breakdown of errors in a sample of 10,000 documents processed by a data-entry department during a one-week period. In addition to the frequencies, the estimated unit cost of fixing each error is listed in the table. Figure 4.14 shows two Pareto charts of these data, one with frequency as the vertical scale and the other with total cost. It can be seen that errors due to misspellings and address changes account for the vast majority of errors. When examined as to what these errors cost the company to fix, the second diagram shows that balance discrepancies (which occur much less frequently than spelling and address errors) are the main problem.

The process of choosing a particular problem, concern, or area of interest to be the horizontal axis of a Pareto chart and then dividing it up into subcategories related to that concern is called **stratification.** It is the repeated application of stratification that allows one to pinpoint process problems. For example, Figures 4.12 and 4.13 show low copper plating to be one of the major causes of re-

TABLE 4.3   Transcription Errors and Estimated Unit Costs from a Sample of 10,000 Documents

Type of error	Frequency (per 10,000)	Unit cost	Total cost
Misspelled name	20.0	$0.20	$4.000
Wrong address	20.0	0.20	4.000
Wrong customer code	3.0	0.30	0.900
Missing file	1.0	5.00	5.000
Balance discrepancy	5.0	1.50	7.500
Billing error	3.0	0.85	2.550
Others (miscellaneous)	4.0	0.05	0.200

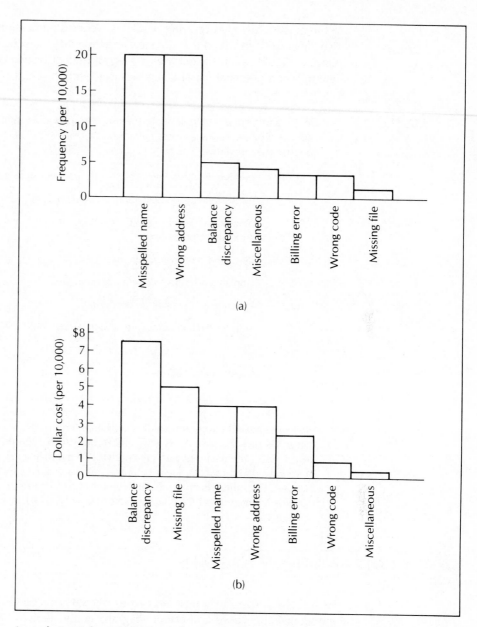

FIGURE 4.14
Pareto charts of data
in Table 4.3

jected circuit boards in Example 4.8. The next step in solving the reject problem is to use stratification to focus on low copper plating as the main concern, making it the horizontal axis of a *new* Pareto chart. The various reasons or causes of low copper plating then become the *new* subcategories, a *new* Pareto chart is constructed, another primary concern is identified, and so forth. Within a few steps, this process usually identifies a concern that has a known remedy. Fixing this problem usually causes a noticeable improvement in all the preceding Pareto charts, since, by the nature of the stratification procedure, attention is continually drawn to the most frequently occurring problems. When stratification is used, situations sometimes arise in which it is not clear exactly what the subcategories of a given concern should be. In those cases, cause-and-effect diagrams can be of help in determining a potential list of categories (see Section 4.7).

EXAMPLE 4.10

Demos and Demos (1989) illustrate the use of stratification in improving patient care in hospitals. Taking *adverse patient outcomes* (APOs) as the primary concern, they identified nine subcategories of APOs:

A   Admissions for adverse results of outpatient management

B   Admissions for complication of problem from previous visit

C   Operation for perforation, laceration, or other injury incurred during an invasive procedure

D   Drug and transfusion reactions

E   Unplanned return to operating room

F   Infection not present on admission

G   Transfer from general care to special care unit

H   Length of care greater than 90th percentile

I   Other complications

Each of these categories is further subdivided. For example, category D (drug and transfusion reactions) could result from D_1: medication errors, D_2: procedure-related errors, and so on. Medication errors (D_1) are divided into categories such as nurse A, nurse B, and so on. When appropriate departments keep track of the data required, Pareto charts of monthly (or weekly) APOs can be examined and problem areas pinpointed. If, for example, the APO chart identifies D as the primary problem, then the Pareto chart of category D is examined, which might identify, say, subcategory D_1 as the main problem.

4.7   CAUSE-AND-EFFECT DIAGRAMS

As a tool for identifying possible causes of quality problems, Dr. Kaoru Ishikawa introduced the **cause-and-effect diagram** in the early 1950s. Since then, this graphical tool has also become known as the **Ishikawa diagram** and the **fish-bone diagram.**

Cause-and-effect diagrams are usually developed in a group setting, in which representatives of various departments participate in 'brainstorming' sessions to find potential solutions to quality problems. One guiding rule of the brainstorming technique is that *all* suggestions, no matter how strange, are acceptable. Afterward, group discussion of the suggestions (avoiding any reference to which suggestions were made by whom) reduces the list to a manageable size.

The cause-and-effect diagram is a graphical method of brainstorming in which the problem to be attacked is first drawn at the 'head' of the diagram as shown in Figure 4.15(a). After suggestions of possible causes of this problem are elicited, additional lines are drawn, one for each 'cause.' To get such a diagram started, it is often convenient to begin with a few basic causes as shown in Figure 4.15(b). In turn, by considering each 'cause' to be the 'effect' of some other list of causes, one iteratively fills out the diagram with more and more levels (i.e., lines). The herringbone appearance of finished diagrams is what gives rise to the alternative name of 'fishbone' diagram. The important thing to remember when constructing the diagram is that the items listed on any branch are considered 'causes' of the next larger branch to which they are connected. Every branch, then, is both a cause and an effect, and the diagram clearly illustrates their interconnections.

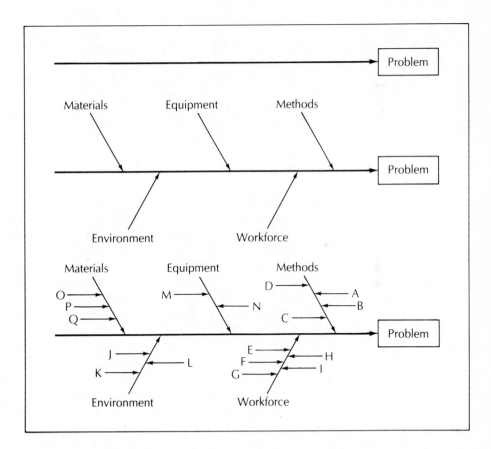

FIGURE 4.15
Construction of a
cause-and-effect
(fishbone) diagram

EXAMPLE 4.11   In Example 4.8, the Pareto analysis showed low copper plating to be the primary cause of rejected circuit boards at final inspection. In a brainstorming meeting with production managers and representatives of the plating and inspection departments, a list of possible causes was identified. The associated cause-and-effect diagram was drawn using the *STATMAN* software package

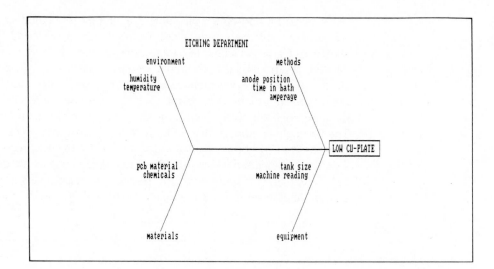

FIGURE 4.16
Cause-and-effect
diagram for
Example 4.11 from
STATMAN software

and is shown in Figure 4.16. The printout lists the secondary causes down the sides of the primary branches. Thus, *anode position, time in bath,* and *amperage* are all possible causes related to the *methods* by which the circuit boards are made. From the completed diagram, some of these causes can then be selected for further study. (*Note:* Although the *STATMAN* software does not allow for more levels of causes in its diagrams, many other quality control software packages do not offer cause-and-effect diagrams as one of their options.)

Cause-and-effect diagrams are most often used when the source of process problems is unknown and some creative thinking is needed to suggest other approaches. This happens frequently—for instance, when Pareto diagrams are used to solve problems. The successive levels of stratification required to track down the roots of a problem with Pareto analysis can get difficult because it is not always readily apparent what subcategories to use for successive Pareto charts. In these cases, *the Pareto and the cause-and-effect diagrams can be used sequentially;* that is, the initial Pareto chart is used to identify the first category to attack, and then a cause-and-effect diagram is used to produce a list of subcategories for the next Pareto chart. Data are collected for this Pareto chart, the largest category is selected for study, the cause-and-effect method is used to help find more subcategories, and so on. Essentially, the main concern in a Pareto chart is used as the 'head' of a cause-and-effect diagram, which identifies subcategories for the next Pareto, and so forth. A schematic of this procedure is shown in Figure 4.17.

4.8   SCATTER DIAGRAMS

In previous sections, with the exception of the cause-and-effect diagram, graphs have been used to describe one characteristic at a time. Histograms show the shape and location of data measured on *one* characteristic, Pareto charts break up data on *one* concern into subcategories, and stem-and-leaf displays and boxplots

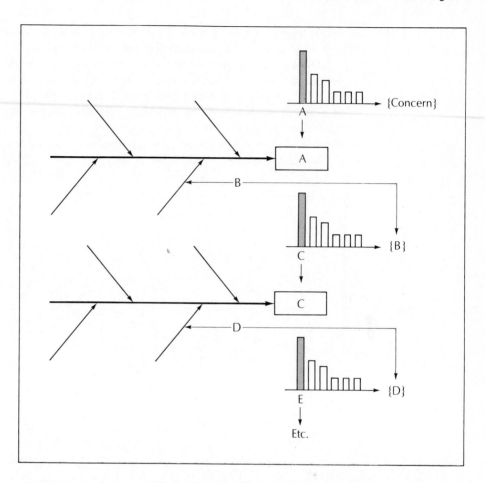

FIGURE 4.17
Pareto charts and
cause-and-effect
diagrams used
iteratively

depict data on *one* variable. The similarity among all these graphs is that they are designed for **univariate data**—that is, measurements made on only one characteristic of each item examined. Thus, although individual metal bolts may have many measurable characteristics (length, diameter, width, weight, strength, etc.), choosing any *one* of these to measure results in univariate data.

To compare more than one characteristic at a time, the histogram (or any univariate graph) of one characteristic can be plotted next to a histogram of the other. This is also a good way to compare the *same* characteristic measured on two or more processes—for example, when more than one machine is used to manufacture a particular part. Differences in process means and variations can be spotted quickly by comparing two histograms, or two stem-and-leafs, or two boxplots.

If two *different* characteristics are measured on each item, the resulting measurement *pairs* are termed **bivariate data.** In general, when many measurements are made on each item, the data are said to be **multivariate.** This section introduces graphs for bivariate data.

With bivariate data, the two characteristics measured can be considered separately as consisting of two sets of univariate data, each of which can be analyzed with histograms or any other univariate procedure. But bivariate (or multivariate) data are richer than this. They also contain information about the *relationship* between the two characteristics. To show the relationship visually requires another type of graph called the **scatter diagram,** or **scatter plot.**

In general, the two characteristics are denoted by X and Y. Suppose that a sample of n items is examined and the resulting bivariate data are (x_1, y_1), (x_2, y_2), (x_3, y_3), ..., (x_n, y_n). A scatter plot of these data is formed by plotting these pairs on a graph as shown in Figure 4.18(a). Relationships between X and Y are evidenced by patterns in the scatter plot. Some typical patterns are shown in Figure 4.19. If desired, the histograms can also be drawn on the axes of the scatter plot to summarize each variable separately [see Figure 4.18(b)].

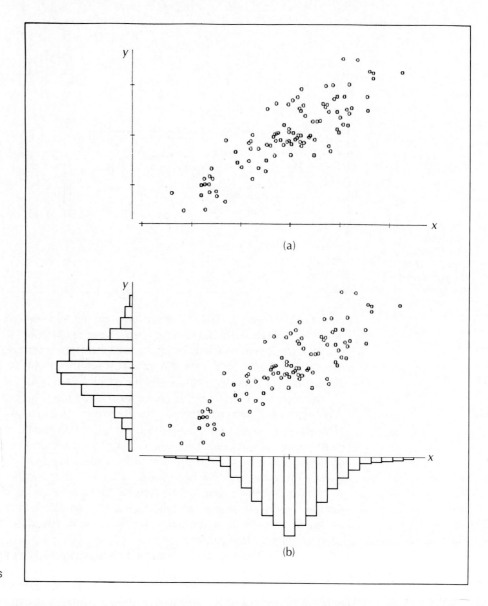

FIGURE 4.18
(a) Scatter plot;
(b) scatter plot and individual histograms of x and y.

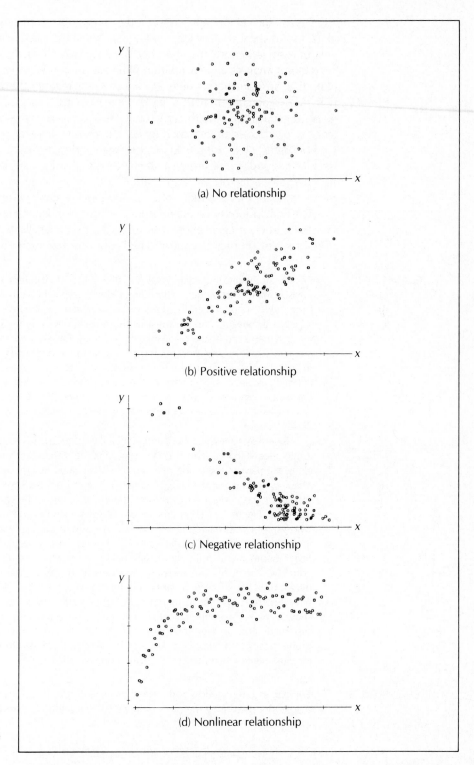

FIGURE 4.19
Examples of bivariate
relationships

Depending on the application, other names are sometimes used for scatter plots. In statistical studies, for example, one's interest may lie in estimating the degree of association or correlation between the two characteristics. Terms such as **correlation analysis** and **correlation plots** are often used when referring to scatter plots that arise in these studies. In another application, the **run chart** (see Section 4.3) is actually a scatter plot of measurements on a single characteristic graphed against the *time* the measurement was taken or the *order* of production. Since two variables are recorded for each item, run charts are indeed just a special kind of scatter plot. In the field of experimental design (see Section 1.4 and Chapter 13), the graphs of a response variable versus various levels of an independent variable (i.e., a factor) are also scatter plots. In the context of designed experiments, though, they are more often called plots of **factor effects.**

The conclusions to be drawn from a scatter plot depend on the patterns found and the particular application. The following examples illustrate two applications of scatter plots in quality control. The list of possible applications is endless.

EXAMPLE 4.12   Coordinate measuring machines (CMMs for short) are used for length measurements that require extreme precision. A typical CMM consists of both a large 'table,' upon which the parts to be measured are situated, and a 'probe,' which is a movable arm (or arms) with a sensing device in its tip for making measurements. To assure measurement accuracy, CMMs are kept in environmentally stable rooms (since changes in temperature and humidity can affect measurements). As the probe moves from one place to another along a part, internal calculations keep track of the exact distance traveled from the start of the measuring process (the 'zero' position) to where the probe reaches its desired location. The difference between the start and finish values is then the desired length.

One way to test the measuring capability of a CMM is to use it to measure a reference standard. This is done by placing the standard at various locations along the path of the probe arm and taking repeated length measurements. Then, by subtracting the known reference length from the CMM's reading, a series of measurement errors is formed corresponding to the various probe positions. Figure 4.20 shows the results of such a study applied to a CMM whose probe was moved in both directions away from the center of the table. In Figure 4.20(a), as the probe moves to the right, the measurement errors show only random variation around 0, as should be the case if the machine is working correctly. The ability of a measuring instrument to hold to true measurements over a distance is called *linearity,* and the CMM appears to exhibit linearity when the probe moves to the right. In Figure 4.20(b), however, the error terms show a marked pattern, indicating that the measurements experience a systematic error as the probe moves to the left. This lack of linearity when the probe moves to the left is undesirable because it could lead to incorrect measurements when the CMM is used on production parts.

EXAMPLE 4.13   Sometimes both vendors and customers collect data on the same parts, with the vendor measuring the parts before shipment and the customer measuring

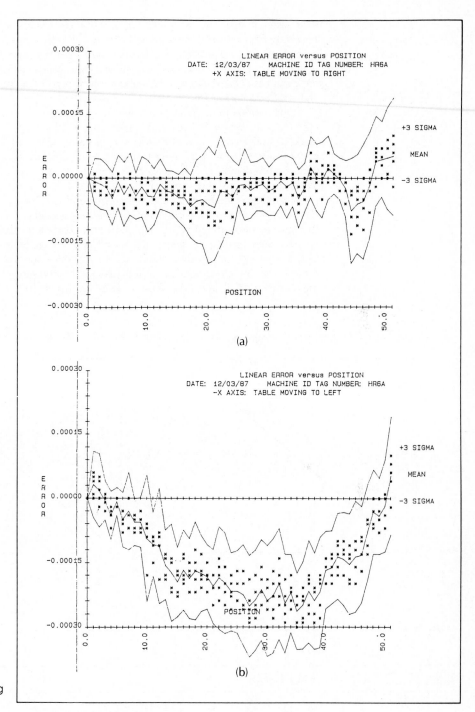

FIGURE 4.20
Scatter plot of
measurement errors
along the axes of a
coordinate measuring
machine

them again at receiving inspection. For precision electronic components such as *hybrids* (which consist of mixtures of circuitry and microprocessors), for example, critical electronic characteristics must meet tight specifications. Small differences in measuring techniques, operators, environmental conditions, and so forth can easily cause disagreement about which parts conform to specification and which do not.

One method for comparing the agreement between a vendor's and customer's testing procedures is to create a scatter plot of the measurements made by both parties on the *same* sample of items. Some of the differences between the pairs of measurements are due to natural process variation (i.e., the parts are not completely identical), and some are due to differences between the measuring instruments (or techniques). Figure 4.21 shows a scatter plot of measurements made by two measuring instruments (sometimes called 'testers') on the same ten parts. If there is no disagreement between the two testers (i.e., they give precisely the same readings for a given part), the scatter plot should approximately follow a 45-degree line as in Figure 4.21(a). However, if the testers differ in their precision and/or accuracy (see Section 3.4), the plot will deviate from the 45-degree line as in Figure 4.21(b). The subject of test and measurement capability is discussed in detail in Chapter 10.

FIGURE 4.21
Scatter plots. (a) Testers agree in precision and accuracy. (b) Testers differ in both precision and accuracy.

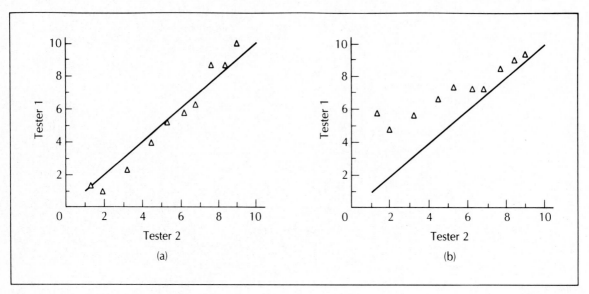

4.9   LOCATION DIAGRAMS

Location diagrams (also called location plots) are among the easiest quality improvement graphs to construct because they require no special skills in doing mathematics, measuring, or plotting data. Nonetheless, they allow very complex relationships in the data to be analyzed. Furthermore, the product itself provides the template for recording the data.

Location diagrams are used to record attributes data, particularly counts of *nonconformities* or *defects*. To construct the graph, a simplified two-dimensional drawing is made of the item to be inspected. Then, as successive items are visually inspected, any defects or nonconformities found are marked on the drawing at the same position as they occurred on the part. In time, the markings on the drawing may show patterns that indicate process problems. Simple though location plots may be, however, they contain information that is available in none of the other graphical methods.

EXAMPLE 4.14

The process of foam molding is used to make soft instrument panels for automobiles. Part of this process includes vacuum forming the vinyl surface of the dashboard and then adding the foam backing. Figure 4.22(a) shows a front-view drawing of a typical dashboard. As each dashboard leaves the mold, it is examined for various aesthetic and functional defects, including voids (air pockets below the vinyl surface), pin holes (front or side), creases on panel, creases on glove box, and other (miscellaneous) flaws. A location chart for voids might appear as in Figure 4.22(b), where marks indicate the locations of voids on several recently inspected dashboards. The concentration of voids near the right side of the diagram indicates a possible problem with the molding process. When presented with this diagram, workers familiar with the molds may recognize the reason for such problems. For example, it could turn out that the concentration of defects occurs near the opening through which the foam is injected. Subsequent cleaning of this opening on a more frequent basis might clear up the void problem in this area.

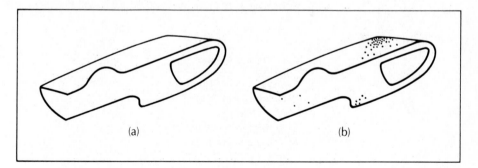

FIGURE 4.22
(a) Dashboard
diagram.
(b) Diagram used to
plot location of
defects.

(a)   (b)

EXAMPLE 4.15

Electronic components (e.g., resistors, capacitors, transistors) are soldered onto printed circuit boards (PCBs) by means of a 'wave' or 'flow' solder process, in which the PCBs are passed over a surface of liquid solder. Soldered boards continue moving through a series of chemical baths before they emerge from the wave solder process. Trained inspectors then visually examine the soldered connections on each board and, on a drawing of the PCB, put red marks over the connections that are incorrectly soldered. These charts are used to direct rework activities (i.e., hand soldering of the defective connections) and are eventually accumulated into a *single* location chart. Concentrations of markings in various regions of the location chart are, to the trained wave solder expert, indicative of various solder problems. For example, many marks near the edges of the chart might indicate that the conveyor system that moves the PCBs over the solder surface is set too fast. Other collections of markings may signal inadequate chemical concentrations, temperature changes, and so forth.

In one variant of the location chart technique, possible defect regions are given coded values or names, and defects are collected on simple checksheets that contain a list of the coded names. This technique is useful for highly complex components where a template of the product may be so complex that plotting the points is out of the question. The advantage of this approach is that data in this form are easily entered into the computer, so that Pareto charts can be constructed and problem areas quickly detected. With printed circuit boards, for example, coding various components and pin numbers ('pins' are the metal connections that emanate from a component) would lead to very precise Pareto charts that could identify major problem areas down to the component and pin level, not just to the less precise visual level found when points are hand plotted on a template of the part.

EXERCISES FOR CHAPTER 4

4.1   Instrument panels for automobiles are made by a process of foam molding in which soft foam material is placed under a harder vinyl skin. At one foam molding press, records are kept of the types of defects found in a group of 50 finished dashboards:

Problem Type	Number of Occurrences
Creases in skin—front	8
Pinholes in skin	11
Voids in foam	25
Creases in skin—glove box	2
Miscellaneous creases	4

Construct a Pareto chart for these data.

4.2   Estimate the frequency of occurrence of the letters in the alphabet by constructing a Pareto chart for the first two paragraphs in this chapter (i.e., use 'a,' 'b,' 'c,' ..., 'z' as the categories). Does the Pareto principle appear to apply in this application? In particular, what are the five most frequently used letters?[1]

---

[1]For a list of the standard occurrences of English letters, see Lucky (1989).

4.3   To analyze the variability in the adhesion of a particular product, the following cause-and-effect diagram was created by a department in the 3M Company:[2]

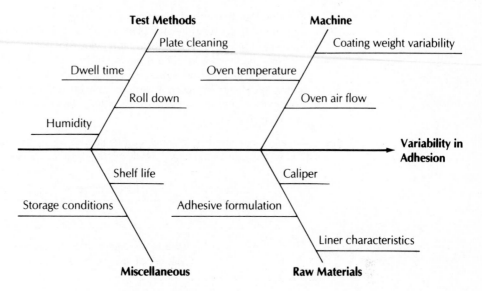

Suppose that adhesive formulation is thought to depend upon chemical concentration which, in turn, is affected by mixing method. Add this information to the cause and effect diagram.

4.4   In a high-volume operation, two identical machines are used to make the same parts. Later in the process, the parts from both machines are collected. As part of a routine inspection, before the parts proceed through the remaining process steps, a sample of 40 parts is selected and an important quality characteristic is measured:

20.3   19.8   19.9   20.3   19.7   20.0   20.2   19.8   20.2   19.9   19.8
19.3   20.4   20.7   19.5   19.6   19.7   20.2   20.2   20.1   18.0   17.6
18.5   18.0   18.0   18.0   17.8   18.4   18.0   17.8   17.9   18.1   18.2
18.2   18.5   18.6   18.3   18.3   18.0   17.7

(a) Create a histogram of these data.
(b) What explanation(s) can you offer for the appearance of the histogram in part (a)?

4.5   Explain how cause-and-effect diagrams and Pareto charts, when applied in a sequential fashion, can be used to discover the root causes of quality problems.

4.6   A majority of the steps in manufacturing printed circuit boards involve some sort of chemical process (e.g., transferring photos of circuit layouts onto the boards, etching surfaces with acid, applying copper plating by electrolyis). To evaluate these steps, small areas on each board are reserved for 'test coupons,' which are very simple circuit and hole patterns. The rest of the board is used for the production circuitry. To evaluate a quality characteristic (e.g., like plating thickness), one can then analyze part of the test coupon, thereby leaving the remainder of the board intact for further process steps. If uniformity in plating thickness across the entire board is a prime concern, describe how this might be monitored with a location

---

[2]From *Quality Progress*, July 1988, p 35.

diagram. Specifically, how would you position test coupons on a rectangular board for use with the location diagram?

4.7 In World War II, bombers returning from missions were examined to determine the location of bullet holes in each plane's body so that their armor could be fortified in the right places to prevent them from being shot down.[3] In essence, each returning plane provided a location diagram for the bullet holes. At first, it was thought that the regions with the highest numbers of holes would show where to fortify the armor. In the end, however, it was concluded that the planes should be fortified in exactly the opposite manner—that is, where the holes did *not* occur on the location diagrams. Why?

4.8 NMRs (nonconforming material reports) provide a natural source of data for Pareto charts. For example, a typical NMR might appear as follows:

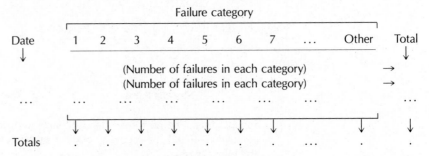

Explain how to create a Pareto chart from an NMR.

4.9 The following data, reprinted from Hunt (1989), list the numbers of rejected aluminum castings by day and defect type for a three-month period:

Cause of Rejection:																		DAY OF MONTH	
	6	7	8	9	10	Wkly. Total	13	14	15	16	17	Wkly. Total	20	21	22	23	24	Wkly. Total	3-Wk. Total
Sand	23	11	22	24	17	97	29	28	18	12	33	120	22	13	16	15	16	82	299
Mis-run	2		2	1	3	8	4		6	4	1	15	1		1	1	1	4	27
Blow																	1	1	1
Shift		15			3	18	1	1		3	7	12						0	30
Cut															1			1	1
Broken	11	2	3	1	4	21	3	3	4	4	3	17	7	4	3	7	4	25	63
Drop	1	1	2	3	1	8	4	3	3	2	1	13	3	4	2	1	2	12	33
Corebreak	1		15	5	2	23	4	5	1	8	3	21	8	1	4	13	12	38	82
Run-out	2			1	1	4	1	2	2	1		6	1			1	3	5	15
Core missing					1	1	1		1		1	3			1	1	1	3	7
Crush											6	6	4	2	3			9	15
Daily Total (Scrap)	40	29	44	35	32	180	47	42	35	34	55	213	46	24	31	39	40	180	573
Production	478	514	500	470	530	2492	430	494	500	505	465	2394	464	500	473	518	570	2525	7411

Create a Pareto chart of these data. What are the primary causes of rejected castings?

---

[3]For further discussion, see *OR/MS Today,* 1991.

4.10   Randomly select 100 words from a standard dictionary and, for each word, record the number of letters, x, and the number of syllables, y, that it contains. Make a scatter plot of y versus x for these data. Does this plot support your prior beliefs about the relationship between the number of syllables and the number of letters in a word?

4.11   The information in a cause and effect diagram can also be arranged in list form, as illustrated below:

Problem
  Cause A
    Effect 1
      Cause a
      Cause b
       .
       .
    Effect 2
      Cause a
      Cause b
       .
       .
  Cause B
    Effect 1
      Cause a
      Cause b
       .
       .

Convert the information in Problem 4.3 into such a list. What are the advantages of this method over the usual cause-and-effect diagram?

4.12   In order to monitor the number of defects found on four-square-foot samples of cloth, an inspector decides to simplify the task by counting the number of defects found on one-square-foot sections of each sample and then multiplying by 4. Describe what a histogram of such data would look like.

REFERENCES FOR CHAPTER 4

Demos, M. P., and N. P. Demos. 1989. "Statistical Quality Control's Role in Health Care Management." *Quality Progress* 22 (no. 8):85–89.

Gunter, B. 1988. "Subversive Data Analysis, Part I: The Stem-and-Leaf Display." *Quality Progress* 21 (no. 9); and "Subversive Data Analysis, Part II: More Graphics, Including My Favorite Example." *Quality Progress* 21 (no. 11):77–78.

Hunt, G. A. 1989. "A Training Program Becomes a 'Clinic'." *Quality Engineering* 2 (no. 1):113–117.

Iglewicz, B., and D. C. Hoaglin. 1987. "Use of Boxplots for Process Evaluation." *Journal of Quality Technology* 19 (no. 4):180–190.

Juran, J. M. 1964. *Managerial Breakthrough.* New York: McGraw-Hill.

Lucky, R. W. 1989. *Silicon Dreams,* p. 106. New York: St. Martin's.

Moore, D. S., and G. P. McCabe. 1989. *Introduction to the Practice of Statistics.* New York: Freeman.

Schmid, C. F. 1986. "Whatever Has Happened to the Semilogarithmic Chart?" *The American Statistician.* 40 (no. 3):238–244.

Tufte, E. R. 1983. *The Visual Display of Quantitative Information.* Cheshire, CN: Graphics Press.

Tukey, J. W. 1977. *Exploratory Data Analysis.* Reading, MA: Addison Wesley; and Chambers, J. M., et al. 1983. *Graphical Data Analysis.* Boston: Duxbury.

Velleman, P. F., and D. C. Hoaglin. 1981. *Applications, Basics, and Computing of Exploratory Data Analysis.* p. 43. Boston: Duxbury.

5

APPLIED PROBABILITY AND STATISTICS

Drawing conclusions from process data involves an assessment of the
confidence one has in statistics generated from those data. Probability
is the tool used to quantify the degree of confidence associated with
process statistics. This chapter reviews basic probability rules and some
of the probability models used most frequently in quality improvement.

CHAPTER OUTLINE

THE RELATIONSHIP BETWEEN PROBABILITY AND STATISTICS

Statistics are numbers generated from data. They are used to summarize particular aspects of a process or product. **Probability**, on the other hand, provides a measure of *confidence* or reliability in the statistics so generated. On an intuitive level, this distinction is readily understood. For example, when checking for defective items, most would agree that finding two defectives in a sample of ten items tested is quite different from finding 200 defectives in a sample of a thousand. Even though the percentage defective (i.e., the statistic) is identical in both samples, one would be very hesitant to infer that the overall process defect rate is around 20% based on the sample of only ten items. By comparison, a 20% defect rate in a sample of a thousand seems much more convincing. Without probability, one could go no further than just the rough idea that larger samples are better. With probability, though, the level of confidence in a statistic can be *quantified.* This quantification allows precise questions about sample sizes and decision risks to be answered.

For a more precise definition, suppose that $x_1, x_2, x_3, \ldots, x_n$ represents a set of n measurements on some quality characteristic, X (using the notation of Chapter 3). Then, a **statistic** is simply any number that can be generated from these data. In particular, the mean, median, standard deviation, variance, trimmed mean, range, percentiles, and order statistics are all examples of statistics that can be generated from $x_1, x_2, x_3, \ldots, x_n$ (see Chapter 3). Similarly, the numbers of defects, nonconformities, defectives, or nonconforming items in a sample of n items are also statistics.

In quality control, and in statistical studies generally, statistics are generated from **samples**; that is, one rarely examines, tests, or measures every item produced or processed. Instead, small representative collections, or samples, of these items are tested or measured. If certain guidelines are followed for picking samples, then the laws of probability can be used to measure the reliability of the resulting statistics one calculates.

This chapter reviews the various concepts of probability and statistics used throughout quality control. It is assumed that the reader has studied the basics of probability elsewhere, however, so the treatment is brief.

SAMPLING

THE REASONS FOR SAMPLING

One way of evaluating a problem is to base decisions on a fraction, or sample, of the thing being evaluated. The rationale for doing this is that samples, if properly chosen, ought to provide an accurate enough representation for making decisions. That being true, more sampling (or even a complete audit) should be unnecessary, since the conclusions drawn from them will generally support the ones drawn from the first sample.

Sampling, then, saves time. Since quality control and improvement efforts take place in production environments, time and other resources must be divided between production, on the one hand, and evaluating the quality of production, on the other. In this setting, sampling methods are of real economic importance because less time spent in evaluation means more time spent in production. Also, if certain production steps must await evaluation or approval before products are allowed to continue through a process, sampling reduces the time it takes to reach such decisions, thereby speeding throughput.

Besides saving time, sampling undeniably saves money. Testing and evaluating 32 items from a lot of 200 simply involves less labor than testing the entire lot. Even the *time* saved by sampling eventually translates into increased production and profits.

Sometimes sampling is unavoidable. **Destructive testing** must be used to evaluate certain quality characteristics. Estimates of the breaking strength and melting points of metal products, the potency of vitamin pills, and the lifetime an electronic component must be based on samples because, for these characteristics, the act of testing destroys the items tested.

Even if testing is nondestructive, it makes sense to sample. In quality control, it is generally accepted that even 100% inspection cannot catch all quality problems (see Section 1.4). In fact, 100% inspection can even be *inferior* to sampling inspection. The explanation for this is that inspection errors begin to creep in when large numbers of items are tested, due to inspector fatigue or differences among inspectors. With samples, more attention can be devoted to each item tested, which should correspondingly result in fewer inspection errors.

More and more production facilities today use some form of automatic inspection and test equipment, usually abbreviated **ATE** (see Sections 1.4 and 3.4). True to their name, these machines require little or no human input and are capable of rapidly inspecting all items produced (i.e., 100% inspection). Since it can be so fast and accurate, automatic inspection might seem to remove some of the arguments given above in favor of sampling.

Attractive as rapid 100% inspection may sound, it is *not* a replacement for sampling. To see this, one must consider the proper *role* of ATE in a quality improvement system. Sampling is used for quick feedback, evaluation, and adjustment of processes. By using ATE only as a *sorting* mechanism, one runs the risk of quickly piling up large numbers of conforming and nonconforming items. In other words, using ATE in this manner is equivalent to using a high-speed *detection* approach to quality. Without proper attention to *prevention,* ATE may only sort good from bad but not *reduce* the proportion of bad items (see Section 2.5).

Thus, the role of ATE is twofold. Used as a tool to perform complex measurements quickly, ATE can play a part in a statistically based improvement program by reducing the time needed to generate *sample* data. Otherwise, when 100% sorting and screening are unavoidable, or even required, the critical issue with ATE changes to one of estimating *misclassification rates* (see Chapter 10).

Besides its obvious economic benefits, sampling is the most important method available for solving problems and understanding what a process is *really* doing. When done correctly, sampling is a method of *actively* generating process data that are not available from company records, log sheets, and other sources. By correctly choosing the **sampling design** (i.e., the exact method by which samples are to be chosen), one can solve problems that company records alone cannot address. It is this ability that gives statistical quality improvement techniques their power and, for this reason, all statistical methods contain specific sampling designs, or recipes, for how samples are to be drawn. Examples are using *random samples* (see Section 5.2), choosing the *rational subgroups* for control charts, using *multistage acceptance sampling,* and *designing statistical experiments* (see Chapter 13), all of which are based on particular sampling schemes.

RANDOM SAMPLING

Devoting some thought to how a sample is generated is the most important step in a statistical analysis, since statistical calculations, no matter how sophisticated, cannot salvage a bad or nonexistent sampling design (see Section 1.4). Samples drawn because they are convenient (e.g., drawing the top few items off a large stack) or because (with good intentions) they are thought to be representative can often turn out to be far from representative of the process or product. To avoid this problem, many sampling schemes involve some sort of **random sampling**.

Random samples are obtained by making sure that every item in the **population** to be sampled has an equally likely chance of being chosen for inclusion in the sample. One popular way to achieve this is to create a list of the items in the population, assign successive positive integers to each item on the list, and then use a *random number generator* to select a 'random' sample of integers of the required size. Random number generators can be found in the form of tables, as functions on hand-held calculators, as commands in programming languages and applications software, or in physical forms (e.g., as in rolling dice). Whatever method is used, the numbers so selected will correspond to numbered items in the population list, which are then considered to form a 'random sample.'

When sampling, one is immediately faced with the decision of whether to sample with or without replacement. **Sampling with replacement** means that each item (number) selected may possibly be selected *again* at a later stage; that is, any number selected for the sample is placed *back* into the population for consideration when the next number is chosen. Sampling with replacement is rarely done in practice. Instead, the more common notion of sampling is to draw *distinct* items from the population, with no repeats allowed. Sampling in this manner is called **sampling without replacement.** Even though these two forms of sampling are different, and the statistics generated from them have slightly different (but related) properties, in most cases (i.e., when the sample size is small compared to the population size) there is little practical difference. Unless otherwise stated, we always assume that sampling is done without replacement.

EXAMPLE 5.1   To draw a random sample of 32 parts from a lot or population of 200 parts, first imagine that each part has been assigned a distinct number from 1 to 200. The integers can be assigned in *any* convenient manner, since 'randomness' is guaranteed by the random selection of these numbers, not by the items themselves. To illustrate, with the *Minitab* software package, 32 random numbers from 1 to 200 were selected as shown in Figure 5.1. To compare the two approaches, Figure 5.1(a) shows a random sample of 32 drawn *without* replacement, and Figure 5.1(b) shows another sample drawn *with* replacement. Notice that in Figure 5.1(b), the number 27 was selected twice. Since the sample size (32) is only about 16% of the population size (200), duplicate values should be fairly unlikely, so sampling with or without replacement should be almost equivalent in this example (in the sense that sampling with replacement also generates *close* to 32 distinct numbers).

```
MTB > set c1
DATA> 1:200
MTB > sample 32 c1 c2
MTB > prin c2

C2
  200   39   95  115  118   87  190   69   70  153  150
  165   98  154   28   31   11   16  196  140   85  143
  129  184  193   60   33    3   57  167   12  182
              (a) Sampling without replacement

MTB > sample 32 c1 c2;
SUBC> replace.
MTB > prin c2

C2
  165  164   27   97  174  110  132  162  137   64  106
    1   38  120  136   27  163   37   55    5   52  154
  170  153  121   62  108  197   68   81   35   66
              (b) Sampling with replacement
```

FIGURE 5.1
Selecting a random
sample using *Minitab*

In practice, random sampling can sometimes seem more like a goal than an actuality. For instance, although one can *imagine* numbering the parts from 1 to 200 in Example 5.1, actually *doing* this may be time-consuming or, if the parts are particularly heavy and stacked in a pile, impossible. Under such circumstances, a little creativity often overcomes such obstacles. If the parts to be sampled are heavy, a small change in the method of stacking (rather than just piling) could make random sampling feasible where it was previously impossible. A different method might be to obtain the random sample by changing the process *location* where the samples are drawn and stacking the sampled items in a separate pile. Even apparent obstacles, such as not knowing the population size in advance, can be overcome. This problem arises, for example, in sampling records from large computer tapes. One solution, given by Kennedy and Gentle (1980), can produce a random sample of any size and does not require knowing the total number of records on the tape. Because of the many benefits of random sampling, any extra effort devoted to achieving (or even approximating) a truly random sample is time well spent. For additional discussion on what constitutes a random sample and some methods for obtaining random samples, see Section 11.2 and the reference by Wright and Tsao (1985).

To conclude, the virtues of random sampling are twofold. First, it assures that representative samples are drawn that add to the credibility of the statistics calculated and to the quality of the decisions made. Second, by random sampling, one produces the conditions necessary for invoking many probabilistic results, which then allow for precise probability statements to be made.

5.3   PROBABILITY CONCEPTS

PROBABILITY MODELS

On an intuitive level, **probability** is the name given to numerical quantities that represent the likelihood with which various events occur. One commonly hears about events that have a '50–50 chance,' a '50% chance,' or 'even odds' of occurring. Each of these equivalent statements is a numerical estimate of the likelihood of the particular event's occurrence. In many cases, probability estimates arise from the long-term frequency with which an event occurs in many identical trials—for example, when a coin is tossed repeatedly to estimate the frequency of 'heads.' In other cases, the event is a one-time occurrence (e.g., a sports team wins or loses a particular game), where exact repetitions of the event are impossible; thus, more subjective probability estimates are required.

Underlying any application of probability is the need to make decisions in uncertain situations. In probability theory, these situations are called **random experiments**—that is, experiments whose outcomes can be itemized or listed but are otherwise unpredictable. Such an experiment, for instance, is the testing of an item to determine whether or not it meets its specifications. The outcome of testing must be that the part meets specifications or that it does not, but precisely *which* of these will occur is unknown until testing is actually performed. Viewing all uncertain situations as 'random experiments' whose outcomes can be listed may sound like a needless abstraction, but it is this very abstraction that makes probability a *tool,* not merely a collection of formulas. By first understanding this tool in the generic setting of 'random experiments,' one has at hand a very general method for attacking unfamiliar problems.

The collection of all possible outcomes to an experiment is called the **sample space** of the experiment. Visualizing what the sample space looks like is vital to correctly estimating probabilities, but doing so requires some practice. For instance, the sample space of the experiment "test five items" is shown in Figure 5.2, where the symbol C is used to denote a conforming item and $\overline{C}$ is a nonconforming one. (Throughout the text, events of interest are denoted by uppercase letters—A, B, C, etc.—and their *nonoccurrence* is indicated by a bar above the letter—$\overline{A}$, $\overline{B}$, $\overline{C}$, etc.) For such a simple experiment, the sample space is surprisingly complex, containing 32 distinct possibilities. To help keep track of all the outcomes in the sample space, **tree diagrams** are often used, since they provide a more structured method of listing outcomes and thereby assure that we haven't missed some of them. Figure 5.3 shows a tree diagram for the "test five items" experiment.

After the sample space is determined, the next step in creating a probability model is to assign numerical estimates of likelihood (i.e., probabilities) to each of the sample space outcomes. There are two rules to follow in this procedure: (1) the probabilities must be nonnegative numbers between 0 and 1, and (2) the sum of all the probabilities must be exactly 1. The first rule reflects the fact that events cannot occur less frequently than 0% of the time or more frequently than 100% of the time. The second rule assures *consistency* in assigning the probabilities, since, when testing an item, it would not make sense to estimate the probability of C to be 0.95 while simultaneously estimating the probability of $\overline{C}$ to be, say, 0.60.

The notation used for the probability of any event, A, is $P(A)$ and is read **"probability of (event) A."** Thus, if it is estimated that conforming items occur 95% of the time, one abbreviates this as $P(C) = 0.95$. Correspondingly, one could also write $P(\overline{C}) = 0.05$ because of the consistency requirement.

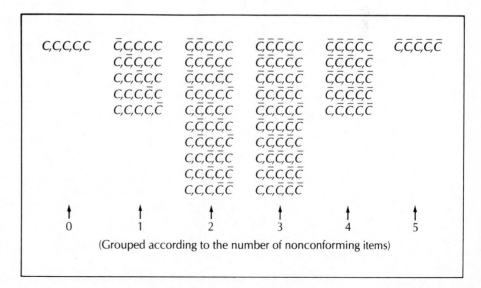

FIGURE 5.2
Sample space when five items are tested for Conformance (C) or Nonconformance (C̄)

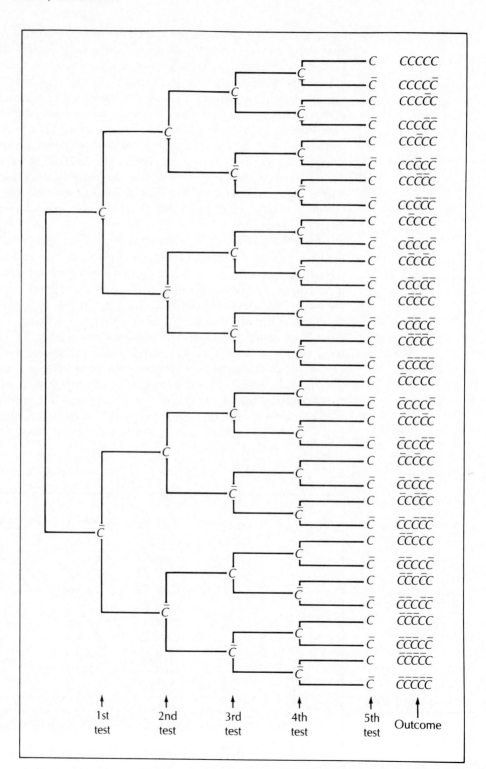

FIGURE 5.3
Tree diagram of the
outcomes when five
items are tested

PROBABILITY LAWS

The sample space outcomes are called **simple events**. In most practical applications, though, one is usually interested not in simple events but in more complex
events consisting of *collections* of simple events. When five items are tested, for
example, the *number* of nonconforming items in the sample of five is of more
practical interest than a particular sequence of five test results. To illustrate, suppose that samples with more than one nonconforming item are of interest. If A denotes the event "one or more nonconforming items," it is clear that A is *not* one of
the simple events shown in Figure 5.2, but is instead a *collection* of many simple
events. In fact, A consists of 31 of the simple events (only the simple event
$\{C, C, C, C, C\}$ is excluded from A). Complex events such as this are called **compound events**, and calculating their probabilities requires being able to *combine*
the probabilities of simple events. Probability 'laws' serve this purpose.

Probabilities of compound events are constructed as follows. First, the probability of any compound event, A, is defined to be the sum of the probabilities of all
the simple events that make up A. Then, 'probability laws' are used to decompose
the probability of compound events into combinations of less complex ones. Complex events are most often decomposed by using the English connectives 'and'
and 'or' to reexpress the events. The event $D =$ "at least two nonconforming
items in five items tested," for example, can be expressed as $D =$ "exactly two
defectives *or* exactly three defectives *or* ...." With the **union** notation, $A \cup B$, to
stand for the phrase "**either A or B, or possibly both, occur**," the event D can be
rewritten as $D = D_2 \cup D_3 \cup D_4 \cup D_5$, where, for $i = 2, 3, \ldots, 5$, D_i denotes the
event "exactly i defectives." Events such as $E =$ "all five items tested conform to
specifications," on the other hand, can be rewritten as $E =$ "the first item is conforming *and* the second item is conforming *and* ...." In this case, we use the **intersection** notation, $A \cap B$, to stand for the event "**both A and B occur**" and
rewrite E as $E = E_1 \cap E_2 \cap E_3 \cap E_4 \cap E_5$, where $E_i =$ "the ith item is conforming." In both examples, the complex events D and E have been decomposed
into some combination of less complex events. To calculate $P(D)$ and $P(E)$, one
uses the probability laws that relate $P(D)$ to each $P(D_i)$, and those that relate $P(E)$
to the $P(E_i)$'s.

There are only two probability laws; all others are special cases of these two:

1. $P(A \cup B) = P(A) + P(B) - P(A \cap B)$ valid for *any* events A and B

2. $P(A \cap B) = P(A \mid B) \cdot P(B)$ valid for *any* events A and B

In the second law, the notation $P(A \mid B)$ is read the "**conditional probability of A
given B**" and represents the probability that A occurs *given* that we know B has
already occurred. When a few more definitions are introduced, these two laws
can be expanded to include frequently occurring special cases.

Complementary events are the 'complements' of each other in the sense that
one, but not both, of them *must* occur. For example, for a single item, the events
$C =$ "conforming item" and $\overline{C} =$ "nonconforming item" are complementary. As
introduced earlier, $\overline{A}$ is used to denote the complement of an event A.

Mutually exclusive events are defined to be events that exclude the occurrence of each other. Thus, since $A \cap B$ cannot occur for any two mutually exclusive events, A and B, we note immediately that $P(A \cap B)$ must equal 0 in such cases. The difference between mutually exclusive events and complementary ones is that mutually exclusive ones need not occur at all in a given experiment, whereas exactly one of a pair of complementary events must always occur. The events $D_i =$ "exactly i defectives," where $i = 2, 3, \ldots, 5$, are all mutually exclusive because, for example, it is impossible to find exactly two defectives *and* exactly three defectives simultaneously in the sample. None of these events are complementary, though, since at any stage, none of the D_i's need occur (e.g., it is possible that no defectives are found).

The final definition needed is that of **independent events**. Two events, A and B, are said to be independent if the outcome of either of them has no effect on the outcome of the other. If A and B are truly independent, then the conditional probability $P(A|B)$ ought to equal $P(A)$; that is, the fact that B has occurred should not change the probability that A occurs. In that case, the second probability law listed above simplifies to $P(A \cap B) = P(A|B) \cdot P(B) = P(A) \cdot P(B)$. Some texts take the equality $P(A \cap B) = P(A) \cdot P(B)$ as the definition of independent events.

The events $E_i =$ "the ith item is conforming" might be considered independent of one another, since they occur on different tests. For example, successive tosses of a coin are independent of one another. The usual way independence arises in practice is when one subjectively decides that two (or more) events *ought* to be independent or when random sampling is used to assure that there is no relationship between successive samples. When five items are tested, if the five items are picked successively from the output of a given process, then the five test results may be highly related. If, instead, the five items are selected using a random sampling technique, there is more assurance of the independence between the test results.

Decomposing complex events into combinations of independent events, mutually exclusive events, and complementary events can greatly simplify probability calculations. With these definitions, the two basic probability laws listed earlier can be expanded to include extremely useful special cases, which in fact are used more frequently than the two general laws themselves:

1. $P(A \cup B) = P(A) + P(B) - P(A \cap B)$ the general addition law
   a. $P(A \cup B) = P(A) + P(B)$ when A and B are mutually exclusive
   b. $P(A_1 \cup A_2 \cup \cdots \cup A_n) = P(A_1) + P(A_2) + \cdots + P(A_n)$ for any collection of mutually exclusive events $A_1, A_2, \ldots, A_n$
   c. $P(A) = 1 - P(\overline{A})$ for any two complementary events

2. $P(A \cap B) = P(A|B) \cdot P(B)$ the general multiplication law
   a. $P(A \cap B) = P(A) \cdot P(B)$ for any two independent events
   b. $P(A_1 \cap A_2 \cap \cdots \cap A_n) = P(A_1) \cdot P(A_2) \cdot \cdots \cdot P(A_n)$ for any collection of mutually independent events $A_1, A_2, \ldots, A_n$

EXAMPLE 5.2

Suppose that a process is known to produce conforming items about 95% of the time and that random sampling is used to select five items to test. The probabilies of the events E = "all five items are conforming" and D = "at least two items are nonconforming" could be calculated as follows.

Since $E = E_1 \cap E_2 \cap \cdots \cap E_5$, where E_i = "ith item is conforming," random sampling assures us that the E_i's are *independent*, so the special case of the multiplicative law gives

$$P(E) = P(E_1) \cdot P(E_2) \cdot \cdots \cdot P(E_5) = (0.95)(0.95)\cdots(0.95) = (0.95)^5$$

$$= 0.7738$$

where the overall rate of 95% conforming is used to estimate each of $P(E_i)$'s.

Calculating $P(D)$ requires more work. From the complementary law, $P(D) = 1 - P(\overline{D})$, where $\overline{D}$ = "at most one nonconforming item." Since $\overline{D}$ can be further decomposed into mutually exclusive events $\overline{D} = D_0 \cup D_1$, where D_i = "exactly i items are nonconforming," notice that D_0 is actually the event E from the previous paragraph. D_1, however, is still a complex event consisting of the five simple events $\{\overline{C}, C, C, C, C\}$, $\{C, \overline{C}, C, C, C\}$, $\{C, C, \overline{C}, C, C\}$, $\{C, C, C, \overline{C}, C\}$, and $\{C, C, C, C, \overline{C}\}$. From the independence of the tests and the multiplicative rule, $P(D_1)$ must, in turn, be the sum of the five probabilities $(0.05)(0.95)^4 + (0.95)(0.05)(0.95)^3 + \cdots + (0.95)^4(0.05) = 0.2036$. Thus, $P(D)$ becomes

$$P(D) = 1 - P(\overline{D}) = 1 - [P(D_0) + P(D_1)] = 1 - [(0.95)^5 + 5(0.95)^4(0.05)]$$

$$= 1 - [0.7738 + 0.2036] = 0.0226$$

Although calculating $P(D)$ is instructive (because it involves the concepts of complementary, independent, and exclusive events in one problem), it is nonetheless too time-consuming to perform such calculations in a production environment. Fortunately, there is a simpler way to do this calculation that requires only a quick glance at some tabled probabilities (see Section 5.5).

5.4   RANDOM VARIABLES, STATISTICS, AND DISTRIBUTIONS

TERMINOLOGY

Although the sample space provides a *framework* for probability calculations, rarely are the probabilities of these simple outcomes of interest themselves. As Example 5.2 illustrates, questions of practical importance are usually represented by *compound* events. Going one step further, we could even state that most probability questions of interest concern events that are *numerical* outcomes to random experiments. For example, questions that are of interest when items are tested for conformance include things like the *number* of nonconforming items, the *number* of conforming items, or perhaps the corresponding *percentages* of such items. As a different example, when the lengths of items in a random sample are measured, it is the average length (a *numerical* outcome) that might be of interest.

The term **random variable** is used to describe a *numerical* value that depends on the particular outcome of a random experiment. Traditionally, uppercase letters near the end of the alphabet are used to represent random variables. Thus, in the last example, one could have studied the random variable X = "the number of nonconforming items in a random sample of five tested," or the random variable Y = "the number of conforming items in five tested," or W = "the percentage defective in five tested." Notice that this terminology agrees with our previous use of uppercase letters to denote quality *characteristics* and lowercase values to denote the actual measurements. Similarly, random variables are classified as *discrete* or *continuous* depending on whether attributes or variables measurements are involved (see Section 1.4).

Random variables have means, variances, and standard deviations of their own that are analogous to the means, variances, and standard deviations calculated for data (see Section 3.3). The mean of a random variable is also called its **expected value**. We denote these quantities as follows:

$\mu_X = E(X)$ mean, or expected value, of the random variable X

$\sigma_X^2 = \text{Var}(X)$ variance of the random variable X

$\sigma_X =$ standard deviation of the random variable X

The notation $E(X)$ is also read "*expected value of X*" and $\text{Var}(X)$ is read "*variance of X.*" μ_X and $E(X)$ may be used interchangeably for the mean of the variable X, and either σ_X^2 or $\text{Var}(X)$ may be used for the variance of X. This dual notation exists because, depending on the application, one form can be more convenient to use than the other. The reader is referred to statistics texts for the formulas used for calculating μ_X and σ_X for any random variable X. For the random variables introduced in this text, the mean and standard deviation formulas are simply *given* as the need arises.

The usefulness of random variables results from knowing their distributions. The **probability distribution** of a random variable X consists of two things: the collection of possible values that X can assume, and the corresponding probabilities with which X assumes these values. For attributes data, the discrete values of X are usually displayed in histogram fashion (see the next example) with the provision that the *areas* of the histogram bars be equal to the *probabilities* of the X values. For variables measurements, a continuous curve, called the density function of the random variable, describes the distribution of probability. The familiar bell-shaped curve associated with the empirical rule (see Section 3.3) is an example of a density function depicting the distribution of a 'normal' random variable (see Section 5.6).

EXAMPLE 5.3

We return to Example 5.2 and let X denote the random variable "number of nonconforming items in a random sample of five tested." Then the possible values of X are 0, 1, 2, 3, 4, and 5 and, from the same methods as in Example 5.2, the corresponding probabilities (rounded to four decimal places) are:

$P(X = 0) = 0.7738$

$P(X = 1) = 0.2036$

$P(X = 2) = 0.0214$

$P(X = 3) = 0.0011$

$P(X = 4) = 0.0000$ (the actual value is 0.000029687)

$P(X = 5) = 0.0000$ (the actual value is 0.000000312)

Since the values of the random variable X correspond to *collections* of simple events, one thinks of the probability in the sample space as being redistributed (or, simply, *distributed*) across the values of X. This distribution is summarized in histogram form in Figure 5.4. Notice also that the events D and E considered in Example 5.2 correspond to the events where $X \geq 2$ (for D) and $X = 0$ (for E). Thus, these probabilities (or *any* probability concerning X) can be found by combining the corresponding bars of the histogram in Figure 5.4. $P(D)$, for example, equals the sum of the bars (probabilities) over the numbers $2, 3, \ldots, 5$. Or, since the sum of the areas (probabilities) must be 1, $P(D)$ could equivalently be calculated by subtracting the bars over $X = 0$ and $X = 1$ from 1 (as was done in Example 5.2).

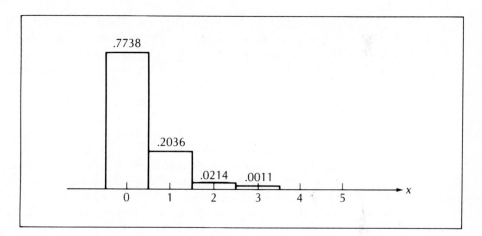

FIGURE 5.4
Histogram of the random variable X of Example 5.3

STATISTICS AS RANDOM VARIABLES

When a *random* sample is taken, any statistic that is subsequently calculated can itself be thought of as a random variable. Indeed, the underlying random experiment created by random sampling is "draw n items at random from a population," making any statistic built from these items a numerical value (i.e., random variable) that depends on the particular outcome of the experiment. Thus, when random sampling is used, the mean, median, variance, standard deviation, range, and other statistics can all be considered random variables.

The import of this is that, as random variables, statistics then have their own distributions, expected values, and variances. Furthermore, these quantities reflect the nature of the population (i.e., the process) being sampled. Under random sampling, the distribution of a statistic is called its **sampling distribution**. In essence, the sampling distribution depicts how the particular statistic would behave *if*

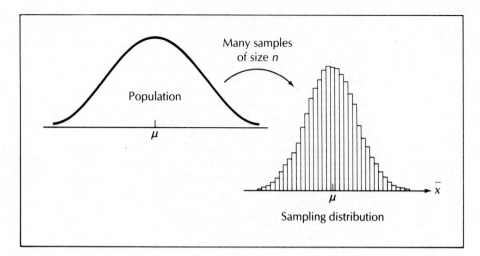

FIGURE 5.5
Sampling distribution
of $\bar{x}$

many samples of the same size were drawn. To illustrate, Figure 5.5 shows how the sampling distribution of the sample mean, $\bar{x}$, arises. (*Note:* In this example, we break with our convention of using uppercase letters for random variables because of the traditional use of lowercase letters to denote the sample mean, standard deviation, etc.)

If a *large number* of random samples of size n were drawn from the population (i.e., process) of interest, and if $\bar{x}$ was calculated for each sample, one would expect each $\bar{x}$ to be slightly different from the rest (since each is based on different data). When all the $\bar{x}$'s are stacked in a histogram, the sampling distribution of $\bar{x}$ is approximated (see Figure 5.5). Since each mean, $\bar{x}$, estimates the true process average, μ, it is also expected that this histogram will be centered around μ. Furthermore, it seems intuitive that the variability of the $\bar{x}$'s should be directly related to the variation in the individual measurements (i.e., the process variation, σ). Note that the same procedure would be followed to construct the sampling distribution of *any* statistic, not just that of the sample mean.

These intuitive ideas are made more exact by the formulas for the expected value and the standard deviation of the random variable $\bar{x}$, which, it can be shown, are related to the process average and process variation by

$$\mu_{\bar{x}} = \mu \quad \text{and} \quad \sigma_{\bar{x}} = \frac{\sigma}{\sqrt{n}} \tag{5.1}$$

Here, μ and σ represent the process mean and standard deviation, respectively, and n is the size of the random sample drawn. These equations not only validate one's intuition, but they go a step further by *quantifying* it. In particular, the second of the two formulas states that larger sample sizes should make estimates of μ more precise, in the sense that the variation in the $\bar{x}$'s decreases in direct proportion to the *square root* of the sample size. Figure 5.6 illustrates the effect that increasing the sample size has on the sampling distribution of $\bar{x}$.

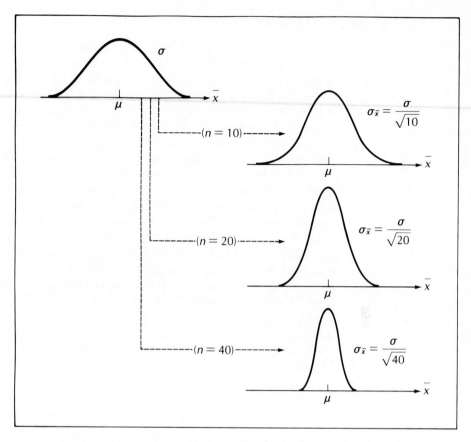

FIGURE 5.6
Increasing the sample
size decreases the
variation in the
sampling distribution
of $\bar{x}$

How is the sampling distribution used? After all, in practice one usually draws only *one* sample, not many samples. The answer is that, by appealing to well-known theoretical results, in many cases the sampling distribution of a statistic is already *known* (either exactly or approximately), so it is not necessary to draw repeated samples in order to approximate it. As an example, the sampling distribution of the mean, $\bar{x}$, is known to exactly follow a normal distribution (see Section 5.6) if the population sampled happens to itself be normally distributed. Even when one has *no* knowledge of the population distribution (as is usually the case), the distribution of $\bar{x}$ may still be *approximated* by a normal curve (see the central limit theorem discussed next) or the closely related t distribution (see Section 5.6). Thus, a sample mean that results from a *single* sample of size n can be compared to a known probability distribution, and it is by doing this that one makes the connection between probability and statistics. Appealing to the empirical rule (see Section 3.3), for example, one could estimate how likely it would be for a given sample mean to fall within 2 standard deviations (i.e., a distance of $\pm 2\sigma/\sqrt{n}$) of the process mean, μ.

THE CENTRAL LIMIT THEOREM

Of all the statistics that can be calculated from a random sample of n process measurements (e.g., the mean, median, standard deviation, range), $\bar{x}$ has a very special property. Roughly stated, the sampling distribution of the mean can often be approximated by a *normal* distribution (see Section 5.6); that is, its sampling distribution should look very much like the bell-shaped curve of the empirical rule as long as the sample size, n, is not too small.

To state this more accurately, for large enough sample sizes, the sampling distribution of $\bar{x}$ should closely follow a particular probability distribution, called the normal distribution. Furthermore, irrespective of the sample size, the mean and standard deviation of this distribution must be given by equation (5.1). This result is known as the **central limit theorem**. Sample sizes required by the central limit theorem turn out to be relatively modest, with an n of 20 to 30 usually being sufficient. Even smaller sample sizes suffice if the population itself is not too skewed.

That the mean and standard deviation should be given by equation (5.1) is not new. Equation (5.1) holds *regardless* of the sample size, whenever random sampling is used. What *is* new is that the shape of the sampling distribution almost always resembles a specific bell-shaped curve and, even more important, nothing need be known about the shape of the population (process) in order to make this statement. The tendency for sample means drawn from any process to follow the normal distribution is used to justify many of the quality control procedures in succeeding chapters.

It almost sounds as though the central limit theorem yields more information than one should rightfully expect, but the following example demonstrates that *some* type of bell-shaped distribution of $\bar{x}$'s *is*, in fact, to be expected.

EXAMPLE 5.4

Suppose that a die is rolled two times and the average, $\bar{x}$, of the two rolls is recorded. This can be viewed as taking a random sample of size $n = 2$ from a 'process' (rolling a die). For such a simple process (i.e., population), the possible measurements are completely known and the process mean can be calculated as $\mu = (1 + 2 + 3 + 4 + 5 + 6)/6 = 3.5$. What about the possible values of $\bar{x}$? By examining all possible pairs of numbers that could occur, one immediately sees that some values of $\bar{x}$ must occur more frequently than others. For example, $\bar{x} = 3$ could result from any of the pairs $\{1, 5\}$, $\{2, 4\}$, $\{3, 3\}$, $\{4, 2\}$, or $\{5, 1\}$, whereas $\bar{x} = 1.5$ can occur in only two ways, $\{1, 2\}$ and $\{2, 1\}$. Since all 36 pairs of numbers are equally likely, one then finds that $P(\bar{x} = 3) = \frac{5}{36}$ and $P(\bar{x} = 1.5) = \frac{2}{36}$, and so on. From these probabilities, the sampling distribution of $\bar{x}$ is shown in Figure 5.7(a). Next, suppose that the die is rolled $n = 3$ three times. By listing all the possible ways that $\bar{x} = 3$ (and other values of $\bar{x}$) could occur and then drawing the resulting sampling distribution, we form Figure 5.7(b).

Three things emerge from this simple example. First, the number of ways that extremely large or small values of $\bar{x}$ (such as $\bar{x} = 1$ or $\bar{x} = 6$) can occur is *limited*. An $\bar{x}$ of 1 can occur only when *each* roll itself results in a 1. But values of $\bar{x}$ close to the middle (i.e., the process mean) can occur in a number of ways, thereby making them more likely. Thus, the sampling distribution *must* exhibit some tendency to pack around the mean. Second, this packing only intensifies as larger samples are used, as can be seen by comparing Figures 5.7(a) and (b). Third, the shape of the normal distribution can already be discerned in Figure 5.7(b), even for such a small sample size as $n = 3$.

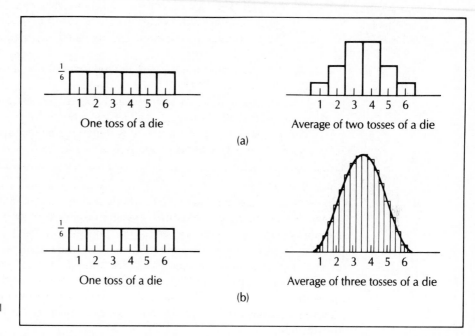

FIGURE 5.7
Illustrating the central limit theorem

5.5   COMMON DISCRETE DISTRIBUTIONS

In this section and the next, the probability distributions most frequently used in quality control are presented. Since the calculations are somewhat different depending on whether attributes (i.e., discrete) or variables (continuous) data are used, the two sections are divided accordingly between discrete and continuous distributions.

THE BINOMIAL DISTRIBUTION

BINOMIAL RANDOM VARIABLES.   The binomial distribution occurs throughout statistics and quality control in myriad, seemingly unrelated, places. Two of its major applications are in the construction of p and np control charts (see Chapter 9) and as an approximation for calculating the operating characteristic curves of acceptance sampling plans (see Chapter 11). There is, however, a general framework underlying all applications of this distribution. To be precise, the binomial distribution can be applied whenever the following conditions hold:

1. There must be a simple two-outcome random experiment involved. The two outcomes are generically called 'success' and 'failure' and are abbreviated by S and F, respectively.
2. This two-outcome experiment is repeated any number, n, of times.
3. The n repeated trials are all *independent* of one another (see Section 5.4).
4. The probabilities of 'success' and 'failure' remain constant from trial to trial. These probabilities are denoted by $p = P(S)$ and $q = P(F)$ where, from the consistency requirement (see Section 5.3), we must have $p + q = 1$.

The terms 'success' and 'failure' are generic and are meant only to distinguish between the two outcomes. No other special meaning is inferred; that is, 'successes' are not necessarily 'good,' nor are 'failures' necessarily 'bad.'

The fact that the above conditions are relatively easily met accounts for the diverse applications of the binomial distribution. When the conditions hold, one can consider a special random variable, X, that *counts* the number of successes in the n trials. Any random variable that arises in this fashion is called a **binomial random variable**, and its associated probability distribution is called a **binomial distribution**.

To illustrate the wide applicability of the binomial distribution, consider the following random variables:

X = the number of nonconforming items in a random sample of five items

Y = the number of defective parts (of the same type) in an assembly containing 50 such parts

U = the number of items that 'pass' when go/no-go gaging is used to inspect a lot of 1,000 items (see Section 3.4)

V = the number of people who respond 'yes' on a yes/no survey of 1,000 people

W = the number of heads that occur when a coin is tossed 100 times

Each of X, Y, U, V, and W could be considered a binomial random variable under the right circumstances. For instance, in testing five items, the two basic outcomes at any trial are C and $\overline{C}$ (using the notation of Section 5.3), and there are $n = 5$ independent trials (assuming random sampling). If it is further given that the proportion of conforming items in the population is about 95%, then one can also state that $p = P(\overline{C}) = 0.05$ and $q = P(C) = 0.95$. Notice in this example that X

counts the number of nonconforming items, so a 'success' on any trial occurs when an item *fails* to meet its specifications.

To completely describe a binomial random variable, two numbers or **parameters** are needed (namely, *n* and *p*). Because of this, the binomial distribution is said to be a two-parameter distribution. These parameters determine the mean and standard deviation of *X*, which are given by

$$\mu_X = np \quad \text{and} \quad \sigma_X = \sqrt{npq} \tag{5.2}$$

Furthermore, *n* and *p* completely determine the probability distribution of *X*, so that any probability concerning *X* can be evaluated once *n* and *p* are specified.

The parameter *n* determines the possible values that *X* can assume—namely, the integers 0, 1, 2, 3, ..., *n*, since in *n* trials we can find anywhere from *no* successes ($X = 0$) up to possibly *all n* successes ($X = n$). Next, the probability associated with any particular number of successes, *k*, is given by

$$P(X = k) = \binom{n}{k} p^k q^{n-k} \quad \text{for } k = 0, 1, 2, 3, \ldots, n \tag{5.3}$$

where the notation $\binom{n}{k}$ stands for the number of distinct samples of size *k* that can be drawn from a group of *n* items. In turn, $\binom{n}{k}$ can be calculated from the formula

$$\binom{n}{k} = \frac{n!}{k!\,(n-k)!} \tag{5.4}$$

where *n*! (read "*n* factorial") is shorthand for the product of the first *n* positive integers. Note that by using the convention that $0! = 1$, equation (5.4) also works in the special cases $\binom{n}{n}$ and $\binom{n}{0}$.

EXAMPLE 5.5

In Example 5.2, let *X* denote the random variable that counts the number of nonconforming items in a random sample of five tested. Given that the overall proportion of conforming items is 95%, the four conditions for applying the binomial distribution are met, so *X* can be considered a binomial variable with parameters $n = 5$ and $p = 0.05$. The probabilities of events *E* and *D* described in Example 5.2 can be expressed in terms of *X* as

$$P(E) = P(\text{all five items are conforming}) = P(X = 0)$$

$$P(D) = P(\text{at least two items are nonconforming}) = P(X \geq 2)$$

Using the binomial formula, equation (5.3), one finds

$$P(E) = P(X = 0) = \binom{5}{0}(0.05)^0(0.95)^5 = \frac{5!}{0!\,(5-0)!}(1)(0.7738) = 0.7738$$

$$P(D) = P(X \geq 2) = 1 - P(X \leq 1) = 1 - [P(X = 0) + P(X = 1)]$$

$$= 1 - \left[\binom{5}{0}(0.05)^0(0.95)^5 + \binom{5}{1}(0.05)^1(0.95)^4 \right]$$

$$= 1 - [0.7738 + 0.2036] = 0.0226$$

which agree with the results of Example 5.2. Notice how much easier the probability calculations become once the binomial distribution is used. In particular, the probabilities result from a simple formula and there is no need to list the sample space outcomes. The next example takes this approach one step further by eliminating the need for formulas altogether and relying instead on tabulated probabilities of the binomial distribution.

BINOMIAL PROBABILITY TABLES.   It is convenient to construct tables of binomial probabilities using equation (5.3). For each sample size, n, the tables are arranged according to the number of successes, k, and the probability of success, p. In addition, because probabilities of the form $P(X \leq k)$ or $P(X \geq k)$ occur much more frequently in practice than do those of the form $P(X = k)$, the tabled values often represent the **cumulative probability $P(X \leq k)$** for given values of n and p. Appendix 3 contains cumulative binomial probabilities for sample sizes ranging from $n = 5$ to $n = 25$. Remember that the tabled values are cumulative probabilities *from the left*, so any binomial probability can be calculated by using one of the following expressions:

$$P(X \leq k) = \text{(in Appendix 3)}$$

$$P(X < k) = P(X \leq k - 1)$$

$$P(X \geq k) = 1 - P(X \leq k - 1)$$

$$P(X > k) = 1 - P(X \leq k)$$

$$P(X = k) = P(X \leq k) - P(X \leq k - 1)$$

where the probabilities on the *right* sides of these equations are found directly from the tables.

EXAMPLE 5.6   In Example 5.5, the binomial random variable X has parameters $n = 5$ and $p = 0.05$. The probabilities of events E and D found from the binomial tables in Appendix 3 are $P(E) = P(X = 0) = 0.7738$ and $P(D) = P(X \geq 2) = 1 - P(X \leq 1) = 1 - [0.9774] = 0.0226$. Thus, by formulating the problem in the context of a binomial distribution, the calculations introduced in Section 5.3 are reduced to a simple arithmetic exercise.

It is always a good practice to sketch the histogram of the binomial distribution. This is easily done by finding its mean $\mu_X = np$ and standard deviation $\sigma_X = \sqrt{npq}$ and then drawing a bell-shaped histogram *centered at μ* extending out approximately $\pm 3\sigma_X$ on either side of μ_X. Under fairly mild conditions (namely, that np and nq both exceed 5), the actual histogram will be very close to this sketch. Figure 5.8 shows typical histograms of binomial distributions for various values of p. For values of p not too near 0 or 1, the histograms have a bell-shaped appearance, but they tend to become less symmetric for values of p close to 0 or 1.

BINOMIAL APPLICATIONS.   We end this section with two examples whose application areas are different, but whose use of the binomial distribution is identical. The reader may observe that both examples could also be approached as simple applications of the multiplicative law for independent events (see Section 5.3).

However, we recommend framing such problems in the context of the binomial distribution because it allows other, harder questions to be answered that would be difficult to approach with probability laws alone.

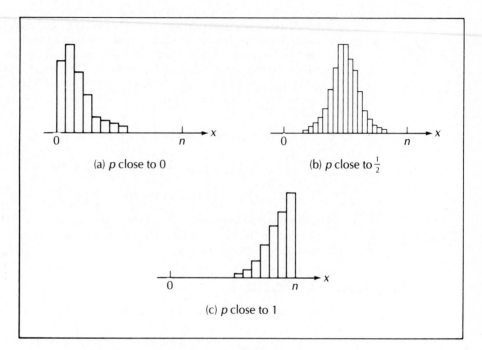

(a) p close to 0

(b) p close to $\frac{1}{2}$

(c) p close to 1

FIGURE 5.8
Histograms of
binomial distributions
for various values
of p

EXAMPLE 5.7   Products or systems are often composed of large numbers of interdependent components that are critical to the proper performance of the finished product. Critical steps in a system (or critical parts in an assembly) are ones that, when improperly functioning, cause the entire system (or assembly) to either malfunction or not meet specifications. In an example showing how vendor quality affects customer quality, Gitlow and Wiesner (1988) consider a product that consists of 50 component parts, any one of which, if defective, could cause the finished product to be defective. Suppose that the defect rate for any of the components is 0.005 (i.e., 0.5%). What proportion of all products made will themselves be defective?

To solve this problem, let X represent the number of defective components in a sample of 50 (i.e., in a given assembled product). Then X appears to satisfy (at least approximately) the requirements of a binomial distribution with parameters $n = 50$ and $p = 0.005$. From this observation, the probability that an assembled product is defective (i.e., has one or more defective parts) is equal to

$$P(X \geq 1) = 1 - P(X = 0) = 1 - \binom{50}{0}(0.005)^0(0.995)^{50} = 0.2217$$

That is, about 22.17% of all assembled products will be defective even though the defect rate for any *individual* component is, it seems, small. The conclusion to be drawn from this example is that quality levels that may appear to be very good (such as a 0.5% rate of defectives) are *not necessarily good enough* when used in large systems.

EXAMPLE 5.8   In a discussion of potential quality problems in the Discovery space shuttle, Hill (1988) uses a binomial random variable to estimate the probability of mission failure. From documented sources, Hill notes that there are 1,232 'critical' parts used in the Discovery, and a malfunction in any one of them could cause a mission failure. Suppose that each of these parts has a fairly low failure rate of 1 failure per 100,000 parts. The probability of mission failure can be modeled using a binomial random variable, X, that counts the number of defective parts in 1,232. Thus,

$$P(\text{mission failure}) = P(X \geq 1) = 1 - P(X = 0)$$

$$= 1 - \binom{1,232}{0}(0.00001)^0(0.99999)^{1,232}$$

$$= 1 - (0.99999)^{1,232} = 0.0122$$

Under these assumptions, then, the Discovery launch had a 1.22% chance of failing, even though the quality of its component parts would normally be considered extremely reliable.

THE POISSON DISTRIBUTION

Another frequently used discrete distribution is the Poisson distribution, whose name is derived from its inventor, the French mathematician S. D. Poisson (1781–1840). The Poisson is used to construct *control limits* for c and u control charts (see Chapter 9), in *compliance sampling,* for *approximating* binomial and hypergeometric distribution *probabilities* in acceptance sampling, and, because of its strong relationship with the exponential distribution, in *reliability analysis.*

Poisson random variables are also used to *count* the number of occurrences of low probability events in situations that involve large numbers of trials. For this reason, some practitioners refer to the Poisson distribution as the "rare events" distribution. More generally, though, a **Poisson random variable,** X, is one that counts the number of occurrences of an event of interest *per unit of measurement.* Like the binomial, a few more conditions must be satisfied before one can justifiably assume that a particular variable follows a Poisson distribution, but these are fairly mild and the interested reader is referred to other discussions on this matter (see Walpole and Meyers 1989).

When a Poisson random variable is described, the unit of measurement can be almost anything, but common choices are time, volume, length, and area. In quality control applications, the units of measure are usually called **inspection units,** and defining them is largely a matter of choice, depending on the particular application. For example, when studying the number of errors that occur in paperwork processes (e.g, billing errors, data entry errors, balance discrepancies, transaction errors), one must select an inspection unit composed of some fixed collection of items to be inspected. Thus, for monitoring data entry errors, samples of 1,000 documents collected at the end of each workshift can be examined, and the total number X of errors found can be recorded. In this case, the inspection unit is '1,000 documents,' and each day one counts the number of errors found in that day's inspection unit. The following examples of Poisson random variables illustrate how free the choice of an inspection unit can be:

X = the number of nonconformities on a finished dashboard (see Example 4.14)

Y = the number of nonconformities on a given *section* of a dashboard

U = the number of errors in 1,000 documents

V = the number of customer complaints per week

W = the number of electronic components that fail after 2,500 hr of use

Notice the distinction between these examples and those of binomial random variables given previously. A binomial random variable counts the number of 'successes' in n trials, whereas a Poisson random variable counts the number of 'successes' *per unit of inspection*. The values of a binomial variable are therefore limited to the integers $0, 1, 2, 3, \ldots n$, but there is no such bound on the possible values of a Poisson variable. In the examples above, U, the number of errors in 1,000 documents, could well exceed 1,000 (e.g., there could be more than one error per document), so U could not possibly be a binomial random variable.

To calculate the probabilities associated with a Poisson random variable requires only one *parameter*—namely, the mean. Traditionally, λ is used to denote the expected value. By specifying λ and appealing to the assumptions underlying the Poisson distribution (Walpole and Meyers 1989), we can completely determine the mean, standard deviation, and probabilities of a Poisson random variable. They are given by

$$\mu_X = \lambda \qquad \sigma_X = \sqrt{\lambda}$$

$$P(X = k) = \frac{e^{-\lambda}\lambda^k}{k!} \qquad \text{for any integer } k \geq 0 \tag{5.5}$$

where e denotes the base of natural logarithms (i.e., $e = 2.7182818\ldots$). Figure 5.9 shows the histogram of a typical Poisson distribution.

As in the case of the binomial distribution, tabled values of the probabilities in equation (5.5) greatly simplify Poisson calculations. Similarly, because most applications involve summing many individual probabilities, Appendix 4 contains the *cumulative* (left-tailed) Poisson probabilities for various values of λ. With the same rules as for the binomial distribution, cumulative tables can be used to calculate any of the probabilities $P(X \leq k)$, $P(X < k)$, $P(X \geq k)$, $P(X > k)$, and $P(X = k)$.

EXAMPLE 5.9   Quick responses to customer problems and inquiries are crucial in forming a customer's perception of service quality. Many service-intensive businesses such as banks, credit card companies, and supermarkets now emphasize prompt service and sometimes offer cash awards to customers when service times are

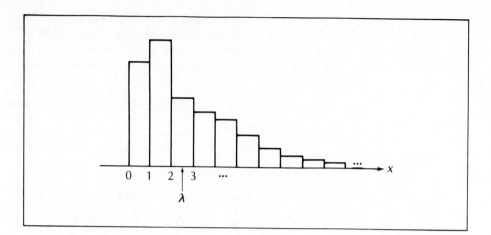

FIGURE 5.9
Histogram of a
typical Poisson
distribution

too long (Diamond 1989). In one example, a bank offers a small payment to customers whose phone calls are not answered within four rings. In order to assign a sufficient number of phone representatives to support such a program, the bank needs to know approximately how many customers are expected to call during a given period of time. Suppose that during peak periods, the bank estimates that there are, on average, five incoming calls per 5-min period and that the majority of customer questions are resolved within a couple of minutes or are referred to another department at the bank. Let X denote the number of calls arriving in a 5-min period. Then a Poisson distribution with $\lambda = 5$ can be used to describe the number of incoming calls. For example, the probability of *more* than five calls arriving in a 5-min period is substantial: $P(X > 5) = 1 - P(X \leq 5) = 1 - 0.616 = 0.384$ (from Appendix 4 with $\lambda = 5$); that is, during peak periods, 38.4% of the time one expects *more* than five incoming calls. When Appendix 4 is used in this manner, other probabilities of interest can be found and reasonable decisions can be made regarding the number of phone representatives to use[1].

The Poisson distribution can also be used to approximate binomial probabilities in cases where the binomial parameter n is large *and p is very small.* Under these circumstances, it can be shown that a *binomial random variable X whose parameters are such that $np \leq 7$ is closely approximated by a Poisson random variable with a parameter of $\lambda = np$.* This approximation is frequently used in acceptance sampling (see Chapter 11).

EXAMPLE 5.10   In Example 5.8, a binomial random variable, X, with parameters $n = 1,232$ and $p = \frac{1}{100,000}$ was used to model the possible number of failures of critical components on the Discovery space shuttle. Since $np = 0.01232$ is less than 7, a

---

[1] If service times are also taken into account, this becomes an easily solved problem in queueing theory (Hillier and Lieberman 1986).

Poisson approximation with $\lambda = 0.01232$ can also be used to find the probability of mission failure:

$$P(X > 0) = 1 - P(X = 0) \approx 1 - \frac{e^{-\lambda}\lambda^0}{0!} = 1 - \frac{e^{-0.01232} \cdot (1)}{1} = 0.012244419$$

Note that this approximation is indeed very close to the exact result, $1 - (0.99999)^{1,232} = 0.01224448$, from Example 5.8.

THE HYPERGEOMETRIC DISTRIBUTION

Up to this point, no attention has been paid to the effect that the population size N might have when the number of 'successes' is counted in random samples of size n. Since most sampling is done *without replacement*, problems can occur when the sample size becomes large in comparison to the population size. For example, when a random sample of $n = 25$ is drawn from a lot of size $N = 50$, the variable X that counts the number of nonconforming items in the sample no longer follows a binomial distribution. The reason is this: as successive items are sampled, the relative percentages of conforming and nonconforming items in the remainder of the lot *change*, so the probability of a 'success' at any trial does not remain constant. Since this violates one of the binomial assumptions, one cannot use binomial probabilities when studying X.

On the other hand, when a sample of $n = 25$ is drawn from a lot of $N = 1,000$, even though the same argument shows that the probability of 'success' changes from trial to trial, the magnitude of this change is extremely small. For all practical purposes, the binomial assumptions are (approximately) satisfied and one can proceed to use the binomial probabilities to study X.

There is, however, a probability distribution that does take into account the size of the population as well as the sample when 'successes' are counted. A random variable that counts the number of 'successes' in a random sample of n taken from a finite population of N items is said to have a **hypergeometric distribution**. We let the number of 'successes' in the population be denoted by r and the number of 'failures' by $N - r$. Then the probabilities for a hypergeometric distribution are given by

$$P(X = k) = \frac{\binom{r}{k} \cdot \binom{N - r}{n - k}}{\binom{N}{n}} \tag{5.6}$$

where the values of X must range between $k = \text{maximum}\{0, n - (N - r)\}$ and $k = \text{minimum}\{r, n\}$.

EXAMPLE 5.11

In Example 5.5, suppose that samples of size 5 are drawn from a lot of size $N = 100$. To calculate the probability that all five items conform to specifications (which is event E of Example 5.5), consider the number of nonconforming items X in the sample to be a hypergeometric random variable. From the overall rate of 5% nonconforming items, suppose that the lot contains five nonconforming items and, therefore, 95 conforming ones. Then

$$P(E) = P(X = 0) = \frac{\binom{5}{0} \cdot \binom{95}{5}}{\binom{100}{5}} = 0.76958$$

For this lot, this is indeed the exact answer, but notice how close it is to the approximate result (0.7738) given in Example 5.5, where sampling was from an ongoing (i.e., infinite) process. This illustrates the general fact that, *when the lot size is large compared to the sample size, the binomial distribution provides a reliable approximation to the correct hypergeometric probabilities.*

How small does the sample size, n, have to be to justify approximating the hypergeometric distribution by the more easily applied binomial distribution? One rule of thumb is $n \le 0.10N$; that is, as long as the sample size does not exceed 10% of the population, the binomial distribution will probably work well.

For finite populations, the hypergeometric distribution provides the true model of the random variable that counts 'successes' in a sample of n items. Often, though, it turns out that the sample size is small compared to the population size, so one rarely needs to calculate probabilities using equation (5.6), since the binomial approximation suffices instead. There are some cases, however, in which the binomial approximation does not work and one must return to equation (5.6) to obtain accurate results. One such case involves compliance sampling, discussed in the following example.

EXAMPLE 5.12

Compliance sampling is used when product safety, consumer protection, and tight standards set by regulatory agencies are of extreme importance (Schilling 1982). In such cases, it is desirable to detect the presence of *any* nonconforming or defective items that may exist in a population. We let X denote the number of nonconforming items in a random sample of n drawn from a population of size N. Then how large should the sample be in order to guarantee a high probability, γ, of detecting at least one nonconforming item? The answer depends, of course, on the prevalence of nonconforming items in the population. Suppose that there are D nonconforming or defective items in the population. Then, to detect at least one defective item with probability γ, the equation $P(X \ge 1) = \gamma$ must be solved for n. Since $P(X \ge 1) = 1 - P(X = 0)$, this is the same as solving $P(X = 0) = 1 - \gamma$ for n. Because the γ is large, the resulting value of n will generally be large, so $P(X = 0)$ must be treated as a hypergeometric probability. Solving $P(X = 0) = 1 - \gamma$ for n is not easy; however, Farnum and Suich (1986) have verified that the following formula provides a very close approximation to the exact solution:

$$n \approx N[1 - (1 - \gamma)^{1/D}] \tag{5.7}$$

For example, if $N = 200$, $D = 4$, and $\gamma = 0.90$, then equation (5.7) indicates that $n = 200[1 - (1 - 0.90)^{1/D}] = 87.53$, or a sample of about 88 should suffice to detect at least one of the four defectives. Farnum and Suich also showed that the exact solution to $P(X = 0) = \gamma$ must lie between $(N - D + 1)[1 - (1 - \gamma)^{1/D}]$ and $N[1 - (1 - \gamma)^{1/D}]$, so the approximation in equation (5.7) is conservative but generally very close to being exact for small values of D.

5.6   COMMON CONTINUOUS DISTRIBUTIONS

THE NORMAL DISTRIBUTION

NORMAL RANDOM VARIABLES.   Having examined some important discrete random variables, we turn to continuous random variables used in quality control. Recall that the distinction between continuous and discrete variables reflects the type of measurements used, either variables or attributes data (see Sections 1.4 and 5.5). For example, the *length* X of a randomly sampled part from a manufacturing process is a continuous random variable, since length itself is a continuous quantity.

Continuous random variables can conceivably be measured to finer and finer degrees of accuracy (see Section 1.4), as is graphically depicted in Figure 5.10. The

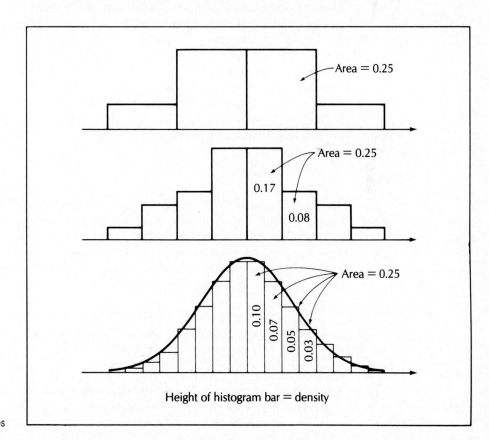

FIGURE 5.10
Histograms of the same data with successively finer measurement scales

successive histograms in Figure 5.10 correspond to a fixed collection of items mea-sured with successively better measuring instruments. In order to make the his-tograms comparable, the convention of making the *areas* of their bars *equal* the corresponding *relative class frequencies* is followed.[2] When this is done (instead of making the frequencies match the *heights* of the bars), the area of any bar in a histogram can be interpreted as the probability with which data fall into that class, and the total area in all the bars must sum to 1.

In Figure 5.10, the histograms begin to approach, in the limit, a smooth curve. Notice that the convention of making the *areas* of histogram bars equal to their relative frequencies causes the *heights* of the bars to correspond to the *density* with which data fall into a given class. Thus, the smooth curve that approximates the heights of the histogram bars is called a **density curve**. Since the areas of the histogram bars represent probabilities, the corresponding areas under the density curve are also interpreted as probabilities.

A small number of density curves occur over and over again in statistical ap-plications, but, by far, the **normal density** is the one most commonly used to de-scribe continuous random variables. Random variables governed by a normal density curve are called **normal random variables,** which then have **normal probability distributions.** The normal distribution acts as a cornerstone in statisti-cal methodology, underlying the probability calculations for what many have called the empirical rule (see Section 3.3), for the central limit theorem (see Sec-tion 5.4), for control charts (Chapters 7 and 9), as well as for approximations to other distributions (especially the binomial). Since so many random variables tend to have a normal distribution, it has become common practice to assume that a majority of continuous process measurements will follow the normal curve. There are, of course, exceptions to this general practice, which are discussed later.

The density curve of a normal random variable, X, that has a mean of μ and a standard deviation of σ is given by the formula[3]

$$f(x) = \frac{1}{\sigma\sqrt{2\pi}} e^{-(1/2)[(x-\mu)/\sigma]^2} \qquad \text{for } -\infty < x < \infty \tag{5.8}$$

This curve is drawn in Figure 5.11. Since two parameters, μ and σ, are needed to define the curve, the normal distribution is a *two-parameter distribution*. Notice from equation (5.8) that any normal curve must be *symmetric* around its mean, μ. It can also be shown that the points on either side of μ, where the curve changes from being concave down to being concave up, must lie at distances of exactly $\pm\sigma$ from μ (this is an aid for sketching a normal curve).

THE STANDARD NORMAL TABLES.   Finding areas (i.e., probabilities) under a normal curve requires the use of tables or software. Fortunately, a separate table for every possible pair of values for μ and σ is not necessary. The only table needed is

---

[2]Relative frequency of a class = class frequency divided by n.
[3]In discussing the normal distribution, we suppress the subscripts on μ_X and σ_X in order to simplify the appearance of the formulas and calculations.

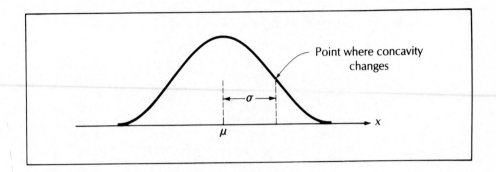

FIGURE 5.11
Graph of normal
curve with mean μ
and standard
deviation σ

the one for the **standard normal distribution,** which is a normal curve with parameters $\mu = 0$ and $\sigma = 1$. It is traditional to use the letter z to denote the standard normal variable. Appendix 2 gives right-tail areas (probabilities) for the standard normal distribution. Probability statements involving *any other* normal curve can always be reduced to equivalent statements about the standard normal curve by a process called *standardization*. Standardization is accomplished by subtracting the mean and dividing by the standard deviation; that is, X is transformed into z by

$$z = \frac{X - \mu}{\sigma} \tag{5.9}$$

For example, to find $P(X > 5)$ for some normal variable with a mean of 3 and standard deviation of 2, we look up the probability $P(z > \frac{5-3}{2}) = P(z > 1)$ in Appendix 2.

EXAMPLE 5.13

The empirical rule (see Section 3.3) is used as a rough approximation for describing any set of process measurements whose histograms appear bell-shaped. In fact, the empirical rule arrives at its probability estimates by assuming that the data follow a normal distribution. Thus, to find the proportion of all process measurements X that fall within 1 standard deviation of the mean, standardization yields

$$P(\mu - \sigma < X < \mu + \sigma) = P\left(\frac{\mu - \sigma - \mu}{\sigma} < z < \frac{\mu + \sigma - \mu}{\sigma}\right)$$

$$= P(-1 < z < 1) = 2 \cdot [0.5 - P(z > 1)]$$

$$= 2(0.3413) = 0.6826$$

or about 68.26%. The reader may verify that similar calculations using the normal table lead to the probabilities associated with 2- and 3-sigma deviations from the mean in the empirical rule.

EXAMPLE 5.14

Table 5.1 contains measurements on the lengths of 100 parts selected from the output of a manufacturing process. Suppose that this process is fairly stable, so that the sample mean of 9.999 in. and the sample standard deviation of

TABLE 5.1   Lengths (in inches) of 100 Parts

9.983	10.000	9.988	10.013
10.020	9.995	9.989	10.011
10.001	9.998	10.009	10.014
9.981	9.995	10.015	9.999
9.994	10.001	10.005	9.987
10.016	10.002	9.999	9.987
9.992	9.981	10.001	9.992
10.023	10.000	10.002	9.987
9.985	9.999	9.979	9.993
10.035	9.996	10.002	10.002
9.960	10.029	9.987	10.021
10.007	10.023	10.017	9.983
9.976	10.014	10.001	10.004
9.998	10.032	10.003	10.012
9.994	9.995	9.988	9.991
9.979	10.010	9.996	10.007
10.034	9.992	10.004	9.980
10.010	9.988	10.012	9.974
9.994	10.032	10.008	9.985
10.049	9.949	10.004	9.984
9.998	9.989	9.972	10.002
10.009	9.977	10.002	10.002
9.975	10.010	10.025	9.985
9.992	10.005	10.001	9.985
10.017	9.978	10.019	10.000

0.0165 in. can be used validly to describe the process. Furthermore, suppose that the lengths are thought to follow a normal curve. If the specification limits for these parts are USL = 10.050 and LSL = 9.950 in., then the proportion of nonconforming items can be calculated using the normal tables as follows. Letting X denote the length of a randomly chosen part, we need to find the proportion, $P(X > \text{USL})$ of parts that exceed the upper specification limit and the proportion, $P(X < \text{LSL})$, that are less than the lower specification limit. With the sample mean and standard deviation, $\bar{x}$ and s, used to approximate μ and σ, respectively, the calculations are:

$$P(X > \text{USL}) = P\left(z > \frac{\text{USL} - \mu}{\sigma}\right) \approx P\left(z > \frac{\text{USL} - \bar{x}}{s}\right)$$

$$= P\left(z > \frac{10.050 - 9.999}{0.0165}\right) = P(z > 3.09) = 0.001$$

(from Appendix 2). Similarly,

$$P(X < \text{LSL}) = P\left(z < \frac{\text{LSL} - \mu}{\sigma}\right) \approx P\left(z < \frac{\text{LSL} - \bar{x}}{s}\right)$$

$$= P\left(z < \frac{9.950 - 9.999}{0.0165}\right) = P(z < -2.97)$$

which, from the symmetry of the normal curve, equals $P(z > 2.97) = 0.0015$. In all, it is estimated that about $0.0010 + 0.0015 = 0.0025$, or 0.25%, of the output of this process will exceed the specification limits.

PARTS-PER-MILLION CALCULATIONS.   Calculations of the type illustrated in Example 5.14 are fundamental to the computation and interpretation of process capability indexes (see Chapter 8). Although the rate of 0.25% nonconforming (and, therefore, 99.75% conforming) estimated in that example might at first seem acceptable, recall that the lesson of Examples 5.7 and 5.8 indicates that this may not, in fact, be good enough. Recognizing this, modern quality improvement programs now strive for much lower rates of nonconforming items than those in Example 5.14.

Switching to lower rates of nonconformance requires changing the base of comparison. Instead of reporting rates on a *percentage* (i.e., parts per hundred) basis, it is now common practice to report them on a *parts per million* basis. One nonconforming (or defective) item in a collection of 1 million items is called one *part per million,* which is abbreviated as 1 ppm. Thus, the 0.25% nonconforming rate in Example 5.14 translates into 2,500 ppm's. For convenience, Table 5.2 shows the equivalent percentage and ppm rates for a range of values that might occur in practice. In order to estimate such small rates, the tables of the normal distribution (which are traditionally limited to z values between ± 3) must be extended to accommodate the larger z values required for ppm calculations. (Appendix 2 includes selected z values up to 5.)

EXAMPLE 5.15   Motorola's "6-Sigma Program," introduced in Example 3.4, has the intended goal of reducing the nonconformance rates of all processes to 3.4 ppm. Actual rates may be even lower than 3.4 ppm, since the program bases its calculations on processes that may have slipped slightly off center by as much as 1.5 sigmas. As described in Example 3.4, the goal of the program is to make the process variation small enough so that 12 sigmas will fit within the specification limits, which means that $12\sigma = \text{USL} - \text{LSL}$, or equivalently $6\sigma = (\text{USL} - \text{LSL})/2$. Then, even if the 2 process center, μ, should slip by as much a 1.5 sigmas off center in one direction—say, to the right so that $\mu = (\text{USL} + \text{LSL})/2 + 1.5\sigma$—the

TABLE 5.2   Table for Converting from Percent Nonconforming to Parts per Million

Percent (%)	Parts per Million (ppm)
10.0	100,000
5.0	50,000
1.0	10,000
0.1	1,000
0.01	100
0.001	10
0.0001	1

proportion of items that would exceed the upper specification (for a shift to the right) would be

$$P(X > USL) = P\left(z > \frac{USL - \mu}{\sigma}\right) = P\left(z > \frac{USL - \left\{\frac{USL + LSL}{2} + 1.5\sigma\right\}}{\sigma}\right)$$

$$= P\left(z > \frac{\frac{USL - LSL}{2} - 1.5\sigma}{\sigma}\right) = P\left(z > \frac{6\sigma - 1.5\sigma}{\sigma}\right)$$

$$= P\left(z > \frac{4.5\sigma}{\sigma}\right) = P(z > 4.5) = 0.00000340$$

(from Appendix 2), which corresponds to 3.40 ppm.

As Examples 5.7 and 5.8 show, there is no absolute scale by which a component's quality can be evaluated. The quality levels required depend on the complexity of the systems built from these components. As a rough guideline for evaluating the quality of components with various ppm levels, Table 5.3 shows the proportions of nonconforming systems (or assemblies) that result whenever one nonconforming subcomponent causes system nonconformance (as in Examples 5.7 and 5.8). This table has quite general use, since complex systems (or assemblies) can often be decomposed into *subsystems* (or assemblies) of the type

TABLE 5.3 Proportion of Nonconforming Systems by Component Nonconformance Rate and Size of System

n ↓	ppm →	100,000	50,000	10,000	1,000	100	10	1
10		0.65132	0.40126	0.09562	0.009956	0.001000	0.0001001	0.0000108
20		0.87842	0.64151	0.18210	0.019812	0.001999	0.0002003	0.0000215
30		0.95761	0.78536	0.26030	0.029570	0.002996	0.0003004	0.0000323
40		0.98522	0.87149	0.33103	0.039232	0.003993	0.0004005	0.0000430
50		0.99485	0.92306	0.39500	0.048797	0.004989	0.0005006	0.0000538
60		0.99820	0.95393	0.45285	0.058267	0.005983	0.0006007	0.0000644
70		0.99937	0.97242	0.50517	0.067642	0.006977	0.0007007	0.0000751
80		0.99978	0.98348	0.55248	0.076924	0.007970	0.0008008	0.0000858
90		0.99992	0.99011	0.59527	0.086114	0.008962	0.0009009	0.0000967
100		0.99997	0.99408	0.63397	0.095212	0.009952	0.0010009	0.0001073
500		1.00000	1.00000	0.99343	0.393636	0.048781	0.0049944	0.0005363
1,000		1.00000	1.00000	0.99996	0.632322	0.095182	0.0099638	0.0010724
1,500		1.00000	1.00000	1.00000	0.777053	0.139320	0.0149082	0.0016081
2,000		1.00000	1.00000	1.00000	0.864813	0.181305	0.0198281	0.0021435
2,500		1.00000	1.00000	1.00000	0.918027	0.221241	0.0247234	0.0026786
3,000		1.00000	1.00000	1.00000	0.950295	0.259230	0.0295942	0.0032135

Note: Tabled values are calculated from $P(X \geq 1) = 1 - \left(1 - \frac{ppm}{10^6}\right)^n$.

addressed by Table 5.3. As an example, suppose that a certain product is composed of two independent subassemblies, A and B, made up of 50 and 20 'critical' parts, respectively. Suppose further that the parts used in assembly A have a nonconformance rate of about 1,000 ppm (i.e., 0.1%), while those of assembly B have a 10,000 ppm (i.e., 1%) nonconformance rate. From Table 5.3, about 4.88% of the A subassemblies and 18.21% of the B subassemblies will be nonconforming. Since these assemblies are assumed to act *independently* of one another, the resulting proportion of products that will be nonconforming can be calculated by using the general addition law and the multiplicative law of Section 5.3:

$$P(\text{product nonconforming}) = P(\text{either system A or B is nonconforming})$$

$$= P(\text{A nonconforming}) + P(\text{B nonconforming})$$
$$- P(\text{A nonconforming}) \cdot P(\text{B nonconforming})$$

$$= 0.0488 + 0.1821 - (0.0488)(0.1821) = 0.2220$$

or about 22.2%.

NORMAL APPROXIMATION TO THE BINOMIAL

Binomial distributions for which p is not too near 0 or 1 tend to have histograms that are closely approximated by a normal density curve. This is why, in Section 5.5, it was recommended that binomial distributions be sketched by drawing bell-shaped histograms around a center of $\mu_X = np$, whose bars are contained within $\pm 3\sigma_X = \pm 3\sqrt{npq}$ of μ_X. As mentioned in that section, the conditions necessary for using this approximation are that *both np and nq exceed 5*. The first condition guarantees that p is not too close to 0, and the second condition assures that p is not too close to 1.

The normal approximation can also be used to approximate binomial probabilities. To do this, in any probability statement involving a binomial random variable X, one simply replaces X by a normal random variable X^* that has the *same* mean and standard deviation as X. Thus, the mean and standard deviation of X^* are np and $\sqrt{npq}$, respectively. Finally, the probability statement is evaluated in the usual fashion, by first standardizing and then using the normal tables in Appendix 2.

In some cases when n is small, the accuracy of the normal approximation can be improved somewhat by incorporating a small adjustment called the **continuity correction**. This adjustment is made because, when one attempts to approximate a *discrete* random variable (the binomial) by a *continuous* one (the normal), a smooth density curve is superimposed over the bars in a histogram, so some corrections must be made to assure that the curve areas correspond to the histogram bars. Figure 5.12 shows a close-up view of this procedure. Essentially, the correction uses the outer boundaries of the histogram bars to establish the correct limits for the normal curve. The next example illustrates this procedure.

EXAMPLE 5.16   Consider a binomial random variable X with parameters $n = 25$ and $p = 0.30$ ($n = 25$ is chosen so that Appendix 3 can be used to check the accuracy of the

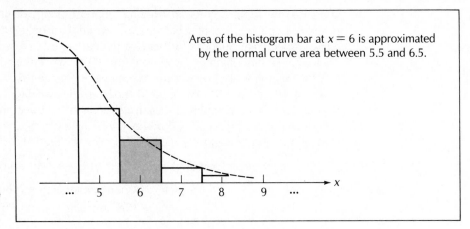

Area of the histogram bar at $x = 6$ is approximated by the normal curve area between 5.5 and 6.5.

FIGURE 5.12
Illustration of the continuity correction for approximating the binomial with the normal distribution

normal approximation), and suppose one wants to find the probability that X exceeds 9. Since $np = 7.5$ and $nq = 17.5$ both exceed 5, the normal approximation can be used. The normal variable X^* that approximates X has a mean of $np = 7.5$ and standard deviation of $\sqrt{25(0.3)(0.7)} = 2.291$; so, from the continuity correction,

$$P(X > 9) = P(X \geq 10) \approx P(X^* > 9.5) = P\left(z > \frac{9.5 - 7.5}{2.291} \right) = P(z > 0.87)$$

$$= 0.1922$$

(from Appendix 2). For comparison, the binomial tables give an exact answer of 0.189, so the normal approximation is off by only 0.003.

With the addition of the normal approximation, there is now little need to use the binomial formula, equation (5.3), to calculate probabilities. Most often, binomial probabilities are found from tables (Appendix 3), from the Poisson approximation (see Section 5.5) if $np \leq 7$, or from the normal approximation if $np > 5$ and $nq > 5$.

STUDENT'S t DISTRIBUTION

Small sample sizes are the norm, not the exception, when on-line statistical quality control methods are used. The resource constraints of the production environment simply demand that one use the minimum necessary sample sizes sufficient to meet quality goals. It was in exactly such a setting that the Student's t distribution was developed in England around 1908 (Student 1908).

'Student' was the pen name of the industrial statistician W. S. Gosset, who developed the basis of what is now known as the Student's t distribution (usually abbreviated as the 't **distribution**') in order to correctly handle small-sample statistical estimates that arose from quality control studies at the Guiness brewery. Gosset's methods and conjectures about this new distribution were eventually given theoretical grounding by another statistician, R. A. Fisher.

The *t* distribution arose from the following observation: to assess how accurately $\bar{x}$ estimates the process mean, the sample standard deviation *s* (see Section 5.4) must be substituted for σ in the expression $\sigma/\sqrt{n}$ (see Section 5.4) and, when the sample size *n* is small, the normal distribution will give misleadingly high estimates of the precision in estimating the process average. The *t* distribution was proposed as the correct one to use for probability calculations when *s* is used to estimate σ. For this reason, many practitioners refer to the *t* distribution as the 'small sample distribution.'

The *t* distribution and normal distribution are closely related and, in fact, become indistinguishable as the sample size *n* increases. This makes sense intuitively, since for large *n*, the sample standard deviation becomes an increasingly accurate estimate of σ, which tends to reduce the inaccuracies introduced by using $s/\sqrt{n}$ in place of $\sigma/\sqrt{n}$. The relationship between these two distributions is shown in Figure 5.13. Notice that *there is a different t distribution for each value of the sample size, n,* and that the *t* distribution always has fatter tails (and is less peaked) than the standard normal distribution, but that the differences between the two curves disappear as *n* increases. As a practical rule of thumb, many practitioners assume that for $n \geq 30$, the differences between the standard normal and *t* distributions are negligible, so that using normal distribution probabilities whenever $n \geq 30$ will produce acceptable results in most applications.

Each *t* distribution is uniquely determined by a parameter called its **degrees of freedom** (abbreviated **d.f.**), which depends on the sample size(s) used. When the process average is estimated based on a single sample of size *n*, for example, the degrees of freedom are given by d.f. $= n - 1$. For other statistical estimates, the degrees of freedom are calculated differently, and the appropriate formulas are given as the need arises. It is not necessary to have a complete understanding of exactly what 'degrees of freedom' are in order to apply the *t* distribution. However, for the interested reader we recommend the discussion found in Mosteller and Rourke (1973), which we have summarized in Section 13.1.

Just as for the normal distribution, calculations based on the *t* distribution require the use of probability tables or software. Appendix 5 lists commonly used tail areas and associated *t* values for the *t* distribution with various values of d.f.

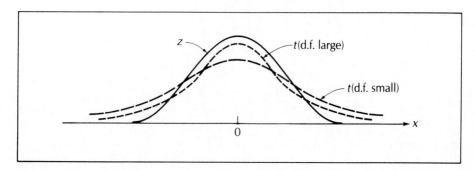

FIGURE 5.13
Relationship between the standard normal (z) and Student's *t* distribution

For example, if d.f. = 10, the table indicates that $t = 1.812$ is the value for which the right tail area $P(t > 1.812) = 0.05$. In any application in which d.f. is large, the standard normal z value can be used in place of the t value.

THE EXPONENTIAL DISTRIBUTION

The exponential distribution is another continuous probability distribution that arises frequently in quality control analyses. In particular, the exponential finds great application in the field of reliability. Reliability has been roughly described by some as the study of quality over time—that is, the study of how long products will last before they eventually become unfit for use. In reliability, the exponential distribution is frequently used to model the behavior of random variables such as the lifetimes of components. Although other distributions can also be used, many government contracts with aerospace suppliers specify that reliability calculations be based on the exponential distribution.

Consider, for example, how one might model the lifetime of a particular type of electronic component. We let X denote the measured lifetime of such a component (i.e., the length of time until the component fails to function). Then Figure 5.14 shows a typical histogram of the lifetimes of a large number of such components. Note that, since X must be nonnegative, the histogram cannot extend into the negative side of the horizontal axis. Furthermore, since 'lifetime' is a continuously measured quantity (see Section 1.4), if lifetimes are measured more and more precisely, the histogram of Figure 5.14 should eventually be well approximated by some density curve. In particular, the density curve shown in Figure 5.14 is called the **exponential density** and has the defining equation

$$f(x) = \begin{cases} \lambda e^{-\lambda x} & \text{for } x \geq 0 \\ 0 & \text{for } x < 0 \end{cases} \tag{5.10}$$

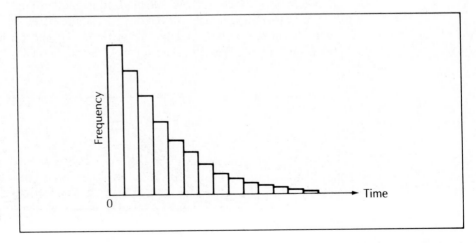

FIGURE 5.14
Distribution of lifetimes of electronic components

If we consider the lifetime, X, of a randomly chosen component as a variable with density curve given by equation (5.10), then X is said to have an **exponential probability distribution.** The mean and standard deviation of X are given by

$$\mu_X = \frac{1}{\lambda} \quad \text{and} \quad \sigma_X = \frac{1}{\lambda} \tag{5.11}$$

where the parameter λ is called the **failure rate.** When lifetimes are modeled, the mean of the exponential distribution is also called either the **MTBF** *(mean time before failure)* or the **MTTF** *(mean time to failure)* of the component. Thus, MTBF = $1/\lambda$; that is, the mean time before failure is the reciprocal of the failure rate.

Finding probabilities for exponential distributions is easier than with most distributions, since there is a simple formula for the right-tail area of the exponential— namely,

$$P(X > a) = e^{-\lambda a} \tag{5.12}$$

By using equation (5.12) along with the fact that $P(X \le a) = 1 - P(X > a)$, it is a simple process to calculate the probabilities associated with exponential random variables.

EXAMPLE 5.17   To assess the failure rate of an electronic component, many components (of the same type) are tested by operating them continuously and recording the time when each fails. This procedure is called **life testing** (Lloyd and Lipow 1984). Suppose that life testing on a certain component shows that the failure rate averages about 1 failure in every 1,000 hr of operation. In this case, $\lambda = 1/10^3 = 10^{-3}$ failures/hr. Assuming that the exponential distribution describes the lifetimes of these components, one can estimate the proportion of the components that will last for more than 2,000 hr by applying equation (5.12) to get

$$P(X > 2,000) = e^{-\lambda(2,000)} = e^{-10^{-3}(2,000)} = e^{-2} = 0.1353$$

that is, about 13.53% of the components should last for 2,000 hr of operation or longer.

When applied to lifetime data, the exponential and Poisson distributions are highly related. Specifically, consider any interval of time from 0 to t and suppose that the *number* of components that fail in this interval can be modeled as a Poisson random variable, Y, with parameter λt (where λ is the average failure rate per *unit* of time). Next, consider the random variable, X, that measures the time to failure of these components. It can be shown that X must follow an exponential distribution with parameter λ. The converse is also true. If X is known to follow an exponential distribution with parameter λ, then Y must be Poisson with parameter λt for any interval of time $[0, t]$.

5.7	INTERVAL ESTIMATION

The formulas for the mean and standard deviation—equations (3.1) and (3.7), respectively—can be applied to *any* set of numbers, but if conclusions or predictions are to be drawn from such calculations, special attention must paid to *how* the raw data are obtained. In particular, when the data arise by some sort of *random* procedure, the mean and standard deviation (or *any* of the measures introduced in Chapter 3) can be thought of as *random variables,* and knowledge of their probability distributions allows one to make predictions and inferences concerning the process. Random sampling is one method of achieving this goal. Another method, used in control chart calculations, is to *assume* that a 'common cause' system is operating (see Section 1.4).

Statistics calculated from random samples are called **estimators** when they are used to estimate unknown **population parameters.** For example, $\bar{x}$ is an estimator of the process average, μ, and s is an estimator of the process standard deviation, σ. Since the numerical value of each estimate naturally varies from sample to sample, an estimator cannot be expected to coincide exactly with the parameter it estimates, but it can be expected to assume any of a range of possible values *around* the parameter. In the language of Section 5.4, the behavior of the statistic is described by its **sampling distribution.** Since the exact forms of the sampling distributions of the statistics used in quality control are known, precise probability statements can be obtained regarding where the population parameters should lie.

The numerical value of an estimator is often called a **point estimate** because it is a single number or 'point' used to estimate a parameter of interest. In practice, however, it is often helpful to obtain an *interval* of possible values in which the population parameter is expected to lie. Estimating a population parameter by using an interval of numbers is called **interval estimation**.

How wide an interval should be used and where should it be centered? The answer depends on knowing the *form of the sampling distribution* (i.e., normal, binomial, Poisson, etc.) of the statistic and the desired *degree of confidence* that the interval estimate should have. For example, the sampling distribution of the estimator $\bar{x}$ usually approximates the shape of a normal distribution. The level of confidence is something that is subjectively chosen. Of course, one usually prefers a high degree of confidence to a low one, but quantifying this desire requires that a specific number between 0% and 100% be selected to represent the level of confidence. Commonly, 90%, 95%, and 99% confidence levels are encountered in practice.

The theory of confidence intervals is not pursued further here. The interested reader may consult general statistics texts for these details (Sachs 1982). Instead, we list some of the most frequently occurring confidence interval formulas in Table 5.4. The notation $z_{\alpha/2}$ (or $t_{\alpha/2}$) used in this table refers to the z value (or t value) for which the probability under the density curve to the right of $z_{\alpha/2}$ (or $t_{\alpha/2}$) exactly equals $\alpha/2$. The confidence level associated with these intervals is then conventionally written as $(1 - \alpha) \cdot 100\%$. Thus, for a 95% confidence interval, one would set $1 - \alpha = 0.95$ and find the tail area of $\alpha/2 = 0.025$. For conve-

TABLE 5.4   Commonly Used Confidence Intervals

Population Parameter	Estimator	$(1 - \alpha) \cdot 100\%$ Confidence Interval
μ (σ known, normal pop.)	$\bar{x}$	$\bar{x} \pm z_{\alpha/2}(\sigma/\sqrt{n})$
μ (σ unknown, normal pop.)	$\bar{x}$	$\bar{x} \pm t_{\alpha/2}(s/\sqrt{n})$ d.f. $= n - 1$
μ (σ unknown, n large)	$\bar{x}$	$\bar{x} \pm z_{\alpha/2}(s/\sqrt{n})$
p (n large)	$\hat{p}$	$\hat{p} \pm z_{\alpha/2}\sqrt{\hat{p}\hat{q}/n}$

Note 1: n denotes the size of the random sample drawn; p denotes the proportion of 'successes' for a binomial random variable; $\hat{p}$ denotes the proportion of successes observed in the sample.
Note 2: The last two intervals are not exact but approximate $(1 - \alpha) \cdot 100\%$ confidence intervals.

nience, Table 5.5 lists some of the most commonly used confidence levels and their associated $z_{\alpha/2}$ values.

It is apparent from the formulas in Table 5.4 that smaller and smaller confidence intervals may be obtained by increasing the sample size n. However, since larger sample sizes are usually achieved only at an increased cost, it is convenient to be able to obtain quick estimates of the minimum necessary sample size for a desired interval width. Exact formulas exist for calculating the minimum sample size, and we again refer the reader to general statistics texts for a discussion of these calculations (Sachs 1982). Instead, for the purposes of obtaining *quick* estimates required in practice, we present the following rule of thumb: given a confidence interval of width W based on a sample of size n, to obtain an interval of width $W/2$ requires a sample of size $4n$. Roughly speaking, *to cut the width of a confidence interval in half, quadruple the sample size.* (In applying this rule, it is assumed that the confidence level is not changed.) Although the rule does not encompass all confidence interval formulas, it *does* apply to the commonly used intervals in Table 5.4 and many others. Thus, if a 95% confidence interval for the

TABLE 5.5   Normal Distribution Factors for Commonly Used Confidence Levels

Confidence Level $(1 - \alpha) \cdot 100\%$	$z_{\alpha/2}$
99%	2.576
98%	2.326
95%	1.960
90%	1.645
80%	1.282

Note: The values of $z_{\alpha/2}$ always coincide with the values of $t_{\alpha/2}$ for d.f. $= \infty$; thus, the last row in the *t* tables can be used as a quick reference for finding $z_{\alpha/2}$ values.

process mean has been calculated from a sample of size 50 and if this interval is not considered small enough (i.e., accurate enough) for our purposes, then we can immediately estimate that it will take a sample of 200 to obtain a 95% confidence interval of half this width. The ability to obtain quick estimates like this can be very useful in the production environment.

We close this section with a discussion of a very frequently occurring special case, one for which the formulas of Table 5.4 are not sufficient. In particular, note that the usual interval estimate for a binomial proportion, given by $\hat{p} \pm z_{\alpha/2} \sqrt{\hat{p}\hat{q}/n}$, is only an *approximate* formula and that it is most accurate when $0.30 \le \hat{p} \le 0.70$ (Fleiss 1981). For more extreme values of $\hat{p}$ (when it is very close to 0 or 1), the formula above breaks down and can give misleading results. In quality control applications, where p is often taken to be the proportion of defective items in a population, values of $\hat{p}$ close to 0 are common. In fact, it is frequently the case that $\hat{p} = 0$. In such cases, the approximate confidence interval formula is of no value. Fortunately, though, the special case $\hat{p} = 0$ is covered by an even simpler formula: if a sample of size n results in $\hat{p} = 0$, then a $(1 - \alpha) \cdot 100\%$ confidence interval for p is given by (Louis 1981)

$$[0, 1 - \alpha^{1/n}] \tag{5.13}$$

EXAMPLE 5.18

In the high-volume manufacture of components such as resistors, springs, and fasteners, large sample sizes are frequently drawn to ascertain the proportion defective in a batch of items. When the quality of such a production process is good, it is very likely that the observed proportion of defective or nonconforming items may be 0. Suppose, for example, that a sample of $n = 100$ items tested on an ATE (see Section 1.4) resulted in all items passing specification, so $\hat{p} = 0$. Even though the sample is entirely nondefective, this does *not* guarantee that the same is true of the batch. Due to the small defect rate, the sample may have just missed some of the defectives in the batch. To estimate the true (unknown) proportion of defectives in the batch, the best we can do is to use expression (5.13). For example, a 95% confidence interval for p is given by $[0, 1 - 0.05^{1/100}] = [0, 0.0295]$. That is, with fairly high (95%) confidence one could say that p is no larger than 0.0295 or, equivalently, that no more than 2.95% of the batch could be defective.

EXERCISES FOR CHAPTER 5

5.1 Suppose a random sample of 100 measurements is obtained from some process, but later it is found that only 25 measurements are needed in a certain calculation. Can the *first* 25 observations in the sample of 100 be considered a random sample of size 25 from the process?

5.2 Some populations and processes are simply too large to list in an economical manner. Discuss how you would go about obtaining a random sample of ten words from a standard dictionary. What is it about your method that assures the sample will indeed be a random one?

5.3   Verify the results of Example 5.4 and Figure 5.7(b) by listing all possible rolls of three dice and the corresponding averages of the three faces showing. As accurately as possible, draw the histogram of the resulting averages.

5.4   After several lots of size 20 each are examined, varying numbers of nonconforming items are found in the lots. From a simple count of the number of lots that are found to contain some (i.e., one or more) nonconforming items, it is estimated that the probability of any lot containing at least one nonconforming item is about 0.50. Assuming that items in the lots are independent of one another and come from a process with a nonconformance probability of p, find an estimate of p.

5.5   Flaws in a certain fabric occur at a rate of about two flaws per square yard.
(a) In a given 1-square-yard section of the material, what is the probability of finding three or more flaws?
(b) What is the probability of finding three or more flaws in a 10-square-yard section of the material?

5.6   A sample of 50 items from a large shipment is inspected and no nonconforming items are found in the sample. Give a 95% confidence interval for the proportion of nonconforming items in the entire shipment.

5.7   From the $\bar{x}$ and s calculated from a random sample of $n = 40$ items, a 95% confidence interval for the process average is generated. This interval turns out to be [3.55, 3.86]. Can you conclude that approximately 95% of the process readings fall within this interval?

5.8   For a process whose measurements can be approximately described by a normal distribution, it is found that about 10% of the measurements fall above 8.5, while about 5% fall below 7.1. What are the process mean and standard deviation?

5.9   For data that follow an exponential distribution, such as the lifetime of some electronic component, it is incorrect to say that about half the lifetimes will be shorter than the mean and half will be longer than the mean lifetime. Such a statement is true, however, for the *median* of any distribution. Define the median of any continuous random variable X to be the particular value $\tilde{x}$ for which half of the area under the density function lies to the left of $\tilde{x}$. Find the median of an exponential distribution with mean μ.

5.10   In an accounting audit of a large batch of bills of lading, a sample of bills is to be drawn and examined for errors. If, in fact, the batch contains exactly two bills with errors, how large a sample should be taken in order to guarantee, with probability 0.95, that at least one of these defective bills will be found?

5.11   In reliability analysis, the probability that a complex system functions correctly is found by examining the probabilities that its subsystems function correctly.[4] In particular, a **series system** is one composed of subsystems $S_1, S_2, S_3, \ldots, S_n$, *all* of which must function if the entire system is to function. The probability that any

---

[4]See Lloyd and Lipow (1984, chap. 9).

system functions correctly is denoted by $P(S)$. Suppose that a particular series system contains four subsystems with the reliabilities: $P(S_1) = 0.95$, $P(S_2) = 0.85$, $P(S_3) = 0.80$, and $P(S_4) = 0.99$. Assuming that all the subsystems are independent of one another, find the reliability of this serial system.

5.12 *(Continuation of the previous problem)* A **parallel system** is one that can fail only if *all* of its subsystems fail to function. Such systems are said to have some *redundancy* built in because they will continue to operate even if a few subsystems have broken down. Suppose that a certain parallel system is composed of three subsystems with reliabilites $P(S_1) = 0.85$, $P(S_2) = 0.90$, and $P(S_3) = 0.9$. Assuming that all the subsystems operate independently of one another, find the reliability of the entire system.

5.13 Fault tree analysis was invented in the 1960s for analyzing the Minuteman launch control system.[5] Fault trees are constructed by selecting a particular undesirable event at the top of the tree and then identifying the various events (called 'faults') that could cause the top event to happen. Drawing lines to show the connection(s) from one level of faults to the next results in a tree diagram such as the one here:

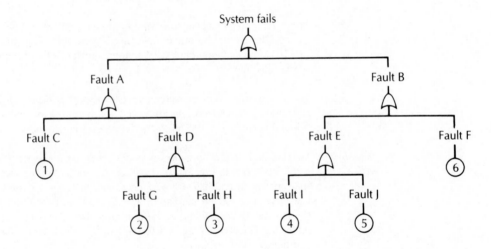

Symbols such as $\triangle$ (which means 'OR') and $\triangle$ (which means 'AND') describe how one fault level affects the next higher level. For example, fault A will occur if *either* fault C or D occurs. The circle symbol represents a 'basic' event, which has no lower-level causes. Assuming that all faults are independent of one another, find the probability of system failure in this tree in terms of the probabilities of the basic events. (*Note:* When it is not reasonable to assume that faults are independent, a recursive fault tree analysis can be used instead.[6])

[5]See Dhillon (1983).

[6]See Page and Perry (1991).

5.14   A standard procedure for testing safety glass is to drop a $\frac{1}{2}$-lb iron ball onto a 12-in. square of glass supported on a frame.[7] The height at which the ball is dropped is determined so that there is a 50% chance of a break. A break-through is considered a 'failure,' while a ball that is stopped by the glass (even if the glass is cracked) is considered a 'success.' Suppose that 100 sheets of safety glass are tested over a ten-day period. If there has been no change in the resin used to manufacture the glass, what is the probability that 60 or more of the sheets will break?

5.15   'Travellers' or 'routing sheets' are documents that accompany a product as it sequences through various production steps. Travellers contain manufacturing instructions pertaining to the particular item. Suppose each of 30 data fields on a certain traveller has a small chance—say, 0.5%—of being filled out wrong or of being wrongly interpreted by an operator. What is the probability that a product built according to such a traveller will contain:
(a) At least one nonconformance?
(b) At least two nonconformances?

5.16   Show that $-(1/\lambda) \ln(\alpha)$ is the formula for the upper $\alpha \cdot 100\%$ percentage point of an exponential distribution with parameter λ.

5.17   For any collection of events $A_1, A_2, A_3, \ldots, A_n$, it can be shown that the inequality

$$P(A_1 \cap A_2 \cap A_3 \cap \cdots \cap A_n) \geq 1 - [P(\overline{A}_1) + P(\overline{A}_2) + P(\overline{A}_3) + \cdots + P(\overline{A}_n)]$$

always holds. This inequality is particularly useful when the events involved each have relatively high probabilities. Suppose that a system consists of ten components connected in series (see Exercise 5.11), with each component having a 0.999 chance of functioning without failure. What lower bound can you put on the reliability of this system?

5.18   For any collection of events $A_1, A_2, A_3, \ldots, A_n$, it can be shown that the inequality

$$P(A_1 \cup A_2 \cup A_3 \cup \cdots \cup A_n) \leq P(A_1) + P(A_2) + P(A_3) + \cdots + P(A_n)$$

always holds. This inequality provides useful information in cases where the events involved have relatively small probabilities. Suppose, for example, that a system consists of five components connected in series (see Exercise 5.11), with each component having a probability 0.01 of failing. Find an upper bound on the probability of a system failure.

5.19   In cases where it is not clear which distribution *best* describes a measured characteristic, the method of 'bootstrapping' can be used. Bootstrapping refers to the procedure of repeatedly *resampling* a set of data *with replacement,* calculating new statistics each time, and using the combined results to assess the sampling distribution of the statistic.[8] For example, suppose that the following 20 measurements were obtained from a process:

5.59	5.67	5.71	5.85	6.26	6.15	5.54	6.14	6.06	5.89
6.06	6.15	5.73	5.91	6.31	5.95	6.25	6.38	5.88	5.85

---

[7]See Gore (1989).

[8]See Diaconis and Efron (1983).

(a) From these data, repeatedly draw samples of size 20 (with replacement) and calculate the mean of each such sample. After doing this many times (the more the better), form a histogram of the resulting means and use it to place 95% confidence bounds on the process mean from which the data were obtained.

(b) Construct a 95% confidence interval for the process mean using the usual *t* distribution formula (which assumes normally distributed process data). How does this result compare to the one in part (a)?

REFERENCES FOR CHAPTER 5

Dhillon, B. S. 1983. *Systems Reliability, Maintainability, and Management*, pp. 229–235. New York: Petrocelli Books.

Diaconis, P., and B. Efron. 1983. "Computer-Intensive Methods in Statistics." *Scientific American* 248: (116–130).

Diamond, S. J. 1989. "Lip Service No Substitute for Genuine Customer Service." *Los Angeles Times,* Oct. 13.

Farnum, N. R., and R. C. Suich. 1986. "Comment on: Determining Sample Size When Searching for Rare Events." *IEEE Transactions on Reliability* R-35 (no. 5):584–585.

Fleiss, J. L. 1981. *Statistical Methods for Rates and Proportions*, p. 15. New York: Wiley.

Gitlow, H. S., and D. A. Wiesner. 1988. "Vendor Relations: An Important Piece of the Quality Puzzle." *Quality Progress,* January.

Gore, W. L. 1989. "Statistical Methods in Plastics Research and Development." *Quality Engineering* 2 (no. 1):81–89.

Hill, H. W. 1988. "Going Back into Space Is Still a Roll of the Dice." *Los Angeles Times,* Sept. 18.

Hillier, F. S. and G. J. Liebermann. 1986. *Introduction to Operations Research,* 4th ed. Oakland, CA: Holden-Day.

Kennedy, W. J., and J. E. Gentle. 1980. *Statistical Computing,* p. 239. New York: Marcel Dekker.

Lloyd, D. K., and M. Lipow. 1984. *Reliability: Management, Methods, and Mathematics,* 2nd ed., Chap. 10. Milwaukee, WI: ASQC.

Louis, T. A. 1981. "Confidence Intervals After Observing No Successes." *The American Statistician* 35 (no. 3):154.

Mosteller, F., and R. E. K. Rourke. 1973. *Sturdy Statistics.* Reading, MA: Addison Wesley.

Page, L. B., and J. Perry. 1991. "Reliability, Recursion, and Risk." *American Mathematical Monthly* 98 (no. 10):937–946.

Sachs, L. 1982. *Applied Statistics.* New York: Springer Verlag.

Schilling, E. G. 1982. *Acceptance Sampling in Quality Control,* Chap. 17. New York: Marcel Dekker.

"Student." 1908. "The Probable Error of the Mean." *Biometrika* 6: 1–25.

Walpole, R. E., and R. H. Meyers. 1989. *Probability and Statistics for Engineers and Scientists*. New York: Macmillan.

Wright, T., and H. Tsao. 1985. "Some Useful Notes on Simple Random Sampling." *Journal of Quality Technology* 17 (no. 2):67–73.

CHAPTER **6**

Control Chart Concepts

This chapter discusses the concepts and philosophy underlying the control chart methodology. It covers the types of control charts available, how to choose the proper chart, the problem of variable subgroup sizes, how to interpret control chart patterns, and how to manage the charting process. The computational aspects of control charts are covered in Chapters 7 and 9.

CHAPTER OUTLINE

6.1 CONTROL CHART THEORY

6.2 TYPES OF CONTROL CHARTS

6.3 SUBGROUPS AND SAMPLE SIZES

6.4 INTERPRETING CONTROL CHART PATTERNS

6.5 MANAGING THE CHARTING PROCESS

6.1   CONTROL CHART THEORY

This chapter introduces the concepts underlying the control chart applications in Chapters 7–9. In deciding which to cover first, the mechanics of control charts or the philosophy of applying them, we have opted for the latter. It is recommended that Chapter 6 be read lightly once before the reader proceeds to the variables control charts of Chapter 7. Afterwards, a rereading of Chapter 6 will enhance the understanding of both chapters.

COMMON CAUSES VERSUS SPECIAL CAUSES

Shewhart's concept of common causes versus special causes was described in Section 1.4. Recall that **common causes** account for the uncontrollable, natural variation present in any repetitive process, whereas **special causes** are those whose effect can be detected and controlled. In the quality control literature, **chance cause** and **assignable cause** are frequently used in place of the terms 'common cause' and 'special cause,' respectively.

Exercising control over special causes requires three things: a mechanism for *detecting* or signaling the presence of special causes, the ability to trace back and *find* the specific cause responsible for the signal, and the ability to *correct* problems once they are found. **Control charts** provide a mechanism for deciding which causes are special and which are not. **Traceability**, however, depends on the quality of the data system used. Sometimes, memory alone suffices to pinpoint a special cause—for example, when one remembers that different operators or materials were used on a certain day. At other times, changes are subtle and less well documented, which clouds the search for special causes. In those cases, cause-and-effect diagrams, Pareto charts, or informal discussions with those directly involved with the process can frequently uncover special causes. In a more proactive vein, these tools can also be used to create a quality *database,* thereby improving the traceability necessary for effective process control (see Chapter 14).

Many control charts resemble the one in Figure 6.1. Successive samples (also called **subgroups**) of size n are taken from the output of a process; then a statistic is calculated for each subgroup and plotted (on the vertical scale) against the subgroup number (horizontal scale). Any statistic of interest can be calculated, but those most commonly used are $\bar{x}$ (sample mean), s (sample standard deviation), R (sample range), $\tilde{x}$ (sample median), p (proportion nonconforming), np (number nonconforming), c (number of nonconformities), and u (number of nonconformities per inspection unit). Statistics that *accumulate* the information in many successive subgroups are also in wide use (see Sections 7.6 and 7.7). Sometimes, when only subgroups of size $n = 1$ are available, the individual process measurement itself is plotted on the chart (see Section 7.5).

Control limits and a **centerline** are calculated and placed on the chart to divide the vertical scale into two regions, one associated with special causes and one with chance causes. Plotted statistics that fall outside (above or below) the control limits are interpreted as signals of possible special causes, whereas points

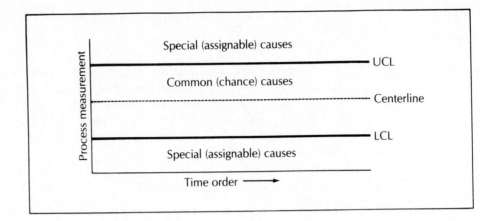

FIGURE 6.1
The control chart

that fall within the control limits are usually (but not always) associated with the *absence* of special causes.

The centerline of a control chart is, in theory, the long-run expected value of the statistic being charted. Thus, when samples of size *n* are drawn and the sample mean, $\bar{x}$, is plotted, the centerline of the resulting $\bar{x}$ chart should approximate μ (the process average), since the *expected value* of $\bar{x}$ is known to equal μ (see Section 5.4). In addition, when it is assumed that the statistic follows (at least approximately) a normal distribution, control limits are placed at a distance of 3 standard deviations on either side of the centerline. One must be careful here, since the standard deviation used is that of the statistic, *not that of the process.* In the $\bar{x}$ chart, for example, the standard deviation of the statistic $\bar{x}$ is equal to $\sigma/\sqrt{n}$, where σ denotes the process variation, so the control limits would *theoretically* be placed at $\mu \pm 3\sigma/\sqrt{n}$. The **upper and lower control limits** are usually abbreviated by UCL and LCL. Figure 6.2 illustrates the relationship between the control chart variation (i.e., $\mu \pm 3\sigma\sqrt{n}$) and the process variation ($\mu \pm 3\sigma$). As long as the sub-

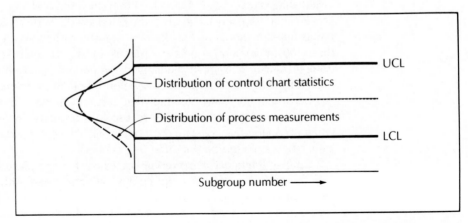

FIGURE 6.2
Control chart
variation and process
variation

group size, *n*, is greater than 1, the spread between the control limits will be smaller than the process spread.

In practice, one never knows the theoretical process mean (μ) and process variation (σ), so *estimates,* calculated from the subgroup data, are used instead. Since there are various ways to estimate a process characteristic (e.g, the standard deviation, see Section 3.3), one must sometimes choose between several estimates for the control limits. For example, in the $\bar{x}$ chart one might base the control limits on an estimate of σ derived from *another* control chart, such as the range chart (see Section 7.2) or the standard deviation chart (see Section 7.3). In either case, the resulting control limits are approximately the same but not necessarily identical.

STATISTICAL CONTROL

When all the points on a control chart are within the 3-sigma control limits *and* when there are no *other* anomalous patterns in the data (see Section 6.4), the process is said to be in a state of **statistical control** or, for short, "in control." Otherwise, the data indicate that the process is "out of control." As Pitt (1978) has pointed out, the terminology "out of control" should be used carefully and correctly when one is talking to those directly involved with the process, since this phrase can sometimes be *misinterpreted* to mean that production workers have lost all control over their processes. Needless to say, such a misunderstanding can be detrimental to the introduction of charting techniques. When control charts are first used on a process, for example, a lack of statistical control is often detected even though the majority of the items produced may be well within specification limits. Conversely, it is also possible to find processes that are in a state of statistical control but, nonetheless, are incapable of meeting specification limits.

The proper interpretation of control chart signals depends on understanding the relationships among the three types of limits associated with any process: specification limits, control limits, and process limits. Rarely do these three limits coincide, nor is it desirable that they should.

Specification limits describe what one *wants* the process to do, process limits indicate how the process is actually behaving, and control limits check for possible correctable problems within the process. **Process limits** indicate where the majority of the process measurements tend to be, so, if we assume the measurements approximately follow a normal distribution, these limits are defined to be 3 sigmas from the process average, μ (see Section 3.3). Thus, if the process variation is σ, then the process limits are $\mu \pm 3\sigma$, since the majority (99%) of the normal curve falls between such limits. Figure 6.3 illustrates some of the possibilities that can occur.

PROCESS CAPABILITY

Only when control charts indicate that a process is in statistical control does it make sense to speak of the process average and standard deviation. Without such control, both the average and standard deviation may change over time, and basing decisions on such volatile measures is risky. The control chart, however,

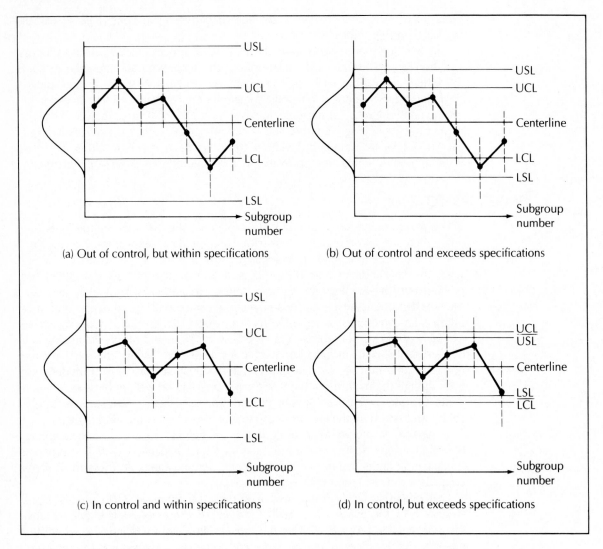

(a) Out of control, but within specifications

(b) Out of control and exceeds specifications

(c) In control and within specifications

(d) In control, but exceeds specifications

FIGURE 6.3
Possible relationships among process limits, specification limits, and control limits

provides a method for stabilizing a process so that its future behavior can be *predicted* with some degree of confidence. In particular, the estimated mean and standard deviation of a controlled process should be reliable measures.

Estimates of the process average, μ, and standard deviation, σ, are derived from the control chart data themselves. There are a number of ways to arrive at the estimates. For the moment, we denote these estimates by $\hat{\mu}$ and $\hat{\sigma}$, regardless of the estimation method used.[1] The estimated *process limits* are then $\hat{\mu} \pm 3\hat{\sigma}$.

---

[1] We follow the standard statistical practice of putting a hat ^ over a parameter to denote an estimate of that parameter.

The **capability** of a process can then be evaluated by comparing these process limits to the specification limits. If the process limits both fall within the specification limits (when the process is 'in control'), then the process is said to be **capable**; otherwise, it is not capable.

This definition of process capability is a rough one. For instance, it implies that a process would be considered capable if its process limits just coincided with the specification limits. From the normal tables, this would mean that about 0.27% (or 2,700 ppm) of the output would be outside of the specifications. Given the current emphasis on achieving nonconformance rates of only a few parts per million, a process whose nonconformance rate is 0.27% would be unacceptable for modern quality improvement programs.

One solution to this problem is to form **capability ratios** (or **capability indexes**) that compare the specification spread (USL − LSL) to the spread between the process limits (i.e., $6\hat{\sigma}$). This way the particular quality standard that a company requires can be stated in terms of acceptable lower limits on the capability ratios. Chapter 8 discusses the calculation and interpretation of capability indexes. Throughout the remainder of Chapter 6, we continue to use the rough definition of process capability mentioned above with the understanding that more precise definitions are available in Chapter 8.

6.2   TYPES OF CONTROL CHARTS

CONTROL CHART CLASSIFICATION

Control charts can be categorized in various ways, according to *type of data, type of control,* and *type of application.* Such classifications can be of help in selecting an appropriate chart to use.

The distinction between variables and attributes has been mentioned in regard to data (see Section 1.4) and random variables (Section 5.4). Similarly, since control charts statistics arise from either variables or attributes data, it follows that the charts themselves can be classified as **variables control charts** or **attributes control charts.** Table 6.1 lists the most commonly used control chart statistics according to whether they use attributes or variables data. Note that CUSUM (cumulative sum of deviations) charts can be applied to either type of data (see Section 7.7).

There are two types of control that the charts can exert: threshold and deviation control. Threshold control is used to detect large shifts in a process, whereas deviation control is used to detect small shifts from a target value. Deviation control is the more recent of the two concepts, originating with the CUSUM procedures introduced by E. S. Page in the 1950s. EWMA (exponentially weighted moving average) charts, also known as geometric moving average charts, are an even more recent addition to the collection of deviation control schemes. Since deviation control monitors very small process shifts, it is best used after threshold control charts have already brought some stability to a process by eliminating the

TABLE 6.1   Control Charts Classified by Type of Data

Variables Control Charts	Attributes Control Charts
$\bar{x}$, sample mean	p, percent nonconforming
$\tilde{x}$, sample median	np, number nonconforming
x, individual measurement	c, number of nonconformities
R, sample range	u, number of nonconformities per inspection unit
s, sample standard deviation	CUSUM, cumulative sum of deviations
MR, moving range	
Narrow-limit gage charts	
CUSUM, cumulative sum of deviations	
EWMA, exponentially weighted moving average	

larger process shifts. All of the charts listed in Table 6.1, with the exception of the CUSUM and EWMA procedures, are threshold charts.

Finally, when charting a process, one is usually interested in tracking a particular characteristic such as the process average, process variation, or some measure of conformance (or nonconformance). Nonconformance is measured either by the number of nonconforming items in a sample or by the number of nonconformities per unit of inspection. Table 6.2 classifies control charts according to these applications.

CONTROL CHART SELECTION

Many factors affect the decision of which control chart to use. Subgroup size, sampling frequency, type of control (threshold or deviation), type of data, the process characteristic, and limitations imposed by the process itself all affect the selection of an appropriate chart. Indeed, most of this chapter is devoted to sorting through these factors and their effect on the charting process.

Unfortunately, there is no standard flowchart for deciding which chart(s) to use, but some guidance is offered in Figure 6.4, which shows yet another classifi-

TABLE 6.2   Control Charts Classified by Type of Application

Characteristic to Control	Chart Type
Process average	$\bar{x}$, $\tilde{x}$, x, CUSUM, EWMA, narrow-limit gage
Process variation	R, s, MR, narrow-limit gage
Proportion nonconforming	p
Number nonconforming	np
Number of nonconformities	c, u, CUSUM

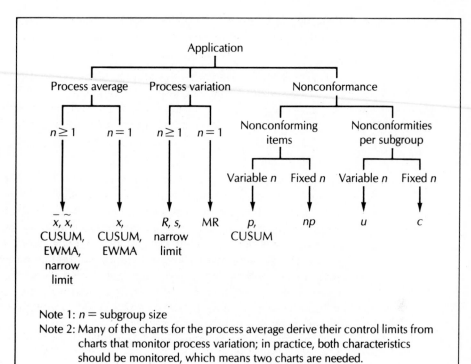

FIGURE 6.4
Control charts
classified by
application type and
subgroup size

cation of control charts, this time according to process characteristic and sample size. Depending on the particular process and the limitations it imposes on the subgroup sizes that are possible, Figure 6.4 may help narrow the list of choices.

6.3   SUBGROUPS AND SAMPLE SIZES

RATIONAL SUBGROUPS

Forming the subgroups of size n from which the control chart is to be formed is not simply a matter of choosing arbitrary samples of n items from the process's output. In fact, the *method* of forming subgroups directly affects the chart's ability or inability to detect process changes. Choosing the subgroups correctly is probably the single most important decision when setting up a control chart (Nelson 1988).

Shewhart solved the problem of forming subgroups by introducing the method of rational subgrouping. **Rational subgroups** are those that contain or enclose only the chance or common causes. Within any rational subgroup, the variation among the readings should be due to only the natural process variation, and there should be little or no opportunity for assignable causes to add to this variation. Forming subgroups that meet this requirement has two advantages:

1. The variation *within* the subgroups can be pooled to give a good estimate of the natural process variation.
2. The presence of special causes can easily be detected, since they are responsible for any large variations *between* the subgroups.

For example, in a labor-intensive process, it is often important to detect differences between different operators or workers. Suppose there are five operators who perform the same process steps. If one item is selected from each operator's work and the resulting five items are combined to form a subgroup of size 5, such subgroups will not serve the purpose of detecting operator-to-operator differences. The variation within such subgroups will be due partly to each operator's performance and partly to the differences between operators. Any differences between operators will effectively be masked, and a control chart based on such groups will rarely show out-of-control signals (except those due to assignable causes *other* than operator differences).

Instead, subgroups should be chosen so as to isolate each operator's performance. Various approaches can be used to achieve this. One method would be to run five separate control charts, one for each operator, so that subgroups represent only one operator's performance. Alternatively, one could use a single chart but rotate which operator's output is used in forming each subgroup. Five items from operator A could be used for the first group, five from operator B for the second, and so on. Cycling through the operators in this manner and, perhaps, adjusting the sampling frequency would give rise to subgroups that are capable of detecting operator differences.

One commonly used method for creating rational subgroups is to choose the sampled items over a fairly *short* time span. For detecting special causes such as differences in raw materials or vendors, this method can work well, since the subgroups have a high probability of representing parts built from only *one* raw material or from a *single* vendor's products. However, even with this approach, attention must still be paid to the types of special causes one wishes to detect. There is a natural tradeoff between forming subgroups over too short a period of time (so that everything except measurement error becomes a special cause) and over too great a span of time (in which case special causes are likely to be present within every subgroup).

In general, subgroups that are selected for control charts do not necessarily come from random samples. Instead, it is more important that samples be chosen to form rational subgroups, so that the resulting control chart is able to pinpoint process problems effectively. Nonetheless, control limits are based on the *assumption* that only a system of 'chance causes' is operating and, in that case, any sample (including that from a rational subgroup) can be considered to be a random sample. In other words, if a process is "in control," then successive items should vary only according to a system of random causes, and one is then permitted to use formulas such as $\sigma/\sqrt{n}$ (which is derived from the assumption of *independent* samples) to construct control limits.

SAMPLE SIZES

What sample size to use is probably the most frequently asked question about any statistical method. Without begging the question, the answer is that the required sample size depends on the size of the underlying process variability, σ, and on the accuracy one wishes to derive from the sample(s). Consider, for example, a hypothetical process whose variability is so small that each item produced is virtually identical to the next one. How large a sample is needed from such a process to accurately predict the next result? Clearly, the answer is $n = 1$. On the other end of the spectrum, imagine sampling from a process that exhibits a large amount of variability from item to item. Basing decisions on samples of size $n = 1$ from this process would not be wise.

In one sense, it should not much matter what sample sizes are used for control charts, since every chart uses 3-sigma control limits, which means the sensitivity of one chart should be about the same as that of another (assuming the normal distribution applies). On the other hand, if the process output does not closely follow a normal distribution, then the normality of control chart statistics must depend to some extent on the central limit theorem (see Section 5.4), which requires *larger* sample sizes in order to assure normality.

In practice, for variables charts, samples of 1 to 25 are used. Usually, n is chosen between 1 and 6. Prior to the computer age, subgroups of size $n = 5$ were a particularly common choice because this facilitated the rapid calculation of subgroup means.[2] Of course, subgroups of size $n = 10$ would make calculations even easier, but at the price of collecting twice as many readings. In general, for variables data, a balance must be struck between the assurance of normality provided by larger sample sizes and the increasing costs associated with larger samples.

Quite often, a particular process imposes its own constraints on the possible subgroup sizes. For example, with processes that take a long time to produce one item, large subgroup sizes require too long a time span, making it too easy for special causes to occur *within* subgroups and thereby diminishing the usefulness of the control chart. As another example, consider continuously changing characteristics such as chemical and environmental measurements that, in manufacturing settings, one tries to control to acceptable ranges or target values. Since the most recent chemical or environmental measurement is most indicative of the true value, using threshold control and subgroup sizes greater than 1 may actually *impair* our ability to detect changes. For such processes, some form of deviation control based on a subgroup size of $n = 1$ may be preferable.

[2]To find the average of any set of five numbers, one needed only to double their sum and move the decimal place one place to the left.

With attributes data, the question is often not one of trying to obtain larger sample sizes, but exactly the opposite, trying to handle extremely large amounts of data. The reason for this is that attributes data are relatively easy to obtain and usually exist in much greater abundance than variables data. For example, at inspection points in a process (e.g., at final inspection), records are kept on much, if not all, of the product passing through the inspection station. Measuring whether items pass or fail tests, counting nonconforming items or nonconformities, or using go/no-go gages generates large amounts of attributes data. In these cases, it becomes important to have charts that can handle the daily *changes* in sample sizes in addition to the magnitudes of the samples. The problem of varying subgroup sizes also affects variables charts. Methods for handling this problem are discussed in the next section.

VARIABLE SUBGROUP SIZES

There comes a time in almost every charting application when the amount of data available for a subgroup does not match that required by the chart. Suppose, for example, that an $\bar{x}$ chart based on subgroups of $n = 5$ has been established for a certain process. What does one do when, during the specified sampling period, fewer than five items are produced, so a subgroup of size five cannot be formed? If this happened only rarely, the subgroup could just be eliminated from the chart, although one then forfeits any information contained in that particular subgroup. Taking a more extreme case, what happens if production output is quite variable, so that subgroups of size less than five occur frequently?

For attribute charts, the problem of variable subgroup sizes is usually the *rule*, not the exception. Most attribute charts are based on production records that consist of audits of production, not on random samples. Formal sampling plans that select a fixed number of items for inspection during each sampling period are used less frequently when attributes data are involved.

There are three solutions to the problem of varying sample sizes:

<table>
<tr><td>METHODS FOR
VARIABLE
SUBGROUP SIZES</td><td>1. Use the **average sample size** to calculate the control limits.
2. Use **variable control limits.**
3. Use **standardized control limits.**</td></tr>
</table>

The **average sample size** approach is best used when sample sizes do not vary *too* greatly. This makes sense because the resulting control limits must apply to all subgroups. As a recommended rule of thumb, for *large* samples, the subgroup sizes should not vary by more than about 25% from the average sample size (*Continuing Process Control* 1989). The average sample size approach works best when applied to attributes charts where samples tend to be large (as long as there are no sample size swings of 25% or more). For variables charts, based on much smaller subgroup sizes, changes of 25% in the subgroup size are not at all unlikely (e.g., a subgroup size of $n = 3$ represents a 25% shift in sample size for an average sample size of $n = 4$).

The average sample size approach makes calculations easier, but with the availability of modern software, this consideration becomes less important and **variable control limits** may be used instead. In this approach, control limits are calculated separately for *every* subgroup. For example, suppose that a certain process has an approximate average of $\hat{\mu} = 10$ and a process standard deviation of $\hat{\sigma} = 2$. For any subgroup of size n, the control limits for an $\bar{x}$ chart are based on $UCL = \hat{\mu} + 3\hat{\sigma}/\sqrt{n}$ and $LCL = \hat{\mu} - 3\hat{\sigma}/\sqrt{n}$. Thus, if the first five subgroup sizes are $n_1 = 2$, $n_2 = 5$, $n_3 = 5$, $n_4 = 4$, and $n_5 = 10$, then the control (variable) limits are

Subgroup number	Subgroup size	Control limits	
i	n_i	UCL	LCL
1	2	5.76	14.24
2	5	7.32	12.68
3	5	7.32	12.68
4	4	7.00	13.00
5	10	8.10	11.90

Notice that the spread between the control limits widens or narrows according to the particular subgroup size. In practice, the estimates $\hat{\mu}$ and $\hat{\sigma}$ are derived from the subgroup data themselves. For an $\bar{x}$ chart, $\hat{\mu}$ is usually taken to be the average of all the subgroup averages, while $\hat{\sigma}$ is approximated from equation (3.9) or (3.11) in Chapter 3. Figure 6.5 shows how variable control limits are graphed on

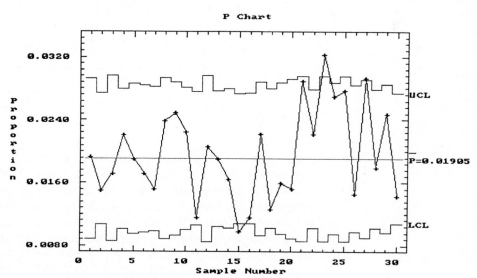

FIGURE 6.5
Example of variable
control limits

a control chart. Of course, the *interpretation* of these limits does not change; the changing widths just take into account the changing amounts of information in each subgroup.

The third approach to establishing control limits is to use **standardized limits.** Standardization refers to the conversion of control chart statistics (and control limits) into 'standardized' form, as described in Section 5.6. Generally, if we let θ denote the quality characteristic of interest, with θ_i denoting the control chart statistic associated with subgroup i, the standardized value of θ_i is given by

$$\text{standardized value} = \frac{\theta_i - \bar{\theta}}{\hat{\sigma}_\theta} \qquad (6.1)$$

where $\bar{\theta}$ is the centerline of the particular control chart and $\hat{\sigma}_\theta$ is an estimate of the standard deviation of the statistic θ_i. The standardized values are assumed to follow (at least approximately) a standard normal distribution, so the control limits are simply $z = +3$(UCL) and $z = -3$(LCL). Thus, the added calculation involved in standardizing the θ_i's is offset by the fact that the control limits are *constant,* regardless of the sample size.

As an example, consider the $\bar{x}$ example given earlier in this section. Here, the control chart statistic is $\bar{x}_i$, the average of the readings in subgroup i, and the estimated process average and standard deviation are $\hat{\mu} = 10$ and $\hat{\sigma} = 2$, respectively. From equation (5.1), the mean and standard deviation of the statistic $\bar{x}_i$ should then be 10 and $2/\sqrt{n_i}$, respectively. Suppose that the subgroup averages corresponding to the samples of size $n_1 = 2$, $n_2 = 5$, $n_3 = 5$, $n_4 = 4$, and $n_5 = 10$ are $\bar{x}_1 = 8.12$, $\bar{x}_2 = 6.75$, $\bar{x}_3 = 10.33$, $\bar{x}_4 = 11.40$, and $\bar{x}_5 = 8.93$. Then the standardized values are as follows:

Subgroup number	Subgroup size	Subgroup mean	Standard value	Control limits	
i	n_i	$\bar{x}_i$	z_i	UCL	LCL
1	2	8.12	-1.33	3	-3
2	5	6.75	-3.63	3	-3
3	5	10.33	0.37	3	-3
4	4	11.40	1.40	3	-3
5	10	8.93	-1.69	3	-3

The standardized control limits are *always* +3 and -3. In essence, standardization shifts the burden of calculation from the control limits to the subgroups and, at the same time, accommodates any variation in sample size from subgroup to subgroup. The only difficulties that arise center around how to interpret the standardized values because they are less intuitively understood than the simple control chart statistics based on means, percentages, and counts. For further discussion of the practice of using standardized control charts, see Nelson (1989).

6.4   INTERPRETING CONTROL CHART PATTERNS

RULES FOR DETECTING SPECIAL CAUSES

Reacting to the information in control charts requires a set of rules for signaling the *probable* existence of special causes. However, irrespective of the rule used, there is always the possibility that false alarms may occur when no special cause is present or, alternatively, that legitimate special causes may go undetected by the chart. These risks are inherent in any statistical inference. Balancing the risk of false alarms against the risk of not detecting problems is handled by deciding on what level of risk (i.e., probability) one will tolerate in a decision procedure.

For control charts, risk levels are determined by the placement of the control limits. In the United States it has been traditional to establish control limits based on 3-sigma limits for the statistic being charted. By assuming that the statistic approximately follows a normal distribution, one then estimates that about 0.0027 (or 0.27%) of the time a point may fall outside the control limits, even when no special cause is present; that is, control charts standardly allow a 0.27% chance of false alarms. In general, the wider the control limits, the less chance there is of detecting special causes, and the sensitivity of the chart may suffer. To increase the sensitivity of a chart, an expanded list of "out of control" rules is used.

Points that fall outside the 3-sigma control limits are one signal of possible special causes. Other rules are based on examining the patterns of points that occur *within* the control limits. The essential idea behind rules for detecting special causes is that a stable process should give rise to only a *random* scatter of points on the control chart.[3] Nonrandom patterns of any kind signal the possible presence of special causes. The following list of rules is commonly used to detect the existence of special causes in control charts. (All rules assume that process data are randomly scattered around the mean; some rules require the additional assumption that the process data follow a normal distribution.)

RULES FOR DETECTING SPECIAL CAUSES	1. One point outside the (3-sigma) control limits. 2. Eight successive points on the same side of the centerline. 3. Six successive points that increase (or decrease). 4. Two out of three points that are on the same side of the centerline, both at a distance exceeding 2 sigmas from the centerline. 5. Four out of five points that are on the same side of the centerline, all four at a distance exceeding 1 sigma from the centerline.

---

[3]In Chapter 12, the definition of 'stable' is expanded to include some processes that are not necessarily random. The rules in this section do not apply to those processes.

There are additional rules (e.g., some practitioners replace 'eight' by 'seven' in rule 2 above), but these five are probably the most frequently used. Nelson (1984) gives a list of these and other rules. Figure 6.6 illustrates some of these rules.

The rules in the list are designed to be comparable in the sense that each has approximately the *same* probability of giving a false signal. In each case, the target value for the false alarm probability is taken to be 0.0027, the same as that imposed by rule 1. Calculating the *exact* probabilities of false signals for each of the rules involves the probability tools of Chapter 5. For example, in rule 2, if we make the assumption that the centerline of a chart is close to the *median* of the data, then there should be about a 50–50 chance that any point falls on one side of the centerline. The probability of eight consecutive points falling on the same side of the centerline—say, above—is then $P(8 \text{ points above}) = P(\text{1st above}) \cdot P(\text{2nd above}) \cdots P(\text{8th above}) = (0.5)^8 = 0.0039$, which is reasonably close to the desired value of 0.0027.

It should also be noted that, when a group of rules are used together, the sensitivity of detection must be greater than that offered by any *single* rule. For example, if two different rules have false alarm rates of α_1 and α_2, respectively, then the false alarm rate associated with using both rules simultaneously is approximately $1 - (1 - \alpha_1)(1 - \alpha_2)$, which, it can be shown, exceeds both α_1 and α_2. Thus, although using more than one rule can enhance the probability of detecting special causes, it also has the effect of slightly increasing the false alarm rate (see also Walker, Philpot, and Clement 1991).

AVERAGE RUN LENGTH

Apart from considering the probability of false signals, there is another method for evaluating a control chart's sensitivity to special causes. This method is based on the **average run length** (ARL), which is the average (or expected) number of subgroups one must examine until the first 'out of control' signal is given. ARLs are calculated by first supposing that the process has shifted off center by a certain number of sigmas, $k\sigma$, and then finding the ARL required to detect such a shift. By graphing the ARL versus k, one can compare the sensitivity of different control chart schemes for detecting shifts in the process average. This is especially useful when comparing different control charts (e.g., $\bar{x}$ charts and CUSUM charts) that could be used on the same data.

The formula for calculating the ARL is given by

$$\text{ARL} = \text{average run length} = \frac{1}{p_d} \tag{6.2}$$

where p_d is the probability that an 'out of control' condition is detected based on examining a *single* subgroup. Notice that the size of p_d depends on the size of the shift that the process has experienced. For example, if *no* shift has occurred in the process, then rule 1 gives a probability of 0.0027 of detecting a shift (a false signal in this case), so the ARL is $1/0.0027 = 370.4$. Thus, for a stable centered process, it would take, on average, about 370 or 371 subgroups for a (false) 'out of control' signal to occur.

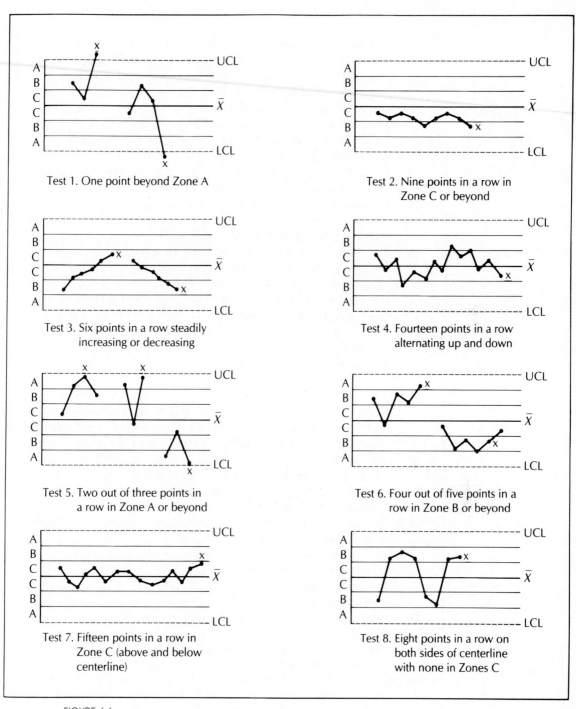

Test 1. One point beyond Zone A

Test 2. Nine points in a row in Zone C or beyond

Test 3. Six points in a row steadily increasing or decreasing

Test 4. Fourteen points in a row alternating up and down

Test 5. Two out of three points in a row in Zone A or beyond

Test 6. Four out of five points in a row in Zone B or beyond

Test 7. Fifteen points in a row in Zone C (above and below centerline)

Test 8. Eight points in a row on both sides of centerline with none in Zones C

FIGURE 6.6

Illustrations of tests for special causes applied to Shewhart control charts (from Nelson, 1984)

(*Continued*)

1. These tests are applicable to $\overline{X}$ charts and to individuals (X) charts. A normal distribution is assumed. Tests 1, 2, 5, and 6 are to be applied to the upper and lower halves of the chart separately. Tests 3, 4, 7, and 8 are to be applied to the whole chart.

2. The upper control limit and the lower control limit are set at 3 sigmas above the centerline and 3 sigmas below the centerline. For the purpose of applying the tests, the control chart is equally divided into six zones, each zone being 1 sigma wide. The upper half of the chart is referred to as A (outer third), B (middle third), and C (inner third). The lower half is taken as the mirror image.

3. When a process is in a state of statistical control, the chance of (incorrectly) getting a signal for the presence of a special cause is less than five in a thousand for each of these tests.

4. It is suggested that Tests 1, 2, 3, and 4 be applied routinely by the person plotting the chart. The overall probability of getting a false signal from one or more of these is about one in a hundred.

5. It is suggested that the first four tests be augmented by Tests 5 and 6 when it becomes economically desirable to have earlier warning. This will raise the probability of a false signal to about two in a hundred.

6. Tests 7 and 8 are diagnostic tests for stratification. They are very useful in setting up a control chart. These tests show when the observations in a subgroup have been taken from two (or more) sources with different means. Test 7 reacts when the observations in the subgroup always come from both sources. Test 8 reacts when the subgroups are taken from one source at a time.

7. Whenever the existence of a special cause is signaled by a test, this should be indicated by placing a cross just above the last point if that point lies above the centerline, or just below it if it lies below the centerline.

8. Points can contribute to more than one test. However, no point is ever marked with more than one cross.

9. The presence of a cross indicates that the process is not in statistical control. It means that the point is the last one of a sequence of points (a single point in Test 1) that is very unlikely to occur if the process is in statistical control.

10. Although this can be taken as a basic set of tests, analysts should be alert to any patterns of points that might indicate the influences of special causes in their process.

FIGURE 6.6
(*Continued*)
Illustrations of tests for special causes applied to Shewhart control charts (from Nelson, 1984)

If, on the other hand, the process has shifted off target, then the probability of detection increases and the ARL decreases accordingly. Suppose, for example, that the process mean has shifted $k = 2$ sigmas off center. Then, for any single subgroup, the probability of finding an 'out of control' signal (using rule 1) in an $\bar{x}$ chart is

$$p_d = P(\text{point above UCL}) = P(\bar{x} > \text{UCL})$$

$$= P\left[z > \frac{\mu + 3\sigma/\sqrt{n} - (\mu + 2\sigma)}{\sigma/\sqrt{n}}\right] = P(z > 3 - 2\sqrt{n})$$

Thus, if subgroup sizes of $n = 5$ are used, $p_d = P(z > -1.47) = 0.9292$, and the corresponding ARL is $1/0.9292 = 1.08$. Thus, a process shift of as much as 2 sigmas would be detected, on average, within about one subgroup after the shift had occurred.

THE RESAMPLING FALLACY

When control charts are used, there is a tendency for some to mistrust the 'out of control' signals and wait for another point or two to go out of control before taking any action on the process. Even though this may seem to be a safe enough procedure, it can significantly delay needed process adjustments.

In fact, it can be shown that if a process has slipped slightly off center and a point has gone beyond one of the control limits, then there is a reasonably high probability that the *next* point will be *within* the control limits, thereby canceling the warning given by the first point. This can *mistakenly* be interpreted to mean that it is correct to ignore the first signal, since the second point is 'in control.'

Disregarding control chart signals in this manner is sometimes called the **'resampling fallacy.'** A process shift will eventually be detected, but unless the shift is fairly large, the practice of resampling may cause it to go unnoticed for many subgroups. The resampling syndrome occurs in many applications. Other examples are found in Pitt (1978). The practice of resampling is sometimes caused by a lack of confidence in the measurement system that generates the data.

PROBABILITY LIMITS

Some countries, notably Great Britain, use another method for establishing control limits, one based directly on the probabilities of false alarms rather than on 3-sigma limits. Both methods have their advantages and disadvantages.

In the **probability limit** approach, the probability of a false alarm is fixed at a particular level, usually 0.002, and control limits are calculated that give rise to this probability. Thus, with a probability of 0.002, control limits are set at k sigmas from the mean, where $P(z > k) = 0.001$. Using the normal tables and solving for k, we place the control limits $k = 3.08$ sigmas on either side of the centerline. This probability is not substantially different from that given by the 3-sigma limits, but one should at least be aware that there is more than one method in use for setting control limits. The impact of the 3-sigma convention may be greater for some charts than for others. In particular, the reader is referred to the related discussion in Section 7.3 concerning control limits for R and s charts.

| 6.5 | MANAGING THE CHARTING PROCESS |

CHOOSING PROCESS CHARACTERISTICS TO MONITOR

For some processes, the decision of which quality characteristics to monitor requires little effort. Such is the case, for example, in the manufacture of fairly simple products such as metal fasteners, where the customer's primary concern is that the fastener exceed certain strength requirements. A variables control chart for the characteristic 'breaking strength' would meet this producer's needs.

On the other hand, a complex product (e.g., a car) offers many opportunities for using control charts, and the decision of exactly which characteristics to chart becomes nontrivial. The list given here offers suggestions for how best to choose the characteristics to monitor. One guiding principle is to *locate the charting activities early* (i.e., upstream) *in the process* (see Section 2.5). Another is to periodically reevaluate the characteristics chosen, since the familiarity with a process gained from charting can itself suggest new characteristics to monitor.

WHERE TO APPLY
CONTROL CHARTS

1. Analyze customer requirements and translate these into product requirements (see Section 2.5) and then into a list of critical quality characteristics.
2. Use Pareto charts and cause-and-effect diagrams to discover the causes of production problems and, therefore, identify important process and/or product characteristics.
3. Use process flow diagrams (see Section 2.4) to identify key upstream inputs and downstream outputs.
4. Work with suppliers to implement on-line charting at the supplier's facility.

COMPUTERIZING THE CHARTING PROCESS

Control charts are constructed and operated by those directly involved in production. Although only a few simple arithmetic calculations are required, along with periodic plotting of points on the charts, the time required for this activity can still be significant. Fortunately, computers now make chart construction painless and instantaneous. The reader is referred to Chapter 14 for a detailed description of what SPC software should do and how to organize a quality data system that makes good use of such software.

For a computerized charting activity, it should be kept in mind what computerization can and cannot do. Although computers certainly eliminate most of the arithmetical and graphical drudgery of chart construction, they do not yet remove the need to understand how charts are to be interpreted.[4] Employees still need training in how to properly react to the signals provided by the charts, and they

---

[4]However, many SPC software packages do check for violations of some of the rules mentioned in Section 6.4.

are still the ones who decide which characteristics to monitor, which charts to use, what sample sizes to use, and so on. Furthermore, decisions must be made on when to revise the control limits and when to discontinue a particular chart. All of these decisions are in the realm of what **expert systems** software may someday be able to do, but until that day it is imperative that production personnel themselves make these important decisions.

REVISING CONTROL LIMITS

By their very nature, control charts are time-dependent. A control chart used on a certain process two months ago may not be applicable today. In practice, two things happen when a process is monitored with control charts. First, as special causes are detected and eliminated, the process variation begins to decrease. Second, the elimination of some special causes may result in changing the process average. In either case, the current centerline and/or control limits will no longer validly describe the process and must be corrected.

For these reasons, every application of control charts should include a plan for revising the charts. If a pronounced process shift has resulted from either process changes or the elimination of some special cause, then data from the *new* process will have to be accumulated and a new centerline and control limits must be calculated. In order to have a running history of these process changes, the new control limits and centerline are often plotted on the same graph as the old chart. Operationally, however, only the newly revised limits are used to signal special causes.

Sometimes there may be no known special cause or process change that can be pinpointed, yet after enough time, the process variation and/or average may nevertheless change. For this reason, it is wise to establish a **periodic revision** cycle; that is, a particular time period (e.g., every week, every month) should be chosen and at the end of each period the most recent process data should be used to update the centerline and control limits. The length of the revision cycle should be chosen with the sole purpose of maintaining the validity of the chart limits. Thus, a very stable process requires a longer revision cycle than a new or otherwise more volatile process.

DISCONTINUING A CONTROL CHART

Control charts are initially used because an important process or product characteristic is experiencing unacceptably high variation. After the chart is in operation for a period of time, two things can happen. In some cases, the presence of special causes may remain a problem and the chart must continue to be operated. In other cases, the identification of a special cause may lead to a process change that reduces the concern with the characteristic being charted. For example, to paraphrase Juran and Gryna (1980), a method may be discovered for *'foolproofing'* the process so that its output becomes precisely predictable and stable and no longer a source of concern to downstream processes. When this happens, there is little or no benefit to be gained from continuing to operate this particular control chart, and it should be discontinued. Resources devoted to this chart can then be more profitably applied to other process variables.

EXERCISES FOR CHAPTER 6

6.1   Are rational subgroups also random samples? Explain why or why not.

6.2   For a stable process, one wants all the points in a control chart to fall within the control limits. However, one does *not* want all the points on the chart to fall close to, or exactly on, the centerline. Why?

6.3   Suppose that for subgroups of size 50 the proportion nonconforming $\hat{p}$ in each subgroup is charted. Furthermore, suppose the process is in control with an overall average nonconformance rate of $\bar{p} = 0.05$.
 (a) If $\bar{p}$ shifts to 0.10, what is the probability that this shift is detected in the next subgroup after the shift occurs? That is, what is the probability that the next $\hat{p}$ falls outside the control limits?
 (b) Generalize the result in part (a) to a shift in $\bar{p}$ of any size, δ.

6.4   In Exercise 6.3, what is the average number of subgroups inspected before an 'out of control' signal is given?

6.5   From an $\bar{x}$ chart, when a process shifts off target by 2 sigmas, the probability of detecting this shift at the *next* subgroup was shown in Section 6.4 to be $p_d = P(z > 3 - 2\sqrt{n})$.
 (a) Generalize this result to a shift of k sigmas.
 (b) Using the result in part (a), plot the average run length (ARL) versus k for samples of size $n = 1, 5, 10,$ and 20.

6.6   How frequently should subgroup data be collected for a control chart? List several factors involved in this decision and give supporting examples.

6.7   The theoretical control limits for an $\bar{x}$ chart are $\mu \pm 3\sigma/\sqrt{n}$ (see Section 6.1). For a given process (i.e., for a fixed μ and σ), these control limits become narrower as the subgroup size increases. Does this mean that the 'out of control' signals will become more likely for control charts based on larger subgroup sizes?

6.8   Three-sigma limits are used with most control charts. In terms of the probability of detecting special causes,
 (a) describe the effect of widening the limits, say, to 3.1-sigma limits;
 (b) describe the effect of narrowing the control limits using, say, 2-sigma limits.

6.9   Show that ARL $= 1/p_d$ (see Section 6.4). (*Hint:* Let $Y =$ number of subgroups until the first 'out of control' signal is given; find the distribution of Y; then find the expected value of Y.)

6.10   Suppose that a measuring instrument used to obtain process data for an $\bar{x}$ and R chart is out of calibration, so that each of its reported measurements is off by $+\delta$ units from the true value. What effect will this have on the signals given by the $\bar{x}$ and R charts?

6.11   A canning machine has 15 different heads, each of which affixes lids to cans passing by on a conveyor system. Two methods have been proposed for collecting subgroup data from this process: (1) taking periodic samples of five cans from the finished cans exiting the machine, and (2) taking periodic samples of 15 cans exiting the machine. Compare these two methods. If one or more of the heads is out of adjustment and causing inadequate sealing, which of the two methods is more likely to detect such a problem?

6.12   When two 'out of control' rules (with false alarm rates α_1 and α_2) are applied simultaneously, the overall false alarm rate is given by $1 - (1 - \alpha_1)(1 - \alpha_2)$. Generalize this statement to the case where any number of rules, k, are used together (assume independence).

6.13   An $\bar{x}$ chart appears as follows:

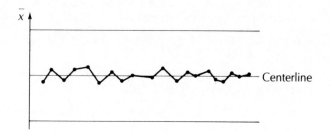

(a) Explain what is wrong with such a pattern.
(b) Give at least two reasons or scenarios for how such a pattern might arise in practice.

6.14   Solder joints that connect electronic components to printed circuit boards are examined and classified as either acceptable or not acceptable by an inspector. The inspector then records the number of unacceptable joints per board.
(a) If all the circuit boards are of the same type, what kind of control chart do you recommend for this process?
(b) When different types of circuit boards are inspected, both the number of unacceptable joints and the total number of joints are recorded for every circuit board. For this type of data, what kind of control chart would you recommend?

6.15   Insurance companies publish standard tables of average heights and weights of Americans. Weight ranges show weights without clothes for people aged 25–29, while heights are measured without shoes. For example, one such table suggests that healthy men who are 5 ft, 11 in. tall with a medium frame should weigh between 152 and 165 lb. For a man 5 ft, 11 in. tall, are these limits best thought of as specification limits, process limits, or control limits?

6.16   Refer to the list of rules for special causes in Section 6.4.
(a) For independent measurements from a normal distribution, calculate the probability of the event in rule 2 (eight successive points on the same side of the centerline).

(b) Under the same assumptions as in part (a), find the probability associated with rule 5 (four out of five points on the same side of the centerline, all four at a distance exceeding 1 sigma from the centerline).

6.17 The list of 'out of control' rules in Section 6.4 cannot be used on standardized control charts.[5] To show this, develop a list of seven (or more) subgroup means and their associated subgroup sizes that has the following property: the unstandardized means show no particular pattern, but the standardized means form an increasing sequence.

REFERENCES FOR CHAPTER 6

Continuing Process Control and Process Capability Improvement. 1989. Statistical Methods Office, Ford Motor Company.

Juran, J. M., and F. M. Gryna. 1980. *Quality Planning and Analysis,* pp. 268–269. New York: McGraw-Hill.

Nelson, L. S. 1984. "The Shewhart Control Chart—Tests for Special Causes." *Journal of Quality Technology* 16 (no. 4):237–239.

Nelson, L. S. 1988. "Control Charts: Rational Subgroups and Effective Applications." *Journal of Quality Technology* 20 (no. 1):73–75.

Nelson, L. S. 1989. "Standardization of Shewhart Control Charts." *Journal of Quality Technology* 21 (no. 4):287–289.

Pitt, H. 1978. "The Resampling Syndrome." *Quality Progress,* April, pp. 27–29.

Walker, E., J. W. Philpot, and J. Clement. 1991. "False Signal Rates for the Shewhart Control Chart with Supplementary Runs Tests." *Journal of Quality Technology* 23 (no. 3):247–252.

[5]See Nelson (1989).

7

Variables Control Charts

The most frequently used control charts for variables data are described in this chapter. Emphasizing the details of their computation, the sections illustrate traditional Shewhart charts as well as newer methods such as CUSUM and EWMA charts.

CHAPTER OUTLINE

7.1   INTRODUCTION

The concepts and philosophy of control charts were introduced in Chapter 6. The next few chapters illustrate the construction and applications of commonly used control charts. Charts for variables data are covered in Chapter 7, followed by capability indexes for variables data in Chapter 8 and charts for attributes data in Chapter 9.

As mentioned in Section 6.1, Chapters 6 and 7 should ideally be covered simultaneously. In order to separate out some of the philosophical issues from the computational ones, we continue to recommend reading first Chapter 6 for an overview of charting and then Chapter 7 for the detailed calculations.

7.2 $\bar{x}$ AND R CHARTS

BACKGROUND

Historically, $\bar{x}$ and R charts (read 'x bar and R charts') were the first variables charts introduced by Shewhart. For this reason, and because $\bar{x}$ and R charts are still among the most frequently used charts, our discussion begins here. A good understanding of $\bar{x}$ and R charts facilitates the study of any chart based on threshold control (see Section 6.2), since other threshold charts use the $\bar{x}$ and R model as the prototype.

With variables data, the two most important process characteristics are the process average and the process variation. Instability in either of these measures is an indication of process problems. Of the many statistics that could be used to estimate the process average, Shewhart initially chose the subgroup mean, $\bar{x}$. To monitor process variation, the sample range, R, was used, since its calculation was so much easier than that of the sample standard deviation. In addition, for the small subgroup sizes normally used in control charts, there is very little statistical information lost by using the range in lieu of the standard deviation (see Section 3.3).

Theoretically, control limits for the $\bar{x}$ charts are given by

$$\text{UCL} = \mu + 3\frac{\sigma}{\sqrt{n}} \quad \text{and} \quad \text{LCL} = \mu - 3\frac{\sigma}{\sqrt{n}} \tag{7.1}$$

where μ and σ denote the long-run process mean and standard deviation, respectively (see Section 6.1). Unfortunately, equation (7.1) cannot be used directly, since both μ and σ must first be estimated from the available process data. Furthermore, the accuracy of the estimates of μ and σ directly affects the accuracy of the control limits in equation (7.1).

To obtain reasonable estimates of μ and σ, the following two-stage procedure is used. *First,* the chart for process variation (the R chart in the present discussion) is brought into statistical control. This assures that the process variation is under control and, consequently, that the centerline is a fairly reliable estimate of the general subgroup range, R. Second, this estimate of R is converted into an estimate of σ—for example, by using equation (3.11), which is then put into equation (7.1) to obtain the *control limits* for the process average (here, the control limits for the $\bar{x}$ chart). The important point to remember is that the *centerline of the R chart governs the control limits of the $\bar{x}$ chart.*

Fortunately, control limits of variation charts (R or s) always turn out to be fairly simple functions of the variation chart's centerline ($\bar{R}$ or $\bar{s}$). For example, when the R chart is used to monitor process variation, its *own* control limits can be shown to be simple multiples of the centerline, $\bar{R}$. This is demonstrated in the discussion to follow.

THE R CHART

To begin the construction of the R chart, some number, k, of successive subgroups of process measurements are used. As suggested in Chapter 6, at least 20 to 25 subgroups should be used. Normally, the same sample size, n, is used in forming each subgroup, although variable sample sizes are also allowed (see Section 6.3). The centerline of the R chart is denoted by $\bar{R}$ and is calculated by averaging the sample ranges $R_1, R_2, R_3, \ldots, R_k$ of these subgroups:

$$\bar{R} = \frac{1}{k} \sum_{i=1}^{k} R_i \tag{7.2}$$

$\bar{R}$ serves as an estimate of μ_R, the mean of the sampling distribution of the ranges (for samples of size n). If the R chart is in control, the sample ranges should vary about their expected value, μ_R, and most of the R_i's should fall within 3 standard deviations (i.e., $3\sigma_R$) of μ_R. Ideally, then, the control limits would be placed at

$$\text{UCL} = \mu_R + 3\sigma_R \quad \text{and} \quad \text{LCL} = \mu_R - 3\sigma_R \tag{7.3}$$

Assuming that the process measurements follow (approximately) a normal distribution, it can be shown that σ_R is estimated by

$$\hat{\sigma}_R = \frac{d_3}{d_2} \bar{R} \tag{7.4}$$

where d_3 and d_2 are constants that depend on the subgroup size, n. Values of d_3 and d_2 can be found in Appendix 1, but, as the ensuing discussion shows, other constants are subsequently used to calculate the control limits for the R chart. Indeed, by substituting $\bar{R}$ for μ_R, and $\hat{\sigma}_R$ for σ_R, equation (7.3) gives approximate control limits of

$$\text{UCL} = \bar{R} + 3\hat{\sigma}_R = \bar{R} + 3\frac{d_3}{d_2}\bar{R} = \left(1 + 3\frac{d_3}{d_2}\right) \cdot \bar{R} = D_4 \bar{R}$$

and

$$\text{LCL} = \bar{R} - 3\hat{\sigma}_R = \bar{R} - 3\frac{d_3}{d_2}\bar{R} = \left(1 - 3\frac{d_3}{d_2}\right) \cdot \bar{R} = D_3 \bar{R}$$

or simply

$$\text{UCL} = D_4 \bar{R} \quad \text{and} \quad \text{LCL} = D_3 \bar{R} \tag{7.5}$$

where $D_4 = 1 + 3(d_3/d_2)$ and $D_3 = 1 - 3(d_3/d_2)$ are conveniently found from Appendix 1.

After these calculations are performed, the R chart is constructed by plotting the k subgroup ranges R_i $(i = 1, 2, \ldots, k)$ versus the subgroup number, i, and drawing in the centerline, $\overline{R}$. From the rules outlined in Chapter 6 for detecting 'out of control' conditions, the chart is examined to see whether or not the ranges are in a state of statistical control. If any 'out of control' conditions are found, and *if assignable causes can be found* to explain these problems, then the subgroups associated with these problems should be eliminated and the calculations in equations (7.2) and (7.5) should be revised accordingly. 'Out of control' subgroups for which *no* assignable causes can be found should not be eliminated.

When the R chart is deemed to be in a state of statistical control, so that the centerline $\overline{R}$ can be considered a reliable estimate of the range (of samples of size n) from a normal population, equation (3.11) can be used to estimate the process standard deviation σ:

$$\hat{\sigma} = \frac{\overline{R}}{d_2} \tag{7.6}$$

where, as before, d_2 is found from Appendix 1. This estimate is then used to calculate the control limits of the $\overline{x}$ chart and to calculate process capability estimates (see Chapter 8).

THE $\overline{x}$ CHART

Turning to the $\overline{x}$ chart, we need estimates of both μ and σ in equation (7.1). We assume that there are k valid subgroups of data,[1] whose subgroup means are denoted by $\overline{x}_1, \overline{x}_2, \overline{x}_3, \ldots, \overline{x}_k$, so the centerline of the $\overline{x}$ chart is calculated by averaging all k subgroup averages:

$$\overline{\overline{x}} = \frac{1}{k} \sum_{i=1}^{k} \overline{x}_i \tag{7.7}$$

The notation $\overline{\overline{x}}$ is used to indicate that the centerline results from *two* averaging operations, one to find the individual subgroup means and another to find their average.

With equation (7.7) used to estimate μ and equation (7.6) to estimate σ, the control limits for the $\overline{x}$ chart, equation (7.1), become

$$\text{UCL} = \mu + 3\frac{\sigma}{\sqrt{n}} \approx \overline{\overline{x}} + 3\frac{(\overline{R}/d_2)}{\sqrt{n}} = \overline{\overline{x}} + A_2\overline{R}$$

and

$$\text{LCL} = \mu - 3\frac{\sigma}{\sqrt{n}} \approx \overline{\overline{x}} - 3\frac{(\overline{R}/d_2)}{\sqrt{n}} = \overline{\overline{x}} - A_2\overline{R} \tag{7.8}$$

---

[1] It is possible that some of the 'out of control' subgroups from the R chart will have been deleted prior to construction of the $\overline{x}$ chart.

where the values of $A_2 = 3/(d_2\sqrt{n})$ can be found in Appendix 1. From equation (7.8) one can see how these control limits for the $\bar{x}$ chart are directly affected by the centerline $\bar{R}$ of the R chart.

EXAMPLE 7.1   The process of making ignition keys for automobiles consists of trimming and pressing raw key blanks, cutting grooves, cutting notches, and plating. Some of the dimensions associated with groove and notch cutting are critical to the proper functioning of the keys. The data in Table 7.1 are measurements on a particular critical groove dimension. Due to the high volume of keys processed per hour, the sampling frequency is chosen to be five (keys) every 20 minutes.

TABLE 7.1   Groove Dimensions (in inches) of Ignition Keys, Subgroups of $n = 5$ Taken Every 20 Minutes

Subgroup i	Measurements					$\bar{x}_i$	R_i
1	0.0061	0.0084	0.0076	0.0076	0.0044	0.00682	0.0040
2	0.0088	0.0083	0.0076	0.0074	0.0059	0.00760	0.0029
3	0.0080	0.0080	0.0094	0.0075	0.0070	0.00798	0.0024
4	0.0067	0.0076	0.0064	0.0071	0.0088	0.00732	0.0024
5	0.0087	0.0084	0.0088	0.0094	0.0086	0.00878	0.0010
6	0.0071	0.0052	0.0072	0.0088	0.0052	0.00670	0.0036
7	0.0078	0.0089	0.0087	0.0065	0.0068	0.00774	0.0024
8	0.0087	0.0094	0.0086	0.0073	0.0071	0.00822	0.0023
9	0.0074	0.0081	0.0086	0.0083	0.0087	0.00822	0.0013
10	0.0081	0.0065	0.0075	0.0089	0.0097	0.00814	0.0032
11	0.0078	0.0098	0.0081	0.0062	0.0084	0.00806	0.0036
12	0.0089	0.0090	0.0079	0.0087	0.0090	0.00870	0.0011
13	0.0087	0.0075	0.0089	0.0076	0.0081	0.00816	0.0014
14	0.0084	0.0083	0.0072	0.0100	0.0069	0.00816	0.0031
15	0.0074	0.0091	0.0083	0.0078	0.0077	0.00806	0.0017
16	0.0069	0.0093	0.0064	0.0060	0.0064	0.00700	0.0033
17	0.0077	0.0089	0.0091	0.0068	0.0094	0.00838	0.0026
18	0.0089	0.0081	0.0073	0.0091	0.0079	0.00826	0.0018
19	0.0081	0.0090	0.0086	0.0087	0.0080	0.00848	0.0010
20	0.0074	0.0084	0.0092	0.0074	0.0103	0.00854	0.0029
						$\downarrow$	$\downarrow$
						$\bar{\bar{x}} = 0.007966$	$\bar{R} = 0.002406$

For convenience, the subgroup means and ranges are also given in Table 7.1. From these, one calculates the grand mean $\bar{\bar{x}} = 0.007966$ and the average range $\bar{R} = 0.002406$. For subgroups of size $n = 5$, the relevant control chart factors for $\bar{x}$ and R charts are $D_4 = 2.114$, $D_3 = 0$, and $A_2 = 0.577$ (see

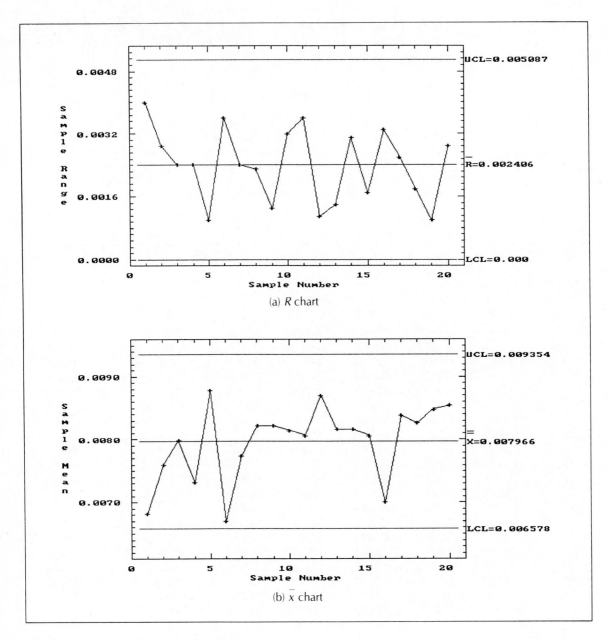

(a) R chart

(b) $\bar{x}$ chart

FIGURE 7.1
Minitab plots of $\bar{x}$ and R charts for Example 7.1

Appendix 1).[2] Starting with the R chart, one finds the control limits are UCL = $D_4 \bar{R}$ = (2.114)(0.002406) = 0.005087 and LCL = $D_3 \bar{R}$ = (0)(0.002406) = 0. A *Minitab* plot of the R chart is shown in Figure 7.1(a).

---

[2] It is useful to remember that $D_3 = 0$ whenever the subgroup size is 6 or less.

Since there do not appear to be any 'out of control' conditions present, we proceed immediately to the construction of the $\bar{x}$ chart. In this case, the control limits are UCL $= \bar{\bar{x}} + A_2\bar{R} = 0.007966 + (0.577)(0.002406) = 0.009354$, and LCL $= \bar{\bar{x}} - A_2\bar{R} = 0.007966 - (0.577)(0.002406) = 0.006578$. A *Minitab* plot of this $\bar{x}$ chart is given in Figure 7.1(b). Although none of the points on the chart fall outside the control limits, the chart does contain a run of eight consecutive points that fall above the centerline (subgroups 8–15). Before using this chart, then, one should search for possible explanations (assignable causes) for this behavior. However, if none can be found, then the chart can be used as it stands. For comparison, both charts have also been plotted using the *NWA Quality Analyst* software (Figure 7.2). This package automatically checks control charts for many of the 'out of control' signals discussed in Section 6.4. Thus, in Figure 7.2, the *NWA* software indicates that a rule violation has occurred at subgroup number 15. Further diagnostics within the *NWA* package mention that it is the 'eight points above the centerline' rule that has been violated.

FIGURE 7.2
NWA Quality Analyst plots of $\bar{x}$ and R charts for Example 7.1

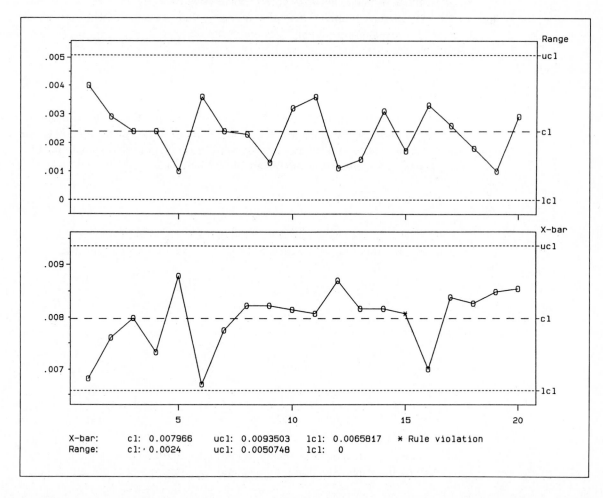

X-bar:	cl: 0.007966	ucl: 0.0093503	lcl: 0.0065817 * Rule violation
Range:	cl: 0.0024	ucl: 0.0050748	lcl: 0

7.3	$\bar{x}$ and s CHARTS

There are several possible substitutes for $\bar{x}$ and R charts. Because there are many different statistics for measuring central tendency, and an equally large number of variation measures, just about any combination of the two can be used to chart the process level and variation. Two widely used combinations are the $\bar{x}$ and s charts and the median and R charts, either of which can be used in lieu of $\bar{x}$ and R charts. Which combination to use is largely a matter of personal preference, the level of computation required, and one's software capabilities, since all the charts are designed to give about the same sensitivity to detecting 'out of control' signals.

The construction of $\bar{x}$ and s charts is the same as for $\bar{x}$ and R charts, in that the variation chart (here, the s chart) is brought into statistical control *first*. The reason is the same as before: the s chart must first be considered reliable, since the control limits for the $\bar{x}$ chart are based on its (the s chart's) centerline.

The s chart is constructed from some number, k, of successive subgroups of process measurements. Each subgroup should have the same number of readings, n. Next, the sample standard deviation of each subgroup is calculated from equation (3.7). The subgroup standard deviations are denoted by $s_1, s_2, s_3, \ldots, s_k$. Their grand average,

$$\bar{s} = \frac{1}{k} \sum_{i=1}^{k} s_i \tag{7.9}$$

forms the *centerline* of the s chart. $\bar{s}$ is an estimate of μ_s, the mean of the sampling distribution of the standard deviations. Following the usual procedure, we place control limits at a distance of $\pm 3\sigma_s$ from the centerline. To estimate σ_s, one must appeal to theoretical results (Ryan 1989) that give

$$\hat{\sigma}_s = \frac{\bar{s}}{c_4} \sqrt{1 - c_4^2} \tag{7.10}$$

where c_4, the same constant used in equation (3.9), is found in Appendix 1. From equation (7.10), the 3-sigma control limits for the s chart are then given by

$$\text{UCL} = \bar{s} + 3\hat{\sigma}_s = \bar{s} + 3\left(\frac{\bar{s}}{c_4} \sqrt{1 - c_4^2} \right) = B_4 \bar{s}$$

and

$$\text{LCL} = \bar{s} - 3\hat{\sigma}_s = \bar{s} - 3\left(\frac{\bar{s}}{c_4} \sqrt{1 - c_4^2} \right) = B_3 \bar{s}$$

or simply

$$\text{UCL} = B_4 \bar{s} \quad \text{and} \quad \text{LCL} = B_3 \bar{s} \tag{7.11}$$

where $B_4 = 1 + (3/c_4)\sqrt{1 - c_4^2}$ and $B_3 = 1 - (3/c_4)\sqrt{1 - c_4^2}$. As before, the constants B_3 and B_4 are conveniently found in Appendix 1.

Plotting the centerline, control limits, and subgroup standard deviations $s_1, s_2,$

$s_3, \ldots, s_k$ together on the same graph completes the s chart. The rules for detecting 'out of control' conditions (see Chapter 6) can be invoked to see whether the process variation seems to be in statistical control or not. As before, any subgroups deemed to be 'out of control' should be eliminated from the data (*only if assignable causes can be found*) and the calculations revised accordingly.

Before proceeding to the $\bar{x}$ chart, we pause to note that the 3-sigma control limits used in the s chart (and the R chart as well) may not have exactly the same sensitivity as the $\bar{x}$ chart because the sampling distribution of the statistic s (based on samples of size n) does not follow a normal distribution.[3] Using 3-sigma control limits with the s chart (or the R chart) causes the false alarm rate to increase somewhat; that is, there is a slightly greater chance that an 'out of control' signal may, in fact, be false. This problem can be fixed by instead using *probability limits* (see Section 6.4) for the control limits of the s chart, but the required calculations do not retain the simplicity of the 3-sigma approach. The reader is referred to other texts for this discussion (e.g., Ryan 1989, Grant and Leavenworth 1988). Regarding the tradeoff between having a slightly increased false alarm rate and using probability limits, we believe that the simpler 3-sigma approach still serves well in practice.

We now turn to the $\bar{x}$ chart, where the subgroup means $\bar{x}_1, \bar{x}_2, \bar{x}_3, \ldots, \bar{x}_k$ and their grand average, $\bar{\bar{x}}$, are calculated as before from equation (7.7). To estimate the process standard deviation, σ, one uses $\bar{s}$ in equation (3.9) to obtain $\hat{\sigma} = \bar{s}/c_4$. The 3-sigma control limits for the $\bar{x}$ chart then become

$$\text{UCL} = \mu + 3\sigma_{\bar{x}} = \mu + 3\left(\frac{\sigma}{\sqrt{n}}\right) \approx \bar{\bar{x}} + 3\frac{(\bar{s}/c_4)}{\sqrt{n}} = \bar{\bar{x}} + A_3\bar{s}$$

and   (7.12)

$$\text{LCL} = \mu - 3\sigma_{\bar{x}} = \mu - 3\left(\frac{\sigma}{\sqrt{n}}\right) \approx \bar{\bar{x}} - 3\frac{(\bar{s}/c_4)}{\sqrt{n}} = \bar{\bar{x}} - A_3\bar{s}$$

where, as usual, the constant A_3 is obtained from Appendix 1.

EXAMPLE 7.2   This example reanalyzes the ignition key data from Example 7.1, this time using $\bar{x}$ and s charts. For convenience, Table 7.2 lists the subgroup standard deviations and means. The means, of course, are identical to those in Table 7.1. From Table 7.2, $\bar{\bar{x}} = 0.007966$ and $\bar{s} = 0.0009672$, and from Appendix 1, one finds $B_3 = 0$, $B_4 = 2.089$, and $A_3 = 1.427$ for subgroups of size $n = 5$. When these results are used in equation (7.11), the s chart has a centerline of $\bar{s} = 0.0009672$ and control limits of UCL = 2.089(0.0009672) = 0.00202 and LCL = 0(0.0009672) = 0. The corresponding $\bar{x}$ chart centerline is $\bar{\bar{x}} = 0.007966$ with control limits of UCL = 0.007966 + 1.427(0.0009672) = 0.00935 and LCL = 0.007966 − 1.427(0.0009672) = 0.00659. The resulting $\bar{x}$ and s charts are plotted with the *SPC1+* software package and, for comparison, with the *SPCII* package in Figures 7.3 and 7.4. Both graphs give the same conclusion as the $\bar{x}$ and R charts of Example 7.1. Notice that the *SPC1+* graph marks possible 'out of control' points on the graph with a + sign.

---

[3] Recall that 3-sigma limits are used for statistics that follow a normal distribution.

TABLE 7.2   Subgroup Means and Standard Deviations for Groove Dimension Data in Table 7.1

	$\bar{x}_i$	s_i
1	0.00682	0.0015881
2	0.00760	0.0011023
3	0.00798	0.0008955
4	0.00732	0.0009418
5	0.00878	0.0003768
6	0.00670	0.0015264
7	0.00774	0.0010831
8	0.00822	0.0009834
9	0.00822	0.0005167
10	0.00814	0.0012361
11	0.00806	0.0012915
12	0.00870	0.0004637
13	0.00816	0.0006309
14	0.00816	0.0012219
15	0.00806	0.0006656
16	0.00700	0.0013248
17	0.00838	0.0010941
18	0.00826	0.0007403
19	0.00848	0.0004207
20	0.00854	0.0012402
	↓	↓
	$\bar{\bar{x}} = 0.007966$	$\bar{s} = 0.0009672$

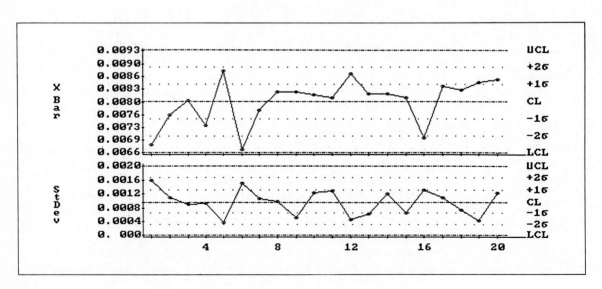

FIGURE 7.3
$\bar{x}$ and s charts of data in Table 7.1 from *SPC1 +* software package

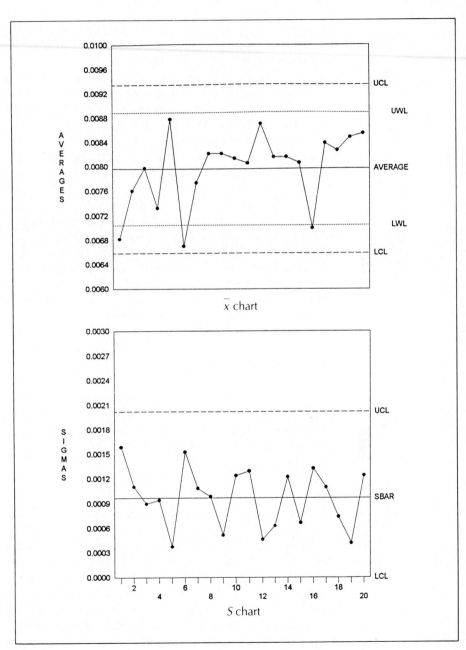

FIGURE 7.4
$\bar{x}$ and s charts of
data in Table 7.1 from
SPCII software
package

7.4   MEDIAN CHARTS

To expedite calculations on the shop floor, subgroup medians can be substituted for means when the level of a process is charted. Recall that the median of a set of n numbers is just the 'middle' value in the data if n is odd and, by convention, the average of the *two* middle values when n is even (see Section 3.3). Thus, to take full advantage of the ease in calculating the median, control charts based on the median generally use *odd* subgroup sizes. Typically, subgroup sizes of 3, 5, 7, 9, or 11 are used, but any odd integer suffices.

Even though the subgroup medians $\tilde{x}_1, \tilde{x}_2, \tilde{x}_3, \ldots, \tilde{x}_k$ are plotted on the median chart, the centerline is still calculated by *averaging* the subgroup means. Other approaches, such as using the *median* of the numbers $\tilde{x}_1, \tilde{x}_2, \tilde{x}_3, \ldots, \tilde{x}_k$, can be used (Grant and Leavenworth 1988, pp. 313–314), but our presentation is based on the traditional choice of $\bar{\bar{x}}$ as the centerline. Another convention commonly used with median charts is to plot *all* the subgroup data on the chart and then connect only their *medians* by line segments. In this way, additional information is available that may be helpful when interpreting the chart. The only cautionary note here is to remember that individual readings always exhibit more variation than do their subgroup medians,[4] so any individual points that fall outside the control limits should *not* be treated as 'out of control' signals. Only the subgroup *medians* are examined using the 'out of control' rules.

If an R chart is used in conjunction with the median chart, then the control limits for the R chart are calculated as in Section 7.2 and the control limits for the median chart are given by

$$\text{UCL} = \bar{\bar{x}} + A_6 \bar{R} \quad \text{and} \quad \text{LCL} = \bar{\bar{x}} - A_6 \bar{R} \tag{7.13}$$

where the constant A_6 is found in Appendix 1. Analogously, if an s chart is used, the control limits for the median chart are

$$\text{UCL} = \bar{\bar{x}} + A_7 \bar{s} \quad \text{and} \quad \text{LCL} = \bar{\bar{x}} + A_7 \bar{s} \tag{7.14}$$

where $\bar{s}$ is the centerline of the s chart from equation (7.9) and A_7 is given in Appendix 1.

There is an implicit tradeoff to be aware of when using the median chart. The ease of calculating the median must be weighed against the fact that it is somewhat less sensitive to extreme values than is the mean (see Section 3.3), so median charts do not immediately react to one or two extreme readings in the data. Although this is a desirable feature in many statistical applications, it is usually *not* desirable in process control because extremely large or small readings are often very important signals of process or measurement system problems.

---

[4] This is also true if the individual values are plotted on an $\bar{x}$ chart.

EXAMPLE 7.3   Consider the data of Table 7.1 again and suppose that a median and R chart are used to monitor the key groove dimension. In this case, the R chart is identical to the one calculated in Example 7.1. In particular, its centerline is $\overline{R} = 0.0024$ so, from equation (7.14), the control limits for the median chart are UCL = $0.007966 + 0.691(0.0024) = 0.00962$ and LCL = $0.007966 - 0.691(0.0024) = 0.0063$. Figure 7.5 shows the median and R charts for these data generated by the *SPC1+* software package. This package follows the usual convention of plotting the individual subgroup readings as well as their medians. As expected, some of the individual values fall outside the control limits, but the *medians* all lie within these limits. In addition, note that Figure 7.5 does not show a violation of the 'eight successive points on the same side of the centerline' rule as did the $\overline{x}$ charts in Figures 7.1–7.4. Thus, unlike the $\overline{x}$ charts of these data, the median chart shows no evidence of any 'out of control' conditions.

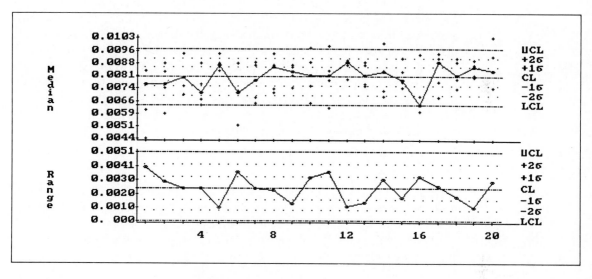

FIGURE 7.5
Median and R charts for the data of Table 7.1

7.5   INDIVIDUALS CHARTS

Frequently, process measurements are not arranged into subgroups before charting; that is, sometimes it is necessary or desirable to monitor successive *individual* measurements. This situation usually arises when data become available only at very slow rates (i.e., there are large gaps of time between successive measurements), or when a continuously varying quantity is monitored (so that the most *current* reading is the best indicator of the process level, and averaging a group of readings only gives a false picture of this level). Typical examples of the former situation are data from low-volume production runs and accounting records such as weekly or monthly customer orders, costs, or returns. Examples of continuously varying quantities are process-related variables such as humidity, temperature, and chemical concentrations.

Various control charts have been developed to monitor ungrouped data. This section looks at the oldest of these methods, the individuals chart. Later sections examine the more recently developed EWMA and CUSUM methods.

An **individuals chart** (also called an **X chart**) is created by plotting the individual process measurements in the order in which they were collected. The resulting chart can be thought of as a control chart for the process average based on subgroups of size 1. Thus, the centerline is just the sample mean, $\bar{x}$, of the data. To place control limits on an individuals chart, however, requires a different strategy than before, since it is no longer possible to calculate the process variation from the subgroups.

One solution to this problem is to *artificially* create subgroups from the data and then calculate the range of each such subgroup. Traditionally, these subgroups are formed from successive pairs of readings; that is, if $x_1, x_2, x_3, \ldots, x_k$ represent the available data for constructing the chart, then the pairs $\{x_1, x_2\}$, $\{x_2, x_3\}$, $\{x_3, x_4\}$, ..., $\{x_{n-1}, x_n\}$ are used to find the ranges $|x_2 - x_1|$, $|x_3 - x_2|$, $|x_4 - x_3|$, ..., $|x_n - x_{n-1}|$. Conceptually, one envisions successive groups being formed by bracketing the first pair of readings and then iteratively sliding the brackets one point to the right to obtain $n - 1$ pairs of readings. For this reason, the resulting ranges are called **moving ranges,** and their average

$$\overline{MR} = \frac{1}{n - 1} \sum_{i=1}^{n-1} |x_{i+1} - x_i| \tag{7.15}$$

is considered to be a good estimate of the range, *for samples of size 2,* drawn from the process. The notation $\overline{MR}$ has become fairly standard because it is suggestive of how the estimate is to be constructed.

Putting $\overline{MR}$ in for R in equation (3.11) then gives an estimate of the process standard deviation:

$$\hat{\sigma} = \frac{\overline{MR}}{d_2} \tag{7.16}$$

Note that since equation (3.11) is based on the assumption that the measurements are normally distributed, this assumption underlies the interpretation of the individuals chart. For subgroups of size $n = 2$, Appendix 1 gives $d_2 = 1.128$, so equation (7.16) can be further simplified to

$$\hat{\sigma} = \frac{\overline{MR}}{1.128} \tag{7.17}$$

Finally, 3-sigma control limits for the process mean can be constructed by

$$\text{UCL} = \bar{x} + 3\hat{\sigma} = \bar{x} + 3\left(\frac{\overline{MR}}{1.128}\right) = \bar{x} + 2.660(\overline{MR})$$

and

$$\text{LCL} = \bar{x} - 3\hat{\sigma} = \bar{x} - 3\left(\frac{\overline{MR}}{1.128}\right) = \bar{x} - 2.660(\overline{MR})$$

$$\tag{7.18}$$

Moving ranges can also be based on subgroups of more than two measurements, but the choice $n = 2$ is so easy to implement that it has become the most common choice. However, it is good to be aware that other choices exist, since some software packages offer these options for individuals charts.

EXAMPLE 7.4   In the process of chrome plating, parts immersed in a chemical bath containing nickel receive a thin plating of the metal when small electric currents are run though the bath. Metals used in plating solutions are referred to as 'electroless nickel,' 'electroless copper,' and so on. Table 7.3 shows the results of measuring the electroless nickel concentration in a bath at the start of each workshift in a chrome plating facility. There are three shifts per day, so the 75 measurements represent a 25-day period. As the plating proceeds, the nickel is depleted. For correct operation, it has been determined that the concentration of electroless nickel in the bath should be about 4.5 oz/gal.

TABLE 7.3   Concentrations (in oz/gal) of Electroless Nickel Measured at the Start of Each Workshift; Three Shifts/Day for 25 Days

(Read across)												
4.8	4.8	4.5	4.5	4.4	4.2	4.4	4.5	5.0	4.2	4.8	4.5	4.4
4.6	4.3	4.5	4.7	4.4	4.5	4.4	4.5	4.7	4.7	4.6	4.4	4.7
4.8	4.6	4.5	4.6	4.3	4.3	4.5	4.8	4.5	4.6	4.4	4.7	4.6
4.5	4.8	4.7	4.5	4.6	4.7	4.7	5.0	4.7	4.8	4.6	4.4	4.8
4.9	4.6	4.3	4.7	4.6	4.8	4.8	4.9	4.9	4.6	4.6	4.8	4.9
4.9	4.7	4.7	4.7	4.8	4.7	4.9	5.2	4.4	4.3			

Grouping these measurements, as would be done in $\bar{x}$ or median charts, does not make sense here because the most recent measurement is the best indicator of the current concentration. Averaging several readings would only delay a signal that the concentration had dropped. Thus, the individuals chart is the appropriate choice. Furthermore, in this application, the object of charting is to *control the process to a target* of 4.5, so $\bar{x}$ is replaced by the value 4.5 in equation (7.18).

Figure 7.6 shows a *Minitab* plot of the individuals chart for these data. *Minitab* allows the user to select the size of the moving subgroups for calculating $\overline{MR}$ and to specify a target value, instead of using $\bar{x}$ as the centerline. Figure 7.6 uses the default subgroup size of $n = 2$ and a specified target of 4.5 for the process mean. It is difficult to discern 'out of control' signals in the chart until about observation 62, when an 'eight points on the same side of the centerline' violation becomes apparent. A search for special causes is needed, but since the chart seems to indicate an upward shift in the concentration, it is apparent that electroless nickel has been added to the bath.

It may have occurred to the reader that an easier way to estimate the underlying process variation would be to use the sample standard deviations of the readings $x_1, x_2, x_3, \ldots, x_k$ instead of the moving ranges; that is, instead of using $\overline{MR}/d_2$,

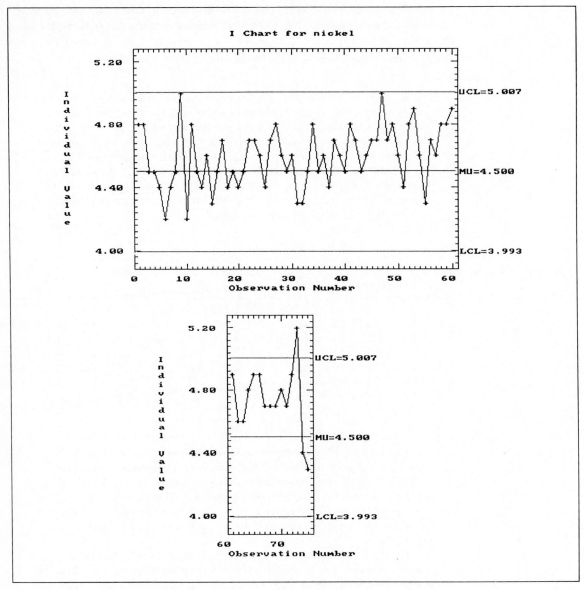

FIGURE 7.6
Minitab plot of an individuals chart for the data of Table 7.3

one could use s/c_4 from equation (3.9). The usual justification for using $\overline{MR}/d_2$ is that it is not so easily influenced by trends in the data as s might be, and that this property therefore makes $\overline{MR}/d_2$ the better estimator. Recently, Cryer and Ryan (1990) have questioned this rationale and have shown that, for normally distributed process data, s/c_4 is a much more efficient estimate of σ than is $\overline{MR}/d_2$. Because of the increased efficiency of s/c_4, they conclude that it, not $\overline{MR}/d_2$, should be used to monitor processes that have *already been brought into statistical control*. When an individuals chart is initially set up, they further recommend that both $\overline{MR}/d_2$ and s/c_4 be calculated and, if they are *not* in reasonable agreement, then an investigation of the possible problem is warranted before the chart is constructed. Operationally, this should be easy to accomplish, since c_4 will be very close to 1 for the usual values of k met in practice. Applying these recommendations to Example 7.4, we find $\overline{MR}/d_2 = 0.19054/1.128 = 0.1689$ and $s/c_4 = 0.20307/0.9966 = 0.2038$. On a percentage basis, these estimates do seem to differ, so further exploration is required. In fact, as seen in the next section, a process shift has probably occurred near the middle of this set of data, which could explain the disparity between the two estimates of σ.

7.6   EWMA CHARTS

The **exponentially weighted moving average (EWMA)** chart was introduced by Roberts in 1959. Instead of EWMA, Roberts originally used the term **geometric moving average chart,** but the chart's name evolved to the current one to reflect the fact that the method of exponential smoothing serves as the basis of EWMA charts. Exponential smoothing, a simple time series technique for forecasting large numbers of time series simultaneously, was developed in World War II by R. G. Brown for the U.S. Naval Operations Evaluation Group (Gardner 1985), and it is still widely used for inventory forecasting.

The EWMA chart can be constructed for any series of subgroup statistics, but it is most commonly used for variables data. EWMA charts are considered to be alternatives to individuals charts (see Section 7.5) or $\overline{x}$ charts (Section 7.2). In addition, the EWMA chart tends to provide quicker responses to shifts in the process average than either the individuals or $\overline{x}$ charts do because each point on an EWMA chart incorporates information from all the *previous* subgroups, not just from the current subgroup. However, to be fair to the individuals and $\overline{x}$ charts, we note that their sensitivity can also be increased if one uses the expanded set of 'out of control' rules from Section 6.4.

To construct the EWMA chart, one begins as usual with k successive subgroup means $\overline{x}_1, \overline{x}_2, \overline{x}_3, \ldots, \overline{x}_k$. If the subgroup size happens to be $n = 1$, then the EWMA chart is considered an alternative to the individuals chart (see Section 7.5). If, instead, $n > 1$, then the EWMA chart is an alternative to the $\overline{x}$ chart. The points on the EWMA chart are denoted by $\hat{x}_1, \hat{x}_2, \hat{x}_3, \ldots, \hat{x}_k$ and are calculated recursively by means of the formula

$$\hat{x}_i = \lambda \overline{x}_i + (1 - \lambda)\hat{x}_{i-1} \tag{7.19}$$

where λ, called the **weighting constant,** is chosen between 0 and 1. (The choice of λ is discussed later in the section.) To initialize the calculations in equation (7.19), a **starting point,** $\hat{x}_0$, is required. If no target value is specified for the process average, then the starting point is taken to be $\hat{x}_0 = \overline{\overline{x}}$. Otherwise, if the process is to be controlled to some target level, μ_0, then one takes $\hat{x}_0 = \mu_0$.

From equation (7.19) it can be seen that each EWMA point is simply a *weighted average* of the subgroup means that *precede* it. More precisely, by repeatedly substituting equation (7.19) into itself, we can write each $\hat{x}_i$ as

$$\hat{x}_i = \lambda \overline{x}_i + \lambda(1 - \lambda)\overline{x}_{i-1} + \lambda(1 - \lambda)^2 \overline{x}_{i-2} + \cdots + \lambda(1 - \lambda)^i \hat{x}_0 \tag{7.20}$$

The fact that the weights $\lambda(1 - \lambda)^i$ in equation (7.20) decrease exponentially gives rise to the name of the EWMA chart.

The weighting constant, λ, controls the amount of influence that previous points have on the *current* EWMA point. From either equation (7.19) or (7.20), one can see that values of λ near 1 put almost all the weight on the current subgroup mean; that is, the closer λ is to 1, the more the EWMA chart resembles the $\overline{x}$ chart. At the other extreme, for values of λ close to 0, the EWMA tends to give a small weight to almost all the past observations and the performance of the chart then parallels that of the CUSUM chart discussed in Section 7.7. Experience has shown that λ's in the range $0.10 < \lambda < 0.30$ generally give good results. The value of λ should not be allowed to get too large or else, as we have seen, the chart is simply a duplicate of the $\overline{x}$ chart.

Control limits for the EWMA chart vary depending on the subgroup number, i, and are calculated from the equations

$$\text{UCL} = \hat{x}_0 + \frac{3\sigma}{\sqrt{n}} \sqrt{\left(\frac{\lambda}{1 - \lambda}\right) \cdot [1 - (1 - \lambda)^{2i}]}$$

and $\qquad\qquad\qquad\qquad\qquad\qquad\qquad\qquad\qquad\qquad\qquad$ (7.21)

$$\text{LCL} = \hat{x}_0 - \frac{3\sigma}{\sqrt{n}} \sqrt{\left(\frac{\lambda}{1 - \lambda}\right) \cdot [1 - (1 - \lambda)^{2i}]}$$

As the subgroup number increases in equation (7.21), the control limits approach the simpler, approximate limits

$$\text{UCL} = \hat{x}_0 + \frac{3\sigma}{\sqrt{n}} \sqrt{\left(\frac{\lambda}{1 - \lambda}\right)} \quad \text{and} \quad \text{LCL} = \hat{x}_0 - \frac{3\sigma}{\sqrt{n}} \sqrt{\left(\frac{\lambda}{1 - \lambda}\right)} \tag{7.22}$$

SPC software packages normally plot the exact control limits in equation (7.21).

The process standard deviation, σ, that appears in equations (7.21) and (7.22) must be estimated from the data. When we average either the subgroup ranges or standard deviations, σ is estimated as $\overline{R}/d_2$ or $\overline{s}/c_4$. In the case where the subgroup size is $n = 1$, σ is estimated via the moving range procedure of Section 7.5.

EXAMPLE 7.5   An individuals chart for the electroless nickel concentration data of Table 7.3 was constructed in Example 7.4. As an alternative to the individuals chart, an EWMA chart is now created. Since the target value for the concentration was set at 4.5 oz/gal, an initial value of $\hat{x}_0 = 4.5$ is used for calculating the EWMA points. With a weighting constant of $\lambda = 0.20$, the EWMA points are

$$\hat{x}_1 = 0.20(4.8) + 0.80(4.5) = 4.56$$

$$\hat{x}_2 = 0.20(4.8) + 0.80(4.56) = 4.61$$

$$\hat{x}_3 = 0.20(4.5) + 0.80(4.61) = 4.59$$

and so forth. Figure 7.7 shows the EWMA chart for these data plotted by the *Minitab* package. Since the subgroup size is $n = 1$, *Minitab* uses the moving

FIGURE 7.7

Minitab plot of an EWMA ($\lambda = 0.20$) chart for the data of Table 7.3

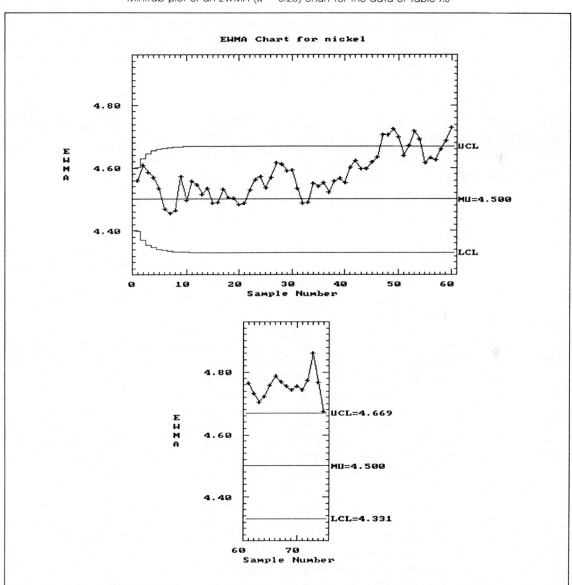

range method to estimate σ when calculating the control limits in equation (7.21). (The default subgroup size of 2 was used in Figure 7.7.) This EWMA chart shows an 'out of control' condition first occurring at about subgroup number 46, followed by many more signals in succeeding groups. Compared with the individuals chart in Figure 7.6, the EWMA chart gives an earlier warning (by about 16 subgroups) of the apparent overfilling of electroless nickel in the plating bath.

Because the points on the EWMA chart are weighted averages of the previous subgroup means, successive EWMA points tend to be highly correlated with one another. As a result, the expanded list of 'out of control' rules in Section 6.4 *cannot* be applied to EWMA charts, since these rules apply to data points that are assumed to be statistically independent. Thus, with EWMA charts, only the 'point outside the 3-sigma limits' rule is used.

7.7   CUSUM CHARTS

BACKGROUND

The **cumulative sum of deviations** (or, more simply, **CUSUM**) chart was introduced by E. S. Page in 1954 as a method for monitoring variables data. Although the CUSUM was intended to monitor individuals data and subgroup means, it has evolved into a method for controlling any subgroup statistic. This section presents the original CUSUM procedure for monitoring subgroup means, which, for a subgroup size of $n = 1$, also applies to individuals data (see Section 7.5). For applications to other types of subgroup statistics, the reader should refer to Wadsworth, Stephens, and Godfrey (1986) or Ryan (1989).

CUSUM charts are *deviation control* charts because they monitor the extent to which a subgroup statistic deviates from a **target value** (see Section 6.2). If the centerline is replaced by a target value, most threshold control charts can also be altered to serve as deviation control charts. However, like EWMA (see Section 7.6), the CUSUM procedure goes one step further by *accumulating* the information from successive subgroups. For threshold control charts, this is analogous to using the extended rules for 'out of control' signals from Section 6.4, which incurs the additional cost of a careful examination of the chart for significant patterns. In essence, by incorporating the previous data into each plotted point, the CUSUM and EWMA procedures allow the user to spot process trends and shifts more quickly and with less inspection of the chart.

It is important to remember that *CUSUM procedures assume that the process standard deviation σ is stable,* and that they use estimates of σ in the chart's construction. For this reason, it is often recommended that the process first be brought into a state of statistical control by using a threshold control chart so that a reliable estimate of σ can be obtained. Once a good estimate of σ is available, the CUSUM procedure can be shown to be much more sensitive to small process shifts (e.g., shifts less than about 2σ) than the corresponding Shewhart chart. For shifts in the process mean that are greater than 2σ (i.e., large process shifts), the Shewhart chart tends to be marginally more sensitive than the CUSUM chart (Ewan 1963).

Given successive subgroup means $\bar{x}_1, \bar{x}_2, \bar{x}_3, \ldots, \bar{x}_s$ and a desired target value μ_0 for the process average, the first step in constructing a CUSUM chart is to create the deviations from target $\bar{x}_1 - \mu_0, \bar{x}_2 - \mu_0, \bar{x}_3 - \mu_0, \ldots, \bar{x}_s - \mu_0$ and then form their partial sums:[5]

$$
\begin{aligned}
&(\bar{x}_1 - \mu_0) \\
&(\bar{x}_1 - \mu_0) + (\bar{x}_2 - \mu_0) \\
&(\bar{x}_1 - \mu_0) + (\bar{x}_2 - \mu_0) + (\bar{x}_3 - \mu_0) \\
&\quad \vdots \\
&(\bar{x}_1 - \mu_0) + (\bar{x}_2 - \mu_0) + (\bar{x}_3 - \mu_0) + \cdots + (\bar{x}_s - \mu_0)
\end{aligned}
\tag{7.23}
$$

When these partial sums are plotted (in place of the subgroup means), the resulting graph is called a CUSUM chart. It should be apparent that an upward shift in the process will result in many positive deviations, causing a positive slope in the CUSUM graph. Similarly, downward process shifts will cause the graph to have a negative slope.

THE TABULAR CUSUM

While on the one hand we want to detect the process shifts when they occur, we do not necessarily want to detect extremely small, inconsequential shifts. Since even the smallest shift in the process average causes the slope of the CUSUM to change, something must be done to 'desensitize' the chart so that only shifts of the desired magnitude are detected. This is accomplished by selecting a **reference value** (denoted by k) that specifies the size of the shift that one desires to detect. It is very common to choose k to be approximately one-half the size of the shift to be detected. Thus, to detect process shifts of magnitude 1σ, a reference value of $k = 0.5\sigma$ is used. Then, instead of calculating the deviations $\bar{x}_i - \mu_0$, one forms a sequence of upper deviations $\bar{x}_i - (\mu_0 + k)$ and lower deviations $\bar{x}_i - (\mu_0 - k)$ whose partial sums are calculated *instead* of those in equation (7.23). The partial sums of the upper deviations (denoted by CUSUM_U) are used to detect upward process shifts, whereas the partial sums of the lower deviations (denoted CUSUM_L) signal downward shifts.

In order to decide when an 'out of control' signal occurs, a **decision interval** (denoted by h) is selected. Whenever CUSUM_U or CUSUM_L exceeds h, an 'out of control' signal occurs. Typically h is selected between $h = 4.5\sigma$ and $h = 5\sigma$. The various combinations of h and k cause the CUSUM to be more or less

---

[5] In this section we use s, not k, to denote the number of subgroups. The letter k is traditionally used to represent the 'reference value' of a CUSUM chart.

sensitive. The choice $k = 0.5$ and $h = 5\sigma$ occurs frequently in practice because it gives rise to a chart with roughly the same ARL as the $\bar{x}$ chart when no process shift has occurred (see Section 6.4). For other choices of h and k and their associated ARLs, we refer the reader to Ewan (1963). The exact rules for calculating CUSUM$_U$ and CUSUM$_L$ are as follows:

THE TABULAR
CUSUM PROCEDURE

1. Starting with CUSUM$_U(0) = 0$, form the upper partial sums CUSUM$_U(i) = $ max$[0,$ CUSUM$_U(i - 1) + \bar{x}_i - (\mu_0 + k)]$ for $i = 1, 2, 3, \ldots, s$.

2. Starting with CUSUM$_L(0) = 0$, form the lower partial sums CUSUM$_L(i) = $ max$[0,$ CUSUM$_L(i - 1) - \bar{x}_i + (\mu_0 - k)]$ for $i = 1, 2, 3, \ldots, s$.

Rules 1 and 2 can be summarized: anytime that CUSUM$_U$ becomes negative, *reset it to 0* and continue; anytime that CUSUM$_L$ becomes negative, *reset it to 0* and continue.

3. An 'out of control' signal is given whenever CUSUM$_U > h$ or CUSUM$_L > h$.

EXAMPLE 7.6

The nickel concentration data of Table 7.3 were analyzed with the individuals chart (Example 7.4) and the EWMA chart (Example 7.5). For comparison, the CUSUM procedure is now applied to these data. As an estimate of the process standard deviation, the sample standard deviation of all 75 readings, $s = 0.203066$, is used. Using the standard choices $k = 0.5\hat{\sigma}$ and $h = 5\hat{\sigma}$, we generate a reference value of $k = 0.5(0.203066) = 0.101533$ and a decision interval of $h = 5(0.203066) = 1.01533$. Table 7.4 shows the calculations for these data following the tabular CUSUM rules listed in this section.

TABLE 7.4   Tabular CUSUM Calculations for the Nickel Concentration Data of Table 7.3

i	x_i	$x_i - (\mu_0 + k)$	CUSUM$_U$	$x_i - (\mu_0 - k)$	CUSUM$_L$
1	4.8	0.198467	0.198467	0.401533	0
2	4.8	0.198467	0.396934	0.401533	0
3	4.5	−0.101533	0.295401	0.101533	0
4	4.5	−0.101533	0.193868	0.101533	0
5	4.4	−0.201533	0	0.001533	0
6	4.2	−0.401533	0	−0.198467	0.198467
7	4.4	−0.201533	0	0.001533	0.196934
8	4.5	−0.101533	0	0.101533	0.095401
9	5.0	0.398467	0.398467	0.601533	0
10	4.2	−0.401533	0	−0.198467	0.198467
11	4.8	0.198467	0.198467	0.401533	0
12	4.5	−0.101533	0.096934	0.101533	0
13	4.4	−0.201533	0	0.001533	0
14	4.6	−0.001533	0	0.201533	0
15	4.3	−0.301533	0	−0.098467	0.098467
16	4.5	−0.101533	0	0.101533	0

17	4.7	0.098467	0.098467	0.301533	0
18	4.4	−0.201533	0	0.001533	0
19	4.5	−0.101533	0	0.101533	0
20	4.4	−0.201533	0	0.001533	0
21	4.5	−0.101533	0	0.101533	0
22	4.7	0.098467	0.098467	0.301533	0
23	4.7	0.098467	0.196934	0.301533	0
24	4.6	0.001533	0.195401	0.201533	0
25	4.4	0.201533	0	0.001533	0
26	4.7	0.098467	0.098467	0.301533	0
27	4.8	0.198467	0.296934	0.401533	0
28	4.6	−0.001533	0.295401	0.201533	0
29	4.5	−0.101533	0.193868	0.101533	0
30	4.6	−0.001533	0.192335	0.201533	0
31	4.3	−0.301533	0	−0.098467	0.098467
32	4.3	−0.301533	0	−0.098467	0.196934
33	4.5	−0.101533	0	0.101533	0.095401
34	4.8	0.198467	0.198467	0.401533	0
35	4.5	−0.101533	0.096934	0.101533	0
36	4.6	−0.001533	0.095401	0.201533	0
37	4.4	−0.201533	0	0.001533	0
38	4.7	0.098467	0.098467	0.301533	0
39	4.6	−0.001533	0.096934	0.201533	0
40	4.5	−0.101533	0	0.101533	0
41	4.8	0.198467	0.198467	0.401533	0
42	4.7	0.098467	0.296934	0.301533	0
43	4.5	−0.101533	0.195401	0.101533	0
44	4.6	−0.001533	0.193868	0.201533	0
45	4.7	0.098467	0.292335	0.301533	0
46	4.7	0.098467	0.390802	0.301533	0
47	5.0	0.398467	0.789269	0.601533	0
48	4.7	0.098467	0.887736	0.301533	0
49	4.8	0.198467	1.086203*	0.401533	0
50	4.6	−0.001533	1.084670*	0.201533	0
51	4.4	−0.201533	0.883137	0.001533	0
52	4.8	0.198467	1.081604*	0.401533	0
53	4.9	0.298467	1.380071*	0.501533	0
54	4.6	−0.001533	1.378538*	0.201533	0
55	4.3	−0.301533	1.077005*	−0.098467	0.098467
56	4.7	0.098467	1.175472*	0.301533	0
57	4.6	−0.001533	1.173939*	0.201533	0
58	4.8	0.198467	1.372406*	0.401533	0
59	4.8	0.198467	1.570873*	0.401533	0
.	.	.	.	.	.
.	.	.	.	.	.
.	.	.	.	.	.

Target $= \mu_0 = 4.5$
Reference value $= k = 0.5\hat{\sigma} = 0.101533$
Decision interval $= h = 5\hat{\sigma} = 1.01533$
* denotes an 'out of control' signal.

The first 'out of control' condition is evidenced at subgroup 49 when $CUSUM_U = 1.086203$ exceeds the decision interval of 1.101533. Normally, this signal would instigate a search for special causes, and *if a special cause was found,* both CUSUMs would be reset to 0 at this point. However, to illustrate the calculations, Table 7.4 continues to show $CUSUM_U$ and $CUSUM_L$ values for a few subgroups past group 49.

THE V MASK

The tabular form of the CUSUM is not the one usually associated with CUSUM *graphs.* The tabular form involves *two* sets of points $CUSUM_U$ and $CUSUM_L$, whereas the CUSUM described at the beginning of this section is based on *one* sequence of partial sums shown in equation (7.23). In addition, as Example 7.6 shows, both $CUSUM_U$ and $CUSUM_L$ must often be reset to 0, so any graph of these quantities would have a very choppy appearance.

To create a smoother looking chart, the rules governing $CUSUM_U$ and $CUSUM_L$ are transformed so that a chart based on the simple deviations from target, $\bar{x}_i - \mu_0$, can be plotted. The control limits for this chart must reflect the reference value k and decision interval h from the tabular CUSUM. To accomplish this, the control limits take a different form than the usual 3-sigma limits placed on either side of a centerline. Instead, the CUSUM graph uses control limits in the form of a "V" as shown in Figure 7.8.

To construct the control limits (called the "**V mask**") at subgroup i, the corner of the V is plotted at a certain distance d *ahead* of subgroup i and the legs of the V are drawn so that they are at a distance of h units above and below the plotted CUSUM point *at position* i (see Figure 7.8). The lead distance (d) and half-angle (θ) of the mask are related to the reference value (k) and decision interval (h) by the formulas

$$ d = \frac{h}{k} \quad \text{and} \quad \theta = \arctan\left(\frac{k}{w}\right) \tag{7.24} $$

where w is a scale factor chosen to make the resulting graph easily readable.

The scale factor works in the following manner: whatever physical distance is used between successive subgroup numbers on the horizontal axis of the graph, *that same physical distance represents w measurement units on the vertical axis.* A scale factor of $w = 2\hat{\sigma}$ usually produces good results when a reference value of $k = 0.5\hat{\sigma}$ is used, since the half-angle then becomes $\theta = \arctan(0.5\hat{\sigma}/2\hat{\sigma}) = \arctan(0.25) \approx 14°$. This means that the angle at the corner of the V is about 28°, making the mask easy to analyze visually.

After the V mask has been constructed at subgroup i, the chart is examined to see whether any part of the graph *prior to subgroup i* falls outside the legs of the mask. If *any* prior point on the graph crosses one of the legs, then an 'out of control' signal is given *at subgroup i*. It must next be determined whether the process has shifted upward or downward by looking at which leg of the mask crossed the CUSUM graph. *If the graph crosses the lower leg, then an upward shift is indicated. Crossing the upper leg signals a downward shift.* These signals may at first

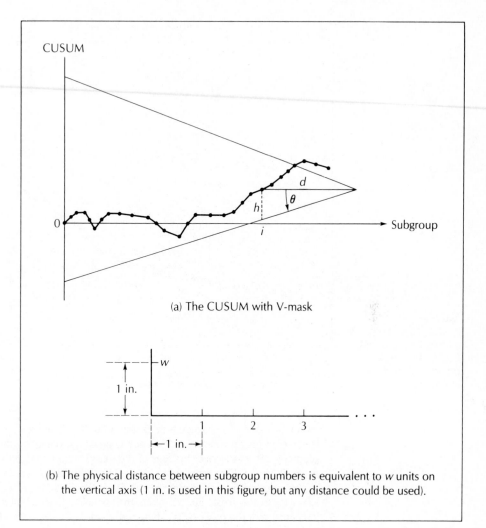

CUSUM

0 ──────────────────────────────────► Subgroup

(a) The CUSUM with V-mask

w

1 in.

1   2   3   . . .

←1 in.→

(b) The physical distance between subgroup numbers is equivalent to *w* units on
the vertical axis (1 in. is used in this figure, but any distance could be used).

FIGURE 7.8
CUSUM graph with
V mask at subgroup *i*

seem backward from the usual threshold control chart scheme, since the *upper* leg of the CUSUM is equivalent to the *lower* control limit on a threshold control chart and, vice versa, the *lower* leg corresponds to the *upper* control limit. With practice, this interpretation becomes more natural.

EXAMPLE 7.7   With the same reference value $k = 0.5\hat{\sigma} = 0.101533$ and decision interval $h = 5\hat{\sigma} = 1.10533$ as in the tabular CUSUM procedure of Example 7.6, Figure 7.9 shows a CUSUM plot of the data in Table 7.3 created by the *NWA Quality Analyst* software. The graph appears to use the common scale factor of $w = 2\hat{\sigma}$, since the angle at the V of the mask is about 28°. Also, each CUSUM point at which an 'out of control' signal occurs is marked by a # symbol. Notice that the # symbols occur at exactly the same subgroups as those identified by

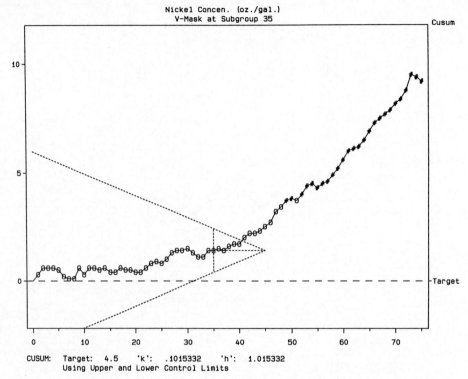

FIGURE 7.9
CUSUM plot of data
in Table 7.3 from *NWA
Quality Analyst*
software; V mask at
subgroup 35

the tabular CUSUM procedure in Example 7.6. The V mask shown in Figure 7.9 is centered over position $i = 35$. Since none of the points on the graph prior to subgroup 35 crosses over the legs of the mask, there is no 'out of control' signal given at $i = 35$. For comparison, Figure 7.10 shows the same chart with the V mask plotted over subgroup $i = 49$. Since the CUSUM graph crosses the lower leg of this mask, an 'out of control' signal is given at subgroup 49, indicating that the process has shifted upward.

As was the case with EWMA charts, successive CUSUM points are highly correlated, so the extended control chart rules from Section 6.4 do not apply to CUSUM charts.

7.8   NARROW-LIMIT GAGE CHARTS

MEASUREMENT BY GAGING

Since early times practitioners have used some sort of gaging to assess product quality. To this day gages play an important role in day-to-day operations, from inspection and testing to process control. Section 3.4 introduced the concept of

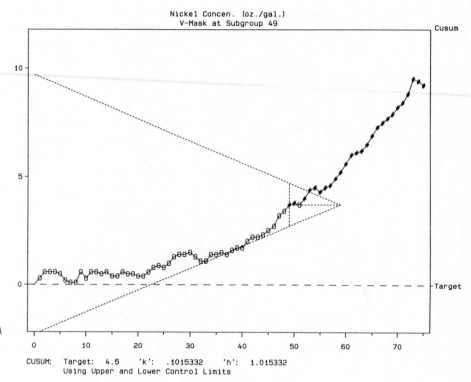

Nickel Concen. (oz./gal.)
V-Mask at Subgroup 49

CUSUM: Target: 4.5 'k': .1015332 'h': 1.015332
Using Upper and Lower Control Limits

FIGURE 7.10
CUSUM plot of data
in Table 7.3 from *NWA
Quality Analyst*
software; V mask at
subgroup 49

gage measurements and listed some of their applications in quality control. We turn now to a particular application, the use of gages for controlling variables measurements. The charts presented here can be considered alternatives to any of the variables control charts discussed in this chapter.

Gages are often used to assess whether or not items conform to their specification limits. When used in this fashion, two gages can check two-sided tolerances, while only one gage is needed for checking one-sided tolerances. As an example, consider the case of a product characteristic that has a two-sided tolerance. One gage is set equal to the upper specification limit (USL), the other is set at the lower limit (LSL), and then finished items are compared to these gages. If the item's dimension falls between the two gages, it is within specifications. Otherwise, if its dimension is smaller than the LSL or larger than the USL, it is classified as nonconforming. The speed with which items can be gaged and classified makes this form of measurement attractive in production environments.

HISTORY OF NLG

Gaging to the specification limit(s) is a very common practice because it is so easy to accomplish and understand. However, this form of gaging is used primarily for the *detection*, not *prevention*, of nonconforming items. In order to use the concept of gaging for prevention-oriented process control (see Section 2.5), a different approach is required in which the gages are set at values *narrower* than the product specification limits. With **narrow-limit gaging** (NLG), many more items will necessarily fall outside the gage limits, but this information gives a much earlier warning of process changes than if gages are set at the specification limits. Of course, the items that exceed these gage limits are not considered conforming or nonconforming, since narrowed limits are not the same as the specification limits. Thus, as with Shewhart charts, the purpose of narrow-limit gaging is to detect special causes, not to assess compliance to specifications.

The first documented use of NLG was by D. J. Desmond in England in 1942, who also coined the term **compressed limit gaging** when referring to the method. Little was published regarding NLG during the war years, but soon afterward, Stevens (1948) was motivated to deliver a paper to the Royal Statistical Society describing the method. Stevens's paper mentions a privately circulated manuscript by Temple in 1943, which stated:

> A control chart for defectives [i.e., as determined by go and not-go gauges] would be very little less sensitive than one using averages of a sample of the same size.

In other words, process control using NLG offers very nearly the same sensitivity of control that Shewhart charts do. In fact, since gaging a few extra items is so economical to do, the NLG method can easily be made more sensitive than the Shewhart charts.

In the United States, the primary proponents of NLG have been Mace, Ott, and Mundel (Mace 1966, Ott and Mundel 1954, and Ott 1975). Since the mid-1950s, research on NLG has consisted of a relative handful of articles, including those by Ladany (1976), Farnum and Stanton (1986, 1991), and Schneider and O'Cinneide (1987). Various versions of the NLG concept have been rediscovered over the years. The most familiar of these is the method of *PRE-Control,* whose development is attributed to Carter, Purcell, Satterthwaite, and Shainin in 1952 (see Shainan and Shainan 1989), which applies the NLG concept to processes whose 3-sigma limits are assumed to be within the allowed specification limits (called capable processes). Other incarnations include *zone control* charts (Jaehn 1987a, 1987b, 1989) and *target control* charts.

NLG CHART CALCULATIONS

Narrow-limit gaging is used only for characteristics that have a *two-sided tolerance.* The two gages create three zones into which product or process measurements can fall: (I) below the lower gage, (II) between the gages, and (III) above the upper gage (Figure 7.11). As a process control measure, the numbers of gaged items that fall in these three regions give information about both the process mean and variability. Most simply, an imbalance in the number of items that fall in the outer two regions, I and III, indicates a shift in the process mean, whereas changing numbers of items in the middle zone, II, point to changes in the process variability (standard deviation). The most important consideration in applying this technique

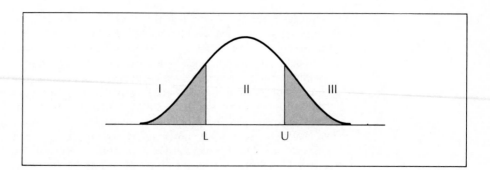

FIGURE 7.11
Placement of
narrow-limit gage
values

is to set the two gages so as to gain the maximum information concerning the values of the process mean and standard deviation.

For a sample of size n, we use the following notation:

n_L = number of items smaller than the lower gage limit, L

n_U = number of items larger than the upper gage limit, U

$f_L = n_L/n$ = fraction of sample below the lower gage

$f_U = n_U/n$ = fraction of sample above the upper gage

z_L = lower $f_L \cdot 100\%$ percentage point of the standard normal distribution. (For $f_L = 0.05$, this is the lower 5% point, which is -1.645.)

z_U = upper $f_U \cdot 100\%$ percentage point of the standard normal distribution. (For $f_U = 0.05$, this is the upper 5% point, which is 1.645.)

For a sample of size n, estimates of the process mean and standard deviation are given by (Farnum and Stanton 1986, 1991, Schneider 1986, and Stevens 1948)

$$\hat{\mu} = U - \frac{z_U(U - L)}{z_U - z_L}$$

$$= L - \frac{z_L(U - L)}{z_U - z_L}$$

(7.25)

and, provided z_U and z_L are both finite (i.e., f_L and f_U are nonzero),

$$\hat{\sigma} = \frac{U - L}{z_U - z_L}$$

(7.26)

EXAMPLE 7.8   To illustrate the method, suppose that a process is to be controlled to a target dimension of $\mu_0 = 1.600$ in. For simplicity, we assume that the process is just capable of maintaining a tolerance of ± 0.009 in. (i.e., the process spread, $\pm 3\sigma$, just fits within the specification limits). The process standard deviation that corresponds to this condition is $\sigma = 0.003$ in. For jointly controlling the average and standard deviation, *gages narrowed to about 0.7 of a standard deviation on either side of the target are recommended.* These limits give probabilities of roughly 25%:50%:25% for regions I, II, and III, respectively. Since $0.7\sigma = 0.7 \times 0.003 = 0.0021$ in., values of about $U = 1.602$ and $L = 1.598$ in. are reasonable to use for the narrow-limit gages. Finally, suppose that it costs no more (time and dollars) to gage a sample of ten articles than it does to measure a sample of five (this is not an unrealistic assumption in most instances), so that a sample (subgroup) size of $n = 10$ is used. The control chart for the mean is run on the difference between the region I and III counts—that is, $d = n_U - n_L$. The control limits for d are selected to be roughly equivalent to the 3σ limits for X charts; that is, they are chosen so that the probability of getting an "out of control" signal from a process that is in control is about 0.27%. Stevens (1948) gives these limits for a sample of size 10 and 25%:50%:25% probabilities for regions I, II, and III, respectively, as $d \geq +7$ and $d \leq -7$. (The test can also be run as a z test, but doing so sacrifices the simplicity and immediacy of using counts.) Suppose a sequence of three samples occurs as shown below:

Sampling period #1:   3 items below the lower gage
                      5 items between the gages
                      2 items above the upper gage

Then, $n_L = 3$, $n_U = 2$, and $d = -1$, and no 'out of control' signal is given.

Sampling period #2:   4 items below the lower gage
                      5 items between the gages
                      1 item above the upper gage

Here, $n_L = 4$, $n_U = 1$, and $d = -3$, and again no alarm is sounded.

Sampling period #3:   8 items below the lower gage
                      1 item between the gages
                      1 item above the upper gage

In this case, $n_L = 8$, $n_U = 1$, and $d = -7$, so an 'out of control' signal does occur for this subgroup. Since both f_L and f_U are nonzero in sample #3, estimates of the process mean and standard deviation can be calculated using tables of the normal distribution:

$$L = 1.598 \qquad U = 1.602$$

$$f_L = \frac{8}{10} = 0.80 \qquad f_U = \frac{1}{10} = 0.10$$

$$z_L = +0.842 \qquad z_U = +1.282$$

$$U - L = 1.602 - 1.598 = 0.004$$

$$z_U - z_L = 1.282 - 0.842 = 0.440$$

$$\hat{\mu} = U - \frac{z_U(U - L)}{z_U - z_L}$$

$$= 1.602 - \frac{1.282 \times 0.004}{0.440} = 1.602 - 0.0117 = 1.590$$

$$\hat{\sigma} = \frac{U - L}{z_U - z_L} = \frac{0.004}{0.440} = 0.009$$

For further reading on the NLG method and its implementation, we refer the reader to Farnum and Stanton (1986, 1991) and Schneider and O'Cinneide (1987).

EXERCISES FOR CHAPTER 7

7.1   For a given process, the control limits of the $\bar{x}$ chart become closer together as the subgroup size n is increased (i.e., the A_2 factor shrinks as n increases; see Appendix 1). Does this mean that $\bar{x}$ charts based on large subgroup sizes are more likely to have some subgroups that fall outside their control limits than charts based on smaller subgroup sizes?

7.2   Instead of constructing $\bar{x}$ and R charts for 30 subgroups of size 4, a friend suggests the alternative of simply finding the averages of all 30 subgroups and then creating an *individuals* chart for these averages. Explain what is wrong with this procedure.

7.3   An R chart is used to monitor the combined output of six identical machines. For the last 25 samples of size 5, the R chart appears as follows:

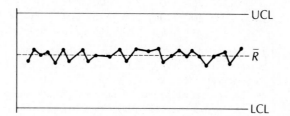

(a)   Does the chart indicate that the process is in statistical control?
(b)   Explain what could be happening to cause the R chart to have this form.

7.4   Prior to shipment, thrust washers supplied to the automotive industry go through a five-step process of sanding, stripping, punching, and baking. The important quality characteristics of each washer are its thickness, inside diameter, and outside diameter.[6] The following data represent 25 subgroups of five washers, with each subgroup sampled from a different production lot:

Lot #	x_1	x_2	x_3	x_4	x_5
1	0.0767	0.0771	0.0774	0.0768	0.0776
2	0.0771	0.0771	0.0776	0.0774	0.0776
3	0.0773	0.0773	0.0772	0.0776	0.0772
4	0.0772	0.0776	0.0779	0.0770	0.0778
5	0.0769	0.0777	0.0776	0.0772	0.0775
6	0.0767	0.0772	0.0773	0.0774	0.0772
7	0.0772	0.0776	0.0773	0.0775	0.0766
8	0.0775	0.0773	0.0770	0.0769	0.0771
9	0.0774	0.0772	0.0773	0.0775	0.0770
10	0.0774	0.0773	0.0777	0.0772	0.0776
11	0.0770	0.0774	0.0774	0.0773	0.0772
12	0.0780	0.0775	0.0767	0.0773	0.0775
13	0.0764	0.0775	0.0776	0.0774	0.0777
14	0.0781	0.0772	0.0772	0.0773	0.0775
15	0.0775	0.0772	0.0776	0.0774	0.0772
16	0.0773	0.0769	0.0776	0.0773	0.0769
17	0.0770	0.0772	0.0775	0.0773	0.0775
18	0.0773	0.0775	0.0778	0.0771	0.0774
19	0.0773	0.0775	0.0774	0.0774	0.0773
20	0.0776	0.0770	0.0771	0.0777	0.0779
21	0.0779	0.0768	0.0769	0.0771	0.0770
22	0.0780	0.0771	0.0776	0.0774	0.0779
23	0.0769	0.0771	0.0773	0.0771	0.0769
24	0.0773	0.0771	0.0780	0.0773	0.0773
25	0.0772	0.0777	0.0773	0.0767	0.0773

---

[6] See Chaudhry and Higbie (1990).

(a) Construct $\bar{x}$ and R charts for these data.
(b) Do the charts constructed in part (a) indicate that the washer stamping process is in statistical control?
(c) Using the centerline from the R chart, obtain an estimate of the standard deviation of washer thicknesses.
(d) Suppose the specification limits for washer thickness are 0.0755 and 0.0795 in. Using the result in part (a), approximately what proportion of the washers have thicknesses that exceed one of the specification limits?

7.5   Galvanized coatings on pipes protect them from rust. A certain coating process for large pipes calls for an average coating weight of 200 lb per pipe.[7] The lower specification limit is 180 lb per pipe, but there is no upper specification, since extra coating material only provides more protection for the pipe. The following data show the coating weights of 30 pipes sampled at a rate of one per shift (read across):

216   202   208   208   212   202   193   208   206   206
206   213   204   204   204   218   204   198   207   218
204   212   212   205   203   196   216   200   215   202

(a) Construct an individuals chart for these data.
(b) Does the chart in part (a) indicate that the coating process is in statistical control?
(c) From the chart in part (a), how is the process performing from the point of view of the customer? How does the producer view these results?

7.6   For the data of Exercise 7.5 , construct a CUSUM chart using $h = 5\hat{\sigma}$, $k = 0.5\hat{\sigma}$, and $w = 2\hat{\sigma}$ (estimate σ from the data). Interpret the resulting chart.

7.7   For the data of Exercise 7.5, construct an EWMA chart using a parameter of $\lambda = 0.20$. Interpret the resulting chart.

7.8   Certain manufactured parts are required to have a length of 0.254 in. Twenty subgroups of three parts each were used to form $\bar{x}$ and R charts for the part lengths. To simplify the data-gathering process, the measurements were reported as deviations from the nominal length in units of 0.001 (e.g., a recorded value of -3 refers to a measured length of 0.251). In this format, the data on the 20 subgroups are given here.

Subgroup number	d_1	d_2	d_3
1	4	0	−2
2	−1	−3	−1
3	−2	4	2
4	−2	−2	1
5	0	−2	2
6	−1	0	2
7	−3	3	3
8	−2	−3	1
9	−3	1	3

---

[7] See Weaver (1990).

10	3	1	1
11	−1	3	0
12	−2	−1	4
13	4	−1	3
14	−3	3	2
15	2	0	3
16	−3	1	−1
17	−2	2	1
18	−3	2	−1
19	−1	−2	0
20	1	−2	−1

Construct $\bar{x}$ and R charts for these data. From the extended list of 'out of control rules' in Section 6.4, are there any indications that this process is not in control?

7.9   Construct a CUSUM chart for the data of Exercise 7.8. Use $h = 5\hat{\sigma}$, $k = 0.5\hat{\sigma}$, and $w = 2\hat{\sigma}$ (estimate σ from the range chart). Interpret the resulting chart and compare these results to those of Exercise 7.8.

7.10   Construct an EWMA chart for the data of Exercise 7.8 Use a parameter of $\lambda = 0.30$. Compare the resulting chart to those in Exercises 7.8 and 7.9.

7.11   In an EWMA chart, explain the effect on the chart of using a value of λ that is very close to 1. What is the effect of using a λ that is close to 0?

7.12   Explain why the extended 'out of control' rules from Section 6.4 do not apply to CUSUM or EWMA charts.

7.13   What is the minimum value of the process coefficient of variation that results in a *positive* lower control limit, LCL $= \bar{\bar{x}} - A_2 \bar{R}$, of an $\bar{x}$ chart? (Use $\bar{R}/d_2$ to estimate the process standard deviation and use $\bar{\bar{x}}$ to estimate the process average.)

7.14   A drilling tool that machines metal parts eventually wears out and periodically must be replaced. If the hole diameters drilled by this machine are monitored on a control chart, describe the type of pattern you would expect to see in the points plotted on the chart.

7.15   Process data that do not closely follow a normal distribution must sometimes be transformed so that they appear normal. One popular transformation used for positive data is to take logarithms of the original data. For a given set of positive measurements, suppose that two $\bar{x}$ charts are constructed, one from the raw data and one from the logarithms (any base) of the raw data. If a point falls beyond the 3-sigma control limits on the $\bar{x}$ chart for the logarithms, must the corresponding point fall outside the control limits on the chart of the raw data?

7.16   In a process that produces molded plastic containers, hourly samples of size 3 were used to create control charts for a critical dimension. For the most recent 20 samples, the measurements were:

Hour	Measurements		
1	.36	.39	.36
2	.33	.35	.30
3	.51	.41	.42
4	.42	.37	.34
5	.39	.38	.38
6	.33	.41	.45
7	.43	.39	.41
8	.41	.32	.32
9	.37	.42	.36
10	.26	.42	.32
11	.36	.32	.36
12	.38	.47	.35
13	.29	.45	.39
14	.44	.38	.43
15	.38	.37	.37
16	.31	.43	.38
17	.39	.49	.35
18	.43	.36	.38
19	.40	.45	.32
20	.40	.40	.32

(a) Construct a range chart for this data. Is there any 'out of control' condition indicated on the chart?

(b) Construct an $\bar{x}$ chart for this data and check for any signs of special causes.

7.17   Each hour a 3-ft length is cut from a continuous extruded sheet of plastic. The weights of these cross sections are used to monitor the uniformity of the extrusion process. The weights (in pounds) of the last 20 cross sections are:

Hour	Weight (lb)
1	169
2	164
3	169
4	178
5	178
6	183
7	181
8	195
9	184
10	179
11	216
12	170
13	168
14	182
15	177
16	164
17	182
18	148
19	176
20	162

Construct an individuals chart for this data. Are there any signs of the presence of special causes?

7.18 Sand is an important component in a process that produces molds for cylinder blocks (Krishnamuoorthi 1990). Foundry workers have determined that the compactibility of the sand is of key importance in making goods molds. Compactibility is measured as the percent reduction in volume in a fixed amount of sand after being compacted with a standard force. Because testing sand samples is time consuming, an individuals chart is used to monitor the compactibility. The following data represent compactibility measurements (in percent) from 30 successive samples taken from the molding process (read across and down):

```
44   39   49   41   38   44   40   43   40   41
33   31   30   46   45   48   45   42   40   45
44   41   49   38   41   40   48   42   36   39
```

(a) Construct an individuals chart for this data.
(b) Does the chart indicate that there are any problems with the sand compactibility?

7.19 CUSUM charts have many applications in chemical industries, in which numerous chemical characteristics must be maintained close to specified target levels. To ensure the chemical purity of a commercial organic chemical, measurements of the level of a certain intermediate chemical material are taken every 4 hr. Data from 22 samples appear below:

Sample number	Level
1	15.3
2	15.7
3	14.4
4	14.0
5	15.2
6	15.8
7	16.7
8	16.6
9	15.9
10	17.4
11	15.7
12	15.9
13	14.7
14	15.2
15	14.6
16	13.7
17	12.9
18	13.2
19	14.1
20	14.2
21	13.8
22	14.6

(a) Given that the target chemical level is 15 and that the process standard deviation is known to be about 1, construct a CUSUM chart for this data. Use $k = 0.5\sigma$ and $h = 5.0\sigma$.

(b) Construct an individuals chart of the data and compare its performance to the CUSUM chart in part (a).

(c) Using a parameter value of $\lambda = .20$, construct an EWMA chart of this data and compare its performance to the CUSUM chart in part (a).

REFERENCES FOR CHAPTER 7

Chaudry, S. S., and J. R. Higbie. 1990. "Quality Improvement Through Statistical Process Control." *Quality Engineering* 2 (no. 4):411–419.

Cryer, J. D., and T. P. Ryan. 1990. "The Estimation of Sigma for an X Chart: $\overline{MR}/d_2$ or s/c_4?" *Journal of Quality Technology* 22 (no. 3):187–192.

Ewan, W. D. 1963. "When and How to Use Cu-Sum Charts." *Technometrics* 5 (no. 1):1–22.

Farnum, N. R., and L. W. Stanton. 1986. "Using Counts to Monitor a Process Mean." *Journal of Quality Technology* 18 (no. 1):22–28.

Farnum, N. R., and L. W. Stanton. 1991. "Narrow Limit Gauging—Go or No Go?" *Quality Engineering* 3 (no. 3):293–307.

Gardner, E. S. 1985. "Exponential Smoothing: The State of the Art." *Journal of Forecasting,* 4:1–38.

Grant, E. L., and R. S. Leavenworth. 1988. *Statistical Quality Control,* 6th ed., pp. 315–318. New York: McGraw-Hill.

Jaehn, A. 1987a. "Improving QC Efficiency with Zone Control Charts." *American Society for Quality Control, Chemical and Processing Industry Division News.* September, pp. 1–2.

Jaehn, A. 1987b. "Zone Control Charts—SPC Made Easy." *Qual. Prog.* 20 (no. 10):51–53.

Jaehn, A. 1989. "All Purpose Chart Can Make SPC Easy." *Qual. Prog.* 12 (no. 2):112.

Krishnamuoorthi, K. S. 1990. "On Assignable Causes That Cannot Be Eliminated—An Example from a Foundry." *Quality Engineering* 3 (no. 1):41–47.

Ladany, S. P. 1976. "Determination of Optimal Compressed Limit Gaging Sampling Plans. *Journal of Quality Technology* 8 (no. 4):225–231.

Mace, A. E. 1966. "The Use of Limit Gauges in Process Control." *Industrial Quality Control.* 8 (no. 4):24–31.

Ott, E. R. 1975. *Process Quality Control.* New York: McGraw-Hill.

Ott, E. R., and A. B. Mundel. 1954. "Narrow Limit Gauging." *Industrial Quality Control* 10 (no. 5):2–9.

Page, E. S. 1954. "Continuous Inspection Schemes." *Biometrika* 41:100–115.

Roberts, S. W. 1959. "Control Chart Tests Based on Geometric Moving Averages." *Technometrics* 1:239–250.

Ryan, T. P. 1989. *Statistical Methods for Quality Improvement.* New York: Wiley.

Schneider, H. 1986. *Truncated and Censored Samples from Normal Populations.* New York: Marcel Dekker.

Schneider, H., and C. O'Cinneide. 1987. "Design of CUSUM Control Charts Using Narrow Limit Gauges." *Journal of Quality* 19 (no. 2):63–68.

Shanin, D., and P. Shanin. "PRE-CONTROL Versus $\overline{X}$ and R Charting: Continuous or Immediate Quality Improvement." *Quality Engineering* 1 (no. 4):419–429.

Wadsworth, H. M., K. S. Stephens, and A. B. Godfrey. 1986. *Modern Methods for Quality Control and Improvement.* New York: Wiley.

Weaver, W. R. 1990. "The Foreman's View of Quality Control." *Quality Engineering* 3 (no. 2):257–280.

8

PROCESS CAPABILITY ANALYSIS

The degree to which process measurements satisfy their specification limits can be quantified in various ways. This chapter discusses several measures of process capability, how these measures are used, and the impact of the measurement system on their interpretation.

CHAPTER OUTLINE

8.1   INTRODUCTION

After special causes have been identified and eliminated, a process is said to be in a state of **statistical control.** One of the by-products of attaining statistical control is that a process then becomes *predictable,* and it makes sense to evaluate its ability to satisfy requirements that are placed on it. If statistical control is *not* achieved, then the process average and standard deviation are unstable and, correspondingly, calculations based on data from such a process are unreliable.

Process capability is judged by comparing process performance with process requirements. Since meeting specification limits is one of the most basic requirements, capability analyses usually involve the specification limits somewhere in their calculations. Thus, before one proceeds with a capability study, it is important to verify that the specification limits have been accurately determined.

To assess capability, it is necessary to use *data* to describe how a process is performing. Usually the data come from the operation of a control chart on the process, so the subgroup values must be translated into estimates of what the entire process is doing. This requires that an *assumption* be made about the type of distribution that the measurements follow. Since many process characteristics tend to follow a normal distribution (see Section 5.6), most capability measures tacitly assume the normal distribution in their definitions. When capability measures are used, it is good to keep in mind that some processes do *not* follow normal distributions and that the data should be examined to determine which distribution is best to use.

8.2   ESTIMATING PROCESS VARIATION

Because statistical control is a prerequisite for capability studies, some form of control chart data is usually available from which to estimate process variation. For variables data, R or s charts are the most frequently used, so estimates of the process standard deviation are generally derived from either $\hat{\sigma} = \overline{R}/d_2$ or $\hat{\sigma} = \overline{s}/c_4$ (see Sections 7.2 and 7.3). These two estimates assume that the underlying process is normally distributed, which is often a good assumption. In cases where other distributions describe the measurements, different methods of estimating σ are needed (see Example 8.2).

Another method of estimating process variation is to ignore the grouping of data in the control chart and to simply calculate the sample standard deviation of the combined data in the subgroups. For example, rather than calculating $\overline{R}/d_2$ from, say, 20 subgroups of size 5, one could calculate s for the combined set of 100 measurements. In theory, *if the process is in control,* the combined-subgroup method is as good as either the $\overline{R}/d_2$ or the $\overline{s}/c_4$ method. However, if a process is not in control, s is usually much larger than $\overline{R}/d_2$ or $\overline{s}/c_4$ because it reflects the shifts *between* the subgroup averages along with the smaller within-subgroup variation. Therefore, some practitioners prefer to use *s* because they believe it gives a truer picture of **process performance** than within-subgroup estimates do. Both methods are illustrated in the ensuing examples.

After the standard deviation and the mean are estimated, it has become traditional to refer to the 3-sigma distance on either side of the mean as the **process spread** (see Section 6.1). The process spread is defined to be $\hat{\mu} \pm 3\hat{\sigma}$, where $\hat{\mu}$ and $\hat{\sigma}$ are estimates obtained from process data. This definition arose from the practice of assuming that all process distributions are essentially *normal* distributions, most of whose probability lies within 3 sigmas of the mean (see Section 3.3). Even though this may not necessarily be the case for every process, we continue to use the phrase 'process spread' to refer to the $\pm 3\hat{\sigma}$ interval around the mean. Doing so simplifies the description of capability indexes and yields simple indexes that apply to most process data arising in practice. Furthermore, since capability measures can be defined and interpreted relative to *any* definition of process spread, the definition chosen does not fundamentally affect the behavior of these indexes.

8.3   ESTIMATING NONCONFORMANCE RATES

With estimates of σ and μ in hand, calculations similar to those of Example 5.14 yield **nonconformance rates**—that is, the proportion of process measurements that fall above the upper specification limit or below the lower limit. In Example 5.14, the measurements were assumed to follow a *normal distribution,* so that nonconformance rates are estimated by

$$\text{proportion above USL} = P(X > \text{USL}) = P\left(z > \frac{\text{USL} - \hat{\mu}}{\hat{\sigma}} \right) \tag{8.1}$$

and

$$\text{proportion below LSL} = P(X < \text{LSL}) = P\left(z < \frac{\text{LSL} - \hat{\mu}}{\hat{\sigma}} \right) \tag{8.2}$$

The proportions in equations (8.1) and (8.2) are usually converted to percentages to facilitate understanding. The shaded regions in Figure 8.1 correspond to nonconformance rates (percentages) for normally distributed data.

If it is thought that the process data follows some distribution *other* than the normal, then equations (8.1) and (8.2) are not appropriate. In such cases, the proportions $P(X > \text{USL})$ and $P(X < \text{LSL})$ are calculated by referring to the particular probability distribution that the process measurement X is thought to have. For example, Figure 8.2 shows process measurements that follow an exponential distribution. In this case, the nonconformance rates are calculated with the aid of equation (5.12).

EXAMPLE 8.1   The ignition key data of Example 7.1 showed only one sign of a potential lack of control—namely, the 'eight points on the same side of the centerline' that

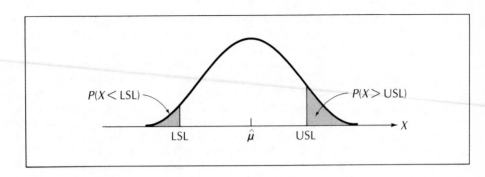

FIGURE 8.1
Percent
nonconforming
(exceeding
specification limits)
for normally
distributed process
data

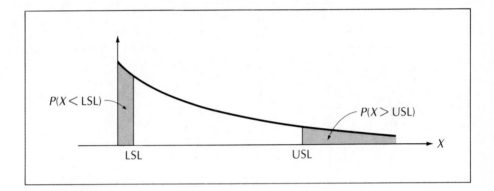

FIGURE 8.2
Percent
nonconforming for
exponentially
distributed process
data

occurred at subgroups 8–15. After an investigation, suppose that no assignable causes are found and that the process is deemed to be in control. Suppose further that the specification limits for the groove dimension are 0.0072 ± 0.0020 in. and that nonconformance rates are to be estimated. The process variation can be estimated by $\hat{\sigma} = \overline{R}/d_2 = 0.002406/2.326 = 0.00103$, where $\overline{R}$ is obtained from the R chart for the data (see Table 7.1). Alternatively, the variation can be estimated from the s chart (Table 7.2) as $\hat{\sigma} = \overline{s}/c_4 = 0.0009672/0.9400 = 0.00103$.[1] Next, if we assume that the measurements follow a *normal* distribution, equations (8.1) and (8.2) yield

$$P(X > USL) = P\left(z > \frac{USL - \hat{\mu}}{\hat{\sigma}}\right) = P\left(z > \frac{0.0092 - 0.007966}{0.00103}\right)$$

$$= P(z > 1.20) = 0.1151$$

or 11.51%, and

$$P(X < LSL) = P\left(z > \frac{LSL - \hat{\mu}}{\hat{\sigma}}\right) = P\left(z < \frac{0.0052 - 0.007966}{0.00103}\right)$$

$$= P(z < -2.69) = 0.0036$$

[1] Normally only one variation control chart is used on a process, so there is usually no question about which estimate to use; also, for stable processes, $\overline{R}/d_2$ and $\overline{s}/c_4$ will be close but may not always be identical.

or, 0.36%. Notice that the process mean is estimated by the centerline of the $\bar{x}$ chart and that the nominal (target) dimension of 0.0072 in. is *not* used to estimate the process average.

The total percentage of nonconforming product is 11.51% + 0.36% = 11.87%, which by most standards is unacceptably high. Thus, statistical control alone does not guarantee good capability. In Figure 8.3, the histogram of the groove dimension data of Table 7.1 clearly shows one reason for the poor capability, a process mean ($\hat{\mu} = 0.007966$) that has shifted above the target dimension 0.0072 (by about $0.74\hat{\sigma}$). However, the problem is even worse because, *even if the process was perfectly centered* (i.e., $\hat{\mu} = 0.0072$), the nonconformance rate would shrink to only 5.24% (assuming that $\hat{\sigma}$ remains at 0.00103). At this point, a study of the process variation is needed, the goal being to institute process changes that will reduce the variance (and thereby increase the capability).

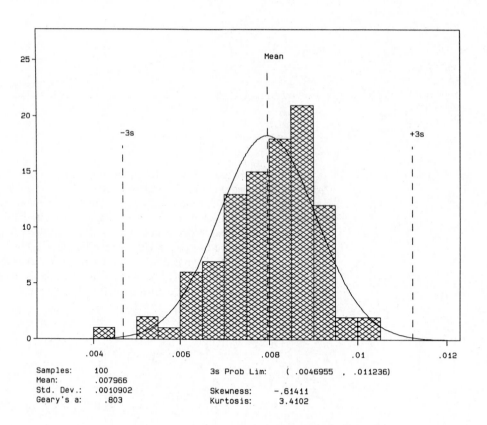

FIGURE 8.3
Histogram of groove dimension data in Table 7.1

Samples:	100	3s Prob Lim:	(.0046955 , .011236)
Mean:	.007966		
Std. Dev.:	.0010902	Skewness:	-.61411
Geary's a:	.803	Kurtosis:	3.4102

EXAMPLE 8.2
In Example 5.17, a certain type of electronic component had an estimated average life of $\hat{\mu}_x = 1{,}000$ hr. Suppose that the process that creates these components is in statistical control and that a capability estimate is required,

with a lower specification limit of 100 hr.[2] The exponential distribution is thought to best describe the process measurements, so equation (5.12) is used to estimate the percentage of nonconforming components;

$$P(X < \text{LSL}) = 1 - P(X > \text{LSL}) = 1 - e^{-\lambda(\text{LSL})} = 1 - e^{-10^{-3}(100)} = 0.0952$$

where $\lambda = 10^{-3}$ is the estimated failure rate from equation (5.11). Note that if the *normal* curve had mistakenly been used to calculate the nonconformance rate, the proportion that falls below the LSL would have been

$$P(X < \text{LSL}) = P\left(z < \frac{\text{LSL} - \hat{\mu}}{\hat{\sigma}}\right) = P\left(z < \frac{100 - 1,000}{1,000}\right)$$

$$= P(z < -0.90) = 0.1841$$

which differs substantially from 0.0952.

One further caution should be noted. Assuring that this process was in statistical control required the use of a control chart. From the control charts, the estimates $\hat{\sigma} = \bar{R}/d_2$ and $\hat{\sigma} = \bar{s}/c_4$ used to estimate process variation are valid only for *normally distributed data* (see Section 3.3), so it would not be appropriate to use them for this example or any other nonnormal process data.

8.4 CAPABILITY INDEXES

THE C_p, C_{pl}, C_{pu}, AND C_{pk} INDEXES

PROCESS POTENTIAL   Nonconformance rates were used to evaluate process capability in Section 8.3. If we use the concept of 'process spread' from Section 8.2, it is possible to go a step further by calculating numerical measures of process capability in the following way. Assume for the moment that process measurements approximately follow a normal distribution, so the process spread should encompass a range of about 6σ. This range can be called the 'actual process spread,' while the distance between the specification limits is the 'allowable process spread.' From these, a **process capability index,** denoted C_p, can be defined by

$$C_p = \frac{\text{allowable spread}}{\text{actual spread}} = \frac{\text{USL} - \text{LSL}}{6\hat{\sigma}} \tag{8.3}$$

where $\hat{\sigma}$ is an estimate of the standard deviation of the process measurements.

The C_p index is interpreted as follows. If $C_p = 1.0$, then the process is said to be 'marginally capable' of meeting its specification limits. Thus, it is desirable that values of C_p exceed 1.0, since then the likelihood is higher that the measurements

---

[2]Lifetimes have a one-sided tolerance; a lower specification limit is necessary because components must operate for some minimum acceptable duration, but no upper limit is needed because one does not mind if components last indefinitely.

will be able to stay within the specification limits. A C_p that exceeds 1.33 (i.e., an 8σ spread fits within the specification limits) is usually considered very good and is commonly used as a goal by many companies. On the other hand, C_p's less than 1.0 imply that a process is not capable of meeting its specifications. Figure 8.4 illustrates typical values of the C_p index along with the associated (normal) distributions of measurements from the process. The simple interpretations in Figure 8.4 can be greatly complicated by the presence of measurement errors, undetected process shifts, and assumptions about the underlying process distribution. *In this section, we discuss capability indexes in their ideal setting, where processes are in control, measurement errors are negligible, and process distributions are approximately normal.* Section 8.5 discusses what happens when these conditions are not met.

The C_p index is one of four measures, originally developed in Japan, which are now routinely used in quality control programs. These indexes derive their usefulness from the fact that they convey much information in a simple fashion. Capability indexes also have the advantage of being *unitless* measures, which allows them to be used to compare two entirely different processes. For example, if copper

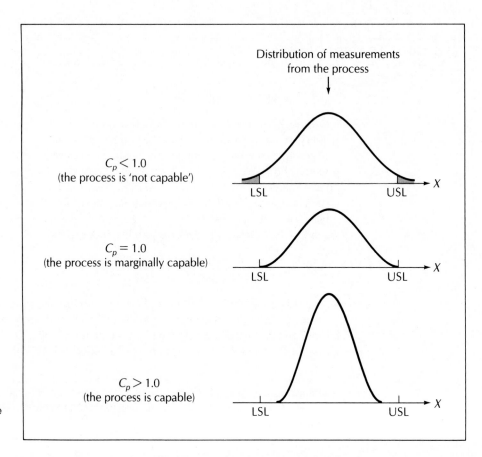

FIGURE 8.4
Interpretation of the process capability index, C_p

plating thicknesses (in inches) from a chemical plating process have a C_p of 0.81, while resistance measurements (in ohms) of electronic components have a C_p of 2.30, then one can conclude that the electronic component process is the more capable of the two, even though their measurement units, inches and ohms, are unrelated.

PROCESS PERFORMANCE.   The C_p index does not take into account the location of the process, only its *potential* for meeting specifications. Figure 8.5 illustrates this by showing two processes, both with C_p's of 2.0, one centered between the specification limits and the other located closer to the upper specification limit. The latter process has the potential to be capable because its C_p is greater than 1.0, but it will realize this potential only if the process average can be moved closer to the center of the specification range.

A closely related index that *does* take the process mean into account is the C_{pk} index:

$$C_{pk} = \text{minimum}\left[\frac{\text{USL} - \hat{\mu}}{3\hat{\sigma}}, \quad \frac{\hat{\mu} - \text{LSL}}{3\hat{\sigma}}\right] \qquad (8.4)$$

For normally distributed measurements, $\hat{\mu}$ is the centerline of the $\bar{x}$ chart and $\hat{\sigma}$ equals either $\bar{R}/d_2$, $\bar{s}/c_4$, or perhaps the combined-subgroup estimate s (described

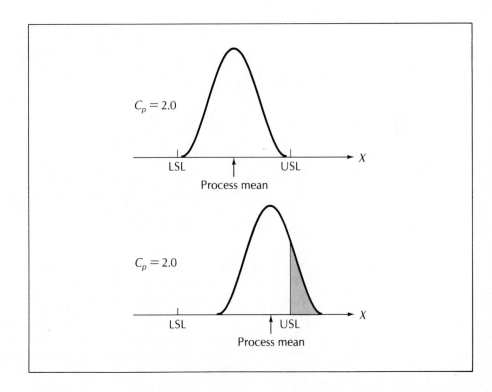

FIGURE 8.5
The C_p index measures process potential but does not consider the process location

in Section 8.2). Other estimates are used for nonnormal distributions. The k in the subscript of C_{pk} refers to the so-called k factor:

$$k = \frac{|(USL + LSL)/2 - \hat{\mu}|}{(USL - LSL)/2} \tag{8.5}$$

which measures the extent to which the process location ($\hat{\mu}$) differs from a target value midway between the specification limits. It can be shown that k always lies between 0 and 1 and that the C_p and C_{pk} indexes are related by the formula

$$C_{pk} = (1 - k)C_p \tag{8.6}$$

Since $0 \leq k \leq 1$, this formula shows that C_{pk} never exceeds C_p and that $C_{pk} = C_p$ *precisely when the process is centered midway between its specification limits.* Together, C_p and C_{pk} give a clear picture of how well a process is performing when compared to its specification limits.

EXAMPLE 8.3

Nonconformance rates for the groove dimension data of Example 7.1 were calculated in Example 8.1, where it was concluded that the process had poor capability and that a shift in the process average had much, but not all, to do with the lack of capability. The same conclusions could have been drawn by a quick inspection of the C_p and C_{pk} indexes for these data. To illustrate, with $\hat{\mu} = 0.007966$, $\hat{\sigma} = 0.00103$, USL $= 0.0092$, and LSL $= 0.0052$ from Example 8.1,

$$k = \frac{|(USL + LSL)/2 - \hat{\mu}|}{(USL - LSL)/2} = \frac{|(0.0092 + 0.0052)/2 - 0.007966|}{(0.0092 - 0.0052)/2} = 0.383$$

so the capability indexes are

$$C_p = \frac{0.0092 - 0.0072}{6(0.00103)} = 0.647$$

$$C_{pk} = (1 - k)C_p = (1 - 0.383)(0.647) = 0.399$$

Neither index exceeds 1.0. That the process mean has shifted away from center is evidenced by the fact that C_p and C_{pk} are not equal. Furthermore, even if the process could be adjusted so that it is centered, the *potential* for the process to have good capability still does not exist, since C_p is less than 1.0.

EXAMPLE 8.4

When SPC software is used, it is important to find out how that software estimates the process standard deviation. Figure 8.6 shows an *SPC1* printout for the groove dimension data of Table 7.1. The estimated standard deviation $\hat{\sigma} = 0.0011$ has been calculated using the combined-subgroup method (see Section 8.2), so the C_p and C_{pk} indexes shown in the figure differ somewhat from those in Example 8.3. If the groove dimension process was not in control, the capability indexes calculated in this manner would have differed markedly from those based on $\bar{R}/d_2$ or $\bar{s}/c_4$.

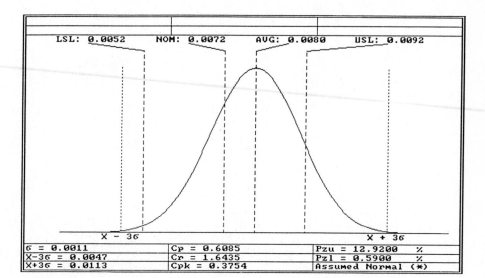

| LSL: 0.0052 | NOM: 0.0072 | AVG: 0.0080 | USL: 0.0092 |

FIGURE 8.6
SPC1 printout of
capability analysis of
data in Table 7.1

σ = 0.0011	Cp = 0.6085	Pzu = 12.9200 %
X−3σ = 0.0047	Cr = 1.6435	Pzl = 0.5900 %
X+3σ = 0.0113	Cpk = 0.3754	Assumed Normal (*)

EXAMPLE 8.5

Some SPC software allows distributions other than the normal to be used to calculate nonconformance rates. Thus, although the capability indexes may be based on the $6\hat{\sigma}$ spread expected for normal distributions, the software uses distributions other than the normal to estimate the percentage of nonconforming items. As an example, Figure 8.7 shows an *SPCII-PC* printout for the groove dimension data of Table 7.1. In this figure, a "best fitting" curve has been used to calculate the nonconformance rates of 11.9538% (above USL) and 0.1982% (below LSL). These rates are somewhat different from those calculated with the assumption of a normal distribution. Notice that the *SPCII-PC* package uses a within-subgroup method for estimatting σ, so the results in the printout agree closely with those found in Example 8.3.

ONE-SIDED TOLERANCES.   The C_p and C_{pk} indexes are used for characteristics with two-sided tolerances—that is, processes with both upper and lower specification limits. Since many characteristics have only one-sided specifications, it is also convenient to have one-sided capability indexes. Such indexes can be defined, and in fact their definitions are contained in that of the C_{pk} index in equation (8.4). For processes that have only a lower specification limit, LSL, the lower capability index C_{pl} is defined by

$$C_{pl} = \frac{\hat{\mu} - \text{LSL}}{3\hat{\sigma}} \tag{8.7}$$

Correspondingly, when only an upper specification exists, we define an upper capability index by

$$C_{pu} = \frac{\text{USL} - \hat{\mu}}{3\hat{\sigma}} \tag{8.8}$$

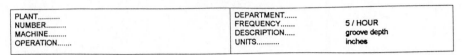

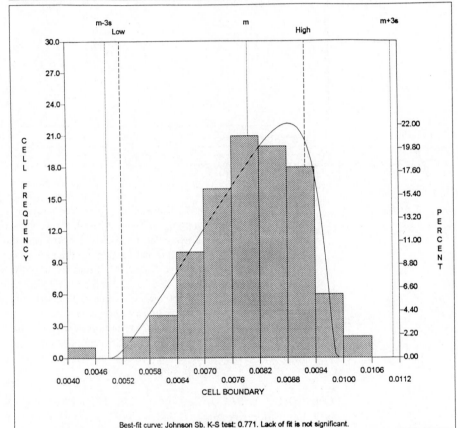

Best-fit curve: Johnson Sb. K-S test: 0.771. Lack of fit is not significant.

SUMMARY DATA	PERFORMANCE STATISTICS	CAPABILITY ANALYSIS (±3.00 sigma)		
ANALYSIS OF GROUPS	LOW SPEC : 0.0052		NON-NORMAL	NORMAL
GROUPS = 20 (100 OBSERVATIONS)	HIGH SPEC : 0.0092	Cp	0.87	0.65
COUNT OF DROPPED GROUPS : 0	AVERAGE(m) : 0.0080	Cr	114%	154%
LARGEST VALUE : 0.0103	m - 3s : 0.0049	Cpm	0.70	0.52
SMALLEST VALUE : 0.0044	m + 3s : 0.0111	Zl(Cpl)	2.88(0.96)	2.69(0.90)
OVERALL RANGE : 0.0059	m - 4.00s : 0.0039	Zu(Cpu)	1.18(0.39)	1.20(0.40)
POPULATION SIGMA : 0.0011	m + 4.00s : 0.0121	Cpk	0.39	0.40
SAMPLE SIGMA : 0.0011	1 VALUES BELOW 0.0052	PROCESS	incapable	incapable
PROCESS SIGMA (s) : 0.0010	9 VALUES ABOVE 0.0092	% LOW	0.1982%	0.3594%
SKEWNESS : -0.6049	10 VALUES OUT OF SPEC	% HIGH	11.9538%	11.5220%
KURTOSIS : 0.3423		% OUT	12.1521%	11.8814%

FIGURE 8.7
SPCII-PC capability analysis of data in Table 7.1

From these definitions and equation (8.4), it is apparent that C_{pk} always equals the smaller of C_{pl} and C_{pu}; that is,

$$C_{pk} = \min[C_{pl}, C_{pu}] \tag{8.9}$$

The reason these indexes use a process spread of $3\hat{\sigma}$ in their denominators is that one-sided capability compares only *one side* of the process to the *single* specification limit.

For data from normally distributed processes, it is convenient to interpret C_{pu} and C_{pl} in terms of their associated nonconformance rates in parts per million (see Section 5.6). For one-sided specifications, the relationships

$$P(X > \text{USL}) = P\left(z > \frac{\text{USL} - \hat{\mu}}{\hat{\sigma}}\right) = P(z > 3C_{pu}) \tag{8.10}$$

$$P(X < \text{LSL}) = P\left(z < \frac{\text{LSL} - \hat{\mu}}{\hat{\sigma}}\right) = P(z < -3C_{pl}) \tag{8.11}$$

are used to convert one-sided capability indexes into nonconformance rates. As an example, a marginally capable process with $C_{pu} = 1.0$ has an upper nonconformance rate of $P(z > 3C_{pu}) = P[z > 3(1.0)] = P(z > 3.0) = 0.00135$, or 1,350 ppm. Table 8.1 lists the ppm rates associated with C_{pu} (or C_{pl}) values ranging from 0.10 to 2.00. Notice that a C_{pu} (or C_{pl}) of 2.0 corresponds to a very low nonconformance rate of about one in a billion, so pursuing capability indexes higher than 2.0 is not warranted in most applications.

EXAMPLE 8.6

Motorola Company's 6-sigma program was described in Examples 3.4 and 5.15. In essence, the program strives to achieve processes with C_{pk} indexes of about 1.5 or greater. The title "6-sigma" has caused some confusion in the literature, since it might imply that the goal is to achieve processes for which 12 sigmas fit within the tolerance range, which would mean that C_{pk}'s of 2.0 are desired. However, a closer reading indicates that the Motorola program allows for some slippage in the process mean, by as much as 1.5 from target. Thus, for a process with a *potential* $C_p = 2.0$, an allowed slippage of 1.5σ translates into an allowed C_{pk} of 1.5. From Table 8.1, nonconformance rates for such processes are no larger than 3.4 ppm (as indicated in the Motorola literature).

PERCENT OF SPECIFICATION USED

A primary concern of quality professionals is to assure that the majority, if not all, of a process's measurements fall within the specification limits. Although this 'goalpost' philosophy (see Section 1.3) has now been replaced by an emphasis on statistically controlled processes and continual reductions in process variation, it still remains useful to compare process behavior with the specifications.

One concept that has been around since *before* the capability indexes is that of the **capability ratio,** which measures the amount of the specification range,

TABLE 8.1   One-Sided Capability Indexes and Their Corresponding Nonconformance Rates (in ppm) for Processes That Follow a Normal Distribution

Capability Index, C_{pu} or C_{pl}	Nonconformance Rate (in ppm), $P(X > USL)$ or $P(X < LSL)$
0.10	382,089
0.20	274,253
0.30	184,060
0.40	115,070
0.50	66,807
0.60	35,930
0.70	17,864
0.80	8,196
0.90	3,467
1.00	1,350
1.10	483
1.20	159
1.30	48
1.33*	33
1.40	13.4
1.50	3.4
1.60	0.793
1.70	0.170
1.80	0.033
1.90	0.006
2.00	0.001

*A capability index of 1.33 is included because of its use in many corporate quality programs.

USL − LSL, that is 'lost' to the process. The wider the process spread, the more a process tends to 'use up' the specification range, and conversely, centered processes with high C_p ratios tend to use much less of the tolerance.

The concern with 'using up the tolerance' originated with considerations about measurement errors and how precise a measuring instrument should be. Measurements that do not have fine enough resolution produce measurement errors that may themselves take up a large portion of the allotted tolerance. It is a mistake to trust any decisions about nonconformance if the measuring instrument itself has poor repeatability or poor resolution (see Section 3.4).

Beyond measurement errors, one can also ask how much of the tolerance is used by the process itself. Even though error variation and process variation can never be disassociated, we continue with the assumptions outlined early in this section by assuming that the magnitude of measurement error is not of concern. The problem of handling measurement and error variation together is addressed in Section 8.5.

To assess the percent of the specification range used, we calculate the proportion of USL − LSL represented by the process spread ($6\hat{\sigma}$). This proportion, $6\hat{\sigma}/(\text{USL} - \text{LSL})$, is called the capability ratio. Thus, the capability ratio is simply the reciprocal of the C_p index:

$$\text{capability ratio} = \frac{1}{C_p} \tag{8.12}$$

When expressed as a percentage, the capability ratio is also called the "percent of specification used" by the process. Notice that the C_{pk} index is not used in this definition. The reason is that one usually wants to know how much of the tolerance range is used in the most ideal circumstances (i.e., when the process is centered), and it is the C_p (not the C_{pk}) index that summarizes the *potential* of a process to meet specifications.

EXAMPLE 8.7

In the *SPC1* printout in Figure 8.6, the reported C_p index is 0.61. As we have seen, capability indexes less than 1.0 indicate that the process has poor capability. For these data, the capability ratio is then $\frac{1}{0.61} = 1.64$; that is, the process measurements (i.e., $6\hat{\sigma}$ spread) use about 164% of the allowed tolerance range. Since the process requires more tolerance (64% more) than is allowed, *even if the process could be perfectly centered at the target dimension of 0.0072 in.*, the process variation would still be too large to fit within the prescribed specification limits. Since a target C_p of 1.33 is widely used, measuring the capability ratio with respect to $8\hat{\sigma}$ is also desirable.

8.5   USING CAPABILITY INDEXES

MEASUREMENT ERROR AND CAPABILITY INDEXES

In Section 8.4 it was assumed that measurement errors can be ignored when capability indexes are interpreted; that is, it was assumed that the measuring instrument has high enough resolution to make measurement error a negligible part of any individual reading. However, intuition tells us that measurement variation must contribute *something* to process variation and that we should try to account for this in our interpretation of capability indexes. In this section, we provide a method for incorporating knowledge of error variation into a capability study.

To begin, recall that any process measurement x_m can be thought of as simply the sum of the 'true' value, x, of the characteristic plus the measurement error, ϵ; that is, $x_m = x + \epsilon$ [see equation (3.13)]. In this expression, the only numerical value available to us is the measurement x_m, not the 'true' value, x, or the error, ϵ. However, if the magnitude of ϵ can be estimated, then it becomes possible to separate the process variation from the error variation. Since it is often reasonable to assume that the measurement error is *statistically independent* of the value being measured, and since the variance of a sum of independent quantities is simply the sum of the component variances, the variation in the measured values, σ_m^2, can be written as

$$\sigma_m^2 = \sigma^2 + \sigma_\epsilon^2 \tag{8.13}$$

where σ_ϵ^2 denotes the variance of the measurement errors and σ^2 is the 'true' process variance. Therefore, we can combine a knowledge of the size of σ_ϵ with σ_m to assess the true process variation. Remember, σ_m is the estimate used to construct capability indexes, not the 'actual' (unknown) process variation σ.

One way of assessing the effect of measurement error is to estimate the percentage of the tolerance 'used' by the error term, ϵ. Studies have shown that errors of measurement often follow approximately a *normal distribution*, so we use $6\sigma_\epsilon$ to represent the range of measurement errors. This range can then be expressed as a percentage of the allowed tolerance range, ULS $-$ LSL, as follows:

$$c = \frac{6\sigma_\epsilon}{\text{USL} - \text{LSL}} \tag{8.14}$$

When multiplied by 100%, c can be thought of as the percent of the tolerance used by the measurement error. The information necessary to estimate c is provided by measuring equipment manufacturers; that is, every instrument is 'rated' to perform at some prescribed level of precision, and this information can be used to find c in equation (8.14).

Rearranging equation (8.14) yields $\sigma_\epsilon = c[(\text{USL} - \text{LSL})/6]$, which, in turn, can be substituted into equation (8.13) to obtain

$$\sigma_m^2 = \sigma^2 + \left[c\left(\frac{\text{USL} - \text{LSL}}{6} \right) \right]^2 \tag{8.15}$$

Next, when the capability index based on the measured data is denoted by C_p^*, equation (8.15) can be used to link C_p^* to the 'true' capability index C_p:

$$C_p^* = \frac{\text{USL} - \text{LSL}}{6\sigma_m} = \frac{\text{USL} - \text{LSL}}{6\sqrt{\sigma^2 + \left[c\left(\dfrac{\text{USL} - \text{LSL}}{6} \right) \right]^2}}$$

which becomes (after dividing top and bottom by USL $-$ LSL)

$$C_p^* = \frac{1}{\sqrt{(1/C_p)^2 + c^2}} \tag{8.16}$$

where C_p denotes the 'true' capability index based on the (unknown) process variation σ. From equation (8.16), we see that if there is no measurement error whatsoever (i.e., $\sigma_\epsilon = 0$), then $c = 0$ and $C_p^* = C_p$. Equation (8.16) also shows that, even if the process variation shrinks to 0 (making C_p arbitrarily large), the measured capability index is *bounded* above; that is,

$$C_p^* \leq \frac{1}{c} \tag{8.17}$$

Thus, the magnitude of the measurement error imposes a limit on how large the reported capability can be. Furthermore, equation (8.16) can be solved for C_p to yield

$$C_p = \frac{1}{\sqrt{[1/C_p^*]^2 - c^2}} = \frac{1}{\sqrt{(CR)^2 - c^2}} \qquad (8.18)$$

where $CR = 1/C_p^*$ is the capability ratio (see Section 8.4). Equation (8.18) can be used to estimate the 'true' process capability after the size of the measurement error is accounted for.

EXAMPLE 8.8   To illustrate the limitations that measurement errors impose on capability indexes, Lucas (1989) described a hypothetical chemical process with specification limits 100 ±30 whose measurements are subject to a variation of $\sigma_\epsilon = 6$. From equation (8.14), $c = 6\sigma_\epsilon/(USL - LSL) = 6(6)/(130 - 70) = 0.60$; that is, the measurement error uses up about 60% of the allowed tolerance range. Equations (8.16) and (8.17) imply that even if the variation in such a process could be reduced to 0 (yielding $CR = 0$), the measured process data would yield a capability index no larger than $1/c = 1.67$. Efforts to achieve a higher C_p^* would be useless. Knowing that such an upper limit exists means that emphasizing higher C_p^* values (e.g., 2.0) would not be wise because they would be unachievable under the current measuring system.

GUIDELINES FOR USING CAPABILITY INDEXES

Much of the popularity of capability indexes is undoubtedly due to their ease of calculation and apparent ease of interpretation. As we have seen, though, their proper interpretation is subject to many conditions. The following list summarizes these conditions and may serve as a checklist for a capability analysis:

1.   *Statistical control*   Before any capability analysis can be performed, the process must be brought into a state of statistical control. Control is established with a variables control chart (see Chapter 7).

2.   *Measurement error*   The magnitude of the errors produced by the measurement system should be ascertained, since the error variation imposes limits on the achievable values of the capability indexes.

3.   *Distribution type*   The exact form of the distribution of process measurements can never be exactly known. It is necessary to *assume* that some distribution (such as the normal) governs the process variability. Some effort should be dedicated to making a sound choice for the underlying distribution.
    a.   Skewed distributions often arise when one-sided specification limits are used (Gunter 1989). The normal distribution does not apply in such situations.

b.   Some processes involve automated sorting, which may produce truncated distributions that underestimate the true process variation. The normal distribution again is not appropriate in this situation (Gunter 1989).

4.   *Sampling variation*   Even for controlled processes, the C_p index is a statistic and is therefore expected to exhibit natural sampling variability from sample to sample. The amount of this variation can be estimated (Kane 1986).

5.   *Software*   If capability analyses are done with SPC software, then it is important to know which method is used to estimate the process variation, σ. Methods based on control chart centerlines ($\bar{R}$ and $\bar{s}$) may give very different results than those based on the standard deviation of the combined data set.

EXERCISES FOR CHAPTER 8

8.1   Why must a process be in statistical control before its capability is measured?

8.2   A process has a C_p of 1.2 and is centered on its nominal value. What proportion of the specification limits are used by this process's measurements?

8.3   Explain how it is possible for all the measurements in a given sample to be within the specification limits, while the same data yield a *nonzero* estimate of the proportion of the process that exceeds specifications.

8.4   Suppose a measuring instrument is very precise but has an unknown offset, δ; that is, each reading from the instrument is exactly δ units higher (or lower if δ is negative) than the true value. What effect does this have on the C_p and C_{pk} indexes generated from such measurements?

8.5   A computer printout shows that a certain process has a C_p of 1.6 and a C_{pk} of 0.9. Assuming that the process is in control, what do these indexes say about the capability of the process?

8.6   A process with specification limits of 5 ± 0.01 has a C_p of 1.2 and a C_{pk} of 1.0. What is the estimated process average $\bar{\bar{x}}$ from which these indexes are derived?

8.7   Show algebraically that C_{pk} can never exceed C_p.

8.8   For a certain process, $\bar{x}$ and R control charts based on subgroups of size 5 have centerlines of 14.5 and 1.163, respectively. Given that the process has specification limits of 12 and 16, calculate C_p, C_{pu}, C_{pl}, and C_{pk}.

8.9   Show that the C_p index of a process that is known to follow an exponential distribution with mean μ is given by $C_p = (U - L)/6\mu$.

8.10   (a)   *(Difficult)* Assuming that process measurements, *X*, follow a normal distribution, show that the following equation always holds for processes with a two-sided tolerance:

$$\text{proportion out of specification} = P(z \geq 3C_{pk}) + P(z \geq 6C_p - 3C_{pk})$$

(b)   What properties of the normal distribution are needed to prove this statement? Does the equation hold for any other distributions?

(c)   Suppose, for a given process, that C_p and C_{pk} are estimated to be 1.4 and 0.90, respectively. Assuming that the process follows a normal distribution, estimate the proportion of the measurements that will be out of specification. (Report your answer in *parts per million*.)

(d)   For a centered process with $C_p = 2.0$, what proportion of the measurements will be out of specification?

8.11   Suppose that a measuring instrument has a rated accuracy of ±1%; that is, its readings are within 1% of the true dimension being measured. When this instrument is used to obtain data on a given process, what is the largest (i.e., best) value that the C_p index could attain? State any assumptions made in obtaining this estimate.

REFERENCES FOR CHAPTER 8

Gunter, B. H. 1989. "The Use and Abuse of C_{pk}, Part 2." *Quality Progress,* March, pp. 108–109.

Kane, V. E. 1986. "Process Capability Indexes." *Journal of Quality Technology* 18 (no. 1):41–52.

Lucas, J. M. 1989. "How Process Industries Are Different." *Chemical and Process Division News.* Milwaukee, WI: ASQC.

9

ATTRIBUTES CONTROL CHARTS

This chapter describes control charts used for processes that generate attributes data. The charts are divided into two groups: those that monitor nonconforming items and those that monitor nonconformities.

CHAPTER OUTLINE

9.1	INTRODUCTION

ATTRIBUTES DATA

Variables data arise from graduated measuring instruments, whereas attributes data usually result from a simple counting procedure (see Section 1.4). To be more precise, **attributes measurements** arise when items are *compared against some standard and are then classified as to whether or not they meet that standard.*

The standard must be carefully defined. For example, in visual inspection, it is easy for different inspectors to have slightly different interpretations of what constitutes a flawed item. This can lead to disagreements concerning actual product quality, and it undermines the usefulness of control chart procedures. (The between-inspector variation is large compared to the within-subgroup variation used to construct control limits, resulting in 'out of control' signals when a different inspector is used.) To avoid such problems, **operational definitions** should be devised, which can be unambiguously and uniformly applied by any inspector (see Section 1.2). The simple preliminary step of creating an operational definition often makes the difference between successful and unsuccessful control chart applications.

NONCONFORMING ITEMS AND NONCONFORMITIES

After a standard is established, one can record whether or not an item meets its specifications. If it doesn't, the item is classified as either **defective** or **nonconforming**. In the past, items that did not meet standards were simply called defectives, but this term was found to be too encompassing because it does not distinguish between salvageable items and totally useless items. Today, the American Society for Quality Control (ASQC) recommends distinguishing between nonconforming units and defective units, with the term 'defective' now reserved for the more serious kinds of nonconformance.

Nonconforming items are those that do not meet specifications, whereas **defective** units are those deemed unsuitable for any intended or foreseeable usage requirements. In other words, defective items are worse than nonconforming ones. For example, a ball bearing might be considered nonconforming if its diameter is not within the specification limits set by design engineers; it might be called defective if it is found to contain large cracks. The reader should be aware that many texts still refer to any item that does not meet its specifications as defective, partly due to past practices and partly in order to shorten the terminology.

Sometimes a standard may *repeatedly* be applied to an item. For example, in the visual inspection of surfaces, standards usually define the magnitude and extent of surface flaws allowed. Since a surface could conceivably have many flaws, inspection consists of counting the number of such flaws, not just classifying the entire surface as nonconforming or not. Similarly, when accounting records are inspected for errors, a single form may contain many small errors. To distinguish this type of counting from simply counting nonconforming *items*, the term **nonconformities** is used to refer to flaws found on a single item. As above, to distinguish levels of seriousness, we can also use the term **defects** to refer to unfixable flaws.

THE ROLE OF THE BINOMIAL AND POISSON DISTRIBUTIONS

Control charts for attributes base their control limit calculations on either the binomial or the Poisson distribution (see Section 5.5). These distributions are appropriate, since both binomial and Poisson random variables *count* the number of occurrences of an event.

The binomial distribution is used for the control limits for the *p* and *np* charts, which track *nonconforming* (or defective) items. The Poisson distribution provides control limits for the *c* and *u* charts, which monitor *nonconformities* (or defects).

SUBGROUP SIZES

A commonly heard adage is that "variables data are better than attributes data," which implies that, given a choice between the two, a variables control chart provides more sensitivity of control than does an attributes chart. This rule of thumb does not apply to all attributes data. It is appropriate only when it is possible to convert variables measurements into attributes measurements by comparing each reading to the *variables* specification limits, and thereby classifying the measurement as conforming or nonconforming. For example, Figure 9.1 depicts the familiar situation of a continuously measurable characteristic (say, the length of a part) and its specification limits. Since it is possible to compare each part's length with the specification limits, the number of nonconforming parts in any sample or subgroup can be obtained. The question then arises: Should one use the original (variables) data to chart the process or would charting the numbers (i.e., attributes data) of nonconforming items per subgroup give equivalent sensitivity of control? The answer is to use the variables data, *not* the associated attributes data. The rationale is that by counting only out-of-specification items, one loses information, which reduces the ability to detect special causes.[1] Increasing the subgroup size can recoup lost information, but the increased sampling can be substantial. For normally distributed data, attributes charts based on out-of-specification counts often require subgroup sizes four to five times larger than that for a variables chart in order to provide the same sensitivity of control (Duncan 1986).

Having said this, we note that there does exist a method for *efficiently* translating variables data into attributes—namely, the narrow-limit gaging procedures described in Section 7.8. Replacing specifications by limits that are closer to the center of the specification region greatly reduces the problem of 'lost' information. Although the numbers of items that fall outside these compressed limits no longer represent the numbers of nonconforming items (with respect to the true

---

[1] For example, an item that is barely larger than the USL would count the same as one that greatly exceeds the USL when one is converting to attributes data.

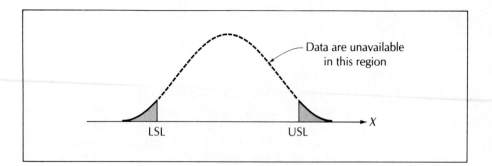

FIGURE 9.1
Much information is
lost when only items
outside the
specifications are
counted

product specifications), they have the virtue of providing increased sensitivity to process changes.

It is also true that a great deal of attributes data do *not* arise by converting variables data into counts. In such cases, the 'variables versus attributes' question described above does not apply, and adequate subgroup sizes must be based on only attributes considerations. Fortunately, obtaining adequate sample sizes is rarely a problem with attributes data because they are usually abundant. The reason is that *classifying* items is almost always easier than *measuring* them, so counted data tend to accumulate faster than variables data would. However, perhaps the main reason for the abundance of attributes data is that so many company records are of an attributes nature (e.g., inspection and test reports, production logs, customer returns data, warrantee records, numbers of backorders, numbers of shipments).

VARIABLE SUBGROUP SIZES

Attributes data drawn from company records are not specifically designed with control chart applications in mind. As a rule, subgroup sizes for attributes charts vary, and control limits must rely on the procedures outlined in Section 6.3.

One such procedure is to use 'standardized' values in place of the original subgroup statistics. For this, some of the 'out of control' rules in Section 6.4 are not applicable, so some care is necessary when searching for signals of special causes (Nelson 1989).

INTERPRETING THE LOWER CONTROL LIMIT

Attributes charts can monitor sample percentages and counts of any sort, but the majority count nonconforming items and nonconformities. Since the fewer nonconforming items generated, the better, a question arises as to how to treat subgroup statistics that fall *below* the lower control limit. On the one hand, such points signal an 'out of control' condition, but on the other, they are very good, since the proportion of nonconforming items (or nonconformities) must have been small for such a signal to occur. In these cases, the recommended procedure is the same as in all control chart applications: search for an assignable cause. The only difference is that, unlike the usual response of eliminating the cause, one may want to take steps to assure that it *continues* to exist, since its general effect is to *reduce* the numbers of nonconformances.

9.2	p CHARTS

CHART CONSTRUCTION

The proportion of nonconforming items in successive subgroups is plotted on a **p chart**. To develop control limits and a centerline, suppose that each subgroup is of the same size, n, and that X denotes the random variable that counts the number of nonconforming items in a subgroup. Assume that all the subgroups come from a statistically controlled process with some 'true' underlying proportion of nonconforming items. Then X can be modeled as a binomial random variable with parameters n and p (see Section 5.5). Converting the number X into a proportion is easily done, and we denote this proportion by $\hat{p} = X/n$. The subgroup proportions $\hat{p}_1, \hat{p}_2, \hat{p}_3, \ldots, \hat{p}_k$ are closely related to the binomial variables from which they arise, and, from well-known properties of the expected value and variance, it can be shown that

$$E(\hat{p}) = E\left(\frac{X}{n}\right) = \frac{np}{n} = p$$

(9.1)

$$\text{Var}(\hat{p}) = \text{Var}\left(\frac{X}{n}\right) = \frac{1}{n^2}\text{Var}(X) = \frac{npq}{n^2} = \frac{pq}{n}$$

so that the standard deviation of the statistic $\hat{p}$ is[2]

$$\sigma_{\hat{p}} = \sqrt{\frac{pq}{n}}$$

(9.2)

In constructing the p chart, each proportion $\hat{p}_i$ is plotted versus its subgroup number i. If the process is in control around an underlying proportion p, then the sample proportions should fall within about 3 standard deviations, or $3\sigma_{\hat{p}}$, of p. To estimate p, the proportions from the k samples are averaged:

$$\bar{p} = \frac{1}{k}\sum_{i=1}^{k} \hat{p}_i$$

(9.3)

with $\bar{p}$ denoting the resulting average. When $\bar{p}$ is substituted for p in equations (9.1) and (9.2), the estimated variation in the fraction defective becomes

$$\hat{\sigma}_{\hat{p}} = \sqrt{\frac{\bar{p}(1 - \bar{p})}{n}}$$

(9.4)

The upper and lower control limits are then

$$\text{UCL} = \bar{p} + 3\sqrt{\frac{\bar{p}(1 - p)}{n}}$$

$$\text{LCL} = \bar{p} - 3\sqrt{\frac{\bar{p}(1 - \bar{p})}{n}}$$

(9.5)

---

[2]Equation (5.2) is used to obtain the mean and variance of X.

Sometimes, because of the small values of $\bar{p}$ that are encountered in practice, LCL can be negative. In those cases, one replaces the LCL by 0.

In the frequently encountered case where the subgroup sizes $n_1, n_2, n_3, \ldots, n_k$ are not all equal, the control chart calculations are modified as follows. When $X_1, X_2, X_3, \ldots, X_k$ denote the numbers of nonconforming items in the subgroups, an estimate of p is found from the *weighted average:*

$$\bar{p} = \frac{X_1 + X_2 + X_3 + \cdots + X_k}{n_1 + n_2 + n_3 + \cdots + n_k} \tag{9.6}$$

We denote the total sample size by N (i.e., $N = n_1 + n_2 + n_3 + \cdots + n_k$), so equation (9.6) could also be written as

$$\bar{p} = \left(\frac{n_1}{N}\right)\hat{p}_1 + \left(\frac{n_2}{N}\right)\hat{p}_2 + \cdots + \left(\frac{n_k}{N}\right)\hat{p}_k$$

which more clearly shows that $\bar{p}$ *is* a weighted average of the subgroup statistics. This is a better estimate than simply averaging the subgroup proportions $\hat{p}_1, \hat{p}_2, \hat{p}_3, \ldots, \hat{p}_k$ because it gives more weight to the larger (i.e., more statistically reliable) subgroups. Note that when all the subgroup sizes are the same, equations (9.6) and (9.3) are identical. Because it is valid for any subgroup sizes (equal or not), equation (9.6) is the most commonly used estimate of p.

From equation (9.6), $\bar{p}$ is substituted into equation (9.2) as before to obtain control limits (except these limits now depend on the particular subgroup size):

$$\text{UCL} = \bar{p} + 3\sqrt{\frac{\bar{p}(1 - \bar{p})}{n_i}}$$

$$\text{LCL} = \bar{p} - 3\sqrt{\frac{\bar{p}(1 - \bar{p})}{n_i}} \tag{9.7}$$

EXAMPLE 9.1

In recent years the Department of Defense has increased its quality requirements for weapons systems and other military hardware (Collins 1988). Aerospace contractors and subcontractors who manufacture such systems must often demonstrate, using control charts, that they are capable of meeting the new requirements. Frequently, such systems use a large number of circuit card assemblies, which consist of printed circuit boards with various electronic components soldered to them. Components are soldered in place by means of a wave solder machine, which passes the boards over a surface of liquid solder. Soldered boards are then connected to test stations, which test the circuits and classify the boards as either conforming or nonconforming. Table 9.1 contains records of the daily numbers of rejected circuit boards for a 30-day period. Since

$$\bar{p} = \frac{14 + 22 + 9 + \cdots + 12}{286 + 281 + 310 + \cdots + 289} = \frac{493}{9{,}155} = 0.054$$

the variable sample size method yields control limits of

$$\text{UCL} = 0.054 + 3\sqrt{\frac{(0.054)(1 - 0.054)}{n_i}} = 0.054 + \frac{0.6781}{\sqrt{n_i}}$$

$$\text{LCL} = 0.054 - 3\sqrt{\frac{(0.054)(1 - 0.054)}{n_i}} = 0.054 - \frac{0.6781}{\sqrt{n_i}}$$

TABLE 9.1   Daily Records of Numbers of Tested and Rejected Circuit Board Assemblies for Example 9.1

Day, i	Rejects	Tested, n_i	Proportion, $\hat{p}_i$	Day, i	Rejects	Tested, n_i	Proportion, $\hat{p}_i$
1	14	286	0.049	16	15	297	0.051
2	22	281	0.078	17	14	283	0.049
3	9	310	0.029	18	13	321	0.040
4	19	313	0.061	19	10	317	0.032
5	21	293	0.072	20	21	307	0.068
6	18	305	0.059	21	19	317	0.060
7	16	322	0.050	22	23	323	0.071
8	16	316	0.051	23	15	304	0.049
9	21	293	0.072	24	12	304	0.039
10	14	287	0.049	25	19	324	0.059
11	15	307	0.049	26	17	289	0.059
12	16	328	0.049	27	15	299	0.050
13	21	296	0.071	28	13	318	0.041
14	9	296	0.030	29	19	313	0.061
15	25	317	0.079	30	12	289	0.042

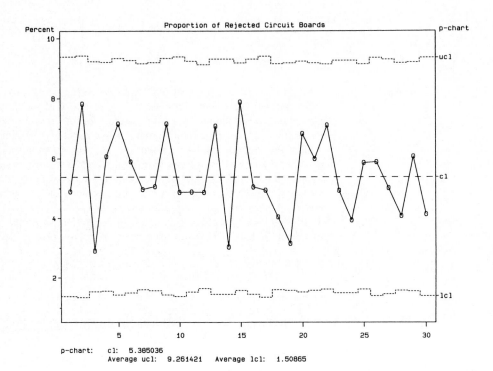

FIGURE 9.2
p chart of data in Table 9.1 from NWA package

With these limits, Figure 9.2 shows the *p* chart for these data created by the NWA software package. From this chart, the process appears to be in control.

Because the subgroup sizes do not vary much around their average value, the appearance of the chart can be simplified somewhat by using the average sample size to calculate control limits. Using an average subgroup size of $\frac{9,155}{30} = 305.17$, which we round off to 305, Figure 9.3 shows the *p* chart of the data as created by a *Minitab* plot. The constant control limits result from using $\overline{p} = 0.054$ and $n = 305$ in equation (9.5). The charts in Figures 9.2 and 9.3 provide relatively the same control over the process because of the stability of the subgroup sizes. In this case, the second chart may be preferred because no other calculations are needed to extend its control limits for future points.

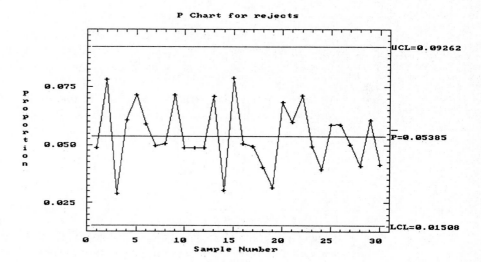

FIGURE 9.3
p chart of data in Table 9.1 from *Minitab* software with a constant subgroup size

SUBGROUP SIZE

Before we construct a *p* chart, it is good to have an estimate of the appropriate subgroup size to use. Setting the subgroup size requires that a reasonable estimate of *p* be available. When *p* is small, larger subgroup sizes are necessary to accurately estimate the subgroup proportions. For larger values of *p*, smaller subgroups can be used. The overriding concern is that the subgroup be large enough to find *some* nonconforming items. With too small a subgroup, the proportions $\hat{p}_i$ can easily be 0 (even if *p* is not 0), which may lead to $\overline{p} = 0$ and control limits of 0 in equation (9.7), rendering the chart useless.

One approach to setting the sample size when *p* is small relies on the fact that a binomial variable *X* can be approximated by a Poisson random variable with parameter $\lambda = np$, where *n* is the subgroup size and *p* is the underlying proportion of nonconforming items (see Section 5.5). In order to assure that *some* nonconforming items are found in a subgroup, one specifies that the probability, γ, of finding at least one such item should be fairly high (e.g., about 0.90 or 0.95). Applying the Poisson approximation, we get

$$\gamma = P(X \geq 1) = 1 - P(X = 0) \approx 1 - e^{-\lambda} = 1 - e^{-np}$$

which can be rewritten as $1 - \gamma = e^{-np}$ and then solved to obtain

$$n = -\frac{\ln(1 - \gamma)}{p} \tag{9.8}$$

where 'ln' denotes the natural logarithm (log base e). Equation (9.8) is easy to apply, but when $\gamma = 0.95$, it is especially simple because $\ln(1 - 0.95) = -2.996$. Rounding this result to -3 and substituting into equation (9.8) yield the simple formula

$$n = \frac{3}{p} \qquad (\gamma = 0.95) \tag{9.9}$$

To illustrate, if one had specified $\gamma = 0.95$ in Example 9.1 and if p was estimated to be about 0.05, then equation (9.9) indicates that subgroup sizes of about $\frac{3}{0.05} = 60$ would be sufficient for the p chart. Thus, to obtain an effective p chart for a process with $p = 0.05$, subgroups should not be allowed to be smaller than 60. Other criteria have also been proposed to find the minimum necessary subgroup size for p charts. For a discussion of these alternatives, we refer the reader to Montgomery (1990).

9.3   *np* CHARTS

Depending on the application, if one has the ability to choose *constant* subgroup sizes, then the p chart of Section 9.2 can be simplified. With a constant base of comparison (i.e., sample size), there is no need to convert nonconforming counts $X_1, X_2, X_3, \ldots, X_k$ into proportions $\hat{p}_1, \hat{p}_2, \hat{p}_3, \ldots, \hat{p}_k$. Thus, in place of the p chart, one can simply plot the counts X_i versus the subgroup number i, thereby forming an ***np* chart.** The common subgroup size n is used to find the centerline and control limits of the np chart.

If $X_1, X_2, X_3, \ldots, X_k$ denote the numbers of nonconforming items in k successive subgroups, then equation (9.6) simplifies to

$$\bar{p} = \frac{X_1 + X_2 + X_3 + \cdots + X_k}{kn} \tag{9.10}$$

When this estimate is used in equation (5.2), the estimated average number of nonconforming items per subgroup becomes $n\bar{p}$, which is taken to be the centerline of the chart.[3] Appealing to equation (5.2) once more for the standard devia-

---

[3] The name 'np chart' derives from the fact that the expected value of a binomial random variable X is np.

tion of *X*, we get the 3-sigma control limits for the *np* chart:

$$UCL = n\overline{p} + 3\sqrt{n\overline{p}(1 - \overline{p})}$$

$$LCL = n\overline{p} - 3\sqrt{n\overline{p}(1 - \overline{p})}$$

(9.11)

EXAMPLE 9.2   In complex systems, items are routed through successions of different processes or procedures before emerging as final products or 'end items.' In 'build to order' systems, it is necessary to route individual orders through slightly different paths from other orders. In either case, a commonly used method for keeping track of an item's progress is to attach paperwork that describes which procedures are to be performed and the associated specifications for each such step. These documents, sometimes called 'travellers,' are created before the order is executed, and it is imperative that they be correct. (Incorrect documents are essentially directions for creating nonconforming products!) To monitor the quality of such paperwork, suppose that periodic samples of 100 travellers are examined for errors, where a nonconforming document is defined to be one that contains at least one error. Table 9.2 shows data from 25 daily samples of 100

TABLE 9.2   Numbers of Documents That Contain Errors

Day, i	Nonconforming documents, X_i	Sample size, n
1	10	100
2	12	100
3	10	100
4	11	100
5	6	100
6	7	100
7	12	100
8	10	100
9	6	100
10	11	100
11	9	100
12	14	100
13	16	100
14	21	100
15	20	100
16	12	100
17	11	100
18	6	100
19	10	100
20	10	100
21	11	100
22	11	100
23	11	100
24	6	100
25	9	100

drawn from completed travellers prior to initiating production. An *np* chart for these data, constructed with the *Minitab* software package, is shown in Figure 9.4. Since the total number of nonconforming documents in the 25 samples is 272, $\bar{p} = 272/(25)(100) = 0.1088$ from equation (9.10). The centerline of the chart becomes $n\bar{p} = 100(0.1088) = 10.88$, with control limits of UCL $= 10.88 + 3\sqrt{10.88(1 - 0.1088)} = 20.22$ and LCL $= 1.54$. Although the chart shows two points that exceed the upper control limit, recall that this chart is a *preliminary* one and that these two subgroups must be investigated before this chart is used as it stands.

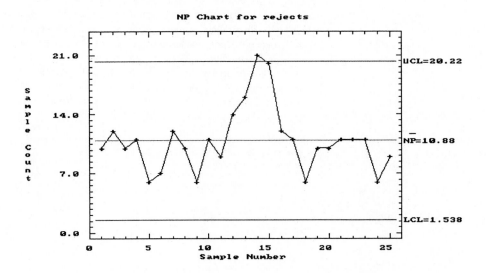

FIGURE 9.4
np chart for data of
Table 9.2 from
Minitab software

Minimum necessary subgroup sizes for *np* charts can be determined with the same criteria used for *p* charts. The reader should refer to Section 9.2, especially equations (9.8) and (9.9), for one method of obtaining this estimate.

9.4   *c* CHARTS

Because the number of nonconformities can vary, sometimes without bound, it is important to establish an **inspection unit** to use with *c* or *u* charts. The inspection unit defines the fixed unit of output that will regularly be sampled and examined for nonconformities. For example, when examining accounting records for errors, one may want to sample 100 records each day. The inspection unit is then 100 records, and the number of errors per 100 records sampled is counted.

Choosing the inspection unit is especially important with 'continuous' processes, such as the production of long rolls of paper, wire, cloth, or metal. To count the number of surface flaws in sheet metal, an inspection unit of, say, 2 square feet could be used. Periodically, then, a 2-square-foot section of the metal surface produced is examined and the number of flaws counted and recorded.

The number of nonconformities per unit (i.e., per inspection unit) is denoted by c. To create a **c chart,** we use a sample of k successive inspection units and estimate the centerline, $\bar{c}$, by

$$\bar{c} = \frac{1}{k} \sum_{i=1}^{k} c_i \qquad (9.12)$$

where c_i is the number of nonconformities in the ith sample. Control limits are based on the fact that the Poisson distribution governs the sampling behavior of statistics that count the number of events per unit (see Section 5.5). Since the mean and variance of a Poisson random variable are equal, and since we have estimated the mean by $\bar{c}$, the variance can also be estimated by

$$\hat{\sigma}_c^2 = \bar{c} \qquad (9.13)$$

The standard deviation is then $\hat{\sigma}_c = \sqrt{\bar{c}}$, so the 3-sigma (i.e., $3\hat{\sigma}_c$) control limits can be approximated by

$$\text{UCL} = \bar{c} + 3\sqrt{\bar{c}}$$
$$\text{LCL} = \bar{c} - 3\sqrt{\bar{c}} \qquad (9.14)$$

As is done with p charts, if the LCL is negative, it is replaced by 0. However, this problem can be avoided if the inspection unit is chosen so that $\bar{c}$ exceeds 9.

EXAMPLE 9.3

One measure of software quality is the error rate per 1,000 lines of code (Dunn 1988). With the abbreviation "K" for the word "thousand," a block of 1,000 lines of computer code is often abbreviated as KLOC (K lines of code). The data in Table 9.3 show the defects per KLOC obtained from daily test logs in a software company. The average number of errors per KLOC is $\bar{c} = \frac{140}{30} = 4.66667$, so the trial upper control limit for the c chart is $\text{UCL} = \bar{c} + 3\sqrt{\bar{c}} = 4.66667 + 3\sqrt{4.66667} = 10.80701$, and because $\bar{c} - 3\sqrt{\bar{c}}$ is negative, the lower limit is taken to be LCL = 0. Figure 9.5 shows a *NWA Quality Analyst* printout of the c chart for these data. Asterisks are used in the plot at subgroups 8, 19, 20, 21, and 23 to indicate 'out of control' rule violations (descriptions of the violations are listed in other *NWA* routines). For instance, at subgroup 8, the 'eight points on the same side of the centerline' violation occurs. At subgroups 19, 20, 21, and 23, no points have fallen below the lower control limit (an impossibility, since LCL = 0 for these data) but other rules have been violated. Before the chart is used, all rule violations have to be investigated and subgroups for which assignable causes can be found should be dropped from the data (and the chart calculations revised accordingly). Probably the most important groups to examine are those from subgroups 17 to 23, where the chart appears to drop dramatically. Assignable causes for data in this region could be very valuable, since these causes have the desirable effect of lowering the error rate.

TABLE 9.3   Number of Errors Per 1,000 Lines
of Code (KLOC)

Day, i	Nonconformities (Errors) per KLOC, c_i
1	6
2	7
3	7
4	6
5	8
6	6
7	5
8	8
9	1
10	6
11	2
12	5
13	5
14	4
15	3
16	3
17	2
18	0
19	0
20	1
21	2
22	5
23	1
24	7
25	7
26	1
27	5
28	5
29	8
30	8

9.5   *u* CHARTS

Sometimes it is neither possible nor convenient to use inspection units of fixed size. For example, an inspection unit of 2 square feet may be specified for counting flaws in a finished car door's surface, but it may be difficult to use this inspection unit in practice. The effort of choosing and measuring a representative 2-square-foot region from the door and then counting the flaws it contains would involve too much production time.

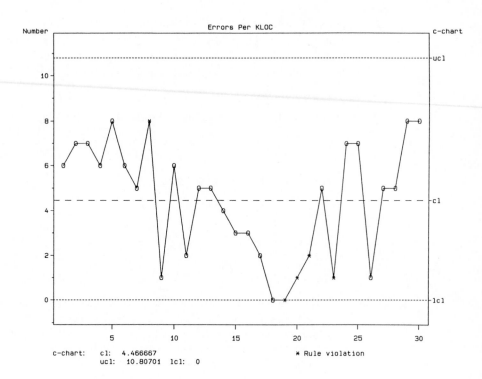

FIGURE 9.5
c chart of data in
Table 9.3 from *NWA
Quality Analyst*
software

In fact, counting flaws on an entire door might be easier and faster. From the door's known surface area (in square feet), the flaws counted could be quickly converted into a rate per 2 square feet (i.e., a rate in terms of the original inspection unit). Thus, a door with 12.8 square feet of surface area would represent $\frac{12.8}{2} =$ 6.4 inspection units. If, say, 2 flaws were found on the door, then the nonconformity rate is estimated as 2 flaws per 6.4 units, or 0.156 flaw per inspection unit. Note that with such naturally occurring inspection samples, the number of inspection units may no longer be an integer and may also vary from sample to sample.

To account for variable numbers of inspection units, *c* charts are replaced by *u* charts. The **u chart** monitors nonconformities per inspection *unit*, when the number of inspection units is allowed to vary from sample to sample. Thus, if subgroup *i* is composed of n_i inspection units,[4] and if a total of c_i nonconformities are found in this subgroup, then the *average* number of nonconformities *per unit* can be estimated as $u_i = c_i/n_i$. For *k* subgroups, the statistics $u_1, u_2, u_3, \ldots, u_k$ are plotted versus the subgroup number to form the *u* chart.

The centerline of the *u* chart is estimated by

$$\bar{u} = \frac{\text{total nonconformities in the } k \text{ subgroups}}{\text{total number of inspection units}} = \frac{c_1 + c_2 + c_3 + \cdots + c_k}{n_1 + n_2 + n_3 + \cdots + n_k}$$

(9.15)

---

[4]Remember, n_i does not necessarily have to be an integer.

For each subgroup, the control limits are derived from the estimate $\bar{u}$ by noting that each $c_i = n_i u_i$ follows a Poisson distribution whose parameter λ_i can be estimated by $\lambda_i \approx \bar{u} n_i$.[5] Thus, the 3-sigma range for the total number of nonconformities c_i in subgroup i is $\lambda_i \pm 3\sqrt{\lambda_i}$, which is approximated by

$$\bar{u} n_i \pm 3\sqrt{\bar{u} n_i} \qquad (9.16)$$

When equation (9.16) is divided through by n_i, the 3-sigma range for u_i is

$$\bar{u} \pm 3 \sqrt{\frac{\bar{u}}{n_i}}$$

which means the subgroup control limits are

$$UCL = \bar{u} + 3 \sqrt{\frac{\bar{u}}{n_i}}$$

$$\qquad (9.17)$$

$$LCL = \bar{u} - 3 \sqrt{\frac{\bar{u}}{n_i}}$$

EXAMPLE 9.4

In Example 9.3, the software error rates per 1,000 lines of code (i.e., per KLOC) were obtained from daily test logs kept for the purpose of tracking error rates. Suppose that the programming department decides to speed up the daily error counting process by simply counting the numbers of errors in finished software modules. (Modules are small task-specific programs that are eventually combined to form a completed software product.) Since a module may consist of any number of lines of code, the reported error rates must be converted to a 'per unit' or 'per KLOC' basis before charting. Table 9.4 lists 25 daily error

TABLE 9.4   Software Error Rates for Completed Modules for a 25-Day Period

Day, i	Completed Modules	Total Lines of Code	Number of Errors	Number of KLOCs, n_i	Nonconformities (Errors) per KLOC, u_i
1	15	7,236	32	7.236	4.422
2	14	7,506	25	7.506	3.331
3	10	6,221	24	6.221	3.858
4	11	5,670	23	5.670	4.056
5	12	6,714	30	6.714	4.468
6	14	7,213	21	7.213	2.911
7	10	4,568	27	4.568	5.911
8	8	3,954	16	3.954	4.047
9	12	7,293	27	7.293	3.702
10	10	4,627	24	4.627	5.187
11	13	6,435	18	6.435	2.797
12	18	7,406	34	7.406	4.591
13	7	3,746	23	3.746	6.140

[5]Since each c_i counts nonconformities.

14	9	6,217	15	6.217	2.413
15	6	5,101	17	5.101	3.333
16	5	5,663	37	5.663	6.534
17	6	5,889	29	5.889	4.924
18	5	4,087	25	4.087	6.111
19	3	3,901	22	3.901	5.640
20	10	5,573	24	5.573	4.306
21	8	4,649	26	4.649	5.593
22	8	4,141	25	4.140	6.037
23	10	5,588	18	5.588	3.221
24	12	6,472	34	6.472	5.253
25	12	7,045	26	7.045	3.691

summaries and the corresponding module lengths (in lines). Figure 9.6 shows an *NWA Quality Analyst* plot of the *u* chart for these data. The daily control limits vary according to the number of KLOCs in the completed modules. Since the chart does not indicate any special causes, we conclude that the error rates seem to come from only a 'common cause' process.[6]

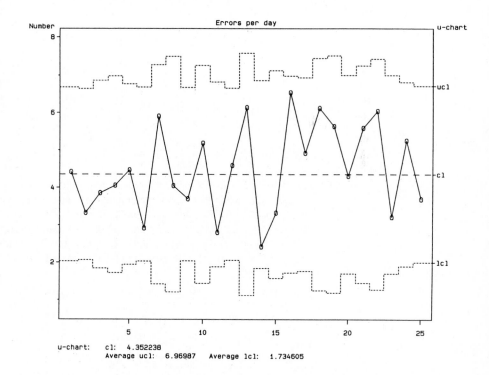

FIGURE 9.6
u chart of data in Table 9.4 from *NWA Quality Analyst*

[6]That is, the programming process appears to be in control.

EXERCISES FOR CHAPTER 9

9.1   The following table contains the number of accidents on the work site across 40 operating divisions of a certain company:[7]

Unit	Number of Accidents			Unit	Number of Accidents		
	March–June	July–October	November–February		March–June	July–October	November–February
1	2	1	2	21	3	4	1
2	1	3	3	22	0	3	1
3	2	4	0	23	1	4	1
4	1	2	4	24	3	1	2
5	1	3	1	25	1	4	4
6	1	1	1	26	1	0	0
7	4	8	8	27	1	0	0
8	0	0	0	28	1	0	0
9	2	1	2	29	0	0	0
10	1	0	2	30	0	0	0
11	3	2	0	31	0	0	12
12	2	6	3	32	1	0	1
13	0	3	1	33	2	3	2
14	0	0	0	34	0	0	0
15	1	0	1	35	0	2	3
16	2	2	4	36	0	0	0
17	0	3	2	37	0	0	0
18	0	0	3	38	0	0	0
19	2	0	4	39	0	1	0
20	2	6	7	40	1	1	1

Construct an *np* chart for these data.

9.2   Off-color flaws in aspirins are caused by extremely small amounts of iron that change color when wet aspirin material comes into contact with the sides of drying containers.[8] The flaws are not harmful but are nonetheless unattractive to consumers. At one Dow Chemical plant, out of every batch of aspirin, a 250-lb sample is taken and the number of off-color flaws is counted. The following table shows the number of flaws per 250-lb sample obtained over a 25-day period:

Sample number	Number of flaws	Sample number	Number of flaws
1	46	14	49
2	51	15	48
3	56	16	59

---

[7]See Bicking (1991).

[8]See Bemowski (1988).

4	57	17	53
5	37	18	61
6	51	19	63
7	47	20	42
8	34	21	45
9	30	22	43
10	44	23	42
11	47	24	39
12	51	25	38
13	46		

Construct an appropriate control chart for these data and examine it for any evidence of lack of statistical control.

9.3 Under what conditions on $\bar{c}$ is the lower control limit of a c chart positive?

9.4 Explain the difference in the actions taken on a process when a point on a p chart exceeds the upper control limit versus the actions taken when a point falls below the lower control limit.

9.5 Under what circumstances does the adage "variables data are stronger than attributes data" apply?

9.6 For a fixed subgroup size, n, what values of $\bar{p}$ lead to a *positive* lower control limit on a p chart?

9.7 For a fixed value of $\bar{p}$, how large does the subgroup need to be to yield a positive lower control limit on a p chart?

9.8 To establish the subgroup size n for a p chart, a probability of 0.90 is specified for finding at least one nonconforming item in any sample of n items. If the process has an average nonconformance rate of about 4%, what subgroup size should be used?

9.9 After passing through a painting operation, individual items are inspected for surface flaws, which are counted and recorded on a c chart. The painting process is currently in statistical control, with an overall average of 0.5 flaw per item inspected. Approximately what proportion of the inspected items will have two or more surface flaws?

9.10 The following data show the number of fabric flaws, c_i, and the number of square feet of material inspected, n_i, for 30 samples of material from a continuous roll of fabric:

i	c_i	n_i	i	c_i	n_i	i	c_i	n_i
1	12	3.9	11	0	8.4	21	21	5.2
2	18	9.0	12	14	6.8	22	6	5.6
3	27	6.7	13	9	4.4	23	16	8.0
4	64	9.2	14	16	5.2	24	27	8.9

(Continued)

(Continued)

i	c_i	n_i	i	c_i	n_i	i	c_i	n_i
5	11	3.6	15	0	7.8	25	21	5.3
6	13	6.7	16	29	9.8	26	12	3.1
7	25	8.3	17	18	8.8	27	19	6.2
8	22	5.6	18	28	7.1	28	14	4.8
9	43	6.1	19	10	3.3	29	42	8.3
10	17	4.2	20	47	5.9	30	19	4.7

(a) From these data, construct a control chart for the number of flaws per square foot of fabric.

(b) Interpret the chart in part (a).

9.11 The following data show the number of nonconforming items per lot for 30 lots, each of size 50 (read across):

```
4  3  0  2  2  2  0  1  1  0  3  2  1  1  0
0  2  4  2  5  0  0  1  1  0  3  2  1  2  4
```

(a) Construct a control chart for the proportion of nonconforming items per lot.

(b) Interpret the chart in part (a).

9.12 Forty consecutive automobile dashboards are examined for signs of pinholes in the plastic molding. The numbers of pinholes found are listed here (read across):

```
6  2  3  2  5  2  2  3  2  4
9  4  0  5  0  6  5  4  2  3
3  1  4  1  7  3  3  5  7  3
6  7  6  4  5  3  8  5  4  3
```

(a) Construct a control chart for the number of pinholes per dashboard.

(b) Interpret the chart in part (a).

REFERENCES FOR CHAPTER 9

Bemowski, K. 1988. "People: The Only Thing That Will Make Quality Work." *Quality Progress,* September, pp. 63–67.

Bicking, C. A. 1991. "The Application of Quality Control to Administrative Problems." *Quality Engineering* 3 (no. 3):413–424.

Collins, F. C. 1988. "Department of Defense Renews Emphasis on Quality," *Quality Progress,* March, pp. 19–21.

Duncan, A. J. 1986. *Quality Control and Industrial Statistics,* pp. 499–501. Homewood, IL: Irwin.

Dunn, R. H. 1988. "Software Quality Assurance: A Management Perspective," *Quality Progress.* 21 (no. 7):55–56.

Montgomery, D. C. 1990. *Statistical Quality Control,* 2nd ed., pp. 131–133. New York: Wiley.

Nelson, L. S. 1989. "Standardization of Shewhart Control Charts," *Journal of Quality Technology* (no. 4):287–289.

MEASUREMENT SYSTEMS

The quality of a measuring instrument or a measurement method affects the reliability of process data and the statistical methods based on these data. This chapter discusses the fundamentals of metrology, calibration, and the estimation of errors produced by measurement systems.

CHAPTER OUTLINE

10.1 INTRODUCTION

As mentioned in the preface, Chapter 10 may be skipped on a first reading of this book; doing so does not diminish one's understanding of the remaining chapters. This chapter is included for those who have had experience in applying statistical methods, especially in manufacturing. In our eagerness to apply statistical methods, data quality is often one of the last things considered, but experience soon teaches that there is a strong relationship between measurement system quality and the quality of the statistical signals derived from the system. For good decision making, one needs to be confident in the measurement system as well as in the statistical procedure used. To this end, Chapter 10 explores basic concepts in the field of **metrology**, the study of measurement.

With few exceptions (noted in Sections 3.4 and 8.5), the statistical methods in Chapters 1–9 presuppose that the measurements used are reliable. In those chapters, it is assumed that the data come from trained operators using well-defined procedures and correctly calibrated instruments of sufficiently high resolution. Assuring the integrity of measurement systems is of fundamental importance in quality control, since reliable measurements support, just as unreliable measurements undermine, the conclusions drawn from analyses of such data.

Reliable measurements, like controlled processes, do not occur naturally. An effort must be made to control the various factors that affect measurement quality. Factors such as differences among operators, differences in measuring procedures, instrument calibration and drift, instrument repeatability and reproducibility, and instrument resolution all affect the quality of the resulting measurements. Typically these factors create uncertainty about the measurements in two ways, by either introducing **systematic biases** or influencing the **variability** in the readings. Once this uncertainty is controlled and estimated, its effect on statistical techniques and conclusions can be assessed.

This chapter examines the above-mentioned factors in some detail. Specifically, it covers the definition of measurements, measurement as a process, calibration and traceability, metrology terminology, statistical control of measurement processes, components of measurement errors, the comparison of inspection instruments, and inspection capability. This may seem like a lot, but the field of metrology is much larger still. Although most of the discussion applies to variables data, Section 10.9 is devoted to attributes data.

10.2 MEASUREMENT: A DEFINITION

MEASUREMENTS

Metrology, the study of measurements, is an important aspect of the quality practitioner's knowledge. Measurements are the window through which we look at products and processes, and it is necessary to know whether the images we see are accurate or, perhaps, somewhat distorted. Whenever conclusions are drawn, practitioners must always be able to respond to the fundamental question: How reliable are the measurements?

To begin the discussion of metrology, it is convenient to have a working definition of the term 'measurement.' From Eisenhart (1963), **measurements** are defined as *"the assignment of numbers to material things to represent the relations existing among them with respect to particular properties."* In other words, physical items have various **properties** (e.g., length, hardness, density), and it is these properties to which one assigns numbers (i.e., measurements). Since numbers are so much easier to manipulate and compare than the physical objects themselves, the process of measurement allows one to substitute simple numerical comparisons for difficult physical ones. Imagine, for example, the difficulties involved with physically sorting 100 steel rods in order of increasing length versus the triviality of sorting their *measured* lengths on a computer.

Intuitively, each measurable property of an object is thought of as having a fixed, constant value called the **true value** of that property. Envision, for example, a steel rod (the object) that has a length (a particular property of the object) of exactly 10 in. One refers to 10 in. as the *true* length of the rod. Thus, even if imperfect measuring instruments and/or techniques are used to measure the rod's length, one still imagines that the underlying "true" length is 10 in.

MEASUREMENT ERRORS

While the concept of "true value" is not without problems, it does provide a basis for discussing and analyzing measurement errors. If, for instance, we are fortunate enough to know the true value, x, of a measured quantity, we then know something about the errors that our measuring system makes, since the difference between the measured and true values must equal the measurement error. To repeat equation (3.13), we have

$$x_m = x + \epsilon \tag{10.1}$$

where x_m is the measured value, x is the "true value," and ϵ is the error of measurement.[1] Equation (10.1) and its extensions lie at the heart of many measurement error calculations.

As stated above, closer inspection reveals some problems with the concept of "true value," problems that must be overcome before measurement errors and measurement systems can be evaluated in a practical fashion. For one thing, it is well known that many properties (e.g., length) of an object change in response to environmental changes, so the object's "true length" may not necessarily be constant. Coupled with the fact that no measuring instrument can have infinite precision, one is led to the conclusion that the "true length" of a measured quantity can never be known (Belanger 1984).

As a workable substitute for the notion of 'true value,' metrologists use the terms **consensus value, generally accepted value,** and **master value.** In many

---

[1]The subscript 'm' is used throughout this chapter to stand for a measured value of some characteristic.

cases, it is through international *agreements* that accepted standards for measured quantities are established (see Section 10.3).

Although it seems that our inability to ever know the "true value" of a measured quantity would render equation (10.1) useless, this is not the case. Instead, equation (10.1) serves as a model of the measurement process, and it can be used in calculations even when the true value, x, is not known. As one example of how this is done, suppose one takes repeated measurements, $x_{m1}, x_{m2}, \ldots, x_{mn}$, of the same quantity. According to equation (10.1), this may be written as

$$x_{m1} = x + \epsilon_1$$

$$x_{m2} = x + \epsilon_2$$

$$x_{m3} = x + \epsilon_3 \tag{10.2}$$

$$\cdot$$
$$\cdot$$
$$\cdot$$

$$x_{mn} = x + \epsilon_n$$

where x is the (unknown) true value, and ϵ_i $\{i = 1, 2, 3, \ldots, n\}$ is the individual measurement error incurred during the ith measurement. Summing these equations and dividing the result by n yield

$$\bar{x}_m = \frac{1}{n} \sum_{i=1}^{n} x_{mi} = x + \frac{1}{n} \sum_{i=1}^{n} \epsilon_i \tag{10.3}$$

If one then makes the assumption that the measurement errors contain no systematic biases (so that they are as likely to be positive as negative), then the average of the n errors appearing in the right side of equation (10.3) ought to become smaller as n increases. We should have

$$\bar{x}_m \approx x \tag{10.4}$$

an approximation that justifies the intuitive notion that averaging many repeated measurements ought to improve the estimate of x.[2] From equation (10.1) used in this manner, many useful results concerning measurement errors can be obtained (see Sections 8.5, 10.6, and 10.7).

EXAMPLE 10.1   The generally accepted standard for *length* has changed many times. Barry (1964) describes the chronology of length standards beginning with early definitions of the cubit (distance from elbow to fingertip) and proceeding through the foot, the inch (distance from thumb joint to thumb tip), the stadia

---

[2]Recall from Section 3.4, however, that such averaging cannot produce precision beyond that provided by the resolution of the measuring instrument.

(the stride of a man), the mile (1,000 strides), and the meter (1 ten-millionth the distance between the earth's equator and a pole).

These highly variable definitions were replaced by successively more precise *generally accepted* definitions. Thus, beginning in 1305, the inch was defined to be equal to the length of three dry, round barley corns placed end to end (Thomas 1974). Later, in 1866, the U.S. Congress decreed that the inch be related to the international standard for the meter (by defining 39.37 in. to equal the meter), a physical artifact housed in the International Bureau of Weights and Measures in Paris. Unfortunately, British and Canadian standards for the inch were not the same as the U.S. standard in 1866. (The American inch equaled 2.540005 cm, whereas the British inch was 2.539995 cm.) Finally, in 1960, international agreement was reached by defining the inch as exactly 2.54 cm and then redefining the meter as 1,650,763.73 times the wavelength in a vacuum of the orange line of the Krypton-86 atom (Thomas 1974, p. 206).

OPERATIONAL DEFINITIONS

It is important to agree on standard definitions because measurements based on these definitions will eventually be used to make manufactured goods, and it is highly desirable that these goods be compatible with those from other manufacturers and other countries. Even the smallest differences in definitions can cause endless confusion for reporting scientific measurements or for trying to integrate manufactured components from several sources.

It is clear that obtaining precise measurements requires close attention to the measurement procedures used. For this reason, it is usually desirable to create **operational definitions** for each type of measurement. These definitions address the various sources of variation in measurements and attempt to control them so that different operators following these procedures have a reasonably good chance of arriving at equally reliable results. Some of the questions that must be addressed include the following (Belanger 1984, pp. 14–15):

FACTORS THAT AFFECT THE QUALITY OF A MEASUREMENT SYSTEM	1. What is the repeatability of the measuring instrument?
	2. Does the time between measurements have an effect on the measurements (i.e., is the measurement system stable over time)?
	3. Do different instruments measuring the same property give different results?
	4. To what extent do different operators add to the magnitude of the measurement error (i.e., reproducibility)?
	5. Does the measurement result depend on the procedure or the sequence of procedures used?
	6. What is the effect of environmental conditions on the resulting measurements?
	7. Is the measurement process in statistical control?

Many operational definitions can be found in the **The American Society for Testing Materials** (ASTM) catalogue of standard test procedures. The ASTM procedures emphasize control over environmental conditions and the exact sequence of measurement steps, as well as the influence of different operators. They further recommend that results be reported along with statements regarding their accuracy, precision, repeatability, and possible systematic errors.

10.3   THE NATIONAL INSTITUTE OF STANDARDS AND TECHNOLOGY (NIST)

Measurement systems have become increasingly more complex and measurement definitions have become more precise throughout history. Correspondingly, it has become necessary to create government agencies that have the responsibility of monitoring and refining these standards and procedures. In 1901, the United States created the **National Bureau of Standards** (NBS) to serve as a repository for the nation's measurement standards and as a laboratory for current research in improving measurement technology. The NBS acted as the ultimate reference to which measurement instruments could be compared and calibrated.

In 1988, the Omnibus Trade and Competitiveness Act caused the NBS to be renamed and to assume expanded responsibilities. Its new name, the **National Institute of Standards and Technology** (NIST), reflects the agency's original role of standards research as well as its new responsibilities in the areas of engineering, physical and chemical measurements, materials science, and computer technology. The NIST is now divided into various centers and divisions charged with supporting these activities (Figure 10.1).

The NIST is a nonregulatory agency that works cooperatively with industry, the government, and academe. The National Measurement Laboratory maintains the nation's physical and chemical standards and provides measurement services necessary for transferring its standard measurements to all other measuring instruments. This laboratory coordinates the nation's measurement standards with those of other nations and also distributes publications outlining measurement and calibrations procedures (*U.S. Government Manual* 1991/1992).

The National Engineering Laboratory is responsible for standards and practices in the areas of electronics, electrical engineering, CAD/CAM, building safety, control of hazardous fires, and chemical and biotechnical engineering. Like the other arms of the NIST, this laboratory provides calibration services and publishes new test methods.

In addition to the usual calibrations and standards services, the Institute for Materials Science and Engineering studies the chemical and physical properties of materials for the purpose of establishing efficient and safe use of materials. It also uses the NIST reactor to develop neutron measurement procedures and oversees a bureau-wide program of nondestructive evaluation and testing.

Finally, the National Computer Systems Laboratory develops federal information-processing standards, interacts with industries in standards development,

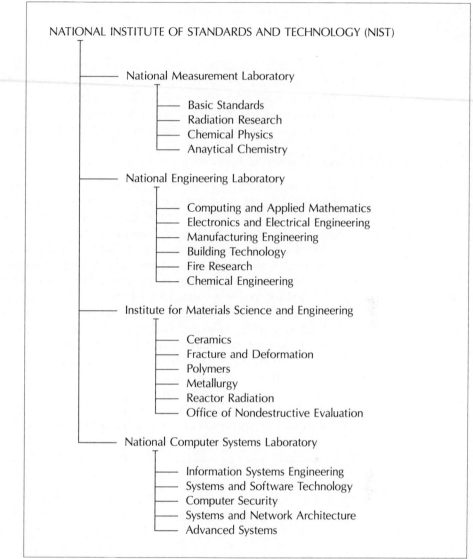

NATIONAL INSTITUTE OF STANDARDS AND TECHNOLOGY (NIST)

- National Measurement Laboratory
  - Basic Standards
  - Radiation Research
  - Chemical Physics
  - Anaytical Chemistry
- National Engineering Laboratory
  - Computing and Applied Mathematics
  - Electronics and Electrical Engineering
  - Manufacturing Engineering
  - Building Technology
  - Fire Research
  - Chemical Engineering
- Institute for Materials Science and Engineering
  - Ceramics
  - Fracture and Deformation
  - Polymers
  - Metallurgy
  - Reactor Radiation
  - Office of Nondestructive Evaluation
- National Computer Systems Laboratory
  - Information Systems Engineering
  - Systems and Software Technology
  - Computer Security
  - Systems and Network Architecture
  - Advanced Systems

FIGURE 10.1
Organization of the
National Institute of
Standards and
Technology (NIST)

conducts research in data processing and telecommunications, and develops test standards and computer security systems.

The NIST is the final authority on measurement accuracy in the United States, although it does look to scientific organizations such as the Los Alamos Scientific Laboratory for direction in establishing many standards. Whenever certifiably accurate measurements are required, the precision of measuring instruments used can be traced back through various calibration levels that always end at the highest level, the NIST (see Section 10.4).

10.4   CALIBRATION

To make use of the ultimate measurement standards maintained or approved by NIST, methods are needed for transferring them to measuring instruments that are used in everyday scientific and manufacturing work. The procedure by which a measurement standard is *transferred* from a higher measurement authority (i.e., a more accurate reference) to a lower one is called **calibration**.

TRACEABILITY TO THE NIST

Assuring the accuracy of measuring instruments used in manufacturing is the responsibility of a company's metrology department. Companies that are not large enough to support a separate metrology department rely on the calibration services of other organizations, called **calibration labs**.

Within the department or lab responsible for calibration, the process of transferring NIST standards to the company's measuring instruments is accomplished by means of a hierarchical system of transfers. First, the company maintains a **primary standard** for each measured quantity. Primary standards are the company's direct link to the NIST. At regularly scheduled times, primary standards are sent to the NIST (or an NIST-approved facility) for recalibration and recertification of accuracy. To assure their accuracy while at the company, primary standards are often kept in environmentally controlled rooms and are not themselves directly involved in the calibration of the company's measurement instruments.

Instead, **secondary standards** are used to calibrate the instruments used in production; that is, primary standards are used to calibrate the secondary standards, which are then used to calibrate the **working standards.** Figure 10.2 shows

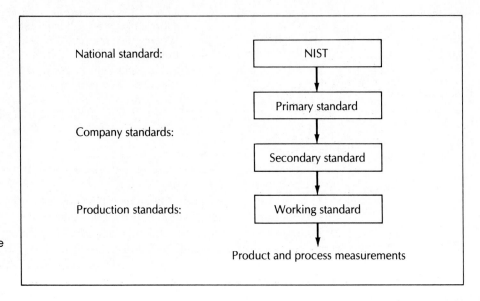

FIGURE 10.2
Traceability of measurements to the National Institute of Standards and Technology (NIST)

the typical hierarchy of calibrations within a company. This process is somewhat similar to that used with microcomputer software, where the purchased copy is reserved as the primary copy and backup copies are used to perform the actual work.

In this manner, a measuring instrument's accuracy can be linked through successive calibrations to the NIST. The ability to establish this connection is called **traceability to NIST** or, more simply, **traceability**. To give another definition, traceability is "the ability to relate individual measurement results to national standards or nationally accepted measurement systems through an unbroken chain of comparisons" (U.S. Dept. of Defense, *Military Standard 45662*).

EXPRESSING CALIBRATION UNCERTAINTIES

Youden (1967) lists three questions that must be answered during the calibration process. First, what is the magnitude of the uncertainty in the 'readings' that the calibrated instrument gives? Second, what kind of data are needed to ascertain the reported uncertainty? Third, how should this uncertainty be summarized and conveyed to those who will use the calibrated instrument? In this section and in Section 10.5, some answers are given to these questions.

Uncertainties in measured quantities are most often reported in terms of the **absolute** or **relative errors** that the instrument is likely to make. Appealing to equation (10.1), one can define these errors as follows:

$$\text{absolute error} = x_m - x$$

$$\text{and} \quad \text{relative error} = \frac{x_m - x}{x} \tag{10.5}$$

Relative errors are usually multiplied by 100 so that they may be more conveniently reported as percentages. For example, a thermometer that reads 30°F when the 'true' temperature is 31°F is said to make an absolute error of −1°F and a relative error of −3.2%. In this example, note that the uncertainties apply only to the thermometer's accuracy near 30°F. To be useful in practice, the reported uncertainties must apply to the full range (or **full scale**) of values that the instrument is intended to measure. Thus, the thermometer in our example might have had a rated accuracy of ±4% over the entire range of temperatures that it is designed to measure.

The **calibration curve** is a convenient device for visualizing calibration uncertainties. This curve depicts the relationship between the 'true' values to be measured (on the horizontal axis) and the measured values (on the vertical axis). Ideally, the true values and the measured values should be related in a linear fashion across the entire operating range (i.e., full scale) of the instrument. Figure 10.3 shows an ideal (linear) calibration curve. The degree to which a measuring instrument approaches this ideal curve is called its **linearity**. In practice, instruments are only approximately linear over their full scale. Figure 10.4, for example, shows both the ideal linear curve and the *actual* calibration curve for a certain instrument.

Measurement uncertainty, relative or absolute, can be easily displayed on the calibration curve. For instance, if a thermometer designed to operate over a certain range has a rated absolute uncertainty of ±1°F, then its calibration curve and

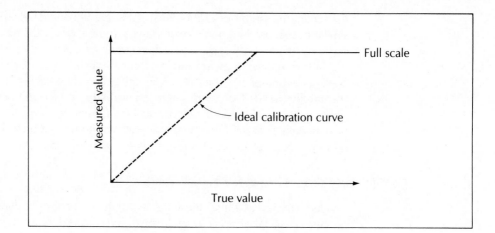

FIGURE 10.3
An ideal (linear)
calibration curve

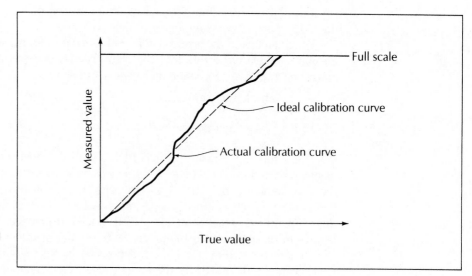

FIGURE 10.4
An instrument's
actual calibration
curve (solid line) and
idealized calibration
curve (broken line)

associated accuracy can be graphed as in Figure 10.5. On the other hand, if the thermometer has a rated relative accuracy of ±1% over its full scale, then the graph appears as in Figure 10.6. In Figure 10.6, note that the absolute error increases as the value to be measured increases. In both cases (absolute and relative), error bands are drawn with the *idealized* curve in the center.

THE PROCESS OF CALIBRATION

The majority of calibrations are performed by means of a *comparison* to a higher reference standard. This comparison can be either direct or indirect. **Direct comparisons** are the simplest to perform. One example of a direct measurement is the

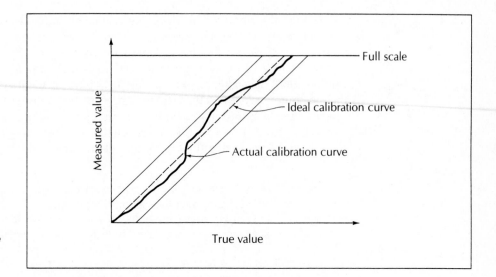

FIGURE 10.5
Expressing absolute
errors on the
calibration curve

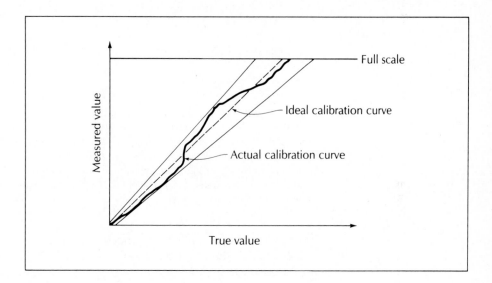

FIGURE 10.6
Expressing relative
errors on the
calibration curve

measurement of an object's length by comparing it with calibrated gage blocks. **Indirect comparisons** involve intermediate mechanisms and calculations that are eventually converted into an instrument reading. Often, measurement instruments based on indirect comparisons rely on physical principles to achieve this conversion. As one example, pressure gages usually measure the effect that an exerted pressure has on a known resistance device (such as a spring) and then use physical laws (e.g., the spring constant and the equation of a spring) to *impute* the pressure that is responsible for the compression measured.

The usual method for certifying the calibration of an instrument is to specify the uncertainty of its readings by a plus-or-minus error (absolute or relative) tolerance. For example, alcohol-filled glass thermometers normally have rated accuracies of $\pm0.5°C$ to $\pm1°C$ over a working range of $-70°C$ to $65°C$ (Sirohi and Radha Krishna 1980). In addition, some instruments have inherent systematic biases in their measured values and these biases, called **offsets**, must also be reported. As a simple example, a scale might be *known* (from calibration) to give measurements that are 5 lb off. Knowing this offset, one can immediately adjust any reading given by the scale to find the true weight of an object.

EXAMPLE 10.2   The NIST maintains 80-piece sets of steel gage blocks spanning a range of 0.100 to 4.0 in. that are available for transferring NIST length standards to laboratory gage blocks (Belanger 1984). From these standards, each laboratory gage block can be assigned an *offset* to be applied to its reported measurements.

Since it is not practical to house thousands of gage blocks in a laboratory, gage block sets are usually small, and they are graduated in a fashion that allows intermediate measurements to be made by simply stacking various combinations of the blocks. For example, one metric set based on the millimeter comes with 112 pieces (Thomas 1974, p. 217):

1 block	$\rightarrow$ 1.0005 mm	
9 blocks	$\rightarrow$ 1.001–1.009 mm	(in increments of 0.001 mm)
49 blocks	$\rightarrow$ 1.01–1.49 mm	(in increments of 0.01 mm)
49 blocks	$\rightarrow$ 0.5–24.5 mm	(in increments of 0.5 mm)
4 blocks	$\rightarrow$ 25–100 mm	(in increments of 25 mm)

A length of 3.0345 mm, which is not included in the set, can be measured by stacking the blocks of lengths 1.0005, 1.004, and 1.03 mm. The end surfaces of gage blocks are very precisely machined so that the process of stacking the blocks (called 'wringing') introduces minimal error into the overall reading. This stackup error is accounted for, and included in, the uncertainty statements that accompany the gage block set.

Once an instrument has been calibrated, it is necessary to certify the calibration and to schedule the next calibration. Certification is usually accomplished by means of a tag or decal affixed to the instrument. These tags commonly specify which reference standard was used, the amount of correction (if any) required, the rated accuracy of the instrument (i.e., its resolution), the date of the calibration, the frequency of calibration, and the next scheduled recalibration. The calibration frequency (the time between successive calibrations) is normally called the **calibration interval.**

EXAMPLE 10.3   MIL-STD-45662A is the document that governs the creation and maintenance of calibration systems used by the government and its subcontractors. The document requires that all measuring and test equipment be listed and given identification numbers. The identification numbers provide calibration traceability over the lifetime of the instrument. Records must also be kept on the type of instrument, its calibration interval, which reference standards are used for calibrating the instrument, the date of calibration, and an NIST number for tracing the reference standard back to the NIST. Calibration intervals are usually established based on how much use the instrument experiences. Intervals can be lengthened or shortened depending on how many times the instrument required (or did not require) adjustments during past calibrations (Sobralske 1989).

10.5   MEASUREMENT PROCESS CONTROL

MEASUREMENT VARIATION

In an ideal measurement system, repeated measurements of the same object should yield identical reported values whose uncertainty is limited only by the rated accuracy (resolution) of the particular instrument. Stated differently, ideal measurements should show no variation or, equivalently, the instrument should have perfect repeatability.

Real measurement systems, of course, do not exhibit perfect repeatability. There are many sources of variation (see Section 10.2), some related to changes in the operating environment and some to changes in the measuring system itself. A more realistic depiction of repeated measurements of a single object is shown in Figure 10.7. Successive observations vary in a (it is hoped) random manner about the nominal value that the system is intended to measure.

Because they are subject to the effects of environmental changes, different operators, different procedures, instrument wear, and other factors, measurement processes have been likened to production processes that must be statistically

FIGURE 10.7
Repeated measurements of the same object exhibit variability

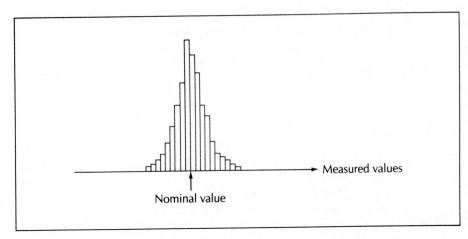

controlled (Eisenhart 1967, p. 165). The production analogy suggests the application of control chart techniques to measurement systems, and indeed such techniques *are* important to assuring the reliability of a measurement system. In fact, Eisenhart (1967, p. 162) stresses that unless a measurement system is brought into a state of statistical control, "it cannot be regarded in any logical sense as measuring anything at all."

CHECK STANDARDS AND CONTROL CHARTS

One procedure common to most measurement control applications is the use of check standards. **Check standards** are objects that are believed to be relatively stable and are periodically measured by the system in order to construct control charts on the system's performance. For example, a 1-in. gage block (the check standard) might periodically be measured by a particular micrometer, with the resulting sequence of readings used to construct control charts. Changes in the patterns (e.g., shifts or slow drifts) on these charts then signal changes in the performance of the micrometer. Such control charts are expressed either in terms of the original readings (measurements) or in terms of the deviations of these measurements from the nominal value (i.e., from the check standard). Figure 10.8 shows control charts for the same set of repeated measurements on a check standard whose length is known to be 1 in.; one chart is based on the original measurements and one is based on their deviations from 1 in. (the nominal value). It should be noted that strict operational definitions should be followed, so that the differences in the measurements can correctly be attributed to random variation or to special causes.

It is advisable to use check standards that are similar to the objects that are measured during normal production (Bishop, Hill, and Lindsay 1987). One reason for this is that no changes need be made in the measuring system to accommodate such check standards. Also, it is convenient (and sometimes necessary) to include the check standards with normal production items so that the standards experience exactly the same conditions found in the production environment.

EXAMPLE 10.4   Belanger (1984, p. 19) reports that check standards have routinely been used in mass calibrations at the NIST since 1963. The data from weighing check standards are used to construct control charts. In a similar fashion, since 1972, the NIST has used check standards to monitor gage block lengths. Not only does this procedure allow for control over the measurement process, but the resulting data can also be used to determine the magnitude of error variation.

Of the various control charts that could be applied to periodic measurements of check standards, the CUSUM (see Section 7.7) is more appropriate than the individuals chart (Section 7.5). First, CUSUM charts are designed to detect *deviations* from a target (in this case, the nominal value of the check standard). Second, even though the individuals chart can also be used to detect deviations from a target value (by making the centerline equal to the target value), the greater sensitiv-

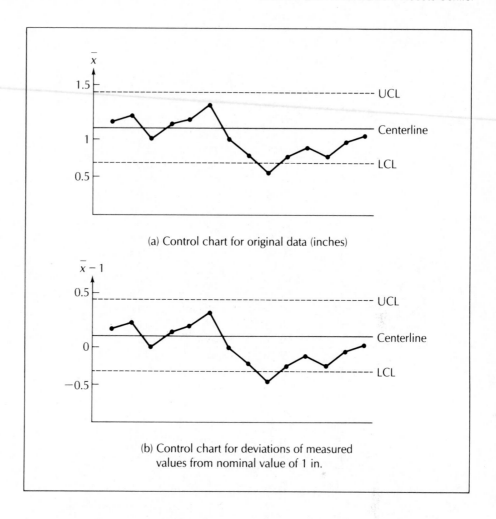

(a) Control chart for original data (inches)

(b) Control chart for deviations of measured
values from nominal value of 1 in.

FIGURE 10.8

ity of the CUSUM chart allows shifts in the measurements to be detected somewhat earlier. Example 10.5 compares both charts.

EXAMPLE 10.5   In a certain calibration laboratory, daily measurements (weights in kilograms) are
taken on a 20-kg check standard used in mass calibrations. Table 10.1 lists the
measurements recorded over a 50-day period. To assess whether or not the
measurement system that produces these weights is stable, both individuals and
CUSUM charts are created. In order to detect shifts away from the nominal
value of 20.00 kg, the centerline of the individuals chart in Figure 10.9 is set at
20.00. It is *not* calculated from the data as it would be in other applications.
Although there are no points outside the 3-sigma limits of the individuals chart,
closer inspection shows that the 'eight points in a row on the same side of the
centerline' rule is violated on day 18 (and later on day 45). This type of behavior, wandering below the centerline and then eventually above it, is indicative of the effect of environmental changes (temperature, humidity, etc.) on

TABLE 10.1   Daily Measurements of a 20-kg
Check Standard

i	x_i	i	x_i
1	19.93	26	19.98
2	20.05	27	20.07
3	19.95	28	20.08
4	20.00	29	20.01
5	20.00	30	20.00
6	19.79	31	20.21
7	19.91	32	20.02
8	19.99	33	19.99
9	20.06	34	19.86
10	19.84	35	20.14
11	19.89	36	20.19
12	19.99	37	19.98
13	19.87	38	20.05
14	19.84	39	20.07
15	19.98	40	20.11
16	19.73	41	20.09
17	19.86	42	20.11
18	19.76	43	20.16
19	20.06	44	20.10
20	19.84	45	20.06
21	20.04	46	20.21
22	20.01	47	20.05
23	20.11	48	20.09
24	19.78	49	20.08
25	19.93	50	20.11

FIGURE 10.9
Individuals chart with
target value of 20 kg
for data of
Table 10.1

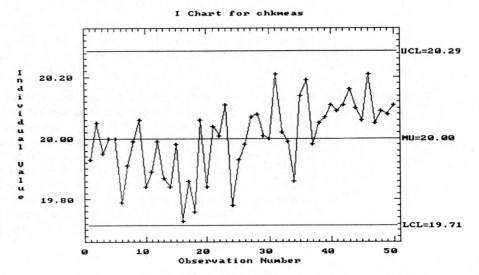

the check standard or on the measuring instrument, which helps to direct the search for special causes. In contrast, a CUSUM chart for these data is shown in Figure 10.10. This chart even more clearly depicts the downward and then upward drift in the measurements and also signals 'out of control' conditions around day 18 (and around day 49).

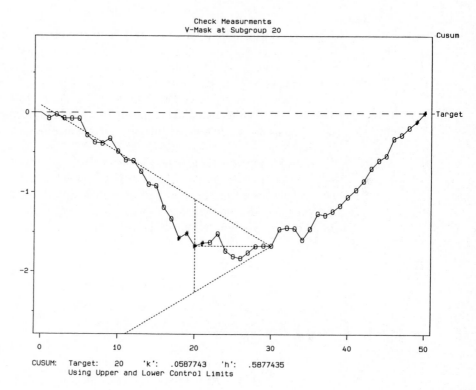

FIGURE 10.10
CUSUM chart with target value of 20 kg for data of Table 10.1

Properly interpreted, the individuals and CUSUM charts are capable of detecting both shifts and slow drifts in the measurements. With most measuring systems, instead of large shifts, it is more likely that instruments slowly drift out of calibration or that slow drifts occur because of environmental changes. We close this section by looking at a statistical test proposed by Youden (1954) for detecting instrumental drift.

A TEST FOR INSTRUMENT DRIFT

Suppose that n successive measurements, $x_1, x_2, x_3, \ldots, x_n$, have been made on a check standard and that one wants to test this sequence for evidence of drift. By 'drifting' measurements, we mean those that consistently move upward, consistently move downward, or slowly wander back and forth (as in Example 10.5). The statistic used to test for such drift is

$$M = \frac{\sum_{i=2}^{n} (x_i - x_{i-1})^2}{\sum_{i=1}^{n} (x_i - \bar{x})^2} \tag{10.6}$$

which was originally developed in 1941 by von Neumann for detecting drifts in time series data (Farnum and Stanton 1989). It can be shown that the statistic M *must lie between 0 and 4 for any set of data.* Values of M close to 2 are indicative of randomness (i.e., no drift) in the readings, whereas values of M closer to 0 signal the presence of a drift in the data. Values of M close to 4 occur in the extremely rare case when the readings swing rapidly back and forth around the mean value. The latter case is rare in measurement system applications and we do not consider it further here.

To decide whether there is no significant drift in the readings (M close to 2) or there is a noticeable drift (M closer to 0) requires the use of Appendix 9. For a given significance level α, one concludes that the drift *is* statistically significant whenever M is less than the tabled value $M_{1-\alpha}$ in Appendix 9. Otherwise, it is concluded that drift is not present in the data.

EXAMPLE 10.6   From the data in Table 9.1, the calculated value of M is

$$M = \frac{\sum_{i=2}^{50} (x_i - x_{i-1})^2}{\sum_{i=1}^{50} (x_i - \bar{x})^2}$$

$$= \frac{(20.05 - 19.93)^2 + (19.95 - 20.05)^2 + \cdots + (20.11 - 20.08)^2}{(19.93 - 20.001)^2 + (20.05 - 20.001)^2 + \cdots + (20.11 - 20.001)^2}$$

$$= 1.340$$

Comparing this value to Appendix 9, we note that, at a significance level of $\alpha = 0.01$, the critical value is $M_{1-0.01} = M_{0.99} = 1.363$ for $n = 50$. Since the observed value of $M = 1.340$ is less than the critical value $M_{0.99}$ from the table, we can conclude (at a significance level of 0.01) that there is some drift in the measurements of Table 10.1. This conclusion agrees with those drawn from our visual inspection of the control charts of these data in Example 10.5.

One caution is in order when using the test for drift based on the statistic M. *The test is sensitive only to drifts in the data; it is not capable of detecting systematic biases (offsets) in the readings.* To see this, suppose that the readings $x_1, x_2, \ldots, x_n$ are all shifted by a fixed offset, c. For the new readings $x_1 + c, x_2 + c, \ldots, x_n + c$, the terms in the numerator of equation (10.6) are *unchanged* because $(x_i + c) - (x_{i-1} + c) = x_i - x_{i-1}$. Similarly, one can see that the terms in the denominator are unchanged, so that offsetting the data by any amount, c, has no effect on M.

10.6 MEASUREMENT ERRORS

The basic model, which defines a measured value x_m to be composed of the true value, x, plus a measurement error, ϵ [see equations (3.13) and (10.1)], is only the starting point for analyzing measurement errors. In order to separate out the distinct *sources* of error, the basic model must be expanded to include precise *definitions* for the different types of measurement errors. In turn, specific data-collection procedures (i.e., *experimental designs*) must be used for estimating the contribution these errors make to the total measurement error. Such models may be thought of as refinements of equation (10.1) in which the error term, ϵ, is successively partitioned into more and more subcategories. Figure 10.11 illustrates this decomposition of the total measurement error.

It is convenient to categorize the various sources of measurement errors as belonging to one of two groups: systematic errors and random errors. Although there is no complete agreement about which group each error source should belong to, the following definitions probably represent the majority view on this subject (Jackson 1987). First, **systematic errors** (also called **offsets** or **biases**) are defined to be the constant values by which, on average, an instrument's readings are off from the true value being measured. Recall the example of the scale in Section 10.4 that is consistently off by 5 lb from an object's true weight. In this case, the scale is said to make a systematic error of 5 lb or, equivalently, it has an offset or bias of 5 lb. The magnitude of an instrument's systematic error is usually specified by the manufacturer or by a calibration lab.

On the other hand, **random errors** are those caused by differences among instruments, differences among operators, instability over time, environmental changes, and different setups. Although random errors cannot be predicted exactly, their magnitudes can be estimated if one follows particular experimental designs when performing measurements and then uses probability distributions (usually, the normal distribution) to generate the estimates.

To begin the analysis of measurement capability, recall the definitions of accuracy, precision, repeatability, and reproducibility given in Section 3.4. **Accuracy** refers to the degree to which repeated measurements of a true value, x, tend to agree with x. Envisioning the repeated measurements histogram form, we can measure accuracy as the difference between x and the middle of the histogram; that is,

$$\text{accuracy} = \bar{x}_m - x \tag{10.7}$$

Figure 10.12 shows a typical histogram of measured values centered around the average $\bar{x}_m$ of these values.

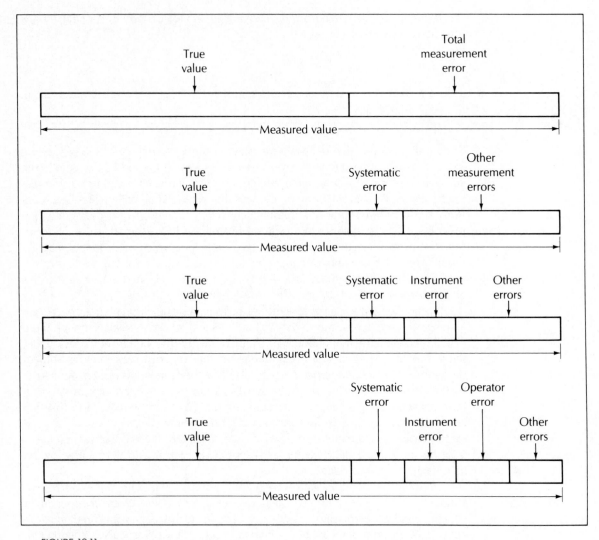

FIGURE 10.11
Separating total measurement error into its components (*Note:* Errors are depicted as positive, but in general they may assume positive or negative values.)

The **precision** of an instrument describes the degree to which repeated measurements tend to agree *with one another.* They do not necessarily have to agree with the true value, *x.* Precision is almost always measured by the sample standard deviation of the measured values:

$$s_m = \sqrt{\frac{1}{n-1} \sum_{i=1}^{n} (x_{mi} - \bar{x}_m)^2} \tag{10.8}$$

where $x_{m1}, x_{m2}, \ldots, x_{mn}$ are n measured values of the same quantity, $x.$ To illustrate, the histograms in Figure 10.13 show measurements made with instruments of

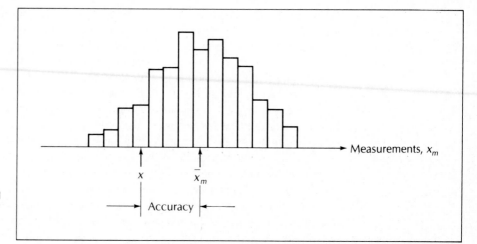

FIGURE 10.12
Measurement
accuracy is the
distance between
the true value, x, and
the average of
repeated
measurements, $\bar{x}_m$

differing precision. From these definitions, one can see that any combination of accuracy (high or low) and precision (high or low) is possible, as depicted in Figure 3.10 in Section 3.4.

It should be noted that some authors refer to the quantity $\bar{x}_m - x$ as the 'bias' in the measurements and reserve the term 'accuracy' to describe the *combined* error caused by this bias along with the variation in the measurements (Eisenhart

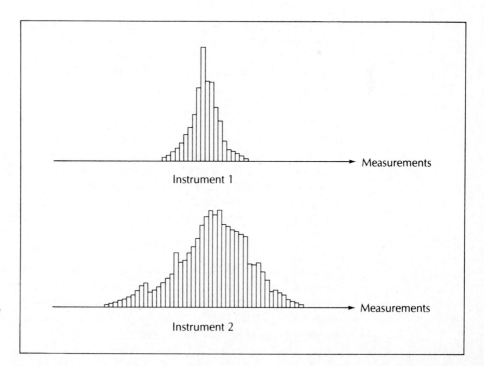

FIGURE 10.13
Repeated
measurements of the
same item made
with instruments of
different precisions

1967, p. 179). However, even these authors must eventually separate out the components of bias and precision defined in equations (10.7) and (10.8).

Equation (10.1) can easily be modified to incorporate additional categories of measurement error. We use the notation found in the AIAG manual on measurement systems for such models (*AIAG Reference Manual* 1990). Recognizing the difficulties involved with the term 'true value' (see Section 10.2), the AIAG manual replaces this term with 'master value,' where a 'master value' is defined to be a measurement made with the most accurate instrument available. The terms 'true value' and 'master value' are used interchangeably in our discussion.

To begin, note that not all products of the same type are identical with respect to their measurable properties; that is, there is always some part-to-part variation. We denote the overall average part measurement by μ. Then each individual part has a true value of $x_i = \mu + \alpha_i$, where α_i is called the **part effect.** In other words, the ith individual part's true value x_i deviates slightly (by an amount α_i) from the overall average part dimension μ. Thus, the basic model in equation (10.1) can be revised to take the part effect into account:

$$x_{mi} = \mu + \alpha_i + \epsilon_i \tag{10.9}$$

where x_{mi} is the measured value of the ith part and ϵ_i is the *remaining* measurement error due to other factors.

It is also possible that an instrument makes some systematic error, b, in its measurements. This bias (hence the use of the letter b) can be included in the model by writing

$$x_{mi} = \mu + b + \alpha_i + \epsilon_i \tag{10.10}$$

where the error term ϵ_i now accounts for all other deviations *except* the offset b and part effect α_i.

Next, the effect of different operators on the resulting measurements can be accounted for by including an **operator effect,** β_j, for the jth operator. This model can be written as

$$x_{mij} = \mu + b + \alpha_i + \beta_j + \epsilon_{ij} \tag{10.11}$$

where ϵ_{ij} now accounts for the remaining measurement errors except for that due to bias, part effects, and operator effects. When several operators are involved, it is also possible to find a lack of consistency among them in how they make their measurements. When such inconsistency exists, there is said to be an **interaction** between operators and parts. Interaction can be modeled by including an 'interaction effect,' γ_{ij}, in equation (10.11) and writing a new model:

$$x_{mij} = \mu + b + \alpha_i + \beta_j + \gamma_{ij} + \epsilon_{ij} \tag{10.12}$$

Obviously, other models can be written to account for additional sources of measurement error, but we stop here. The interested reader may refer to the AIAG manual and Duncan (1986) for more complicated models. From the expanded models in equations (10.9)–(10.12), the repeatability and reproducibility of a measurement system can now be estimated.

10.7   ESTIMATING MEASUREMENT ERROR

ERROR VARIANCE

The magnitude of the error made by a measurement system increases with its complexity. One intuitively expects, for example, that measurement errors from a simple system (one instrument, one operator, one part measured) will be smaller than those encountered with complex systems (several instruments, several operators, several parts measured). Fortunately, these differing levels of complexity can be modeled with equations such as those in the previous section, equations (10.9)–(10.12), and when combined with the right type of data, the resulting measurement error can be quantified.

There are two primary sources of measurement errors, those caused by systematic biases in the measurement system and those due to small random variations caused by environmental changes, operator variation, and instrument resolution. Systematic bias, also referred to as accuracy (see Section 10.6), is usually estimated during the process of calibration with a higher reference standard. Normally, either the instrument manufacturer or a calibration lab performs the necessary repeated measurements on check standards to ascertain the instrument's accuracy (see Example 3.5). Thus, it is assumed that the systematic bias is known, so it is not considered further.

The magnitude of random errors is evidenced by their *variation,* not their average. It is usually assumed that random errors are as likely to be positive as negative, so that their expected value is 0. Furthermore, it is reasonable to assume that the various sources of error are independent of one another (see Section 5.3). Applying these assumptions to equations (10.9)–(10.12) and using the fact that the variance of a sum of independent random variables equals the sum of the individual variances, we can write the following equations:

<table>
<tr><td align="center">Measurement Model</td><td align="center">Measurement Variation</td><td></td></tr>
<tr><td align="center">$x_{mi} = \mu + \alpha_i + \epsilon_i$</td><td align="center">$\longrightarrow \quad V(x_{mi}) = \sigma_\alpha^2 + \sigma_\epsilon^2$</td><td align="right">(10.13)</td></tr>
<tr><td align="center">$x_{mi} = \mu + b + \alpha_i + \epsilon_i$</td><td align="center">$\longrightarrow \quad V(x_{mi}) = \sigma_\alpha^2 + \sigma_\epsilon^2$</td><td align="right">(10.14)</td></tr>
<tr><td align="center">$x_{mij} = \mu + b + \alpha_i + \beta_j + \epsilon_{ij}$</td><td align="center">$\longrightarrow \quad V(x_{mij}) = \sigma_\alpha^2 + \sigma_\beta^2 + \sigma_\epsilon^2$</td><td align="right">(10.15)</td></tr>
<tr><td align="center">$x_{mij} = \mu + b + \alpha_i + \beta_j + \gamma_{ij} + \epsilon_{ij}$</td><td align="center">$\longrightarrow V(x_{mij}) = \sigma_\alpha^2 + \sigma_\beta^2 + \sigma_\gamma^2 + \sigma_\epsilon^2$</td><td align="right">(10.16)</td></tr>
</table>

where $V(x_{mi})$ and $V(x_{mij})$ denote the variances of the measured values, σ_α^2 represents the part-to-part variance, σ_β^2 is the variance between operators, σ_γ^2 is the variance caused by the part–operator interaction, and σ_ϵ^2 is the remaining error variation that is not accounted for by the other factors in the particular model. Notice that a constant offset b in equations (10.14)–(10.16) does not affect the variance of the measured values (although it certainly affects their *mean* value). Equations (10.13)–(10.16) may be easily remembered by noting that the offset adds nothing to the variance and that (under the assumption of independence) the variances of the error categories simply add up to the total error variance; that is, V(total error) $= V$(category 1) $+ V$(category 2) $+ \cdots + V$(category k).

REPEATABILITY AND REPRODUCIBILITY

The concepts of repeatability and reproducibility are not uniquely defined in the quality control literature. To some these terms mean determining the precision of a single measuring instrument, to others they refer to the largest distance expected between two repeated measurements, and to still others they address the consistency of readings that emanate from a single measurement laboratory. At the heart of all these views, however, lies the broader definition that repeatability and reproducibility refer to the *variability* encountered when several repeated measurements of the same characteristic are made. Sometimes the measurements are made on a single part by one operator using the same instrument. At other times measurements include the additional variation encountered when several operators, labs, or instruments are used. Once this variability is quantified, other measures of repeatability or reproducibility can be derived, often by the use of a simple multiplicative factor. For simplicity, in the remainder of this section we use the term 'repeatability' when referring to any study conducted for the purpose of estimating the various sources of measurement error.

Repeatability studies that employ only *one* operator are modeled by either equation (10.9) or (10.10). Since equation (10.9) is a special case of equation (10.10) (in which the bias b is 0), the more general model, equation (10.10), is used in the ensuing discussion. Furthermore, because only one part is involved in the study, the part effect α_i can be considered to be *constant* during the study. Therefore, the subscript is dropped and we simply use α to denote the part effect of the particular part selected for the study. Using equation (10.10), we can write n repeated measurements of this part as

$$x_{m1} = \mu + b + \alpha + \epsilon_1$$

$$x_{m2} = \mu + b + \alpha + \epsilon_2$$

$$x_{m3} = \mu + b + \alpha + \epsilon_3$$

$$\cdot$$
$$\cdot$$
$$\cdot$$

$$x_{mn} = \mu + b + \alpha + \epsilon_n$$

Summing these equations and dividing the resulting sum by n gives

$$\bar{x}_m = \mu + b + \alpha + \bar{\epsilon} \qquad (10.17)$$

where $\bar{x}_m$ is the average of the measured values and $\bar{\epsilon}$ denotes the average of the corresponding (unknown) measurement errors. Using equation (10.17) to calculate the sample variance s_m^2 of the measured values, we get

$$s_m^2 = \frac{1}{n-1} \sum_{i=1}^{n} (x_{mi} - \bar{x}_m)^2$$

$$= \frac{1}{n-1} \sum_{i=1}^{n} [(\mu + b + \alpha + \epsilon_i) - (\mu + b + \alpha + \bar{\epsilon})]^2$$

$$= \frac{1}{n-1} \sum_{i=1}^{n} (\epsilon_i - \bar{\epsilon})^2 = s_\epsilon^2$$

That is, the sample variance of the measured values is also an estimate of the variance of the error terms. Taking square roots, we can then use s_m as an estimate of the variation in the measurement errors.

To convert s_m into a measure of repeatability, more precise definitions are needed. The simplest of these reflects the view that repeatability is just the *precision* in the measurements. When we make the additional assumption that the error term ϵ_i follows a normal distribution, then most (99.73%) of the errors lie no more than $3s_m$ from their mean (which is assumed to be 0). In other words, the majority of measurement errors should fall within $3s_m$ of the true value (after accounting for any bias b) or, more simply said, the repeatability should be about $\pm 3s_m$.

To be technically correct, the majority of the errors should lie in a range of $\pm 3\sigma_\epsilon$, not $\pm 3s_m$, because s_m is only an *estimate* of the true (but unknown) error variance σ_ϵ. Because of this, it is important to assure that s_m is as precise an estimate as possible. Since precise estimation of a population standard deviation generally requires larger sample sizes than those needed for estimating a population mean, we recommend against using very small samples in repeatability studies. If desired, exact sample size formulas for estimating σ_ϵ can be used (Sachs 1982), but as a rough rule of thumb, we recommend using samples of 25 or more.

One definition in common use is that repeatability (or reproducibility) is the maximum difference that, with high probability, can be expected between any *two* instrument readings (or between two measurement labs). From this definition, it can be shown that the repeatability is given by the formula

$$k\sqrt{2}s_m \tag{10.18}$$

where the factor k depends on the probability level specified. Tabled values of k along with a more detailed discussion of this form of repeatability/reproducibility can be found in Mandel and Lashof (1987).

EXAMPLE 10.7   In a repeatability study, an operator took 25 repeated measurements of a certain characteristic on a single part selected from the daily production run. The measurements, along with their sample mean and standard deviation, are given in Table 10.2. Since the standard deviation of the 25 measurements is $s_m = 0.096$, the repeatability of the instrument is estimated to be around $\pm 3(0.096) = \pm 0.288$. Thus, the instrument used to make these measurements introduces an error of no more than about 0.288 in each reading it gives.

TABLE 10.2   25 Repeated Measurements of a Single Part from One Instrument and One Operator

Repetition, i	Measurement, x_{mi}	Repetition, i	Measurement, x_{mi}
1	9.92	14	9.90
2	10.05	15	9.88
3	9.99	16	9.82
4	9.85	17	9.91
5	9.90	18	10.05
6	10.00	19	9.87
7	9.99	20	10.05
8	9.98	21	9.94
9	10.17	22	9.75
10	9.97	23	9.89
11	9.97	24	9.85
12	10.02	25	10.12
13	10.00		
$\bar{x}_m = 9.95$		$s_m = 0.096$	

PART-TO-PART VARIATION

Multiple measurements made by one operator on only one part paints an optimistic picture of an instrument's repeatability. Would it not be better, perhaps, for an operator to take repeated measurements of *several* different parts in an attempt to simulate how the instrument would behave in the production environment? In fact, repeatability studies can be conducted in this manner. The only additional problem that must be addressed is how to separate part-to-part variation (σ_α^2) from measurement error variation (σ_ϵ^2). This section describes two procedures for estimating σ_ϵ^2 when measuring several parts.

As in the previous section, we continue to examine repeatability studies involving only one operator, so equation (10.10) again serves as the measurement model. The method used to obtain separate estimates of σ_α^2 and σ_ϵ^2 is to make more than one measurement on each of k parts. Starting with the smallest possible number of repeated measurements, we suppose that an operator makes *two* measurements, call them x_{mi} and x'_{mi}, on each of k parts. Using equation (10.10), one can then make Table 10.3.

When we make the assumption that the errors ϵ_i and ϵ'_i are independent, it can be shown (Jaech 1985, p. 50) that the sample variance s_d^2 of the differences d_i provides an estimate of σ_ϵ^2:

$$\sigma_\epsilon^2 \approx \frac{s_d^2}{2} \qquad (10.19)$$

TABLE 10.3   Two Measurements of Each of k Parts

Part Number, i	First Measurement, x_{mi}	Second Measurement, x'_{mi}	Difference, $d_i = x_{mi} - x'_{mi}$
1	$\mu + b + \alpha_1 + \epsilon_1$	$\mu + b + \alpha_1 + \epsilon'_1$	$\epsilon_1 - \epsilon'_1$
2	$\mu + b + \alpha_2 + \epsilon_2$	$\mu + b + \alpha_2 + \epsilon'_2$	$\epsilon_2 - \epsilon'_2$
3	$\mu + b + \alpha_3 + \epsilon_3$	$\mu + b + \alpha_3 + \epsilon'_3$	$\epsilon_3 - \epsilon'_3$
.	.	.	.
.	.	.	.
.	.	.	.
k	$\mu + b + \alpha_k + \epsilon_k$	$\mu + b + \alpha_k + \epsilon'_k$	$\epsilon_k - \epsilon'_k$

With this estimate in hand, we first use equation (10.14) to estimate σ_α^2:

$$(s_m)^2 \approx V(x_{mi}) = \sigma_\alpha^2 + \sigma_\epsilon^2$$

and
$$(s'_m)^2 \approx V(x'_{mi}) = \sigma_\alpha^2 + \sigma_\epsilon^2$$

where s_m and s'_m are the sample standard deviations of the x_{mi} and x'_{mi} readings, respectively. Averaging these two equations (an operation referred to as 'pooling the variances') gives

$$\frac{(s_m)^2 + (s'_m)^2}{2} \approx \sigma_\alpha^2 + \sigma_\epsilon^2$$

which, along with equation (10.19), yields

$$\sigma_\alpha^2 \approx \frac{(s_m)^2 + (s'_m)^2}{2} - \frac{s_d^2}{2} \tag{10.20}$$

Thus, it is possible to obtain estimates of both the part-to-part variation, equation (10.20), and the measurement error variance, equation (10.19).

This method can be extended to the case of more than two measurements per part, but the resulting calculations quickly become burdensome. The interested reader should see the original work on this subject by Grubbs (1948). Many of the refinements of Grubbs's work and that of others are summarized in the book by Jaech (1985) and the survey article by Grubbs (1973).

A slightly different procedure for analyzing pairs of measurements on each part is based on the R chart (see Section 7.2). By considering the k pairs of readings as k subgroups, we can construct an R chart with $\overline{R}/d_2$ used to approximate σ_ϵ. As before, the estimate of σ_ϵ is then used in equation (10.20) to approximate the part-to-part variance:

$$\sigma_\alpha^2 \approx \frac{(s_m)^2 + (s'_m)^2}{2} - \left(\frac{\overline{R}}{d_2}\right)^2 \tag{10.21}$$

One of the virtues of the R chart method is that it can readily be extended to handle any number of repeated measurements per part. The only change in the calculations involves looking up the appropriate d_2 factor associated with the number of repeated measurements. One cautionary note to add is that the R chart method relies on the assumption that the measurement errors follow a normal distribution, an assumption not required by the previous method with equation (10.19).

EXAMPLE 10.8

In an effort to estimate the repeatability of an instrument in the production environment, an operator took two measurements on each of 25 different production parts. The data are displayed in Table 10.4 along with the differences, d_i, and the ranges, R_i, of each pair of readings. Following the first method presented in this section, we use equation (10.19) to estimate the measurement error variance, $\sigma_\epsilon^2 \approx s_d^2/2 = (0.03168)^2/2 = 0.00050181$. From this and the standard deviations s_m and s_m', the part-to-part variation is estimated to be

$$\sigma_\alpha^2 \approx \frac{(s_m)^2 + (s_m')^2}{2} - \frac{s_d^2}{2} = \frac{(0.209)^2 + (0.198)^2}{2} - 0.00050181 = 0.04094$$

Converting both variances to their corresponding standard deviations gives $\sigma_\epsilon \approx 0.0224$ and $\sigma_\alpha \approx 0.202$. We invoke the usual assumption that both the measurement error and the part effect follow approximately normal distributions, so $6\sigma_\epsilon$ approximates the span of the errors (i.e., the repeatability), while $6\sigma_\alpha$ approximates the span of the part characteristics. The ratio of these two numbers is often used as a quick comparison to show how large the instrument variation is compared to the part-to-part variation. In this example, $6\sigma_\epsilon/6\sigma_\alpha = \sigma_\epsilon/\sigma_\alpha = 0.0224/0.202 = 0.1108$; that is, the measurement error is about 11.1% of the product variation.

Next, analyzing these data by the R chart method, we obtain the estimate of σ_ϵ from equation (7.6), so $\sigma_\epsilon \approx \overline{R}/d_2 = 0.0250/1.128 = 0.0222$. For these data, the two methods are in very close agreement.

OPERATOR VARIATION

The final repeatability study we examine takes operator variation, part-to-part variation, and instrument error into account. It is based on equation (10.12) and generates estimates for part-to-part variance σ_α^2, operator variation σ_β^2, operator consistency (also called the part–operator interaction) σ_γ^2, and instrument error σ_ϵ^2.

The statistical method used to separate the individual error components is called the **analysis of variance** (abbreviated **ANOVA**). ANOVA techniques are essentially sampling plans that provide specific directions for collecting data and formulas for extracting the variance estimates from those data. The ANOVA approach also allows one to perform statistical tests for deciding which factors truly affect the measurement errors and which do not.

The sampling plan used with equation (10.12) is shown in Figure 10.14. In the language of ANOVA studies, this plan is called a 'two-factor crossed design with replication.' The design requires that each of k different parts be measured n times by each of c operators. The two 'factors' in this design are the 'operators' and the

TABLE 10.4   Two Measurements on Each of 25 Different Parts

i	x_{mi}	x'_{mi}	$d_i = x_{mi} - x'_{mi}$	$|d_i| = $ range
1	9.966	9.925	0.041	0.041
2	10.319	10.298	0.021	0.021
3	9.874	9.878	−0.004	0.004
4	10.014	9.975	0.039	0.039
5	10.009	10.027	−0.018	0.018
6	10.488	10.430	0.058	0.058
7	10.249	10.208	0.041	0.041
8	10.064	10.060	0.004	0.004
9	9.913	9.908	0.005	0.005
10	9.836	9.901	−0.065	0.065
11	10.018	9.967	0.051	0.051
12	10.165	10.172	−0.007	0.007
13	9.707	9.696	0.011	0.011
14	10.211	10.184	0.027	0.027
15	9.931	9.959	−0.028	0.028
16	10.081	10.052	0.029	0.029
17	9.981	10.014	−0.033	0.033
18	9.787	9.847	−0.060	0.060
19	10.352	10.372	−0.020	0.020
20	9.995	10.004	−0.009	0.009
21	9.655	9.661	−0.006	0.006
22	9.926	9.937	−0.011	0.011
23	9.925	9.934	−0.009	0.009
24	9.653	9.638	0.015	0.015
25	10.040	10.053	−0.013	0.013
$s_m = 0.209$		$s'_m = 0.198$	$s_d = 0.03168$	$\bar{R} = 0.0250$

'parts.' The model also estimates the interaction between these two factors (see Section 10.6), which in this application can be thought of as 'operator consistency' (see the *GAGE∗STAT* software manual).

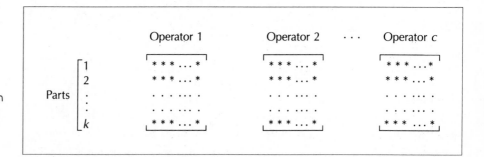

FIGURE 10.14
Data-collection plan for a two-factor crossed design with repeated measurements

A further distinction concerns whether or not the group of operators and the collection of parts are to be considered *populations* or simply *samples* from some larger groups. In the majority of repeatability studies, it seems more reasonable to opt for the latter interpretation, and we do so in the remainder of this section. That is, it is assumed that the operators (and the parts) are only *samples* of the operators (and parts) that could be used in production. Under this interpretation, the model is said to be a 'random effects' model. If, for some reason, the operators and parts are thought to be populations, then the model is called a 'fixed effects' model. The primary difference between the models arises when statistical tests are performed on the factors, a topic we do not pursue here. For further discussion on this matter, we refer the reader to Dixon and Massey (1983).

The total variation among the measured values x_{mij} is

$$SST = \sum (x_{mij} - \bar{\bar{x}})^2$$

where the summation extends over all the values (there are ckn of them), $\bar{\bar{x}}$ is their grand average, and the abbreviation SST is read 'sum of squares, total variation.' It is SST that is partitioned into components to account for the effects of operators, parts, and their interaction. Table 10.5 summarizes the calculations involved.

Each entry in the column of 'mean square' in Table 10.5 estimates a certain combination of the error variances σ_ϵ^2, σ_α^2, σ_β^2, and σ_γ^2 (these combinations are

---

TABLE 10.5   ANOVA Calculations for the Two-Factor Crossed Design with Repeated Measurements

	Operator 1	Operator 2	$\cdots$	Operator c	
Parts 1	* * * ... *	* * * ... *		* * * ... * $\rightarrow$	$\bar{P}_1$
2	* * * ... *	* * * ... *		* * * ... * $\rightarrow$	$\bar{P}_2$
$\vdots$	. . . .	. . . .		. . . .	$\vdots$
k	* * * ... *	* * * ... *		* * * ... * $\rightarrow$	$\bar{P}_k$
Operator means $\rightarrow$	$\bar{O}_1$	$\bar{O}_2$	$\cdots$	$\bar{O}_c$	Part means

	Sum of squares	Degrees of freedom	Mean square	Mean square is an estimate of
SS (operators) $= nk\sum_{j=1}^{c}(\bar{O}_j - \bar{\bar{x}})^2$		$c - 1$	$\dfrac{\text{SS (operators)}}{c - 1}$	$\sigma_\epsilon^2 + n\sigma_\gamma^2 + nk\sigma_\beta^2$
SS (parts) $= nc\sum_{i=1}^{k}(\bar{P}_i - \bar{\bar{x}})^2$		$k - 1$	$\dfrac{\text{SS (parts)}}{k - 1}$	$\sigma_\epsilon^2 + n\sigma_\gamma^2 + nc\sigma_\alpha^2$
SS (interaction) =				
SSM $-$ SS (operators) $-$ SS (parts)		$(c - 1)(k - 1)$	$\dfrac{\text{SS (interaction)}}{(c - 1)(k - 1)}$	$\sigma_\epsilon^2 + n\sigma_\gamma^2$
SS (instrument) = SST $-$ SSM		$ck(n - 1)$	$\dfrac{\text{SS (instrument)}}{ck(n - 1)}$	σ_ϵ^2

Notes: SSM $= n\sum_{i,j}(\bar{R}_{ij} - \bar{\bar{x}})^2$, where $\bar{R}_{ij}$ denotes the average of the n measurements made by the jth operator on the ith part.

SST, the total sum of squares, is the sum of all $c \cdot k \cdot n$ squared deviations from the grand mean $\bar{\bar{x}}$. The grand mean $\bar{\bar{x}}$ is simply the average of all $c \cdot k \cdot n$ measurements.

listed in the rightmost column of the table). For example, the mean square for operators, denoted 'MS (operators),' approximates the quantity $\sigma_\epsilon^2 + n\sigma_\gamma^2 + nk\sigma_\beta^2$, while 'MS (interaction)' approximates $\sigma_\epsilon^2 + n\sigma_\gamma^2$. From these, it is a simple matter to solve for the various variance components, as the following steps illustrate:

$$\text{MS (operators)} \approx \sigma_\epsilon^2 + n\sigma_\gamma^2 + nk\sigma_\beta^2$$

$$\text{MS (interaction)} \approx \sigma_\epsilon^2 + n\sigma_\gamma^2$$

$$\text{MS (operators)} - \text{MS (interaction)} \approx nk\sigma_\beta^2$$

$$\sigma_\beta^2 \approx \frac{\text{MS (operators)} - \text{MS (interaction)}}{nk}$$

In a similar fashion the other variance components can be estimated. These estimates are summarized in Table 10.6.

TABLE 10.6   Estimates of Part-to-Part Variation σ_α^2, Operator Variation σ_β^2, Part–Operator Interaction σ_γ^2, and Instrument Variation σ_ϵ^2

Variance Component	Estimated by
σ_α^2	$\dfrac{\text{MS (parts)} - \text{MS (interaction)}}{cn}$
σ_β^2	$\dfrac{\text{MS (operator)} - \text{MS (interaction)}}{kn}$
σ_γ^2	$\dfrac{\text{MS (interaction)} - \text{MS (instrument)}}{n}$
σ_ϵ^2	MS (instrument)

Note: Mean squares (MS) are calculated as in Table 10.5.

EXAMPLE 10.9

In a repeatability study, 25 parts were selected from daily production and each of three operators was asked to make two measurements on each of the selected parts. Data from this study appear in Table 10.7. Using the notation introduced in this section, we find $k = 25$, $c = 3$, and $n = 2$. To estimate the components of measurement error, the *Minitab* software package was used to perform the calculations required by Table 10.5. From the ANOVA table in the printout in Figure 10.15, we obtain the following estimates:

$$\sigma_\alpha^2 \approx \frac{\text{MS (parts)} - \text{MS (interaction)}}{cn} = \frac{0.214207 - 0.003523}{3 \cdot 2} = 0.03511$$

$$\sigma_\beta^2 \approx \frac{\text{MS (operators)} - \text{MS (interaction)}}{kn} = \frac{0.005249 - 0.003523}{25 \cdot 2}$$

$$= 0.000034$$

$$\sigma_\gamma^2 \approx \frac{\text{MS (interaction)} - \text{MS (instrument)}}{n} = \frac{0.003523 - 0.000505}{2}$$

$$= 0.00151$$

$$\sigma_\epsilon^2 \approx \text{MS (instrument)} = 0.000505$$

TABLE 10.7   Repeatability Study Involving Several Parts ($k = 25$), Several Operators ($c = 3$), and Repeated Measurements ($n = 2$) per Part

Part	Operator 1		Operator 2		Operator 3	
	x_{m1}	x'_{m2}	y_{m1}	y'_{m2}	z_{m1}	z'_{m2}
1	10.21	10.26	10.15	10.19	10.16	10.17
2	9.83	9.79	9.80	9.76	9.77	9.80
3	10.05	10.06	9.92	9.92	10.10	10.09
4	10.01	10.02	10.00	9.97	9.98	9.97
5	9.80	9.85	9.84	9.82	9.78	9.79
6	10.12	10.17	10.12	10.10	10.10	10.09
7	10.23	10.26	10.26	10.23	10.24	10.29
8	10.19	10.18	10.13	10.13	10.16	10.15
9	10.00	10.04	9.98	10.02	10.06	10.02
10	9.71	9.70	9.68	9.72	9.73	9.74
11	10.24	10.28	10.10	10.14	10.20	10.22
12	10.06	10.01	10.07	10.08	10.11	10.12
13	10.02	10.06	10.02	10.03	10.08	10.09
14	9.87	9.89	10.03	10.05	9.87	9.87
15	9.83	9.92	9.95	9.95	9.88	9.94
16	10.36	10.33	10.27	10.30	10.35	10.34
17	10.15	10.10	10.16	10.18	10.26	10.21
18	9.84	9.86	9.85	9.83	9.86	9.89
19	10.58	10.59	10.50	10.47	10.47	10.50
20	9.98	10.02	10.04	10.00	9.99	10.02
21	10.03	10.02	9.99	9.97	9.97	9.97
22	9.70	9.66	9.76	9.70	9.78	9.76
23	10.15	10.14	10.09	10.05	10.08	10.13
24	9.98	10.00	9.98	9.98	9.95	9.98
25	10.22	10.21	10.15	10.19	10.17	10.16

Although the variance components can always be calculated from the ANOVA table, most software packages also have the capability of generating these estimates. Notice that in Figure 10.15 the components are included in the lower part of the printout. Taking square roots, we estimate that the standard deviations of the components are

$$\sigma_\alpha \approx 0.1874$$

$$\sigma_\beta \approx 0.0058$$

$$\sigma_\gamma \approx 0.0389$$

$$\sigma_\epsilon \approx 0.0225$$

which shows that the part-to-part variation accounts for the majority of the variation in the measurements. This is a good sign, since, in capable measurement systems, the various sources of measurement errors should be small so as not to significantly interfere with the reliability of the measured values.

```
Analysis of Variance for measurements

Source           DF       SS          MS         F       P
PARTS            24    5.140976    0.214207   60.80   0.000
OPERATOR          2    0.010497    0.005249    1.49   0.236
PARTS*OPERATOR   48    0.169104    0.003523    6.98   0.000
Error            75    0.037850    0.000505
Total           149    5.358428



Source            Variance  Error  Expected Mean Square
                  component  term   (using unrestricted model)
  1 PARTS          0.03511     3    (4) + 2(3) + 6(1)
  2 OPERATOR       0.00003     3    (4) + 2(3) + 50(2)
  3 PARTS*OPERATOR 0.00151     4    (4) + 2(3)
  4 Error          0.00050          (4)
```

FIGURE 10.15
Analysis of variance
(ANOVA) printout for
the data of
Table 10.7

10.8 INSTRUMENT COMPARISONS

Calibrating an instrument is the most reliable method for ensuring the accuracy of its measurements. Calibration, though, addresses only an instrument's measurement error and does not normally take into account errors of repeatability or reproducibility introduced by different users of the instrument. In turn, conducting full-scale repeatability studies for every instrument and measurement lab can be time-consuming and costly, so an interim step is needed. One solution is to use a relatively simple method for *comparing* two or more instruments, a method capable of detecting any significant differences in the measurements generated by the instruments. Based on the outcome of such a test, one can then decide whether or not to pursue an in-depth repeatability study.

Instrument–instrument comparisons are frequently needed when products measured by a vendor are subsequently remeasured by a customer's receiving inspection department. If the customer's and vendor's measurements are not in agreement, it is possible that 'good' products will be rejected or 'bad' products will be accepted (by the customer). One way of preventing such problems is first to make sure both parties are using the same operational procedures for obtaining the measurements. Then a simple test (such as described in this section) can compare the customer's and vendor's measurements on the *same* sample of items.

EXAMPLE 10.10

Brown (1991) describes the confusion caused by dissimilar procedures used for measuring the length of forged aluminum bars. Although the manufacturer's measurements of the bars were indeed accurate and showed each bar to be well within the allowed tolerance, the customer's inspection of the same bars resulted in a rejection rate of nearly 50%! The discrepancy was due to different measurement *procedures,* not a difference in the measurement instruments. The bars were supposed to be measured in a controlled environment with an ambient temperature of 72°F, a requirement to which the vendor had adhered but the customer had not. The customer had measured the length of the bars in an ambient temperature of 95°F. The resulting thermal expansion increased the lengths enough for some bars to be out of tolerance. Note that in this example the discrepancies could have been avoided if the same operational procedure had been rigorously followed by both parties when making the measurements.

This section considers instrument comparisons involving *variables* measurements. Attributes measurements, on the other hand, require a different approach (see Section 10.9). For variables data, the accuracy and precision of each instrument are compared to those of the other instrument(s). The most desirable outcome is for the instruments to have relatively the same accuracy (as measured by the means) and the same precision (as measured by the variances). Failing that, the comparison should indicate in what manner the instruments disagree: (1) different means/equal variances, (2) different variances/equal means, or (3) different means/ different variances.

To compare two instruments, suppose that k parts are measured *once* by each instrument. The data format for this study is identical to that of Section 10.7, and, for convenience, it is repeated in Table 10.8. Next, on a part-by-part basis, the readings from the two instruments are added (forming the sums $s_1, s_2, \ldots, s_k$) and then subtracted (forming the differences $d_1, d_2, \ldots, d_k$). From these sums and differences, a statistical test was developed by Maloney and Rastogi (1970), who showed that the correlation coefficient, r, between $s_1, s_2, \ldots, s_k$ and $d_1, d_2, \ldots, d_k$ could be used to detect differences in the variances of the two instruments.[3] Differences in accuracies (i.e., the means) are examined by the usual 'matched pairs' test found in introductory statistics texts. The following procedure summarizes both tests:

COMPARING VARIANCES (PAIRED DATA)	1. Calculate the correlation coefficient (Pearson's r) between the sums $s_1, s_2, \ldots, s_k$ and the differences $d_1, d_2, \ldots, d_k$. 2. Apply the usual test of significance to r; that is, calculate $t = (r\sqrt{k-2})/\sqrt{1-r^2}$ and, at significance level α, conclude that $\sigma_x \neq \sigma_y$ if $|t| > t_{\alpha/2}$ [where $t_{\alpha/2}$ is the upper $\alpha/2 \cdot (100\%)$ percentage point of the t distribution with $k - 2$ degrees of freedom].

---

[3]Note that the frequently used F test for comparing variances from two independent samples does not apply in this setting, since the same parts are measured by each instrument; that is, the samples are not independent.

<table>
<tr><td rowspan="2">COMPARING
MEANS (PAIRED
DATA)</td><td>1.</td><td>Calculate the mean $\bar{d}$ and the standard deviation s_d of the differences $d_1, d_2, \ldots, d_k$.</td></tr>
<tr><td>2.</td><td>Apply the usual paired difference test to $\bar{d}$; that is, calculate $t = \bar{d}\sqrt{k}/s_d$ and conclude that $\mu_x \neq \mu_y$ if $|t| > t_{\alpha/2}$ [where $t_{\alpha/2}$ is the upper $\alpha/2 \cdot (100\%)$ percentage point of the t distribution with $k - 1$ degrees of freedom].</td></tr>
</table>

TABLE 10.8   Comparing the Precision and Accuracy of Two Instruments

Part, i	Instrument 1, x_{mi}	Instrument 2, y_{mi}	Sum, $s_1 = x_{mi} + y_{mi}$	Difference $d_i = x_{mi} - y_{mi}$
1	x_{m1}	y_{m1}	$s_1 = x_{m1} + y_{m1}$	$d_1 = x_{m1} - y_{m1}$
2	x_{m2}	y_{m2}	$s_2 = x_{m2} + y_{m2}$	$d_2 = x_{m2} - y_{m2}$
.	.	.	.	.
.	.	.	.	.
.	.	.	.	.
k	x_{mk}	y_{mk}	$s_k = x_{mk} + y_{mk}$	$d_k = x_{mk} - y_{mk}$

**Testing for a Difference in Precision
Between Instrument 1 and Instrument 2**

1. Calculate the correlation coefficient between $s_1, s_2, \ldots, s_k$ and $d_1, d_2, \ldots, d_k$:

$$r = \frac{\sum s_i d_i - \dfrac{1}{k}\left(\sum s_i\right)\left(\sum d_i\right)}{\sqrt{\left[\sum s_i^2 - \dfrac{1}{k}\left(\sum s_i\right)^2\right] \cdot \left[\sum d_i^2 - \dfrac{1}{k}\left(\sum d_i\right)^2\right]}}$$

2. Conclude, at significance level α, that there is a difference in the precisions of the two instruments if $|t| > t_{\alpha/2}$, where $t = r\sqrt{k - 2}/\sqrt{1 - r^2}$ and $t_{\alpha/2}$ is the upper $(\alpha/2) \cdot 100\%$ point of the t distribution with $k - 2$ degrees of freedom.*

**Testing for a Difference in Accuracy
Between Instrument 1 and Instrument 2**

1. Calculate the mean $\bar{d}$ and standard deviation s_d of the differences $d_1, d_2, \ldots, d_k$.
2. Conclude, at significance level α, that there is a difference in the accuracies of the two instruments if $|t| > t_{\alpha/2}$, where $t = \bar{d}\sqrt{k}/s_d$ and $t_{\alpha/2}$ is the upper $(\alpha/2) \cdot 100\%$ point of the t distribution with $k - 1$ degrees of freedom.*

*d.f. $= k - 2$ for the test of precisions; d.f. $= k - 1$ for the test of accuracies.

EXAMPLE 10.11   To compare the accuracy and precision of one instrument against another, $k = 25$ parts are selected and measured once by each instrument. The measurements are listed in Table 10.9, which also includes the summary calculations required for the tests outlined in Table 10.8. To test for a difference

TABLE 10.9 Comparing the Precision and Accuracy of Two Instruments

Part, i	Instrument 1, x_{mi}	Instrument 2, y_{mi}	Sum, s_i	Difference, d_i
1	20.00	19.95	39.95	0.05
2	20.01	20.02	40.03	−0.01
3	19.99	20.06	40.05	−0.07
4	19.97	19.95	39.92	0.02
5	20.02	20.03	40.05	−0.01
6	19.99	20.10	40.09	−0.11
7	20.02	19.95	39.97	0.07
8	19.99	19.94	39.93	0.05
9	19.99	19.98	39.97	0.01
10	19.98	20.01	39.99	−0.03
11	19.98	19.94	39.92	0.04
12	19.97	19.94	39.91	0.03
13	19.99	19.99	39.98	0.00
14	20.00	19.98	39.98	0.02
15	20.02	20.02	40.04	0.00
16	20.01	19.97	39.98	0.04
17	20.00	19.96	39.96	0.04
18	20.02	19.81	39.83	0.21
19	19.96	19.89	39.85	0.07
20	20.01	20.13	40.14	−0.12
21	20.01	19.92	39.93	0.09
22	20.00	19.91	39.91	0.09
23	20.00	20.01	40.01	−0.01
24	20.01	20.01	40.02	0.00
25	19.97	20.07	40.04	−0.10

Correlation between s_i and d_i is $r = -0.878$; $\bar{d} = 0.0148$ and $s_d = 0.0704$.

in the standard deviations of the readings from the two instruments, the t statistic

$$t = \frac{r\sqrt{k-2}}{\sqrt{1-r^2}} = \frac{-0.878\sqrt{25-2}}{\sqrt{1-(-0.878)^2}} = -8.797$$

is calculated and (at a preselected significance level of $\alpha = 0.05$) compared to the critical value $t_{\alpha/2} = t_{0.025} = 2.069$ found from Appendix 5 (using $k - 2 = 23$ degrees of freedom). Since $|t| = 8.797$ exceeds the critical value for this test, it is concluded that the precisions (as measured by the standard deviations of the readings) differ between the two instruments. To test for possible differences in accuracy, the statistic

$$t = \frac{\bar{d}\sqrt{k}}{s_d} = \frac{(0.0148)\sqrt{25}}{0.0704} = 1.051$$

is calculated and compared (using the same significance level of $\alpha = 0.05$) to the critical value $t_{\alpha/2} = t_{0.025} = 2.064$ of Appendix 5 (recall that the degrees of freedom for this test are $k - 1 = 24$). Since $|t| = 1.051$ does not exceed this critical value, it is concluded that there is no significant difference in the average readings generated by the two instruments.

Much work has been done to extend instrument–instrument comparison tests from the original work by Maloney and Rastogi. Blackwood and Bradley (1989, 1991) combined the two tests outlined in this section into a single test for simultaneously detecting differences in the precision or accuracy of two instruments. Other authors have treated the case of more than two instruments. For further reading on multi-instrument comparisons, we again recommend the survey article by Grubbs (1973) and the text by Jaech (1985).

10.9 INSPECTION CAPABILITY

MISCLASSIFICATION RATES

Attributes data arise when inspection instruments or human inspectors classify items by comparing them to given standards. Even though standards are often *stated* in terms of variables measurements, the resulting classifications (i.e., attributes data) are materially different from variables data and must be treated separately. Traditionally, human inspectors have been responsible for generating most attributes data, but with the increasing dependence on ATE (automatic test equipment), much inspection data are now machine generated. Therefore, in the remainder of this section, the term 'inspector' refers to any inspection instrument (human or otherwise) whose evaluation of products and processes is given in terms of attributes measurements. To evaluate inspection capability, we take the same approach as in Section 10.8 by focusing first on a single inspector and then considering the comparison of two inspectors.

The task facing a single inspector is depicted in Figure 10.16. We let the letter C denote a conforming item and let $\overline{C}$ denote a nonconforming one. The inspector examines a group of items and classifies each as either conforming or nonconforming. Since the inspector's *decision* about an item may or may not agree with the item's *true* disposition (i.e., C or $\overline{C}$), the additional notation T and $\overline{T}$ is used to denote the inspector's decision (T means the inspector or tester classifies the item as conforming; $\overline{T}$ indicates a classification as nonconforming).

There are two kinds of mistakes an inspector can make: conforming items may mistakenly be classified as nonconforming, and nonconforming items may incorrectly be classified as conforming. We refer to the *rates* with which an inspector makes these errors as the **performance characteristics** of the inspector. Using the notation of Section 5.3, we can write these rates as *conditional* probabilities:

$P(\overline{T} \mid C) =$ probability of classifying a conforming item as nonconforming

$P(T \mid \overline{C}) =$ probability of classifying a nonconforming item as conforming

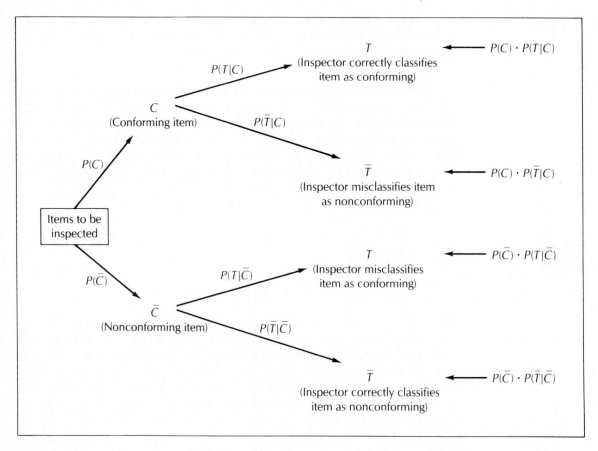

FIGURE 10.16
Conforming items (C) and nonconforming items ($\overline{C}$) are classified as conforming (T) or nonconforming ($\overline{T}$) by an inspector

Note that the rates with which *correct* decisions are made can then be found from the law of complementary events (see Section 5.3); that is, $P(T|C) = 1 - P(\overline{T}|C)$ and $P(\overline{T}|\overline{C}) = 1 - P(T|\overline{C})$. In Figure 10.16, $P(C)$ and $P(\overline{C})$ denote the overall rates with which items submitted to inspection are conforming and nonconforming, respectively. The conditional probability formula from Section 5.3 allows these rates to be combined with the performance characteristics of the inspector to find the rates associated with each of the four branches of the tree diagram. As an example, consider the uppermost branch of Figure 10.16. The probability that an item is conforming *and* is classified as such by the inspector is $P(C \cap T)$, which is found by multiplying the probabilities along the branch; that is, $P(C \cap T) = P(C) \cdot P(T|C)$. These calculations are needed in the analysis that follows.

Consider next what happens when an inspector is used in practice. The most important thing one wants to know is how reliable the inspector is. In other words, what is the probability that items *classified* as conforming (nonconforming) really *are* conforming (nonconforming)? In terms of conditional probabilities, we want to estimate $P(C|T)$ and $P(\overline{C}|\overline{T})$. This can be done by concentrating on the *complements* of these probabilities—namely, $P(C|\overline{T})$ and $P(\overline{C}|T)$. $P(C|\overline{T})$ is called a **false positive** in the sense that the inspector says the item is nonconforming while, in truth, it *is* conforming. Similarly, $P(\overline{C}|T)$ is a **false negative,** since the inspector has called an item conforming even though it is truly nonconforming. We borrow the terminology of false positives and negatives from the study of medical screening tests, where a false positive refers to the case of a patient testing 'positive' for a disease when, in fact, the patient is free of the disease (Fleiss 1981). In the language of inspection, nonconforming items correspond to patients with a disease, and conforming items correspond to those free of the disease.

To estimate the **misclassification rates** $P(C|\overline{T})$ and $P(\overline{C}|T)$, we again appeal to the conditional probability formula and write

$$P(C|\overline{T}) = \frac{P(C \cap \overline{T})}{P(\overline{T})} \tag{10.22}$$

$$P(\overline{C}|T) = \frac{P(\overline{C} \cap T)}{P(T)} \tag{10.23}$$

Both the numerators and denominators in these equations are found by summing the appropriate branches in Figure 10.16. For example, to find $P(\overline{T})$, one sums the probabilities associated with both branches that end with a $\overline{T}$; thus, $P(\overline{T}) = P(C) \cdot P(\overline{T}|C) + P(\overline{C}) \cdot P(\overline{T}|\overline{C})$. Substituting these results into equations (10.22) and (10.23), we arrive at the misclassification rates

$$P(C|\overline{T}) = \frac{P(C) \cdot P(\overline{T}|C)}{P(C) \cdot P(\overline{T}|C) + P(\overline{C}) \cdot P(\overline{T}|\overline{C})} \tag{10.24}$$

$$P(\overline{C}|T) = \frac{P(\overline{C}) \cdot P(T|\overline{C})}{P(C) \cdot P(T|C) + P(\overline{C}) \cdot P(T|\overline{C})} \tag{10.25}$$

Notice that in equations (10.24) and (10.25) we have succeeded in 'turning around' the conditional probabilities in the sense that the conditioning events T and $\overline{T}$ in $P(C|\overline{T})$ and $P(\overline{C}|T)$ are exchanged for calculations that involve conditioning events C and $\overline{C}$. Readers with a background in probability theory will notice that equations (10.24) and (10.25) are simple applications of Bayes theorem.

In order to find the misclassification rates, a study must first be conducted to estimate the performance characteristics $P(\overline{T}|C)$ and $P(T|\overline{C})$. To do this, a special group of items must be inspected, items whose true disposition (conforming or nonconforming) is already known with certainty. As a practical matter, the items should be submitted for inspection in a random manner to simulate how the inspection operation normally operates. Since the exact disposition (C or $\overline{C}$) of each item in this group is known, estimates of $P(\overline{T}|C)$ and $P(T|\overline{C})$ can be obtained. We let n_c and $n_{\overline{c}}$ denote the numbers of known conforming and nonconforming items submitted for inspection. Then the results of this study can be summarized as in Figure 10.17.

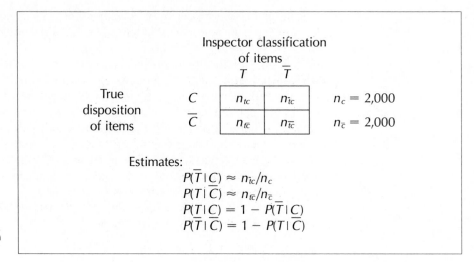

FIGURE 10.17
Data format for
estimating inspection
performance

Notice that the performance characteristics are unique to the inspector but that the misclassification rates depend on *both* the performance characteristics and the underlying rates of conforming and nonconforming items, $P(C)$ and $P(\overline{C})$, submitted for inspection. Because of this, misclassification rates vary depending on the relative proportions of conforming and nonconforming items in the population.

In most applications, the nonconformance rate $P(\overline{C})$ should be fairly small. Unfortunately, as $P(\overline{C})$ becomes smaller, it is well known that the false positive rate $P(C\,|\,\overline{T})$ becomes larger (Diamond 1990; Fleiss 1981, p. 7; Yasuda 1989). Security alarm systems for automobiles provide a good example of the false alarm problem. In the realm of quality control, the better a process becomes, the more likely it is that any items the inspector classifies as bad are, in fact, good. Of course, this problem may be overcome by improving inspection performance, but the degree of improvement required can be substantial. Example 10.12 illustrates the calculation of misclassification rates and the problem of high false positive rates.

EXAMPLE 10.12

To characterize an inspection device, $n_c = 2{,}000$ known conforming items and $n_{\overline{c}} = 2{,}000$ known nonconforming items are inspected by the device. The results of the inspection are summarized in Figure 10.18. From these data, the performance characteristics of the device are estimated to be

$$P(\overline{T}\,|\,C) \approx \frac{n_{\overline{t}c}}{n_c} = \frac{40}{2{,}000} = 0.02$$

$$P(T\,|\,\overline{C}) \approx \frac{n_{t\overline{c}}}{n_{\overline{c}}} = \frac{60}{2{,}000} = 0.03$$

$$P(T\,|\,C) = 1 - P(\overline{T}\,|\,C) \approx 1 - 0.02 = 0.98$$

$$P(\overline{T}\,|\,\overline{C}) = 1 - P(T\,|\,\overline{C}) \approx 1 - 0.03 = 0.97$$

To evaluate how the device will behave in practice, the nonconformance rate $P(\overline{C})$ is needed. Since $P(\overline{C})$ is rarely known with certainty, it is more informative

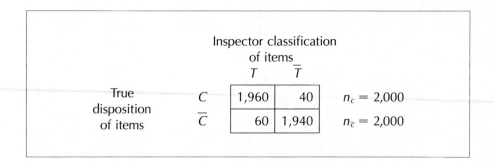

FIGURE 10.18
Characterizing the
performance of an
inspection device

to select an interval of possible values that $P(\overline{C})$ could assume and then calculate the misclassification rate for *each* value of $P(\overline{C})$ in the interval. For the purposes of this example, suppose that nonconforming items could range from as many as 1 per 100 to as few as 1 per million. After equations (10.24) and (10.25) are applied to these rates, the misclassification probabilities are tabulated in Table 10.10. Notice how high the false positive rate is across the whole range of possible values of $P(\overline{C})$. If this device is to be used in practice, a decision has to be made as to whether any problems or costs associated with false positives can be ignored or whether an effort should be made to improve the capability of inspection.

In the latter case, equations (10.24) and (10.25) can be used to determine just how much of an improvement is needed to lower the false positive rate significantly. To illustrate, suppose that with some additional costs the inspection performance can be improved approximately tenfold so that $P(\overline{T}|C) \approx 0.001$ and $P(T|\overline{C}) \approx 0.001$. The misclassification rates associated with this improved level of inspection are listed in Table 10.11. The false positive rate has decreased at each level of $P(\overline{C})$, but not as much as intuition might have suggested. By experimenting with different values of $P(\overline{T}|C)$ and $P(T|\overline{C})$, one can explore the tradeoff between higher levels of inspection performance (with their associated higher costs) and the costs of false positives.

TABLE 10.10   Misclassification Rates for the Inspection Device of
Example 10.12

		Misclassification rates	
Nonconformance rate $\downarrow$ $P(\overline{C})$		False positive $\downarrow$ $P(C|\overline{T})$	False negative $\downarrow$ $P(\overline{C}|T)$
1/100	= 0.01	0.671186	0.0003091
1/500	= 0.002	0.911416	0.0000613
1/1,000	= 0.001	0.953699	0.0000306
1/10,000	= 0.0001	0.995173	0.0000031
1/100,000	= 0.00001	0.999515	0.0000003
1/1,000,000	= 0.000001	0.999951	0.0000000

Note: Misclassification rates are calculated from equations (10.24) and (10.25) from $P(\overline{T}|C) \approx 0.02$ and $P(T|\overline{C}) \approx 0.03$.

TABLE 10.11   Improving the Inspection Performance Improves the
Misclassification Rates

Nonconformance rate $\downarrow$ $P(\overline{C})$	Misclassification rates	
	False positive $\downarrow$ $P(C\,|\,\overline{T})$	False negative $\downarrow$ $P(\overline{C}\,|\,T)$
$1/100$ $= 0.0$	0.090164	0.0000101
$1/500$ $= 0.002$	0.333111	0.0000020
$1/1{,}000$ $= 0.001$	0.500000	0.0000010
$1/10{,}000$ $= 0.0001$	0.909165	0.0000001
$1/100{,}000$ $= 0.00001$	0.990109	0.0000000
$1/1{,}000{,}000 = 0.000001$	0.999002	0.0000000

Note: Misclassification rates are calculated from equations (10.24) and (10.25) with improved rates of $P(\overline{T}\,|\,C) \approx 0.001$ and $P(T\,|\,\overline{C}) \approx 0.001$.

COMPARING TWO INSPECTORS

Two questions arise when one compares the capabilities of two (or more) inspectors. First, is there a significant difference in the performance characteristics of the inspectors? Second, if a difference does exist, how should the inspection data from both inspectors be combined to give a better view of the process? From an operational point of view, the second question seems more important, especially when human inspectors are involved. Since it is very unlikely that two human inspectors will always produce identical inspection results, testing for such differences often only confirms this intuitive fact. Remedial action (e.g., training) is possible in cases where large differences exist between inspectors, but even after such corrections are made, there will continue to be some discrepancies in their classifications of inspected items. In this section, we describe statistical methods for comparing inspection performance and for combining inspection data from two inspectors.

TESTING FOR DIFFERENCES.   One approach to testing for differences between two inspectors is to compare their performance characteristics, $P(\overline{T}\,|\,C)$ and $P(T\,|\,\overline{C})$. Suppose, for example, that one wants to compare $P(\overline{T}\,|\,C)$ for two inspectors. Starting with samples of n_1 and n_2 *known* conforming items, inspector 1 examines all n_1 items in the first sample, while inspector 2 examines all n_2 items in the second sample. (From a statistical standpoint it is best to have $n_1 = n_2$, but constraints of the production environment may make this impossible in some cases.) The results are then recorded as in Figure 10.19, where X_1 and X_2 denote the numbers of nonconforming items found by the inspectors 1 and 2, respectively. Next, the proportions $\hat{p}_1 = X_1/n_1$ and $\hat{p}_2 = X_2/n_2$, which estimate $P(\overline{T}\,|\,C)$ for each inspector, are calculated and tested using the standard "independent samples z test" for testing a difference of two proportions:

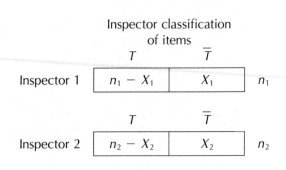

Inspector classification
of items

FIGURE 10.19
Data format for
comparing the
performance
characteristics
$P(\bar{T}|C)$ of two
inspectors

Note: The n_1 and n_2 items inspected are *known,* prior to inspection, to be conforming items.

TESTING FOR A
DIFFERENCE
BETWEEN TWO
PROPORTIONS
BASED ON DATA
FROM TWO
INDEPENDENT
SAMPLES

1.   Form the "pooled estimate" $\hat{p} = (X_1 + X_2)/(n_1 + n_2)$ and the statistic $z = (\hat{p}_1 - \hat{p}_2)/\sqrt{\hat{p}(1 - \hat{p})\left[\frac{1}{n_1} + \frac{1}{n_2}\right]}$.

2.   Conclude, at significance level α, that the proportions p_1 and p_2 are different if $|z| > z_{\alpha/2}$ [where $z_{\alpha/2}$ is the upper $\alpha/2 \cdot (100\%)$ percentage point of the standard normal distribution].

In a similar fashion, a separate test using only *nonconforming* items can be conducted to detect possible differences between the $P(T|\bar{C})$ rates for both inspectors.

Testing for differences in performance characteristics, as described above, is usually done in a nonproduction setting. This raises a concern about the effect that examining batches of entirely conforming (or nonconforming) items might have on the quality of the inspection results. It is easy to imagine, for example, how seeing 100 nonconforming items in a row might sway one's opinion about the 101st item. In an attempt to handle this problem, one can instead submit *any* batch of items to the inspectors and then *subsequently* analyze the items to determine their true disposition (C or $\bar{C}$), from which the performance characteristics could then be estimated. If this approach is adopted, reliable estimates of the performance characteristics require much larger samples than before because the proportion of nonconforming items, $P(\bar{C})$, which is normally much smaller than $P(C)$, results in smaller numbers of nonconforming items being examined. In other words, with this approach, the only way to increase the number of nonconforming items inspected to a reasonable level is to increase the total number of items inspected.

Turning to a different concern, we note that even if reliable estimates of $P(T \mid \overline{C})$ and $P(\overline{T} \mid C)$ can be obtained, they eventually have to be translated into estimates of the misclassification rates. This requires knowing the rates $P(C)$ and $P(\overline{C})$ with which conforming and nonconforming items are submitted to inspection, and the problem is that these rates may vary over time. It is desirable, then, to have methods that do not depend on knowing $P(C)$.

One approach, instead of submitting controlled groups of all conforming (or all nonconforming) items to the inspectors, is to simply select *any* group of n production items and ask both inspectors to examine them. This way, both inspectors can be compared (since they examine the *same* group of n items), and we feel comfortable that they perform their inspections in an environment similar to that experienced in production. Unlike the previous method, however, the procedures described next do not require that we establish the *true* disposition (i.e., conforming or nonconforming) of each of the n items inspected.

Suppose that n production items are first submitted to inspector 1 and then the *same* n items are submitted to inspector 2. Figure 10.20 shows a convenient format for displaying the results of this inspection. For any item, the inspectors are said to "agree" if they both rate the item as conforming (T) or if both rate it as nonconforming ($\overline{T}$). Otherwise, they are said to "disagree" on the item. Furthermore, the *direction* of the disagreements can be either fairly evenly split between $(\overline{T}, T)$ and $(T, \overline{T})$ or heavily one-sided (e.g., if one inspector consistently classifies items as T, while the other classifies them as $\overline{T}$ *whenever they disagree*). The following procedure provides a test for the second of these two forms of disagreement between inspectors:

TESTING FOR DISAGREEMENT BETWEEN TWO INSPECTORS— PAIRED DATA	1. For a set of n items inspected by each of two inspectors, count the number of pairs $(\overline{T}, T)$ and $(T, \overline{T})$ in which the inspectors *disagreed* on the classification of an item. Denote these numbers by b and c, respectively. 2. Calculate the statistic $\chi^2 = (b - c)^2/(b + c + 1)$ and conclude, at significance level α, that the inspectors disagree in the 'direction' of their classifications if $\chi^2 > \chi^2_\alpha$ [where χ^2_α is the upper $\alpha \cdot (100\%)$ point of the chi-square distribution with 1 degree of freedom]. {*Note:* Because the chi-square with 1 degree of freedom is equal to the square of a standard normal variable, $\chi^2 > \chi^2_\alpha$ is equivalent to $[(b - c)^2/(b + c + 1)]^{1/2} > z_{\alpha/2}$, where $z_{\alpha/2}$ is the upper $\alpha/2 \cdot (100\%)$ point of the standard normal distribution.}

In statistical terms, this procedure compares two sets of *paired* attributes data (the pairing occurs because the *same* items are examined by both inspectors) and goes under the name of the McNemar test in the literature (Marascuilo and McSweeney 1977).

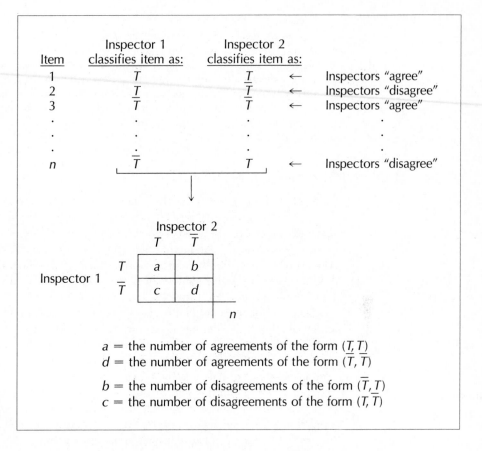

FIGURE 10.20
Data format for
counting
disagreements
between two
inspectors when the
same set of *n* items is
examined

EXAMPLE 10.13

Two auditors have to examine 100 accounting records for possible errors.
Records that contain any errors are classified as nonconforming ($\overline{T}$); otherwise,
they are considered to be conforming (T). Figure 10.21 summarizes the results of
this 200% inspection. To test for a possible difference in the inspectors' work,
the χ^2 statistic is calculated:

$$\chi^2 = \frac{(b - c)^2}{b + c + 1} = \frac{(12 - 2)^2}{12 + 2 + 1} = \frac{100}{15} = 6.67$$

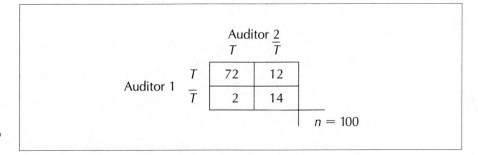

FIGURE 10.21
100 accounting
records inspected for
errors by each of two
auditors

Recalling that the critical value χ_α^2 is just the square of $z_{\alpha/2}$, we find $\chi_{0.05}^2 = z_{0.025}^2 = (1.96)^2 = 3.84$ for a significance level of $\alpha = 0.05$. Since $\chi^2 > \chi_\alpha^2$ for these data, it can be concluded that there does exist a (statistically) significant difference in the *disagreements* between the inspectors. In particular, disagreements of the form $(T, \overline{T})$ occur more often than those of the form $(\overline{T}, T)$; that is, when the inspectors disagree, it is usually because inspector 2 thinks an item is nonconforming while inspector 1 believes it is conforming. This could mean that inspector 2 is using a more stringent definition of 'nonconforming' than inspector 1.

COMBINING INSPECTION DATA

Up to now, the tests in this section have been designed to detect possible differences in inspection capability. In practice, however, we generally expect to find *some* disagreements between two inspectors, regardless of how good each inspector is, and in this frequently occurring situation, the question that naturally arises is whether or not their inspection results can somehow be combined to give a better *estimate* of the number of nonconforming items in the items inspected. The following procedure provides such an estimate.

Suppose that two inspectors examine the same set of n items and that inspector 1 classifies X_1 of them as nonconforming, while inspector 2 finds X_2 nonconforming items. Usually, there is a reasonable amount of agreement between the inspectors; that is, many of the *same* items are classified as nonconforming by both inspectors. Let d denote the number of items that *both* inspectors agreed are nonconforming. From this information, it is possible to estimate the *true* number of nonconforming items in the batch as well as the number of nonconforming items that are mistakenly missed by both inspectors. Polya (1976) has shown that the total number of nonconforming items among the n items inspected is estimated by

$$\hat{n}_{\bar{c}} = \frac{X_1 \cdot X_2}{d} \tag{10.26}$$

Furthermore, the number of nonconforming items *missed* by the two inspectors is estimated by[4]

$$\hat{m}_{\bar{c}} = \frac{(X_1 - d) \cdot (X_2 - d)}{d} \tag{10.27}$$

EXAMPLE 10.14   While reasons were investigated for the discrepancies in the two auditors' inspection results in Example 10.13, it was determined that auditor 2 was, in fact, using more acceptable criteria for what constituted an accounting 'error' than was auditor 1. Furthermore, the two errors found by auditor 1 were deemed to be genuine (auditor 2 had not caught them), as were the 14 errors that both auditors agreed upon. Thus, the total number of errors found by

---

[4]Feller (1950) attributes these estimates to a method originated by the physicist Rutherford.

auditor 1 was $X_1 = 2 + 14 = 16$, while auditor 2 found $X_2 = 12 + 14 = 26$. They agreed on $d = 14$ of these errors. From equations (10.26) and (10.27), the number of errors that were *missed* by both inspectors is estimated to be about $\hat{m}_{\bar{c}} = (X_1 - d) \cdot (X_2 - d)/d = (16 - 14) \cdot (26 - 14)/14 = 1.71$, while the total number of errors in the 100 records is estimated by $\hat{n}_{\bar{c}} = X_1 \cdot X_2/d = 16 \cdot 26/14 = 29.71$.[5] According to these estimates, the auditors *together* performed fairly well, missing only perhaps one or two errors in the entire batch.

EXERCISES FOR CHAPTER 10

10.1   Can calibration improve the precision of a measuring instrument?

10.2   Averaging several repeated measurements of an object does tend to reduce measurement error, but it does not reduce the error variation to 0. Explain.

10.3   Many instruments cannot be brought into exact calibration (e.g., a standard gage block that is longer than its stated value). Explain how the process of calibration enables such instruments to be used with accuracy.

10.4   Explain the role of the NIST in assuring measurement accuracy and precision.

10.5   In Section 10.7 and equation (10.21) it was noted that $\bar{x}$ and R charts can be used to estimate measuring instrument precision, as well as part-to-part variation.[6] Suppose that 30 parts are each measured twice and that the resulting pairs of readings are used to form the subgroups for $\bar{x}$ and R charts.
(a) For these data, explain why the R chart describes only the measuring instrument variation and does not include any of the part-to-part variation.
(b) Explain why the majority of the points on the corresponding $\bar{x}$ chart should be *outside* the control limits if the measuring instrument is of adequate precision.
(c) Explain why finding a majority of the points on the $\bar{x}$ chart within the control limits is an indication of a poor measurement system.

10.6   A scale gives weight measurements that are accurate to within 2% of the true weight of an object; that is, virtually all (99%) of the repeated measurements of a single object fall within 2% of the object's true weight. Two different objects are to be weighed and their total weight reported. Assuming that all weighings are independent of one another, how should you proceed? Should the two objects be weighed simultaneously and the total recorded, or should each object be weighed separately and the two weights summed? Justify your answer by considering which method has the smaller error variation.

---

[5]Note that only one of $\hat{n}_{\bar{c}}$ or $\hat{m}_{\bar{c}}$ need be calculated, since they are related by the equation $\hat{n}_{\bar{c}} = (X_1 + X_2 - d) + \hat{m}_{\bar{c}}$.

[6]For more on this subject, see Anthis, Stanula, and Hart (1991).

10.7 The web thickness of an aluminum alloy part used in the construction of airliners is measured by two inspectors. Ten different parts are used and the thickness (in inches) is measured twice by each inspector.

Part		Inspector 1	Inspector 2
1	Measurement 1	0.135	0.130
	Measurement 2	0.135	0.130
2	Measurement 1	0.120	0.115
	Measurement 2	0.120	0.120
3	Measurement 1	0.125	0.125
	Measurement 2	0.130	0.130
4	Measurement 1	0.140	0.120
	Measurement 2	0.125	0.125
5	Measurement 1	0.130	0.125
	Measurement 2	0.130	0.115
6	Measurement 1	0.135	0.125
	Measurement 2	0.135	0.130
7	Measurement 1	0.125	0.125
	Measurement 2	0.130	0.125
8	Measurement 1	0.120	0.125
	Measurement 2	0.125	0.120
9	Measurement 1	0.125	0.120
	Measurement 2	0.130	0.115
10	Measurement 1	0.125	0.120
	Measurement 2	0.125	0.125

From these data, estimate the part-to-part variation, the inspector variation, and the instrument variation.

10.8 Make two photocopies of the first page of Chapter 10. On one copy, circle each occurrence of the letter 'f.' Give the other copy to a friend to do the same. Work independently and examine each word only one time (i.e., perform only a 100% inspection). Afterward, compare the two sheets by counting the number of f's you found, the number that your friend found, and the number you *both* found (i.e., those occurrences where you both circled the exact same letter in a word).
(a) Using these data, estimate the number of f's on the entire page.
(b) Estimate the number of f's that were missed by *both* you and your friend.

10.9 Human visual inspection of solder joints on printed circuit boards (PCBs) can be very subjective. Part of the problem stems from the numerous ways in which a joint can be nonconforming (e.g., pad nonwetting, knee visibility, voids, to name but a few) and the *degree* to which a joint may exhibit any of these problems. Consequently, even highly trained inspectors tend to disagree when examining the same PCB. The accompanying data show the results of two inspectors who examined a number of solder joints for the particular problem of 'pad nonwetting.'

	Number of defective solder joints found
Inspector I	724
Inspector II	751
Common to both	316

(a) Estimate the total number of solder joints that have the 'pad nonwetting' problem on the PCBs that were examined.

(b) Estimate the number of solder joints with this problem that were missed by *both* inspectors.

REFERENCES FOR CHAPTER 10

Anthis, D. L., R. J. Stanula, and R. F. Hart. 1991. "The Measurement Process: Roadblock to Product Improvement?" *Quality Engineering* 3 (no. 4):461–470.

Barry, B. A. 1964. *Engineering Measurements,* pp. 2–4. New York: Wiley.

Belanger, B. 1984. "Measurement Assurance Programs, Part I: General Introduction." *NBS Special Publication 676-I,* U.S. Department of Commerce.

Bishop, L., W. J. Hill, and W. S. Lindsay. 1987. "Don't Be Fooled by the Measurement System." *Quality Progress,* December, p. 37.

Blackwood, L. G., and E. L. Bradley. 1991. "An Omnibus Test for Comparing Two Measuring Devices." *Journal of Quality Technology* 23 (no. 1):12–16.

Bradley, E. L., and L. G. Blackwood. 1989. "Comparing Paired Data: A Simultaneous Test for Means and Variances." *American Statistician* 43:234–235.

Brown, B. J. 1991. "Precision Measurement with Thermal Expansion." *Quality Progress* 24 (no. 2):65–68.

Diamond, S. J. 1990. "Home Security Needn't Be So Alarming." *Los Angeles Times,* November 2.

Dixon, W. J., and F. J. Massey. 1983. *Introduction to Statistical Analysis,* 4th ed., chaps. 10, 15. New York: McGraw-Hill.

Duncan, A. J. 1986. *Quality Control and Industrial Statistics,* 5th ed. Homewood, IL: Irwin.

Eisenhart, C. 1963. "Realistic Evaluation of the Precision and Accuracy of Instrument Calibration Systems." *Journal of Research,* vol. 67C, no. 2, National Bureau of Standards.

Eisenhart, C. 1967. "Realistic Evaluation of the Precision and Accuracy of Instrument Calibration Systems." Special Publication 300, vol. 1, National Bureau of Standards.

Farnum, N. R., and L. W. Stanton. 1989. *Quantitative Forecasting Methods,* pp. 73–76. Boston: PWS-Kent.

Feller, W. 1950. *An Introduction to Probability Theory and Its Applications,* vol. 1, p. 170. New York: Wiley.

Fleiss, J. L. 1981. *Statistical Methods for Rates and Proportions,* New York: Wiley.

Grubbs, F. E. 1948. "On Estimating Precision of Measuring Instruments and Product Variability." *Journal of the American Statistical Association* 43:243–264.

Grubbs, F. E. 1973. "Errors of Measurement, Precision, Accuracy and the Statistical Comparison of Measuring Instruments." *Technometrics* 15(no. 1):53–66.

Jackson, D. 1987. "Instrument Intercomparison and Calibration." *Proceedings of the 1987 Measurement Science Conference,* January 29–30, Irvine, CA, pp. III A(1)–III (A)21.

Jaech, J. L. 1985. *Statistical Analysis of Measurement Errors,* New York: Wiley.

Maloney, C. J., and S. C. Rastogi. 1970. "Significance Tests for Grubbs Estimators." *Biometrics* 26:671–676.

Mandel, J., and T. W. Lashof. 1987. "The Nature of Repeatability and Reproduceability." *Journal of Quality Technology* 19(no. 1):29–36.

Marascuilo, L. A., and M. McSweeney. 1977. *Nonparametric and Distribution-Free Methods for the Social Sciences.* Monterey, CA: Brooks/Cole.

Measurement Systems Analysis Reference Manual. 1990. Southfield, MI: Automotive Industry Action Group.

Polya, G. 1976. "Probabilities in Proofreading." *American Mathematical Monthly* 83 (no. 1):42.

Sachs, L. 1982. *Applied Statistics,* p. 249. New York: Springer-Verlag.

Sirohi, R. S., and H. C. Radha Krishna. 1980. *Mechanical Measurements.* New York: Wiley.

Sobralske, B. 1989. "Measuring Up to MIL-STD-45662A." *Quality Progress,* September, pp. 51–52.

Thomas, G. G. 1974. *Engineering Metrology.* New York: Wiley.

The United States Government Manual. 1991/1992. Washington, DC: National Archives and Records Administration.

U.S. Department of Defense. 1980. MIL-STD-45622, *Military Standard Calibration Systems Requirements.* Washington, DC.

Yasuda, G. 1989. "Big Market Seen for Bomb Detectors." *Los Angeles Times,* August 15, Part IV, p. 7.

Youden, W. J. 1954. "Instrumental Drift." *Science* 120(no. 3121):627–631.

Youden, W. J. 1967. "Precision Measurements and Calibration: Statistical Concepts and Procedures." Special Publication 300, Vol. 1, pp. 133–138, National Bureau of Standards.

ACCEPTANCE SAMPLING

The concepts and calculations of acceptance sampling plans are presented along with a discussion of the modern role of acceptance sampling in quality programs. In addition to the general theory of acceptance sampling, the MIL-STD-105E sampling system (for attributes data) and the MIL-STD-414 sampling system (for variables data) are covered.

CHAPTER OUTLINE

| 11.1 | INTRODUCTION |

Acceptance sampling is another statistical tool for assessing product quality. It is similar to other statistical methods in that it makes inferences about a process based on samples of items from the process. However, it differs from the control chart approach in its method of selecting samples (random samples versus rational subgroups) and in its traditional point of application in the production process.

Historically, acceptance sampling has been applied at either final inspection (by the producer) or incoming inspection (by the customer). Although there is no *statistical* reason to use it this late in production, final and incoming inspections are certainly convenient locations because most acceptance sampling plans are based on the inspection of batches (or 'lots') of items, and lot formation is normally done late in the production cycle. The tendency to use acceptance sampling late in production has led to the criticism that it is primarily a detection-oriented approach to quality (see Section 2.5). In many ways this criticism is valid, and it is for this reason that modern quality programs stress process control techniques over acceptance sampling. However, acceptance sampling still has an important role to play, one that control chart methods are not designed to handle. Section 11.3 examines the complementary roles played by acceptance sampling and control charts in modern quality control.

The mechanics of acceptance sampling are fairly simple. At the end of production, products are grouped together into **lots** before they are shipped to the customer. Normally, the lot size (the number of items that comprise a lot) is related to the physical size of the product. Thus, items such as bolts or fasteners are more economically shipped in large lots, whereas television sets usually come one per lot. Lots are identified by means of **lot numbers,** which facilitate traceability. Acceptance sampling proceeds by taking a random sample from a particular lot and either *measuring* a particular quality characteristic or *counting* the number of sampled items that do not meet specifications. With variables data, a lot is rejected whenever its sample mean exceeds some specified limits. For attributes data, a lot rejected when the number of nonconforming items in the sample exceeds some specified number. Lots that pass the inspection are then shipped (if the manufacturer does the inspection) or put into stock (if the customer does the inspection). Because of their relative simplicity, acceptance sampling plans based on attributes data tend to be used much more frequently than plans based on variables data.

Schilling (1982) traces the roots of both acceptance sampling and control charts to the same source, the Inspection Engineering Department of Bell Laboratories in the early 1920s. Control chart methodology arose from Shewhart's now famous 1924 memo describing the *p* chart; the fundamentals of acceptance sampling were developed during 1925–1927. After that, acceptance sampling was applied sporadically until World War II, when it was incorporated into military standards. The use of these plans in military and defense contract work helped promote widespread use of acceptance sampling, which has continued to this day. One of these plans, MIL-STD-105E, is in such wide use that we devote a section of this chapter to it. For more information on the history of acceptance sampling, the reader is referred to the summaries given in Duncan (1986) and Schilling (1982).

11.2 SAMPLING

REASONS FOR SAMPLING

By their nature, statistical methods involve sampling. Control charts use rational subgrouping, surveys use stratified and/or cluster sampling, and experimental designs use sampling plans for estimating the effects of process variables. Similarly, acceptance sampling is based on *random* sampling of lots.[1]

As summarized in Section 5.2, sampling (instead of 100% inspection) is used for a variety of reasons. From the point of view of data quality, a thorough inspection of a few sampled items often provides more reliable information than can be obtained by 100% inspection. As Vardeman (1986) points out, with 100% inspection, each item receives *some* attention, but the extent of this examination is necessarily less than the detailed analysis possible when a sample is inspected.

RANDOM SAMPLING

Control charts are formed from the data in *rational subgroups,* not from random samples per se (see Section 6.3). In charting, randomness is brought in by the *assumption* that a process is influenced by only a system of 'common causes.' Under this assumption, any collection of items from the process (including those in a rational subgroup) can be considered to be statistically independent. In turn, the sampling distributions of control chart statistics ($\bar{x}$, R, s, p, c, u, etc.) from independent samples are well known and provide the necessary estimates of control limits for the charts (see Section 5.4). If some points on the chart exhibit an 'out of control' condition, then one concludes that the assumption of a 'chance cause' system is wrong and that some 'special causes' have influenced the system.

Acceptance sampling, on the other hand, requires that random sampling be done at the outset; that is, the samples selected for inspection are drawn in as 'random' a manner as possible. Random sampling does the same thing for acceptance sampling that the assumption of a 'chance cause' system does for control charts: it assures the independence of the items selected and thereby makes probability calculations possible (see Section 5.2).

Obtaining a random sample requires the use of random number generators of some sort because samples based on expert judgment are rarely random in the statistical sense. To understand the mechanics of drawing a random sample, the reader should briefly review Sections 5.2 and 5.3. In addition, we mention a few

---

[1] In the case of continuous processes, where no lots are formed, acceptance sampling is based on a fraction of the items produced.

simple rules that are of great help in practice. These rules, along with their proofs and many examples, may be found in Wright and Tsao (1985):

<table>
<tr>
<td>AIDS FOR
GENERATING
RANDOM SAMPLES</td>
<td>

1. The *complement* of a random sample of *n* from a lot of size *N* is itself a random sample of size *N − n* from the lot.
2. Any random *subsample* of a random sample from a lot is also a random sample from the lot.
3. Any random *subsample* from the *complement* of a random sample of a lot is itself a random sample from the lot.
4. After a random sample of *n* is drawn, any random sample from its *complement* can be added to it to form a larger random sample from the lot.

</td>
</tr>
</table>

11.3   ACCEPTANCE SAMPLING VERSUS CONTROL CHARTS

The move from product inspection to process control is a fundamental tenet of modern quality philosophy. It is widely recognized that control over processes should be as immediate as possible, with inspection moved upstream in production and prevention-based techniques used in favor of detection-based methods (see Section 2.5). The justifications for this shift in emphasis are uncomplicated and easily summarized by catch phrases, such as "You can't inspect quality into a product" and "Inspection isn't perfect."

As the number of arguments to support prevention-based methods grows, so grows the number of criticisms of inspection-based methods. Because of its widespread usage, acceptance sampling has received much of this criticism. In this section, we consider acceptance sampling in this new light and examine the role it plays in modern quality control.

In a very real sense, the goal of acceptance sampling is to reduce the need for acceptance sampling (Schilling 1984). Acceptance sampling is best used in situations either where control charts cannot be applied or where 'special causes' still abound. In particular, acceptance sampling is of value (1) for processes that are not yet 'in control,' (2) as a means of protection against gross production mistakes, (3) for products where safety or liability is of paramount concern, (4) for volatile processes with uneven quality, (5) for correcting problems in already created lots in order to meet specified lot quality levels, and (6) for low-yield processes or first-article inspection (Vardeman 1986).

Although it may be the desired goal, even Shewhart and Deming (1939) recognized that the statistical control of processes "cannot be reached in one day" and, furthermore, that "it [statistical control] cannot be reached in the production of a product where only a few pieces are manufactured." In general, the amount of acceptance sampling used should gradually decrease, "depending on the maturity of the product or process involved and the success (or lack thereof) in achieving statistical control."[2]

To the above list of applications we add the broad area of service-related applications, where monitoring via control charts is sometimes impossible. As an example, Mudryk (1988) lists several reasons for using acceptance sampling (and not control charts) to control the processing quality of large surveys conducted by *Statistics Canada*. This list is paraphrased as follows:

<table>
<tr><td>REASONS
FOR USING
ACCEPTANCE
SAMPLING</td><td>

1. Process stability cannot be assumed, nor is it always attained in the long run.
2. Assignable causes for errors are not always known or knowable, since the operations are labor-intensive.
3. Processes cannot be readily stopped and adjusted for assignable causes, even when these causes are known.
4. Large 'between-operator' variation is common, and running individual control charts for each operator that are updated after each observation is operationally difficult to do.

</td></tr>
</table>

Notice how general this list is. It could be applied to many paperwork-intensive or labor-intensive operations and is not limited solely to the survey application described by Mudryk. Another observation is that many service products are produced precisely at the moment of delivery to the customer (e.g., the creation of bank transactions, medical treatments, travel arrangements), which offers little or no time for the type of feedback provided by the control chart.

Part of modern quality emphasis also includes achieving nonconformance rates of a few parts per million (ppm) instead of rates measured in percentages (i.e., nonconforming items per hundred). Can acceptance sampling plans help achieve quality levels in the ppm range? For the most frequently used acceptance sampling plans (based on attributes data), the answer is no. To show why, recall that even if *no* nonconforming items are found in a random sample of n items, then a $(1 - \alpha) \cdot 100\%$ confidence bound for the nonconformance rate in the population is given by $1 - \alpha^{1/n}$ (see Section 5.7). To achieve confidence bounds on the order of a few ppm then requires huge sample sizes, well beyond those used in common acceptance sampling programs. To illustrate, suppose that a confidence level of 95% (so that $\alpha = 0.05$) is specified and that one wants an upper bound on the nonconformance rate of 1 ppm. This means that $1 - \alpha^{1/n} = 1/10^6 = 0.000001$, which, when solved for n, becomes $n = \ln(\alpha)/\ln(1 - 1/10^6) = \ln(0.05)/\ln(0.999999) = 2,995,729.3$. Sample sizes used in acceptance sampling are nowhere near this size, again affirming that acceptance sampling is best used to control quality for processes that operate with much larger nonconformance rates, such as those listed earlier in the section.

---

[2] See Schilling (1990, pp. 181–191).

Variables control charts are the preferred method of monitoring processes that operate at low ppm nonconformance levels. Of course, control charts also use very small sample sizes, even smaller than those used in acceptance sampling plans. However, if we *assume* that a process follows a certain distribution (usually, the normal distribution), control chart statistics can be converted into estimates of nonconformance rates at *any* level, including the ppm level. It should be noted, though, that process measurements can never be *proven* to follow a particular probability distribution, so that by using control chart data in this fashion, one is implicitly performing what Deming calls an 'analytic' study (see Section 3.1).

11.4   LOT FORMATION

Acceptance sampling can be performed either on a lot-by-lot basis or on a continuous stream of product, but the most frequently used plans are based on lot sampling. Of the various types of lots that can be formed (e.g., manufacturing lots, shipping lots), **inspection lots** are used in acceptance sampling. Many times, the inspection lot is predetermined or dictated by the way the product is handled or shipped. At other times, it is possible to have some influence on the size and composition of the inspection lot. When this is possible, it is desirable to know *how* the lots should be formed. There are two basic principles of (inspection) lot formation:

GOALS IN FORMING
INSPECTION LOTS

> 1. Within-lot homogeneity is desirable.
> 2. If homogeneous, then larger lots are better than smaller lots.

The primary reason for wanting large lots is that they allow larger sample sizes, and large sample sizes allow for a more reliable determination of lot acceptability. But a large lot size is valuable only if each lot is also homogeneous. Homogeneity within a lot means that the items should be as similar as possible in their date and manner of production. Mixing lots of distinctly different quality levels can actually result in a *decrease* in the quality of the lots passed by an acceptance sampling plan (see Duncan 1986, pp. 180–181). It is apparent that the goals of homogeneity and increased lot size are somewhat at odds with each other and that lot formation is necessarily a tradeoff between these goals.

11.5   SINGLE-SAMPLE ATTRIBUTES PLANS

INTRODUCTION

Lot-based sampling with attributes data is the simplest and most commonly used form of acceptance sampling. An **attributes acceptance sampling plan** is characterized by a sample size, n, and an **acceptance number,** c (c is always a nonnegative integer). Operationally, a *random* sample of n items is drawn from a given lot and, after the number of nonconforming items in the sample is compared to the acceptance number, the entire lot is either rejected or accepted.[3] Specifically, if X denotes the number of nonconforming items found in the sample, then a lot is *rejected* if X exceeds c ($X > c$) or *accepted* if X does not exceed c ($X \leq c$). Notice that the lot size, N, does *not* appear as part of the sampling plan (see Section 11.6).

In practice, one never knows the quality of a particular lot in advance. Certainly, if the lot is composed entirely of conforming items, then a sampling plan will always accept or 'pass' the lot (since X will always be 0 in such cases).[4] Similarly, lots that contain only nonconforming items will always be rejected. However, most lots submitted for inspection fall in between these two extremes, and the proportion of nonconforming items, p, in each such lot is unknown and varies from lot to lot. To get around this problem, acceptance sampling plans are evaluated by calculating the probability of acceptance $P_a = P(X \leq c)$ over the entire range of possible values of p. By plotting P_a versus p ($0 \leq p \leq 1$), one can see how the plan will react to any situation (i.e., to any level of nonconforming items in a lot). For this reason, the graph of P_a versus p is called the **operating characteristic curve** (or, more simply, the **OC curve**) of the particular sampling plan. As one would expect, P_a cannot increase as p increases, so OC curves generally decrease as p increases from 0 to 1. As an illustration, Figure 11.1 shows OC curves for two common sampling plans.

After looking at the OC curve for a particular plan, one might decide that it does not offer the desired degree of control over lot quality. For example, a sampling plan with a large P_a for lots of low quality is deemed inadequate. On the other hand, it is possible for a plan that gives very good protection against accepting low-quality lots to reject inordinately large numbers of high-quality lots. In either case, one needs to choose a different plan (i.e., choose different values for n and c).

Acceptance sampling plans can be **tightened** by increasing n and/or decreasing c. Similarly, plans can be **loosened** by decreasing n and/or increasing c. By tightening or loosening plans in this manner and then examining the OC curve(s) of the resulting plans, one can eventually determine values of n and c that provide an acceptable degree of protection at all values of p.

The tradeoff involved can be illustrated on the OC curve. The probability that a high-quality lot is rejected by a plan is called the **producer's risk** (since the producer must then take the lot back, even though it is of acceptable quality). High-quality lots are those with small values of p. However, from the consumer's point of view, low-quality lots should not be accepted by the plan. Low-quality lots are those with large values of p. The probability of accepting a low-quality lot is referred to as the **consumer's risk.** Deciding which values of p correspond to high-quality lots and which correspond to low-quality lots is, of course, a subjective

---

[3] Alternate terminology found in the literature refers to accepting and rejecting lots as 'sentencing' the lot or determining the 'lot disposition.'

[4] In making this statement, we assume that no misclassification errors are made (see Section 10.9).

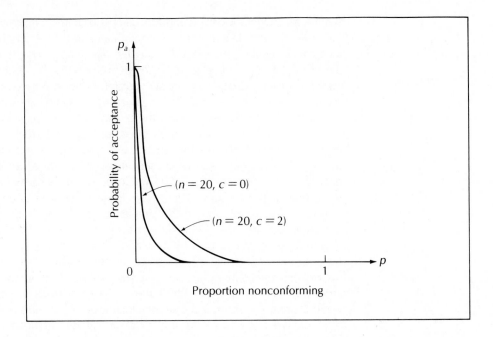

FIGURE 11.1
Operating
characteristic (OC)
curves

decision. The usual approach is to agree on a **producer quality level** (PQL) and a **consumer quality level** (CQL) so that P_a is high when $p = $ PQL and P_a is low for lots with $p = $ CQL.

Figure 11.2 shows a typical OC curve together with the PQL and CQL and their associated producer's and consumer's risks. As an example, if PQL and CQL are set at 0.02 and 0.10, respectively, then one might specify that P_a should exceed 0.95 for lots with $p = $ PQL and P_a should not exceed 0.02 for lots with $p = 0.10$. In essence, this procedure establishes two points on the OC curve and, with the correct choice of n and c, it should be possible to select a plan whose OC curve comes close to passing through these two points.

PROBABILITY OF LOT ACCEPTANCE

To calculate the probability P_a that a lot is accepted, one must first determine the *manner* in which the sampling plan will be used. Essentially, there are two distinct ways to use acceptance sampling, as a screening device for judging an individual lot or as a device for screening a continuing stream of lots. Notationally, sampling plans for screening single lots are called **Type A** plans, and plans for screening a series of lots are said to be of **Type B.** Since the majority of acceptance sampling is done in settings where lots are to be inspected on an ongoing basis, Type B plans are the more frequently used of the two. Knowing whether the plan is Type A or Type B is essential to correctly calculating P_a and, therefore, to finding the OC curve of the plan.

Probability calculations for Type A plans employ the hypergeometric distribution (see Section 5.5). Since only a single lot of size N is to be examined, there are

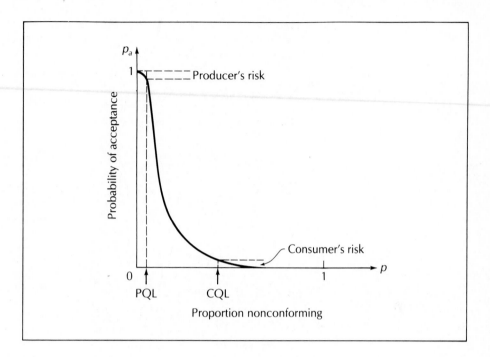

FIGURE 11.2
Producer's and
consumer's risks

only $N + 1$ possible values of p for which P_a can be calculated. That is, since there can be only $0, 1, 2, \ldots$, or N nonconforming items in the lot, the only possible values for p are $0, 1/N, 2/N, \ldots, (N - 1)/N$, or 1. For each such value of p, the composition of the lot (its numbers of nonconforming and conforming items) is known, as is the lot size N, so that $P_a = P(X \le c)$ can be found by direct application of hypergeometric probabilities. Note that the resulting OC curve must necessarily be a step function, since values of p *between* $0, 1/N, 2/N, \ldots, (N - 1)/N$, or 1 are unachievable. Example 11.1 illustrates these calculations.

The situation is quite different for Type B plans. Here, the proportion of nonconforming items, p, in the *process* from which the lots are derived can assume *any* value between 0 and 1. Thus, the sampling plan can be thought of as offering protection against *any* level of nonconforming items *in the process,* not just for $N + 1$ levels as is the case with Type A plans. In addition, each lot formed can be thought of as a collection of items (from the process) from which a random sample of n is to be drawn. Thus, sampling in the Type B setting is viewed as random sampling from an infinite population, which means that the binomial distribution should be used (see Section 5.5). Example 11.1 compares the OC curves for Type A and Type B sampling plans.

EXAMPLE 11.1   For inspecting an individual lot of $N = 10$ items, a Type A sampling plan with $n = 5, c = 1$ is used.[5] Because there are just 10 items in the lot, P_a can be

---

[5] We have chosen such a small value of N in order to illustrate the calculations in this example.

calculated only at the points $0, 0.1, 0.2, \ldots, 0.9$, and 1. Since $P_a = P(X \leq 1) = P(X = 0) + P(X = 1)$, the hypergeometric probabilities corresponding to these values of p are found as follows:

p	$P(X = 0)$	$+$	$P(X = 1)$	$= P(X \leq 1)$
0	$\binom{10}{5} \cdot \binom{0}{0} \Big/ \binom{10}{5}$		{not possible} $=$	$1 \quad + \quad 0 \quad = 1$
0.1	$\binom{9}{5} \cdot \binom{1}{0} \Big/ \binom{10}{5}$	$+$	$\binom{9}{4} \cdot \binom{1}{1} \Big/ \binom{10}{5} =$	$126/252 + 126/252 = 1$
0.2	$\binom{8}{5} \cdot \binom{2}{0} \Big/ \binom{10}{5}$	$+$	$\binom{8}{4} \cdot \binom{2}{1} \Big/ \binom{10}{5} =$	$56/252 + \ 70/252 = 0.5$
0.3	$\binom{7}{5} \cdot \binom{3}{0} \Big/ \binom{10}{5}$	$+$	$\binom{7}{4} \cdot \binom{3}{1} \Big/ \binom{10}{5} =$	$21/252 + 105/252 = 0.5$
0.4	$\binom{6}{5} \cdot \binom{4}{0} \Big/ \binom{10}{5}$	$+$	$\binom{6}{4} \cdot \binom{4}{1} \Big/ \binom{10}{5} =$	$6/252 + \ 60/252 = 0.2619$
0.5	$\binom{5}{5} \cdot \binom{5}{0} \Big/ \binom{10}{5}$	$+$	$\binom{5}{4} \cdot \binom{5}{1} \Big/ \binom{10}{5} =$	$1/252 + \ 25/252 = 0.1032$
0.6	{not possible}		$\binom{4}{4} \cdot \binom{6}{1} \Big/ \binom{10}{5} =$	$0 \quad + \quad 6/252 = 0.0238$
0.7	{not possible}		{not possible} $=$	$0 \quad + \quad 0 \quad = 0$
$\downarrow$			$\downarrow$	$\downarrow \qquad \downarrow \qquad \downarrow$
1.0	{not possible}		{not possible} $=$	$0 \quad + \quad 0 \quad = 0$

Because of restrictions imposed by the small lot size N, P_a must be 0 for values of p that are 0.7 or greater. The OC curve for this plan is shown in Figure 11.3. Broken lines are used to emphasize the fact that P_a can be calculated only at the points $0, 0.1, 0.2, \ldots, 0.9$, and 1.

Suppose, instead, that the $n = 5$, $c = 1$ plan is to be used for the ongoing screening of lots at a company's receiving inspection department. In this setting, the plan is of Type B and its associated OC curve is found by using the binomial distribution. Thus, for any p, $P_a = P(X \leq 1) = P(X = 0) + P(X = 1) = (1 - p)^5 + \binom{5}{1}p^1(1 - p)^4 = (1 - p)^4(1 + 4p)$. For comparison with the Type A plan, the OC curve for the Type B plan is also drawn in Figure 11.3 (solid line). The two curves differ somewhat because, for the Type A plan, a sample $n = 5$ is relatively large compared to a lot size of $N = 10$. For plans where the lot size is much larger than the sample size (say, $N \geq 0.10n$), two things happen: the Type A OC curve is plotted at many more points ($N + 1$ to be exact), giving it a smoother appearance, and the Type A and Type B probabilities become almost indistinguishable, since the hypergeometric distribution is closely approximated by the binomial when $N \geq 0.10n$ (see Section 5.5).

As can be seen in Example 11.1, probabilities generated by the binomial distribution are generally good approximations to OC curves for Type A plans (and *exact* values for Type B plans). For this reason, we consider only the more frequently encountered Type B scenario in the remainder of this chapter.

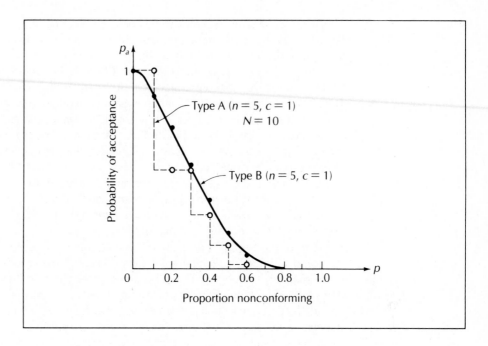

FIGURE 11.3
Type A and Type B
OC curves for
Example 11.1

The parameters *n* and *c* necessarily dictate the values of P_a for a given plan. Furthermore, the *shape* of OC curves with $c = 0$ are different from those with $c > 0$. Plans with $c = 0$ are aptly called **zero acceptance plans** and give rise to OC curves described by simply $P_a = (1 - p)^n$, which are *always* convex (i.e., opening upward). On the other hand, plans with $c > 0$ have a characteristic 'shoulder' near $p = 0$. Figure 11.4 compares these two shapes. A statistical comparison is given in Section 11.6.

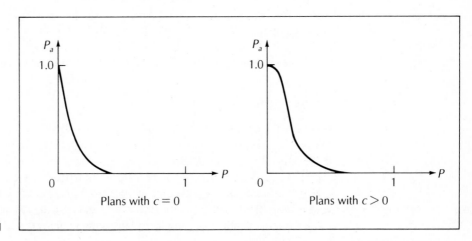

FIGURE 11.4
OC curves for plans
with c = 0 and c ≥ 1

Acceptance sampling can also be applied when one is counting *nonconformities* per item instead of nonconforming items. In such applications, it is common to express the number of nonconformities found either as a rate per item inspected or as the corresponding rate per 100 items inspected. For example, MIL-STD-105E uses the nonconformities per 100 items convention (see Section 11.9). This is analogous to what is done when nonconforming items are counted: $p = 0.10$ is equivalent to 10 per hundred, or 10%. The convention allows the familiar OC curve construction to be used for these plans. For example, suppose that a sample of $n = 5$ accounting records are inspected for errors, and the total number of errors found is expressed as an error rate per 100 records. If a total of three errors are found in the five records inspected, then the error rate can be expressed as either 0.6 error per record or 60 errors per 100 records. Similarly, 12 errors found (which is possible—a record could have more than one error) might be expressed as 240 errors per 100 records.

Calculating the OC curve for plans that count nonconformities requires the Poisson distribution (see Section 5.5). If we keep in mind the fact that the acceptance number c refers to the total number of nonconformities allowed in the n items inspected, then the values of p (interpreted as 'errors per item') are first converted to the corresponding rate per n items, followed by an application of the Poisson probabilities. In the example above, to use an $n = 5$, $c = 2$ plan for counting errors per accounting record, a value of $p = 0.10$ nonconformities per record is first translated into the equivalent rate of 0.5 nonconformities per 5 records. After that, the lot acceptance probability is calculated as

$$P_a = P(X \le 2) = P(X = 0) + P(X = 1) + P(X = 2)$$

$$= \frac{e^{-\lambda}\lambda^0}{0!} + \frac{e^{-\lambda}\lambda^1}{1!} + \frac{e^{-\lambda}\lambda^2}{2!}$$

$$= e^{-\lambda}\left(1 + \lambda + \frac{\lambda^2}{2}\right) = e^{-0.5}\left(1 + 0.5 + \frac{0.5^2}{2}\right) = 0.9856$$

In a similar fashion, P_a can be calculated at other values of p to produce the OC curve of the plan. Figure 11.5 compares the OC curve for the $n = 5$, $c = 2$ plan based on nonconformities to the same plan based on counting nonconforming items. The curves are very nearly identical for small values of p because the Poisson is an excellent approximation to the binomial for small p (see Section 5.5).

ACCEPTANCE SAMPLING SYSTEMS

Even though OC curves are relatively easy to calculate, it would be very inefficient if these calculations had to be repeated each time one needed to select a sampling plan. Furthermore, it would probably be necessary to calculate the OC curves of *several* plans, since it is hard to estimate in advance the exact values of n and c needed in order to achieve specific producer and consumer risk levels. In order to avoid this problem, and in an effort to make acceptance sampling plans even more versatile, large numbers of sampling plans with similar properties are grouped to-

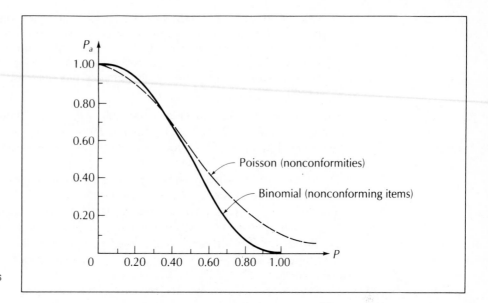

FIGURE 11.5
Sampling plans with
$n = 5$ and $c = 2$ for
monitoring
nonconformities
(Poisson) or
nonconforming items
(binomial)

gether into **sampling systems** whose OC curves are published and included as part of the system. MIL-STD-105E is one such system of plans. Selecting the appropriate plan from a sampling system is then reduced to the relatively simple task of perusing the plan's published OC curves to see which one gives the desired level of protection.

Collecting together plans with specified consumer's and producer's risks is one way of forming a sampling system, but there are others. In fact, each system is based or 'indexed' on a particular measure of performance, of which there are several. The performance characteristics most frequently used are the acceptable quality level (AQL), the lot tolerance percent defective (LTPD), the average outgoing quality limit (AOQL), and the average total inspection (ATI). These measures are briefly described in this section and discussed in detail in later sections. Selecting an appropriate plan becomes a matter of choosing the desired value for a particular performance characteristic and then using the published tables to find values of n and c that will deliver this level of performance.

AQL sampling systems are those collections of sampling plans indexed on the **acceptable quality level** (AQL).

AQL is the quality level for which, for the purposes of sampling inspections, is the limit of a satisfactory process average. The process average is the average percent defective or average number of defects per hundred units (whichever is applicable) of product submitted by the supplier for original inspection. Original inspection is the first inspection of a particular quantity of product as distinguished from the inspection of product which has been resubmitted after prior rejection. A sampling plan and AQL are chosen in accordance with risk assumed. Use of a value of AQL for a certain defect or group of defects indicates that the sampling plan will accept the great majority of the lots or batches provided the process average level of percent defective (or defects per hundred units) in these lots or batches be no greater than the designated value of AQL. Thus, the AQL is a designated value of percent defective (or defects per hundred units) for which lots will be accepted most of the time by the sampling procedure used (MIL-STD-105E, 1989).

In other words, the AQL specifies a particular value of the producer's quality level shown in Figure 11.2. Following the usual convention, AQL can equivalently be specified in terms of the proportion defective (i.e., nonconforming) or percent defective.

To use an AQL system, one first specifies a value for the AQL (say, AQL = 0.01) and then uses the system's published tables to select values of n and c that will achieve this AQL. This implicitly means that P_a should be high when $p = $ AQL. Accordingly, the sampling system must also specify the value (or range of values) that it uses for P_a when $p = $ AQL. It has become conventional in most sampling systems to choose $P_a = 0.95$ at $p = $ AQL, although any value of P_a could be specified if desired.

LTPD sampling systems are indexed on the **lot tolerance percent defective** (LTPD). The LTPD is simply a specific numerical value for the consumer's quality level shown in Figure 11.2. By convention, the notation LTPD usually refers to the point on the OC curve at which $P_a = 0.10$. Thus, to assure that most lots containing 20% nonconforming items ($p = 0.20$) are rejected by a sampling plan, one specifies an LTPD of 0.20 (equivalently, LTPD = 20%). A sampling plan with this LTPD then has an OC curve for which $P_a = 0.10$ for lots with $p = 0.20$; that is, 90% of such lots (where $p = 0.20$) are rejected by the plan. Of course, the probability of acceptance can be set at any value, not just at $P_a = 0.10$, but the majority of LTPD sampling systems are based on $P_a = 0.10$. One sampling system indexed by LTPD is the Dodge-Romig system (Grant and Leavenworth 1988, chap. 13).

AOQL sampling systems are indexed on the **average outgoing quality limit** (AOQL). This performance characteristic applies to only a certain form of sampling inspection in which rejected lots subsequently receive 100% screening and any nonconforming items found are replaced by good items. This form of sampling and subsequent 'correction' of rejected lots is called **rectifying inspection** (see Section 11.7).

Under rectifying inspection, there is a natural upper bound on the average percent nonconforming that will be shipped to the customer. It is easy to understand *why* such a bound exists: most high-quality lots will be accepted (so the percent nonconforming in these lots will be low), and most low-quality lots will be rejected and rectified (so the percent nonconforming in these lots should then be close to 0). Thus, plotting the **average outgoing** (i.e., shipped) **quality** (AOQ) versus the percent nonconforming, p, in the lots submitted for inspection results in a curve resembling the one in Figure 11.6.

The maximum point on this curve is called the average outgoing quality limit, or AOQL. The **AOQL** represents the worst possible percent nonconforming that can be expected in lots shipped under the particular sampling plan. AOQL tables are valuable additions to sampling systems. Both MIL-STD-105E and the

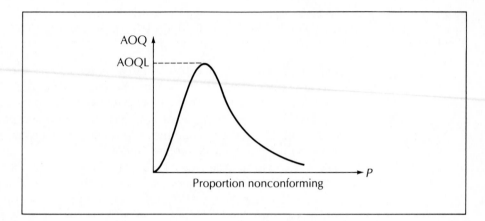

FIGURE 11.6
Average Outgoing
Quality (AOQ) curve
for sampling with
rectifying inspection

Dodge-Romig system include AOQL tables and/or graphs to help the user choose the right plan when rectifying inspection is employed.

Since the number of items inspected is always greater under rectifying inspection than it is without 100% screening of rejected lots, the total amount of inspection work required becomes a factor in choosing the sampling plan. Each rejected lot, for example, adds $N - n$ additional items to the inspector's burden. Of course, as the lot quality becomes worse (i.e., as p increases), more and more lots are rejected and the total amount of inspection should increase. In general, the average total inspection (ATI) to be expected for any lot quality level, p, can be plotted versus p. Figure 11.7 shows a graph of a typical ATI curve. The ATI graph is used to determine the additional costs associated with rectifying inspection.

FIGURE 11.7
Average Total Inspection (ATI) curve for plans with rectifying inspection

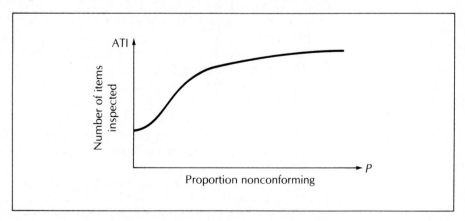

11.6	COMMON QUESTIONS ABOUT ACCEPTANCE SAMPLING

Some questions about acceptance sampling tend to arise over and over again, especially among those first encountering the subject. The two most frequently encountered problems stem from an intuitive feeling that sample sizes should somehow be related to lot sizes and that plans with acceptance numbers of 0 should naturally be better than those with nonzero acceptance numbers. A third question that must be added to the list concerns whether or not acceptance sampling is even worthwhile when a process is known in advance to be in a state of statistical control. The last question has been the source of much of the criticism of acceptance sampling in recent years, and for this reason we include it with the two simpler questions.

Turning first to the determination of sample sizes, we find the somewhat surprising answer is that the lot size, N, actually has very little to do with the choice of the sample size, n. In fact, as long as the lot size is reasonably large compared to the sample size, so that $0.10N \geq n$, the hypergeometric and binomial distributions are very nearly the same (recall Section 5.5). Therefore, since the binomial distribution involves only the parameters n and p (and *not* N), the lot size is not needed for constructing the OC curve. More simply said, for large lots, sampling without replacement (hypergeometric distribution) is about the same as sampling with replacement (binomial distribution), since there is a very slim chance that the same item would ever be selected twice when sampling with replacement.

Instead of the lot size, it is the absolute magnitude of the sample size itself that has the greatest influence on the discriminating power of a sampling plan. Thus, a plan with $n = 10$ could be used on lots of size $N = 100$ or $N = 1,000$ with almost exactly the same effect (i.e., the OC curves would be virtually indistinguishable). In contrast, plans with $n = 10$ and $n = 25$ would provide very different levels of protection when used with lots of size $N = 1,000$ (or any other large lot size).

Part of the confusion regarding possible relationships between lot size and sample size stems from the way that the popular MIL-STD-105E sampling system is designed. In that system, sample sizes *are* selected based on tables that relate sample size to lot size. Furthermore, the sample sizes given by these tables tend to increase as the lot size increases, giving the impression to some that sample sizes might be some percentage of the lot size. In fact, the tables were designed to address the fact that larger sample sizes have better discriminating power and that this power is needed most for the larger lots (where an incorrect decision on lot quality can have more serious consequences). Also, on closer inspection, the percentage of the lot represented by the sample size decreases quickly in these tables as the lot size increases, so the tables are not at odds with the results of the previous paragraphs.

We turn next to the choice of the acceptance number. Recall that in Section 11.5, plans with $c = 0$ were called zero acceptance plans. Such plans can be appealing to customers who naturally do not want *any* nonconforming product in a lot and who intuitively feel that zero acceptance plans ought to offer the best protection in this regard. Although it *is* true that using $c = 0$ gives the strictest possible plan among all plans with the same fixed sample size, plans that use slightly larger sample sizes and nonzero values of c may, in fact, perform better. The problem with using $c = 0$ is that the good lots tend to get thrown out along with the bad; that is, there is generally a high probability that 'good' lots (with small values of p) will not be accepted by the plan. Plans with $c \geq 1$, on the other hand, tend to have a 'shoulder' near $p = 0$, which allows more of the 'good' lots to pass inspection. These plans make it easier to decrease *both* the producer's and consumer's risks. To alleviate concerns about using plans that allow a certain number of nonconforming items to appear in the sample, users should be reminded that with *any* plan (even zero acceptance plans), the fact that no nonconforming items appear in a sample is no guarantee at all that the lot itself is free of nonconforming items. Instead, the correct approach to selecting a sampling plan should be first to identify the performance characteristics of greatest concern and then to select plan(s) that achieve these goals. Further discussion of zero acceptance plans can be found in Baker (1988).

Finally, we come to the most difficult of the three questions: Does acceptance sampling do any good when applied to processes that are known to be in statistical control? Besides the usual argument that acceptance sampling may not embody the preventative approach to quality (see Section 11.3), one of the main arguments against using acceptance sampling for controlled processes is based on a 1943 theorem of Mood (1943). This theorem simply states that if a controlled process is *known* to produce a fixed proportion, p, of nonconforming items, then the number of nonconforming items found in a sample can supply no additional information on the proportion nonconforming in the population. (In statistical jargon, the number of nonconforming items in the sample is independent of the number remaining in the lot.)

Taken at face value, the theorem has been used by some to argue that sampling is of no value for controlled processes (where p is approximately constant). A simpler version of the argument goes as follows. Suppose, for a *controlled* process with $p = 0.04$, that every lot of size, say, $N = 50$ contains *exactly* the same number of nonconforming items—say, two per lot (so $p = 0.04$ for *every* lot, which accurately reflects the process nonconformance rate). When a sampling plan is applied to this stream of lots, some are rejected and some are accepted (since P_a is not 0), yet the quality of the accepted lots is identical to that of the rejected lots; that is, both accepted and rejected lots have $p = 0.04$. The conclusion is that sampling has no effect on improving the quality of the accepted lots.

To put these arguments in the proper perspective, let us be more specific about what Mood's theorem does and does not say. First, and most apparent to the practitioner, is the fact that all lots are *not* identical. Even under the best of circumstances (e.g., controlled processes), the proportion nonconforming varies somewhat from lot to lot. Mood's theorem does not apply in this setting (since p varies from lot to lot) and, as one would expect, acceptance sampling *will* cull out the lower quality lots (Vardeman 1986). For a convincing numerical illustration of this fact, we refer the reader to Schilling (1982, pp. 2–4).

Second, as David (1986) points out, Mood did not intend the interpretation some have ascribed to his work and, in fact, it *is* possible to make inferences regarding a process nonconformance rate from the information in a sample. Instead, the theorem applies in the very special circumstance where one *knows in advance* what the underlying value of *p* is.

Although its interpretation can be somewhat subtle, we do not believe Mood's theorem to be as important a factor in the decision to use acceptance sampling as are the factors outlined in Section 11.3. The important factors in the decision should involve whether or not acceptance sampling is needed to protect against gross mistakes (for liability reasons or for processes not yet in control) and, for controlled processes, whether acceptance sampling is even capable of detecting small nonconformance rates, especially those in the parts per million range.

11.7 RECTIFYING INSPECTION

Rectifying inspection is a form of acceptance sampling in which any nonconforming items found in rejected lots are replaced before such lots are shipped to the customer. Under rectifying inspection, it makes sense to define (for any nonconformance rate *p*) the **average outgoing quality** (AOQ) to be the average proportion of nonconforming items in the lots shipped. Notice that, for any value of *p*, AOQ accounts for *both* the accepted lots shipped and the rejected lots that are subsequently 100% screened prior to shipment. Furthermore, as is usually the case in sampling inspection, any nonconforming items found in a sample of *n* from a lot of size *N* are also replaced by good items, regardless of the decision to reject or accept the lot. Thus, the proportion nonconforming in lots *accepted* by a plan should be about $p \cdot (N - n)/N$ for any particular value of *p*. Figure 11.8 summarizes the key elements in rectifying inspection.

Averaging over all lots that have nonconformance rate *p* gives an average outgoing quality of

$$\text{AOQ} = p\left(\frac{N - n}{N}\right)P_a \tag{11.1}$$

Note that the AOQ is just a weighted average of the proportion nonconforming in rejected lots (which should contain *no* nonconforming items after rectification) and the proportion nonconforming in accepted lots [which equals $p(N - n)/N$ as previously described]. The weights used are the probabilities of rejection and acceptance, respectively. When the sample size is small compared to the lot size, the factor $(N - n)/N$ will be close to 1, so a very good approximation to the AOQ is given by

$$\text{AOQ} \approx pP_a \tag{11.2}$$

Plotting AOQ versus *p* gives rise to the AOQ curve associated with the particular sampling plan (see Figure 11.6). It is also informative to include another line on the AOQ graph to show the effect of *not* using rectifying inspection. With ordinary (no rectification) acceptance sampling, only lots that pass inspection are

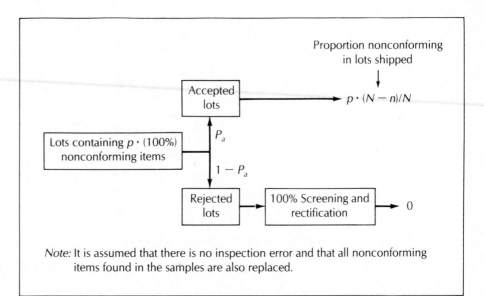

FIGURE 11.8
Proportion of nonconforming items in lots shipped under rectifying inspection

shipped, so any lot containing $p \cdot 100\%$ nonconforming items that is accepted contains $p[(N - n)/N] \cdot 100\%$ nonconforming items when shipped (recall that any nonconforming items in the sample are normally replaced). Usually, the factor $(N - n)/N$ can be ignored, so that the average outgoing quality is about $p \cdot 100\%$ for a lot that has $p \cdot 100\%$ nonconforming items to begin with. This means that the average outgoing quality curve is just the 45° line for plans *without* rectification (see Figure 11.9).

The point p at which the AOQ curve reaches its maximum can be found by elementary calculus. Since a plan with sample size and acceptance number c

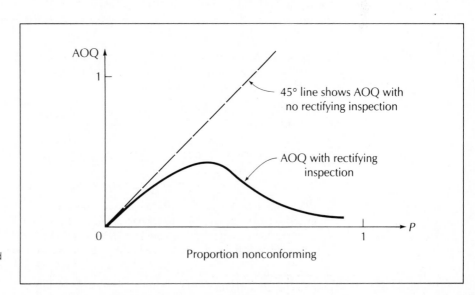

FIGURE 11.9
AOQ curves with and without rectifying inspection

gives rise to an AOQ of $p[(N - n)/N]P_a$, the maximum point can be found by setting the derivative of this expression equal to 0 and solving for p.[6] We use p_{max} to denote this value of p. The following steps illustrate the procedure:

$$\frac{d}{dp}(AOQ) = \frac{d}{dp}\left[p\left(\frac{N-n}{N}\right)P_a\right] = \left(\frac{N-n}{N}\right) \cdot \frac{d}{dp}[p \cdot P(X \leq c)]$$

$$= \left(\frac{N-n}{N}\right) \cdot \frac{d}{dp}\left[p \cdot \sum_{k=0}^{c}\binom{n}{k}p^k(1-p)^{n-k}\right]$$

$$= \left(\frac{N-n}{N}\right) \cdot \frac{d}{dp}\left[\sum_{k=0}^{c}\binom{n}{k}p^{k+1}(1-p)^{n-k}\right]$$

$$= \left(\frac{N-n}{N}\right) \cdot \sum_{k=0}^{c}\binom{n}{k}[p^{k+1}(n-k)(1-p)^{n-k-1}(-1) + (1-p)^{n-k}(k+1)p^k]$$

which, with a little simplification, becomes

$$\frac{d}{dp}(AOQ) = \left(\frac{N-n}{N}\right) \cdot \sum_{k=0}^{c}\binom{n}{k}p^k(1-p)^{n-k-1}[(k+1) - (n+1)p] \qquad (11.3)$$

Setting equation (11.3) equal to 0 and solving for p result in some trivial roots (such as negative roots and the root $p = 1$) as well as the desired root $p = p_{max}$. It should be noted that, after the highest power of $(1 - p)$ is eliminated from equation (11.3), solving the resulting equation amounts to finding the roots of a polynomial of degree $c + 1$ (where c is the acceptance number of the plan), so a computer is generally required. Alternatively, special tables are available for approximating p_{max} (Schilling 1982, pp. 386–387).

The value of AOQ when $p = p_{max}$ is called the **average outgoing quality limit**, or AOQL (recall Figure 11.6). Caution should be exercised when interpreting the AOQL, since it is *not* the maximum proportion nonconforming that could be shipped under the particular plan. It is possible that some lots shipped may have quality levels slightly worse than the AOQL. However, if we average over all lots where $p = p_{max}$, the AOQL is the maximum that this *average* can assume.

EXAMPLE 11.2

To find the AOQL associated with the sampling plan $n = 20$, $c = 1$ (assuming rectifying inspection is used), these values are put in equation (11.3), which is then set equal to 0 and solved:

$$\frac{d}{dp}(AOQ) = \sum_{k=0}^{1}\binom{20}{k}p^k(1-p)^{20-k-1}[(k+1) - 21p]$$

$$= \binom{20}{0}p^0(1-p)^{20-1}[1 - 21p] + \binom{20}{1}p^1(1-p)^{20-2}[2 - 21p]$$

$$= (1-p)^{18}[(1-p)(1-21p) + 20p(2-21p)] = 0$$

When the term $(1 - p)^{18}$ is eliminated, p_{max} is one of the roots of the expression in brackets, which simplifies to solving the quadratic $-399p^2 + 18p + 1 = 0$. The two roots of this equation are $p = 0.0775$ and $p = -0.0324$. Obviously, only the positive root makes sense in this application, so we conclude that

---

[6] Regarding P_a as a function of p.

$p_{max} = 0.0775$. To find the AOQL, $p_{max} = 0.0775$ is substituted into the AOQ formula to yield

$$AOQL = p_{max}\left(\frac{N - n}{N}\right)P_a = p_{max} \cdot P(X \le 1) \cdot \left(\frac{N - n}{N}\right)$$

$$= (0.0775)\left[\binom{20}{0}(0.0775)^0(1 - 0.0775)^{20}\right.$$

$$\left. + \binom{20}{1}(0.0775)^1(1 - 0.775)^{19}\right] \cdot \left(\frac{N - n}{N}\right)$$

$$= 0.0414 \cdot \left(\frac{N - n}{N}\right)$$

If we make the usual assumption that the lot size is large compared to the sample size, then the AOQL is approximately 0.0414, or 4.14%.

Screening rejected lots necessarily adds to the inspection burden. Each lot inspected requires that a sample of n be inspected and, beyond that, each rejected lot contributes $N - n$ additional items. For planning and resource allocation, it is natural to want an estimate of the total inspection burden that a given plan will incur. In particular, it would be desirable to have such estimates given on a per-lot basis, so that total effort can easily be found by just multiplying the average per-lot inspection burden by the number of lots to be inspected.

Since the probability of rejecting a lot is $1 - P_a$, the average number of items inspected per lot, or the **average total inspection** (ATI) per lot, can be estimated by

$$ATI = n + (N - n) \cdot (1 - P_a) \tag{11.4}$$

That is, *every* lot accounts for n inspected items, while the $(1 - P_a) \cdot 100\%$ of the lots that are rejected add another $N - n$ items to the task. Since P_a ranges from $P_a = 1$ (at $p = 0$) down to $P_a = 0$ (at $p = 1$), the ATI must always start out at the point $ATI = n$ (at $p = 0$) and end up at the point $ATI = N$ (when $p = 1$). Figure 11.10 shows the typical ATI curve that passes through these two points.

ATI curves are included with many sampling systems to aid the user in selecting a plan. For zero acceptance plans, the ATI is particularly simple and we include it here for the reader's convenience. For plans with $c = 0$, the lot acceptance rate is just $P_a = (1 - p)^n$, so the average total inspection associated with such plans is found from equation (11.4) to be

$$ATI = n + (N - n) \cdot [1 - (1 - p)^n] \qquad \text{for } c = 0 \tag{11.5}$$

11.8   DOUBLE AND MULTIPLE ATTRIBUTE SAMPLING

INTRODUCTION

Certain situations arise during the operation of single-sample attributes plans that suggest other sampling procedures might be even more effective. It often turns

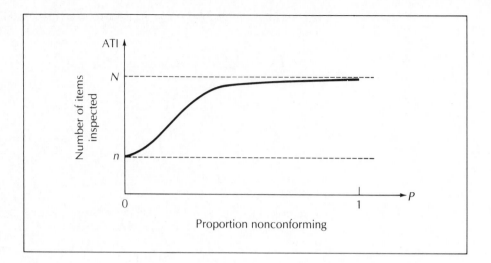

FIGURE 11.10
All ATI curves pass
through ATI = n and
ATI = N

out, for example, that when low-quality lots are inspected, the acceptance number c is exceeded after only a few of the sampled items are inspected. At that point, the lot disposition is clear, the lot is to be rejected, and there is no real need to examine any more items. Putting an end to further inspection as soon as X exceeds c is called **curtailment**. Curtailed inspection obviously saves resources, but quite often sampling is not curtailed so that more information on the lot quality can be obtained. Even without curtailed inspection, though, the thought remains that there might exist a more efficient sampling procedure capable of taking such situations into account.

Since smaller sample sizes suffice to reject the low-quality lots, one idea might be simply to decrease the sample size, n. By doing this, however, the *overall* amount of protection offered by the plan must decrease (i.e., the OC curve rises), so that the desired levels of producer and consumer risks will not be attainable.

One approach that *does* meet the goal of using smaller samples for low-quality lots while maintaining the desired level of control over high-quality lots is to use multiple sampling. **Multiple sampling plans** involve two or more *stages* of sampling, where at each stage a decision is made to accept, reject, or proceed to the next sampling stage and take additional samples before sentencing the lot. Clearly, this procedure addresses the problem described in the previous paragraphs. Lots of low quality will be rejected after the first sample, lots of high quality will be accepted after the first sample, and the remaining lots will require a bit more sampling inspection before they are sentenced. Figure 11.11 illustrates the operation of multiple sampling plans and the much-used special case of double sampling.

Double sampling attributes plans are specified by choosing five numbers, the sample sizes n_1 and n_2 for the two sampling stages along with the respective acceptance and rejection numbers (c_1, c_2, r_1, and r_2) for each stage. Although there *appear* to be six parameters here, in fact only five are freely chosen, since the rejection number at the second stage is exactly determined once c_2 is given (as ex-

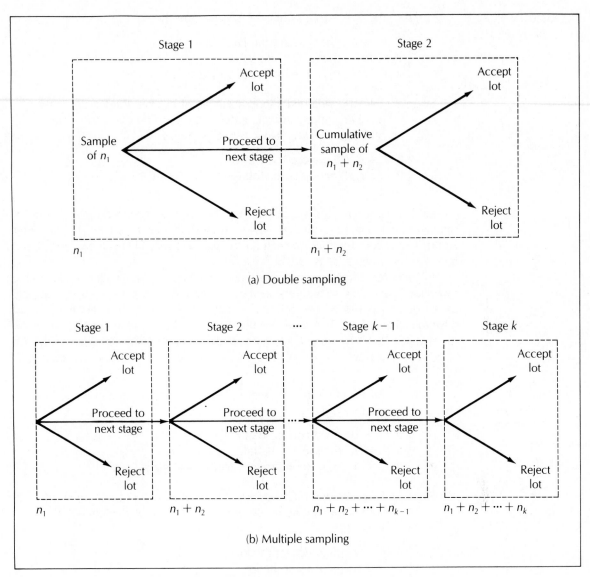

FIGURE 11.11
Double and multiple acceptance sampling

plained shortly). To be specific, we let X_1 denote the number of nonconforming items found in the first sample and, if a second sample is needed, let X_2 denote the number of nonconforming items it contains. The total number of nonconforming items found in both samples is then denoted by $Y = X_1 + X_2$. We use the same notation as for single sampling plans, so c_1 and c_2 represent the acceptance numbers for the two stages. Rejection numbers r_1 and r_2 are also needed at each stage, and the operation of a double sampling plan then proceeds as follows:

DOUBLE SAMPLING
BY ATTRIBUTES

Stage 1: Take a random sample of size n_1 from the lot.

> If $X_1 \leq c_1$, then accept the lot.
> If $X_1 \geq r_1$, then reject the lot.
> If $c_1 < X_1 < r_1$, then proceed to Stage 2.

Stage 2: Take an additional random sample of n_2 from the lot, count the number of nonconforming items X_2, and find the *total* number of nonconforming items in both samples, $Y = X_1 + X_2$.

> If $Y \leq c_2$, then accept the lot.
> If $Y \geq r_2$, then reject the lot. (*Note:* $r_2 = c_2 + 1$.)

Since a decision *must* be made on or before the second sample (in double sampling), the rejection number r_2 is restricted so that $r_2 = c_2 + 1$. The other natural restrictions that must be met for the plan to be feasible can be summarized by the inequalities $c_1 \leq r_1 \leq c_2$.

It is apparent that this type of sampling procedure need not be limited to double sampling. The extension to the general case of multiple sampling is easy. Using the same notation as before, we choose the acceptance and rejection numbers c_i and r_i for sampling stage i. X_i denotes the number of nonconforming items found in the ith sample. Then the *cumulative* number of nonconforming items as of the ith stage, $Y_i = X_1 + X_2 + \cdots + X_i$, is used to sentence the lot at the ith stage:

(k STAGE) MULTIPLE
SAMPLING BY
ATTRIBUTES

Stage 1: Take a random sample of n_1 from the lot and let X_1 denote the number of nonconforming items found.

> If $X_1 \leq c_1$, then accept the lot.
> If $X_1 \geq r_1$, then reject the lot.
> If $c_1 < X_1 < r_1$, then proceed to the next stage.

Stage 2: Take an additional random sample of n_2, count the number of nonconforming items X_2, and form the cumulative count $Y_2 = X_1 + X_2$.

> If $Y_2 \leq c_2$, then accept the lot.
> If $Y_2 \geq r_2$, then reject the lot.
> If $c_2 < Y_2 < r_2$, then proceed to the next stage.
> .
> .
> .

Stage k: Take an additional random sample of size n_k, count the number of nonconforming items X_k, and form the cumulative count $Y_k = X_1 + X_2 + \cdots + X_k$.

> If $Y_k \leq c_k$, then accept the lot.
> If $Y_k \geq r_k$, then reject the lot. (*Note:* $r_k = c_k + 1$.)

As in double sampling, no more than k stages are allowed, so the restriction $r_k = c_k + 1$ is incurred at the *last* stage. There are no such restrictions on r_i and c_i at any of the preceding stages.

Although double sampling $(k = 2)$ is just a special case of multiple sampling, we continue to treat the case $k = 2$ separately in the remainder of the chapter. Double and multiple sampling plans can also be collected into sampling systems in the same manner that single sampling plans are, by indexing on some performance characteristic. MIL-STD-105E, for example, includes the option of selecting single, double, or multiple attributes sampling plans indexed on the AQL.

OC CURVES FOR DOUBLE AND MULTIPLE SAMPLING

The end result of using a multiple sampling plan is the same as that for any acceptance sampling plan: a lot is either accepted or rejected. Thus, the probability of lot acceptance, P_a, can be calculated and plotted against the lot quality, p, to form the operating characteristic curve (OC) associated with the plan. However, with multiple sampling, the calculations are a little more involved, since a lot's acceptance (or rejection) could occur at any of the k sampling stages. In this section, we illustrate these calculations for OC curves of double sampling plans. Multiple sampling OC curves are more complex but are handled in essentially the same fashion.

The parameters that describe a double sampling plan are $\{n_1, c_1, r_1\}$ and $\{n_2, c_2, r_2 = c_2 + 1\}$. If we summarize the plan by means of the tree diagram in Figure 11.12, lot acceptance occurs when either $X_1 \leq c_1$ or (if a second sample is needed) $Y = X_1 + X_2 \leq c_2$. Since these two eventualities are mutually exclusive, we simply add their respective probabilities to find P_a. The first of these components, $P(X_1 \leq c_1)$, is found in the usual fashion by summing binomial

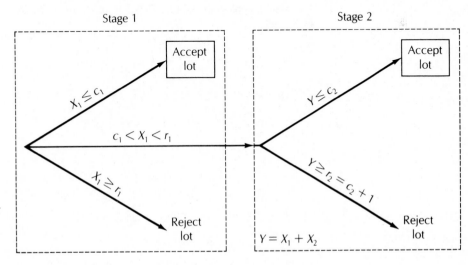

FIGURE 11.12
The two ways that lot acceptance can occur in double sampling

Stage 1

$X_1 \leq c_1$ → Accept lot

$c_1 < X_1 < r_1$

$X_1 \geq r_1$ → Reject lot

Stage 2

$Y \leq c_2$ → Accept lot

$Y \geq r_2 = c_2 + 1$ → Reject lot

$Y = X_1 + X_2$

probabilities. The second, $P(Y \le c_2)$, is simply the product of the appropriate branches in the tree diagram:[7]

$$P(Y \le c_2) = P(Y \le c_2 \mid c_1 < X_1 < r_1) \cdot P(c_1 < X_1 < r_1)$$

To simplify the expressions that follow, we use the notation $b(s, n)$ to denote the binomial probability of finding s nonconforming items in n items inspected; that is, $b(s, n) = \binom{n}{s} p^s (1 - p)^{n-s}$. Then, for any value of p, P_a can be written as

$$
\begin{aligned}
P_a &= P(X_1 \le c_1) + P(Y \le c_2) \\
&= P(X_1 \le c_1) + P(Y \le c_2 \mid c_1 < X_1 < r_1) \cdot P(c_1 < X_1 < r_1) \\
&= P(X_1 \le c_1) + P(X_2 \le c_2 - X_1 \mid c_1 + 1 \le X_1 \le r_1 - 1) \\
&\quad \cdot P(c_1 + 1 \le X_1 \le r_1 - 1)
\end{aligned}
$$

or

$$P_a = \sum_{s=0}^{c_1} b(s, n_1) + \sum_{s=c_1+1}^{r_1-1} \left[b(s, n_1) \cdot \sum_{j=0}^{c_2-s} b(j, n_2) \right] \tag{11.6}$$

EXAMPLE 11.3

Suppose that the double sampling plan with parameters $\{n_1 = 10, c_1 = 0, r_1 = 2\}$ and $\{n_2 = 15, c_2 = 2, r_2 = c_2 + 1 = 3\}$ is being considered as a possible replacement for the single sampling plan $\{n = 12, c = 1\}$. To compare the two plans, their OC curves are calculated. With the notation introduced in this section, the single sampling plan yields, for any p,

$$P_a = \sum_{s=0}^{1} b(s, 12) = (1 - p)^{12} + \binom{12}{1} p^1 (1 - p)^{11} = (1 - p)^{11}(1 + 11p)$$

The more complex double sampling plan has a lot acceptance probability given by equation (11.6):

$$
\begin{aligned}
P_a &= \sum_{s=0}^{0} b(s, 10) + \sum_{s=1}^{1} \left[b(s, 10) \cdot \sum_{j=0}^{2-s} b(j, 15) \right] \\
&= b(0, 10) + b(1, 10) \cdot \{b(0, 15) + b(1, 15)\} \\
&= \binom{10}{0} p^0 (1 - p)^{10} + \binom{10}{1} p^1 (1 - p)^9 \\
&\quad \cdot \left\{ \binom{15}{0} p^0 (1 - p)^{15} + \binom{15}{1} p^1 (1 - p)^{14} \right\} \\
&= (1 - p)^{10} + 10p(1 - p)^9 \cdot \{(1 - p)^{15} + 15p(1 - p)^{14}\}
\end{aligned}
$$

or $\quad P_a = (1 - p)^{10} \cdot \{1 + 10p(1 - p)^{13}(1 + 14p)\}$

---

[7] Recall Section 5.3 on the conditional probability formula.

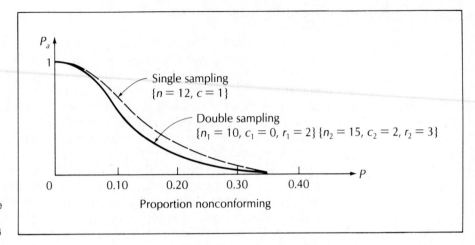

FIGURE 11.13
OC curves for double
and single sampling
plans of Example 11.3

To compare the two plans, P_a is calculated for a few relevant values of p and then graphed as in Figure 11.13:

p	P_a (double sampling)	P_a (single sampling)
0	1.00000	1.00000
0.01	0.99485	0.99383
0.02	0.97793	0.97689
0.03	0.94885	0.95135
0.04	0.90885	0.91906
0.05	0.85999	0.88164
0.10	0.56139	0.65900
0.20	0.15224	0.27488
0.30	0.03252	0.08503
0.40	0.00626	0.01959

The OC curves appear to be very close over the range $0 \le p \le 0.05$, so they provide about the same producer's risk in that range. Beyond $p = 0.05$, the OC curve for double sampling diverges sharply below that for single sampling until about $p = 0.30$, when the curves again become fairly close. We conclude that for AQLs below 0.05, the double sampling plan provides essentially the same degree of protection as the single sampling plan and provides even greater protection against lots of lower quality (i.e., when $p > 0.05$).

If it is assumed that a supplier is capable of delivering lots of quality $p \le 0.05$ the majority of the time, then the double sampling plan should reduce the overall numbers of items inspected (since a sample of size 10 usually suffices to sentence such lots). For lots of lesser quality, it is not quite clear whether double sampling reduces or increases the inspection burden. This question is answered in the next section.

AVERAGE SAMPLE NUMBER (ASN)

In double and multiple sampling, the sample size required to reach an 'accept' or 'reject' decision varies. Sometimes the decision can be reached from the first sample of n_1 items; sometimes more sampling is necessary. Since the amount of inspection required is always an important factor to consider, finding the *average* sample size is of interest for multiple sampling plans. In the literature, this quantity is normally called the **average sample number** (ASN).

For a double sampling plan with parameters $\{n_1, c_1, r_1\}$ and $\{n_2, c_2, r_2 = c_2 + 1\}$, the minimum possible sample size is n_1, since every lot must go through at least the first sampling stage. Because an additional sample of n_2 is needed only if $c_1 < X_1 < r_1$, the long-run average size of the second sample should be about $n_2 \cdot P(c_1 < X_1 < r_1)$. Together, the average sample number for the double sampling plan, for any value of p, is then $\text{ASN} = n_1 + n_2 \cdot P(c_1 < X_1 < r_1)$, or

$$\text{ASN} = n_1 + n_2 \cdot \sum_{s=c_1+1}^{r_1-1} b(s, n_1) \tag{11.7}$$

where, as before, $b(s, n)$ denotes the binomial probability of finding s nonconforming items in a sample of n. Plotting the ASN curve versus p helps to determine the amount of inspection effort one can expect when using the plan.

EXAMPLE 11.4   In Example 11.3, the double sampling plan $\{n_1 = 10, c_1 = 0, r_1 = 2\}$, $\{n_2 = 15, c_2 = 2, r_2 = 3\}$ compared favorably with the single sampling plan $\{n = 12, c = 1\}$ over the range $p \leq 0.05$. To see how much inspection effort the double sampling plan requires over the entire range of possible values of p, its ASN curve is compared with that of the single sampling plan. Since the single sampling plan takes a fixed sample of $n_1 = 12$ from *every* lot, its ASN curve is plotted simply as a horizontal line at ASN = 12 (Figure 11.14). For the double sampling plan, equation (11.7) gives

$$\text{ASN} = n_1 + n_2 \cdot \sum_{s=c_1+1}^{r_1-1} b(s, n_1)$$

$$= 10 + 15 \cdot \sum_{s=1}^{1} b(s, 10) = 10 + 15 \cdot b(1, 10)$$

$$= 10 + 15 \cdot \binom{10}{1} p^1 (1 - p)^9 = 10 + 150p(1 - p)^9$$

By plotting this ASN on the same graph with the ASN of the single sampling plan in Figure 11.14, we see that the double sampling plan requires more inspection, on average, than the single sampling plan over the range $0.02 \leq p \leq 0.28$. For fairly high-quality lots (say, $p \leq 0.02$) and for very low-quality lots (say, $p > 0.28$), double sampling requires less overall inspection.

ASN curves for multiple sampling plans with more than two stages are calculated in the same manner as those for double sampling. Successive sample sizes are weighted by the probabilities of proceeding from one stage to the next. At

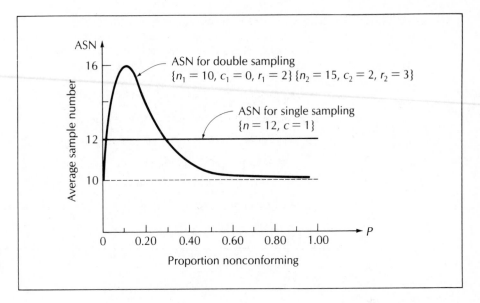

FIGURE 11.14
Average Sample
Number (ASN) curves
for double and single
sampling plans of
Example 11.4

stage i, we denote the probability of proceeding to stage $i + 1$ by P_i. For example, in three-stage sampling, one proceeds to the second stage when $c_1 < X_1 < r_1$, so $P_1 = P(c_1 < X_1 < r_1)$. Similarly, the third stage is reached whenever the cumulative number of nonconforming items at the *second* stage, $Y_2 = X_1 + X_2$, satisfies $c_2 < Y_2 < r_2$, so $P_2 = P(c_2 < Y_2 < r_2)$. With this notation, the average sample number for a k-stage multiple sampling plan becomes

$$\text{ASN} = n_1 + P_1\{n_2 + P_2[n_3 + \cdots + P_{k-1}(n_k) + \cdots]\} \tag{11.8}$$

AVERAGE TOTAL INSPECTION (ATI)

Double and multiple sampling can be applied along with rectifying inspection of all rejected lots. In this case, **average total inspection** (ATI) per lot is a performance characteristic of interest just as it was for the single sampling case in Section 11.7.

Some additional notation helps to simplify the presentation. We denote the probability of lot acceptance at stage i by P_{a_i}. Then the average total inspection per lot (under rectifying inspection) can be shown to equal (Schilling 1982)

$$\text{ATI} = n_1 P_{a_1} + (n_1 + n_2)P_{a_2} + (n_1 + n_2 + n_3)P_{a_3} + \cdots$$
$$+ (n_1 + n_2 + \cdots + n_k)P_{a_k} + N(1 - P_{a_1} - P_{a_2} - P_{a_3} - \cdots - P_{a_k}) \tag{11.9}$$

For the frequently used case of double sampling, equation (11.9) becomes

$$\text{ATI} = n_1 P_{a_1} + (n_1 + n_2)P_{a_2} + N(1 - P_{a_1} - P_{a_2}) \tag{11.10}$$

EXAMPLE 11.5   Suppose that the double sampling plan $\{n_1 = 10, c_1 = 0, r_1 = 2\}, \{n_2 = 15, c_2 = 2, r_2 = 3\}$ of Example 11.3 is to be used in conjunction with rectifying inspection. It was shown in that example that

$$P_{a_1} = (1 - p)^{10}$$

$$P_{a_2} = 10p(1 - p)^9 \cdot [(1 - p)^{15} + 15p(1 - p)^{14}] = 10p(1 - p)^{23}(1 + 14p)$$

From equation (11.10), the average total inspection is

$$\text{ATI} = n_1 P_{a_1} + (n_1 + n_2) P_{a_2} + N(1 - P_{a_1} - P_{a_2})$$

$$= 10(1 - p)^{10} + 25(10p)(1 - p)^{23}(1 + 14p)$$

$$+ N[1 - (1 - p)^{10} - 10p(1 - p)^{23}(1 + 14p)]$$

For lots of size $N = 200$, the ATI curve for this plan is graphed in Figure 11.15.

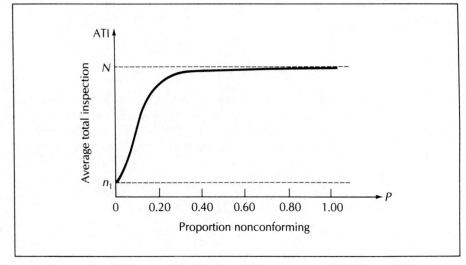

FIGURE 11.15
Average total
inspection curve for
double sampling plan
of Example 11.3 (lot
size $N = 200$)

11.9 THE MIL-STD-105E SAMPLING SYSTEM

HISTORICAL DEVELOPMENT

Military Standard 105E (MIL-STD-105E) is a *system* of attributes acceptance sampling plans that was initially developed as a set of sampling inspection tables for Army Ordnance during World War II. These tables passed through various levels of enhancement from 1942 through 1949 until the formal MIL-STD-105A system was first published in 1950. Successive modifications (105B–105C) occurred until 1963, when the revised MIL-STD-105D was released. Version 105D became the modern standard and reigned from 1963 through 1989. On May 10, 1989, the system was updated to MIL-STD-105E, which is the most current version.

During the postwar years, the widespread use of MIL-STD-105D for government work eventually led to the adoption of this system by national and international standards organizations. In the United States, it was adopted with some minor additions in 1973 as the American National Standards Institute's **ANSI/ASQC Z1.4** standard. For international distribution, it was adopted by the International Organization for Standardization and given the name **ISO 2859.**

Because of its formal adoption by government agencies, sampling inspection occupied a substantial position in quality assurance from about 1950 until the late 1970s, when the Japanese philosophy of continuous process improvement began to play a larger role in the United States. Because the idea of ever-decreasing nonconformance rates (continuous improvement) is at odds with the concept of fixed nonconformance levels inherent in systems based on *acceptable* quality level (AQL), sampling inspection came under heavy criticism in many circles. As a result, the role of acceptance sampling was reexamined during the 1980s and thus there was a reduced emphasis on sampling inspection for mature processes that are in a state of statistical control. The shift away from sampling inspection has naturally led to decreased reliance on MIL-STD-105E, but this system has been (and remains) in such wide use that the informed quality practitioner must still be versed in its application.

OVERVIEW OF MIL-STD-105E

It should be noted at the outset that 105E differs in only a few small ways from 105D. Practitioners versed in 105D do not have to relearn any new techniques. Essentially, 105E represents a rewriting of 105D in which the language has been modernized, in particular emphasizing that any AQL has its associated risks (of misclassifying lots) and that the word 'acceptable' in AQL does not imply that the customer (usually the government) is allowing the purchase of defective products.[8] In addition, the tables in 105E have been recalculated, resulting in some small corrections, and some of the language in the switching rules has been clarified. A good overview of the changes in 105E is Mundel (1990).

AQL LEVELS.   MIL-STD-105E is a system that contains single, double, and multiple acceptance sampling plans indexed on the AQL. The levels of AQL contained in the plans vary from a low of 0.01% up to 10%, and from above 10% up to 1,000%. AQLs of 10% or less are used for counting nonconforming items; AQLs of 10% and above are used for counting nonconformities per 100 units (recall Section 11.5).

There is a repeated pattern in the listed AQLs. Between any two AQLs that are successive powers of 10, there are exactly four AQLs, each selected so that it represents a fixed multiple of the AQL immediately less than it. For example, between an AQL of 10^{-2} and an AQL of 10^{-1}, there are AQLs of 0.015, 0.025, 0.040,

---

[8] In 105D, the AQL is defined to be "the maximum percent defective that, for purposes of sampling inspection, can be considered satisfactory as a process average." When the dependent clause was ignored, this sentence was often misinterpreted to mean that there was a 'satisfactory' level of defectives that customers would agree to purchase.

and 0.065; between 10^{-1} and 10^0, the AQLs are 0.15, 0.25, 0.40, and 0.65; and so forth. Given any AQL—say, p_0—this particular pattern arises very simply from the design requirement that there be a constant multiple r for which the AQLs are p_0, rp_0, r^2p_0, r^3p_0, r^4p_0, and r^5p_0. Since p_0 and r^5p_0 are successive powers of 10, their ratio must be exactly 10, which means $10 = r^5p_0/p_0 = r^5$ and, therefore, $r = 10^{1/5} = 1.5849$. All AQLs in MIL-STD-105E are computed using the factor 1.5849 but then rounded to two significant figures for simplicity. This method of selecting AQLs (and sample sizes) was chosen in order to keep P_a relatively constant for a fixed acceptance number c and varying AQL–sample size combinations. For a detailed discussion of the history of this procedure, see Duncan (1986, p. 243).

MIL-STD-105E's definition and clarification of the concept of AQL were given in Section 11.5. The standard states, in addition, that *"the selection or use of an AQL shall not imply that the contractor has the right to supply any defective unit of product."*[9] In other words, the plans in MIL-STD-105E are designed to accept lots in which $p = $ AQL with high probability, but that does *not* mean that a supplier can add nonconforming items to a lot while keeping its overall proportion of nonconforming items below the AQL.

Some of the language used in MIL-STD-105E is still indicative of an earlier era in quality control before the distinction between 'defective' and 'nonconforming' (or between 'defect' and 'nonconformity') was made. The modern distinction between these terms, however, does not affect the operation of MIL-STD-105E. Indeed, MIL-STD-105E originally recognized that there could be various levels of departure from specifications, and it introduced definitions for 'critical defects,' 'major defects,' and 'minor defects' along with what amount to operational definitions for distinguishing between them.

SWITCHING RULES.   One of the unique features of MIL-STD-105E is that it does not simply provide one sampling plan to achieve a given AQL; instead, it prescribes a collection of plans and rules for switching between these plans to take into account shifts in the quality of lots submitted for inspection. To provide incentives to the supplier, there are three inspection levels: **Tightened Inspection, Normal Inspection,** and **Reduced Inspection.** When MIL-STD-105E is first put online, the Normal Inspection mode is chosen. The plans dictated by Normal Inspection are used until too many successive lots are rejected (prompting a switch to Tightened Inspection) or until many successive lots are accepted (allowing a switch to Reduced Inspection). Thus, if the quality of submitted lots begins to decline, Tightened Inspection is invoked and subsequent lots are examined even more closely, providing an incentive for the supplier to address the problem while simultaneously
protecting the consumer from unusually bad lots. Conversely, if the lot quality improves, Reduced Inspection rewards the supplier with less stringent inspection and

---

[9] This limitation was rewritten in 105E. Formerly, in 105D, it was stated that the supplier did not have the right to 'knowingly' ship any defective units.

assures that the customer receives these higher quality lots. The rules that govern when to switch from one inspection mode to another are as follows:

Switching Rules

Required switch		Condition for switching
From Normal to Tightened	⟶	▪ After 2 out of 2, 3, 4, or 5 consecutive lots have been rejected on original inspection[10]
From Tightened to Normal	⟶	▪ After 5 consecutive lots have been accepted under Tightened Inspection on original inspection
From Normal to Reduced	⟶	▪ When 10 consecutive lots have been accepted on original inspection, *and* ▪ the total number of nonconforming items in these 10 lots does not exceed the applicable limit in Table VIII of the system, *and* ▪ production is steady, *and* ▪ it is approved by a responsible authority
From Reduced to Normal	⟶	▪ When a single lot is rejected, *or* ▪ a lot is marginally acceptable (i.e., the number of nonconforming items falls between the Accept number and the Reject number of the Reduced Inspection plan), *or* ▪ production becomes irregular, *or* ▪ other conditions warrant that Normal Inspection be reinstituted

---

Within each of these inspection modes there are several inspection levels. There are three General Inspection Levels, denoted by I, II, and III, as well as four Special Inspection Levels, S-1, S-2, S-3, and S-4. These levels determine the relationship between the lot size and the sample size and are prescribed by a responsible authority. *Unless otherwise specified, General Inspection Level II is used.* The difference between the levels is the degree of discrimination offered. Level I offers less discrimination than does Level II, which offers less discrimination than Level III; that is, the OC curves of plans at these levels *fall* as one moves from Level I through to Level III. To illustrate, Figure 11.16 shows the OC curves of

---

[10]In 105D, this rule was not clear about whether or not resubmitted lots could be considered. 105E makes it clear that only 'original' (not previously inspected) lots can be used.

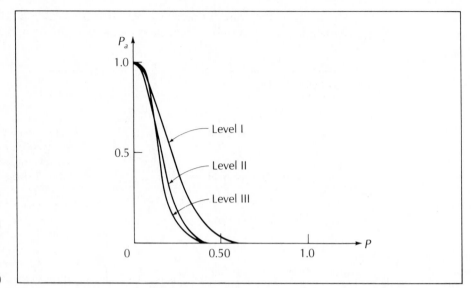

FIGURE 11.16
OC curves for Normal
Inspection, General
Inspection Levels I, II,
and III (assuming
$N = 100$, AQL = 6.5%)

Levels I, II, and III plans for a given lot size and AQL under the Normal Inspection mode. The Special Inspection Levels (S-1, S-2, S-3, and S-4) are used whenever destructive or otherwise expensive testing is involved, which necessitates smaller sample sizes (accompanied by a corresponding decrease in discriminating power).

OC CURVES.   MIL-STD-105E specifies sample sizes and Accept/Reject numbers so that there is a high probability of lots of AQL quality or better being accepted. It is interesting to note, though, that for a fixed AQL, P_a does not remain constant as the lot size changes. Instead, for plans with *nonzero* acceptance numbers ($c \geq 1$), P_a ranges from about 0.91 to 0.99 as the lot size increases. For zero acceptance plans ($c = 0$), P_a is approximately 0.88. This variation in P_a was designed into the system for the purpose of *increasing* the probability of acceptance for lots of AQL quality *as the lot size increases*. The reasoning behind this decision was that the consequences of rejecting lots become more serious (for the supplier) as the lot size increases, so that an effort should be made to increase P_a (for lots of AQL quality) in these circumstances. Since MIL-STD-105E chooses samples sizes that increase with the lot size, P_a (at $p =$ AQL) also increases as the sample size increases. This feature can be a little disconcerting in practice because even moderate changes in lot sizes affect the probability of lot acceptance (but it still stays in the range of 0.91 to 0.99). In order to be sure of the protection a given plan offers, it then becomes necessary to refer to the published OC curves given in MIL-STD-105E for the particular plan.

IMPLEMENTATION

There are two fundamental steps in the operation of MIL-STD-105E: (1) selecting the appropriate plans from the published tables in the standard, and (2) using the switching rules when the lot quality changes for better or worse. Unfortunately, it is easy to forget the second step and just use the standard to select a *single* plan, which is then used regardless of changes in lot quality. Usually, this happens when the Normal Inspection, Level II tables are used to select a plan and then, for one reason or another, that plan is used permanently. The problem with doing this is that if the lot quality begins to deteriorate, the consumer ends up accepting more low-quality lots than would otherwise be accepted if the switching rules had been used.

The obvious solutions to this problem are not necessarily good. For example, merely using Normal Inspection, Level III or perhaps Tightened Inspection can result in unacceptably high rejection rates for the supplier. There are, however, two general approaches that do solve the problem of isolated lots. One method is to use special tables created by Schilling and Johnson (1980) that yield individual sampling plans designed to match the performance of MIL-STD-105E under the switching rules. Another approach is to abandon MIL-STD-105E altogether and use a system of plans such as in the ANSI/ASQC Q3–1988 standard designed specifically to handle isolated or infrequent lots. Since MIL-STD-105E was not originally designed with such situations in mind, we do not pursue these alternate applications further. In the remainder of this section, it is assumed that the usual conditions for applying MIL-STD-105E exist; that is, a steady stream of lots is to be inspected and the switching rules are to be used.

In order to avoid a confusion of subscripts, MIL-STD-105E uses the very simple notation 'Ac' to stand for an 'acceptance number' and 'Re' to stand for a rejection number. For single sampling plans, a lot is accepted if the number of nonconforming items, X, is less than or equal to the 'Ac' number given in the tables, and it is rejected if X exceeds or equals the 'Re' number. A majority of the time, 'Re' is exactly one more than 'Ac,' so this procedure is identical to that introduced in Section 11.5. In some cases (Reduced Inspection, double sampling, and multiple sampling), 'Ac' and 'Re' may differ by *more* than one, but there is no change in the decision procedure: Accept if $X \leq$ 'Ac,' Reject if $X \geq$ 'Re.' (The point to remember here is that 'Ac' is *included* in the acceptance region and 'Re' is included in the rejection region.)

The steps in implementing MIL-STD-105E can be summarized as follows:

OPERATION OF THE MIL-STD-105D SAMPLING SYSTEM	1. Determine the AQL (based on an agreement between the producer and the customer). 2. Decide on an inspection mode and level (unless otherwise specified, use Normal Inspection, Level II to start the system). 3. Determine the lot size. 4. Use the table of **Sample Size Code Letters** (see Appendix 11) to select the appropriate code letter. 5. Decide on the type of sampling procedure: single, double, or multiple sampling.

(Continued)

(Continued)

> 6. Use the table corresponding to the sampling procedure selected in step 5 and the inspection mode/level from step 2 to find the sample size(s) and acceptance and rejection numbers for the plan. In cases where a plan does not exist for a given lot size and AQL combination, be careful to follow the arrows in the tables to the nearest available plan.
>
> 7. Begin by using the plan selected in step 6 and keep a record of acceptances and rejections so that the switching rules can be applied. When switching becomes necessary, determine the revised inspection mode and level and repeat steps 4–6 to find the revised sampling plan.

Many of the MIL-STD-105E tables are given in Appendix 11 of this text. Although these tables have their own specific numbering system within MIL-STD-105E, we recommend that the user simply follow the table *titles* when selecting the sampling plan. The following examples illustrate the steps in selecting a plan from MIL-STD-105E.

EXAMPLE 11.6

Unless otherwise specified, one always starts MIL-STD-105E under Normal Inspection, Inspection Level II. In this setting, suppose that lots of size 200 are to be inspected to an AQL of 1.5%. What kind of sampling plans are appropriate? Examining the table of **Sample Size Code Letters** in Appendix 11, we find the code letter for Level II Inspection is 'G' (since 200 falls in the 151–280 lot size range). For single sampling plans, we then proceed to the **Single Sampling Plans for Normal Inspection (Master Table)** in Appendix 11 and find that code letter G corresponds to a sample size of $n = 32$, an 'Ac' number of 1, and a 'Re' number of 2. Thus, the single sampling plan is to draw a *random* sample of 32 items from the lot and accept the lot only if one or fewer nonconforming items are found. If double sampling is used instead, we use the **Double Sampling Plans for Normal Inspection (Master Table)** in Appendix 11 to find that the first and second sample sizes are both specified to be 20, with 'Ac' = 0, 'Re' = 2 for the first sample and 'Ac' = 1, 'Re' = 2 for the *cumulative* sample. In the notation of Section 11.8, this is the plan $\{n_1 = 20, c_1 = 0, r_1 = 2\}$, $\{n_2 = 20, c_2 = 1, r_2 = 2\}$. Similarly, for multiple sampling, the **Multiple Sampling Plans for Normal Inspection (Master Table)** in Appendix 11 specifies $k = 7$ stages of sampling with a sample size of 8 at each stage. Note that an acceptance number is not permitted for the first two of the seven stages. It is mandatory either that a lot be rejected at one of these stages or that it pass on to at least a third stage of inspection. For the latter two plans, one can refer to the **Average Sample Size Curves for Double and Multiple Sampling** in Appendix 11 to more closely gage the level of sampling that might be needed for these plans.

EXAMPLE 11.7

As demonstrated at the end of Section 11.3, one cannot hope to use acceptance sampling when the nonconformance rate becomes very small. This is evidenced in the MIL-STD-105E tables by the fact that some combinations of lot sizes and AQLs simply do not have an appropriate sampling plan and, in those cases, one must follow the directional arrows in the tables to the nearest available plan. When doing this, it is very important to realize that the sample size must *change* accordingly (to match the code letter of the nearby plan). For example, suppose that the AQL in Example 11.6 was changed from 1.5% to 0.065%. As before, Normal Inspection, Level II would specify the code letter G, but upon examining the **Single Sampling Plans for Normal Inspection (Master Table)** in Appendix 11, we would find only an arrow directing downward to the next available plan. This plan corresponds to code letter L and specifies that the sample size be changed to 200 with an acceptance number of 0 (and, therefore, a rejection number of 1). But this means that the sample size equals the lot size! In other words, depending on the limitations imposed by the lot size and the size of the AQL, the plan selected may involve a great deal of sampling (sometimes even 100% inspection). In this example, it is also informative to verify that the zero acceptance plan $n = 200$, $c = 0$ indeed gives $P_a = (1 - 0.00065)^{200} \approx 0.88$ at the AQL of 0.065% (i.e., $p = 0.00065$).

EXAMPLE 11.8

The overall acceptance rates of MIL-STD-105E plans can be evaluated from the published OC curves for each code letter. The curves are constructed for the single sampling case, but OC curves for double and multiple sampling are matched as closely as possible to them. As an example, the OC curves associated with Code Letter D appear in Figure 11.17. To facilitate reading these curves, the tables that appear below the graphs give a range of acceptance probabilities and the corresponding values of p at which these probabilities are obtained. For example, if the AQL was 1.5%, the table indicates that P_a would equal 90% for $p = 1.31\%$, and so on.

EXAMPLE 11.9

Another performance characteristic of interest is the average outgoing quality limit (AOQL) associated with a plan. Recall that the AOQL is calculated to show the maximum proportion nonconforming, on average, that can be expected when a sampling plan is used with rectifying inspection (see Section 11.7). MIL-STD-105E contains tables of the AOQL factors by inspection mode, AQL, and code letter. For AQL = 1.5% and lot size 200, the single sampling plan in Example 11.6 corresponds to code letter G with a sample size of 32. To find the AOQL associated with this plan, we refer to the table of **Average Outgoing Quality Limit Factors for Normal Inspection** in Appendix 11 to find a factor of 2.6% for that AQL and code letter. The values listed in the table must be multiplied by $(N - n)/N$ to get the exact value of AOQL that takes sample size and lot size into account (recall Example 11.2). Thus, for $N = 200$ and $n = 32$, we find an exact AOQL of $0.026[(200 - 32)/200] = 0.02184$, or 2.184%.

EXAMPLE 11.10

To illustrate the MIL-STD-105E switching rules, let us suppose that lots of size 600 are to be inspected to an AQL of 1.0%. To start the system, we use Normal Inspection, Level II, which yields a code letter of J. The single sampling plan for this code letter is found in the table of **Single Sampling Plans for Normal Inspection (Master Table)** in Appendix 11 and specifies $n = 80$, 'Ac' = 2,

CHART D – OPERATING CHARACTERISTIC CURVES FOR SINGLE SAMPLING PLANS

(Curves for double and multiple sampling are matched as closely as practicable)

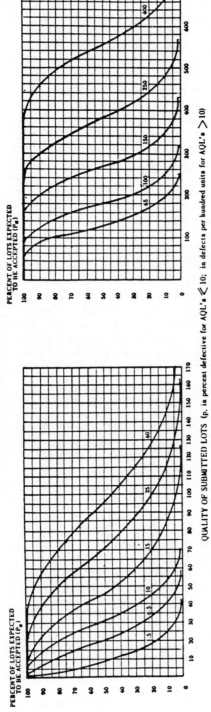

PERCENT OF LOTS EXPECTED TO BE ACCEPTED (P_a)

QUALITY OF SUBMITTED LOTS (p, in percent defective for AQL's $\leqslant$ 10; in defects per hundred units for AQL's $>$ 10)

Note: Figures on curves are Acceptable Quality Levels (AQL's) for normal inspection.

TABLE X-D-1 – TABULATED VALUES FOR OPERATING CHARACTERISTIC CURVES FOR SINGLE SAMPLING PLANS

The first three value columns are p (in percent defective). The remaining columns are p (in defects per hundred units) for the Acceptable Quality Levels (normal inspection). Columns marked × carry an X (no plan) in the original header.

P_a	1.5	6.5	10	1.5	6.5	10	15	25	40	65	×	100	×	150	×	250	×	400	×
99.0	0.126	1.97	6.08	0.126	1.86	5.45	10.3	22.3	36.3	43.8	59.6	76.2	93.5	129	157	215	244	355	386
95.0	0.639	4.64	11.1	0.641	4.44	10.2	17.1	32.7	49.8	58.7	77.1	96.1	116	156	186	249	281	399	432
90.0	1.31	6.88	14.7	1.32	6.65	13.8	21.8	39.4	58.2	67.9	87.8	108	129	171	203	268	301	424	458
75.0	3.53	12.1	22.1	3.60	12.0	21.6	31.7	52.7	74.5	85.5	108	130	153	199	234	303	339	468	504
50.0	8.30	20.1	32.1	8.66	21.0	33.4	45.9	70.9	95.9	108	133	158	183	233	271	346	383	521	558
25.0	15.9	30.3	43.3	17.3	33.7	49.0	63.9	92.8	121	135	163	190	217	272	312	392	432	577	617
10.0	25.0	40.6	53.8	28.8	48.6	66.5	83.5	116	147	162	193	222	252	309	352	437	479	631	672
5.0	31.2	47.1	60.0	37.4	59.3	78.7	96.9	131	164	180	212	243	274	334	378	465	509	665	707
1.0	43.8	59.0	70.7	57.6	83.0	105	126	164	200	218	252	285	318	382	429	522	568	732	776
Acceptable Quality Levels (tightened inspection)	2.5	10	×	2.5	10	15	25	40	65	100	×	150	×	250	×	400	×	×	×

FIGURE 11.17
OC curves for MIL-STD-105E code letter D

'Re' = 3. Suppose, now, that this plan has been put into operation and that the following history of accepted (A) and rejected (R) lots is recorded:

$$\begin{bmatrix} \text{Lot} & \to & 1 & 2 & 3 & 4 & 5 & 6 & 7 & 8 & 9 & 10 \\ \text{Decision} & \to & A & A & R & A & A & A & A & R & A & R \\ & & & & & & & & & & & \downarrow \\ & & & & & & & & & & & \text{Switch} \end{bmatrix}$$

Since the rule for switching to Tightened Inspection is that two out of five consecutive lots must be rejected, the switch to tightened inspection does not occur until lot 10 in this case. The tightened plan is found from the tables of **Single Sampling Plans for Tightened Inspection (Master Table),** which specifies $n = 80$, 'Ac' = 1, and 'Re' = 2. Notice that the sample size does not change (this is *always* the case when switching from Normal to Tightened) but that the acceptance/rejection numbers are both reduced, thereby tightening the plan. Under this tightened plan, suppose that the next sequence of lots inspected is as follows:

$$\begin{array}{lccccccccc} \text{Lot} & \to & 11 & 12 & 13 & 14 & 15 & 16 & 17 & 18 \\ \text{Decision} & \to & A & A & R & A & A & A & A & A \\ & & & & & & & & & \downarrow \\ & & & & & & & & & \text{Switch} \end{array}$$

In this case, the switch back to Normal inspection occurs because five consecutive lots were accepted as of lot 18. At this point, the Normal Inspection plan $\{n = 80, \text{'Ac'} = 2, \text{'Re'} = 3\}$ again becomes operative. Finally, suppose that the next group of lots inspected has the following history:

$$\begin{array}{lccccccccccccc} \text{Lot} & \to & 19 & 20 & 21 & 22 & 23 & 24 & 25 & 26 & 27 & 28 & 29 & 30 \\ \text{Decision} & \to & A & R & A & A & A & A & A & A & A & A & A & A \\ \text{Number} \\ \text{nonconforming} & \to & 0 & 3 & 0 & 1 & 0 & 0 & 0 & 2 & 0 & 0 & 0 & 0 \\ & & & & & & & & & & & & & \downarrow \\ & & & & & & & & & & & & & \text{Switch} \end{array}$$

Although ten consecutive lots have been accepted as of lot 30, before a switch to Reduced Inspection can be made, the total number of nonconforming items in these ten lots must be compared to the limit number found in the table of **Limit Numbers for Reduced Inspection** in Appendix 11. To use this table, the accumulated sample size over the ten lots must be found (in this case, it amounts to $80 \cdot 10 = 800$ items). Corresponding to the accumulated sample size and the AQL, the limit number of four is obtained from the table. This means that the total number of nonconforming items in the ten accepted lots cannot exceed four if a switch is to be to Reduced Inspection. Since the ten lots contained a total of three nonconforming items (one from lot 22, two from lot 26), the switch to Reduced Inspection is allowed in this case. To find the appropriate Reduced Inspection plan, we refer to the tables of **Single Sampling Plans for Reduced Inspection.** Code letter J now gives a plan with $n = 32$ (a substantial reduction from $n = 80$), 'Ac' = 1, and 'Re' = 3. Notice the gap between 'Ac' and 'Re' for this plan. The gap comes into play when the decision is made whether or not to switch back to Normal Inspection. Under Reduced Inspection, any rejected lot causes a shift back to Normal Inspection, but any *accepted* lot whose number of nonconforming items falls between 'Ac' and 'Re' *also* causes a switch back to Normal Inspection.

COMPLIANCE TESTING

In Chapter 5 (Example 5.12), a form of acceptance sampling called compliance sampling was introduced. **Compliance sampling** is a stringent form of inspection used when consumer safety, product liability, or tight regulatory standards are considered to be of extreme importance. In general, compliance testing can be used whenever serious consequences could result from accepting lots that contain nonconforming items.

Compliance sampling plans are usually single-sample zero acceptance plans (defined in Section 11.5). Even though finding no nonconforming items in a sample provides no guarantee that a lot is itself free of nonconforming items, it is very difficult to convince responsible parties to use anything except $c = 0$ when product safety and liability are paramount. As a result, the plans used in compliance testing (with $c = 0$) can be very harsh from the supplier's point of view.

It seems apparent that fairly large sample sizes should be required to guarantee the levels of lot quality sought in the compliance sampling setting. This is indeed the case and, because the sample can represent a large portion of the lot, the hypergeometric distribution is used instead of the binomial distribution for calculating probabilities of lot acceptance (see Section 5.5).

Suppose, then, that one wants to assure that lots of size N containing D nonconforming (or, defective) items will be rejected by a zero acceptance plan. One way to run the test procedure is to find the minimum sample size n such that the plan has a high probability, γ, of detecting at least one of the nonconforming items in the lot. Letting X denote the number of nonconforming items in a *random sample* of n items drawn from the lot, we want to find the minimum n for which $P(X \geq 1) = \gamma$, where X has a hypergeometric distribution.

Schilling (1978) solved this problem by providing a table of sampling fractions for assuring that lots with D nonconforming items have a 90% chance of being rejected by the plan (i.e., $\gamma = 0.90$). Specifically, these tables give the *fraction, f,* of the lot that must be sampled to meet the 90% probability requirement. With a bit of additional work, the table can also be used for values of γ other than 0.90. To illustrate, if one wants a compliance plan that protects (at level $\gamma = 0.90$) against lots of size $N = 200$ that contain $D = 10$ nonconforming items, Schilling's table would require a sampling fraction of $f = 0.21$. In other words, the zero acceptance plan would require a sample size of $n = fN = 0.21 \cdot 200 = 42$.

As a convenient alternative to using these tables to find sample sizes for compliance testing, Farnum and Suich (1986) have shown that for a hypergeometric variable X to satisfy $P(X \geq 1) = \gamma$, the minimum sample size n *must* fall in the interval

$$(N - D + 1)[1 - (1 - \gamma)^{1/D}] \leq n \leq N[1 - (1 - \gamma)^{1/D}] \tag{11.11}$$

In essence, Schilling's table is equivalent to using the right-hand inequality in equation (11.11) to find a conservative estimate for n (conservatively large, that is).

Equation (11.11) has several interesting features. First, it requires only a simple calculation to approximate the sample size needed for a compliance sampling plan; that is, no tables are needed. Second, equation (11.11) can immediately be applied for any value of γ, not just $\gamma = 0.90$. Third, the width of the interval in equation (11.11) depends only on D and is completely independent of N, since

$$\text{width} = N[1 - (1 - \gamma)^{1/D}] - (N - D + 1)[1 - (1 - \gamma)^{1/D}]$$

$$= (D - 1)[1 - (1 - \gamma)^{1/D}]$$

Using the data from the previous example, we substitute $N = 200$, $D = 10$, and $\gamma = 0.90$ into equation (11.11) to yield

$$(200 - 10 + 1)[1 - (1 - 0.90)^{1/10}] \leq n \leq 100[1 - (1 - 0.90)^{1/10}]$$

$$(191)[0.205672] \leq n < (200)[0.205672]$$

$$39.3 \leq n \leq 41.3$$

When rounded (to be conservative, sample sizes are always rounded upward) to integer values, the required sample size must lie between 40 and 42. As noted above, the width of this interval (two units) would not change at all if the lot size was changed. For example, with $D = 10$ and $\gamma = 0.90$ again but N changed to 400, equation (11.11) indicates that the minimum sample size must fall in the interval 81 to 83 (after rounding off), an interval of the same width (two units) as before.

| 11.11 | VARIABLES ACCEPTANCE SAMPLING PLANS |

INTRODUCTION

Prior sections of this chapter have shown how attributes data (specifically, the number of nonconforming items in a sample) are used for acceptance sampling. Since many processes generate variables data, it is also desirable to have acceptance sampling plans based on variables measurements. This section presents the fundamentals of acceptance sampling by variables. Section 11.12 then discusses the most frequently used variables sampling system, MIL-STD-414, and its national and international counterparts, ANSI/ASQC Z1.9–1980 and ISO 3951.

At the outset, it should be noted that most acceptance sampling is done using attributes, not variables, data. There are a few reasons for this. First, variables measurements are almost always more expensive to obtain than attributes data because measuring instruments and trained operators are needed for variables measurements. The expense involved includes the time needed to make the measurements as well as the economic costs. As Grant and Leavenworth (1988, p. 537) point out, these costs multiply rapidly if more than one characteristic is measured and used to determine lot acceptance because a separate plan is required for *each* characteristic. By comparison, only *one* attributes plan is needed if several characteristics are examined to determine the conformance or nonconformance of each sampled item. Another complication with variables plans is that they usually assume that the underlying process generating the measurements follows a normal distribution. This means that, prior to implementing a variables plan, some effort should be expended to ascertain whether process measurements are approximately normal. Finally, variables plans require a bit more computational effort than do attributes plans, an unattractive feature to some prospective users.

On the positive side, variables data are generally more powerful than attributes data (based on the same sample sizes), so variables sampling plans usually involve smaller sample sizes than attributes plans for the same levels of protection. As an example, for lots of size 500, MIL-STD-105E requires samples of size 50 (Normal Inspection, Level II), whereas for the same level of protection, the variables sampling system MIL-STD-414 requires samples of about 25 (Normal Inspection, Level IV).[11]

In addition to reducing sample sizes, working with variables data usually gives a much deeper picture of a process's behavior than is possible with attributes data. Variables data can also uncover problems or deficiencies with the measuring system being used.

AN OVERVIEW OF ACCEPTANCE SAMPLING BY VARIABLES

FORM 1 AND FORM 2 PLANS   Like attributes plans, variables plans operate by comparing the summarized data from random samples to calculated 'acceptance' values. As in attributes plans, a lot is then accepted or rejected depending on whether or not the sample statistic exceeds the acceptance value. Whereas attributes plans count the number, X, of nonconforming items in a random sample from a lot, variables plans summarize *measurements* by calculating the sample mean, $\bar{x}$, of a random sample from a lot.

Variables data also differ from attributes data in that specification limits are available for each process characteristic. It is the position of the sample mean with respect to the specification limit(s) that determines whether a lot is likely to be of acceptable quality.

Statistics used in variables plans combine the sample mean, the process variation (whether it is known or must itself be estimated), and the specification limit(s) by forming ratios of the form

$$\frac{USL - \bar{x}}{\sigma}, \quad \frac{\bar{x} - LSL}{\sigma}, \quad \frac{USL - \bar{x}}{s}, \quad \frac{\bar{x} - LSL}{s}, \quad \text{or} \quad \frac{USL - \bar{x}}{\bar{R}}, \quad \frac{\bar{x} - LSL}{\bar{R}}$$

$$(11.12)$$

depending on whether the process standard deviation σ is known or, when it is unknown, depending on how one chooses to estimate the variation. When σ is unknown, either the sample standard deviation s is used or the sample is broken into *subsamples* of size 5, thereby allowing one to calculate the average range $\bar{R}$ of these subsamples. In the latter case, it is not necessary to go the extra step of

---

[11] In MIL-STD-414, Normal Inspection, Level IV is the equivalent of Normal Inspection, Level II for MIL-STD-105E.

forming the estimate $\overline{R}/d_2$, since the published acceptance values automatically take the d_2 factor into account.

For example, suppose that a variables sampling plan is to be used on a process whose standard deviation σ is known and which has only an upper specification limit, USL. Such a plan requires that random samples of a specified size n be taken from each lot. The same characteristic on each of the n items sampled is then measured, and the sample mean $\overline{x}$ is used to form the statistic $(USL - \overline{x})/\sigma$. After an AQL level is determined, an acceptance value k is found in a published table of such values and a lot is accepted whenever $(USL - \overline{x})/\sigma \geq k$. The lot is accepted as long as the sample mean is a 'reasonable' number of standard deviations below the USL.

Alternatively, the statistic $(USL - \overline{x})/\sigma$ could be used to estimate the proportion $\hat{p}_U$ of product whose measured values fall above the upper specification limit. If we compare $\hat{p}_U$ to a **maximum acceptable proportion nonconforming,** M, a lot is accepted whenever $\hat{p}_U \leq M$. Values of M, which depend on the specified AQL and sample size, are found in published tables in the same way that k values are found in the preceding paragraph.

To distinguish between these two methods, variables acceptance sampling plans are classified as either **Form 1** or **Form 2.** Form 1 plans compare statistics such as those in equation (11.12) to tabulated constants, k, with the lot disposition depending on which side of k the statistic falls. Form 2 procedures first convert the sample data into estimates of the proportions $\hat{p}_U$ and $\hat{p}_L$ of product exceeding the upper or lower specification limits and then compare these proportions to a maximum allowable proportion M found in tables accompanying the sampling plan. Tables of acceptance values are calculated for each variation measure (σ, s, or $\overline{R}$) and for both acceptance procedures (Form 1 and Form 2).

CALCULATING ACCEPTANCE VALUES. To illustrate how acceptance values are calculated, consider the simplest type of variables plan, one where σ is known and a single-sided specification limit (say, an upper spec) is used. To develop the acceptance value k for a Form 1 plan, one needs first to specify an AQL, just as is done in many attributes plans. Suppose, for example, that an AQL of 1% is chosen. Since one normally wants the majority of AQL lots to be accepted by the plan, let us further specify that there should be a probability of around 95% of accepting lots of AQL quality.

If we make the *assumption* that the process measurements X follow a normal distribution, then lots with AQL = 0.01 will satisfy the equation

$$P(X > USL) = AQL = 0.01 \tag{11.13}$$

Converting X to the standard normal variable z, equation (11.13) becomes $P[z > (USL - \mu)/\sigma] = 0.01$, where μ is the mean value of measurements from lots with AQL quality. From the normal table (Appendix 2), we find $P(z > 2.326) = 0.01$, from which it follows that

$$\frac{USL - \mu}{\sigma} = z_{0.01} = 2.326 \tag{11.14}$$

From the criterion that acceptable lots are those for which $(USL - \bar{x})/\sigma \geq k$, 95% of all lots with AQL quality should satisfy this inequality; that is,

$$0.95 = P\left(\frac{USL - \bar{x}}{\sigma} \geq k\right) \tag{11.15}$$

We add and subtract μ/σ to convert equation (11.15) into

$$0.95 = P\left(\frac{USL - \bar{x}}{\sigma} \geq k\right) = P\left(\frac{USL - \mu}{\sigma} + \frac{\mu - \bar{x}}{\sigma} \geq k\right)$$

$$= P\left(\frac{\bar{x} - \mu}{\sigma} \leq \left(\frac{USL - \mu}{\sigma} - k\right)\right)$$

$$= P\left[\frac{\bar{x} - \mu}{\sigma}\sqrt{n} \leq \left(\frac{USL - \mu}{\sigma} - k\right)\sqrt{n}\right]$$

$$= P\left[z \leq \left(\frac{USL - \mu}{\sigma} - k\right)\sqrt{n}\right] \tag{11.16}$$

From the fact that the standard normal value $z_{0.05} = 1.645$ corresponds to a right tail area of 5% (and, therefore, a left tail area of 95%), equation (11.16) simplifies to

$$\left(\frac{USL - \mu}{\sigma} - k\right)\sqrt{n} = z_{0.05} = 1.645 \tag{11.17}$$

Because this equality holds only for lots of AQL quality—that is, lots for which equation (11.14) holds—$(USL - \mu)/\sigma = z_{0.01} = 2.326$ can be substituted into equation (11.17) to yield $(2.326 - k)\sqrt{n} = 1.645$. In turn, this equation can easily be solved for k:

$$k = 2.326 - \frac{1.645}{\sqrt{n}} \tag{11.18}$$

It is easy to generalize equation (11.18). If z_p denotes the standard normal value corresponding to an upper tail area of p, we have

$$k = z_{AQL} - \frac{z_{1-P_a}}{\sqrt{n}} \tag{11.19}$$

where P_a is the specified probability of accepting lots of AQL quality.

Note that equations (11.18) and (11.19) work for *any* sample size, n. To determine a unique sample size, an additional condition must be imposed on the OC curve of the variables plan. One such condition is to specify some quality level, call it the RQL, that should be *rejected* by the plan; that is, suppose that lots of RQL quality are to be accepted with a specified low probability, P_a'. Figure 11.18 illustrates the OC curve with this additional requirement included.

From an argument like the one that leads to equation (11.19), it can be shown that this additional condition results in the equation

$$k = z_{RQL} + \frac{z_{P_a'}}{\sqrt{n}} \tag{11.20}$$

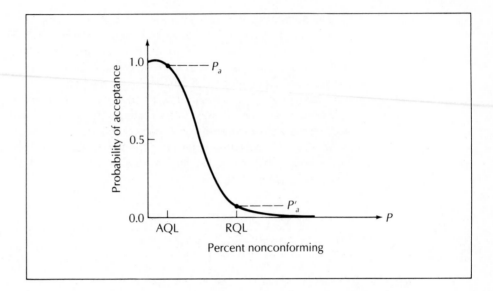

FIGURE 11.18
OC curve of a
Variables Sampling
Plan (AQL =
acceptable quality
level; RQL =
rejectable quality
level)

Equations (11.19) and (11.20) represent a system of two equations in two un-knowns, k and n. The simultaneous solution of the two equations yields

$$n = \left(\frac{z_{1-P_a} + z_{P'_a}}{z_{AQL} - z_{RQL}} \right)^2 \qquad (11.21)$$

The value of n given by equation (11.21) is generally not an integer and must be rounded. Then the rounded value of n can be substituted into either equa-tion (11.19) or (11.20) to solve for k, which, because a rounded value of n is used, gives two slightly different values of k. Various options are possible for handling the two k values: take the average of the two k values, or just select *one* of equa-tion (11.19) or (11.20) to solve for k.

As an example of the foregoing procedure, suppose that we add the condition $P'_a = 0.025$ for lots of quality RQL = 0.10 to our original example of $P_a = 0.95$ for lots of quality AQL = 0.01. Equation (11.21) then yields a sample size of

$$n = \left(\frac{z_{0.05} + z_{0.025}}{z_{0.01} - z_{0.10}} \right)^2$$

$$= \left(\frac{1.645 + 1.960}{2.326 - 1.282} \right)^2 = 11.92$$

Since the sample size must be an integer, 11.92 must be rounded to 12. Putting $n = 12$ in equations (11.19) and (11.20) gives the two k values:

$$k = \frac{z_{0.01} - z_{0.05}}{\sqrt{12}} = 2.326 - \frac{1.645}{\sqrt{12}} = 1.851$$

and

$$k = \frac{z_{0.10} + z_{0.025}}{\sqrt{12}} = 1.282 + \frac{1.960}{\sqrt{12}} = 1.848$$

To implement the sampling plan, either an average k value of 1.8495 could be used or one might simply decide to use the value $k = 1.851$ in order to strictly adhere to the requirement that $P_a = 0.95$ for the AQL lots. Since it is generally the case that the two estimates of k are close, the choice between these options does not have a material effect on the functioning of the plan.

The type of argument illustrated above can be repeated for cases where σ is unknown and either s or $\overline{R}$ is used to estimate the process variation. The approach is the same, but the analysis is somewhat more difficult because the sampling variability of s or $\overline{R}$ must also be included in the development. For readers interested in exactly how this extra variability is handled, we recommend the excellent treatment in Duncan (1986, chap. 12).

11.12 THE MIL-STD-414 SAMPLING SYSTEM

INTRODUCTION

MIL-STD-414, first published in 1957, was designed for acceptance sampling with variables data. Like MIL-STD-105E, it is an acceptance sampling *system* in the sense that it includes several sampling plans, several levels of inspection, and a set of rules for switching between the inspection levels. MIL-STD-414 was developed as a substitute for 105E in those cases where smaller sample sizes are desired. This is especially important, for example, when sampled items are evaluated by destructive testing.

MIL-STD-414 is matched fairly closely to 105E. It is an AQL system with the same inspection levels (reduced, normal, tightened) as 105E. Whereas 105E and its national and international counterparts are closely matched, MIL-STD-414 differs somewhat from its national and international counterparts, ANSI/ASQC Z1–1980 and ISO 3951.

OVERVIEW OF MIL-STD-414

Like most variables sampling plans, MIL-STD-414 is based on the assumption that process measurements approximately follow a normal distribution. Before the plans in this system are used, then, some sample data should first be examined to see whether the normality assumption is reasonable for the application at hand.

As Schilling (1982) points out, MIL-STD-414 can easily be considered to be three separate sampling systems, each depending on the particular method used to estimate the process variation. The system includes separate sets of sampling plans for the cases where (1) σ is known, (2) σ is unknown and is estimated by the sample standard deviation s, and (3) σ is unknown and is estimated by $\overline{R}/d_2$. Form 1 or Form 2 procedures can be used with any of the three systems, with one exception: only Form 2 plans are allowed for process data that have two-sided specification limits. Figure 11.19 shows the possible combinations of variation estimates and Forms 1 and Form 2 available in MIL-STD-414. Form 2 plans are the most popular

Specification Limit(s)	Variation Estimate Used	Form Available
One-sided specification ⟵→	standard deviation method ⟶	1 and 2
	range method ⟶	1 and 2
	variance known ⟶	1 and 2
Two-sided specification ⟵→	standard deviation method ⟶	2
	range method ⟶	2
	variance known ⟶	2

FIGURE 11.19
Variation estimates
and forms for
MIL-STD-414

among all these choices because the switching rules are *only* given in terms of percent nonconforming (which has already been calculated in Form 2 plans).

Since MIL-STD-414 is an AQL system, one must decide beforehand on an AQL to use. In order to narrow down the range of AQLs possible with variables measurements, MIL-STD-414 allows only a small number of AQLs, ranging from 0.04 through 15.0. The incremental steps between these endpoints are established in the same manner as for 105E (see Section 11.10). Note that the AQL levels in MIL-STD-414 represent a *subset* of those in 105E. 105E, for example, allows AQLs smaller than 0.04 and larger than 15.0. Table 11.1 shows the AQL conversion table used in MIL-STD-414. Suppose, for example, an AQL of 1.30% is desired. From Table 11.1, the AQL of 1.5% would be used to select one of the MIL-STD-414 plans.

As in 105E, the lot size and inspection levels are used to determine the required sample sizes in MIL-STD-414. Table 11.2 gives the sample size Code Letters

TABLE 11.1   AQL Conversion Table

For specified AQL values falling within these ranges	Use this AQL value
———— to 0.049	0.04
0.050 to 0.069	0.065
0.070 to 0.109	0.10
0.110 to 0.164	0.15
0.165 to 0.279	0.25
0.280 to 0.439	0.40
0.440 to 0.699	0.65
0.700 to 1.09	1.0
1.10 to 1.64	1.5
1.65 to 2.79	2.5
2.80 to 4.39	4.0
4.40 to 6.99	6.5
7.00 to 10.9	10.0
11.00 to 16.4	15.0

TABLE 11.2 Sample Size Code Letters[a]

Lot Size		Inspection Levels				
		I	II	III	IV	V
3 to	8	B	B	B	B	C
9 to	15	B	B	B	B	D
16 to	25	B	B	B	C	E
26 to	40	B	B	B	D	F
41 to	65	B	B	C	E	G
66 to	110	B	B	D	F	H
111 to	180	B	C	E	G	I
181 to	300	B	D	F	H	J
301 to	500	C	E	G	I	K
501 to	800	D	F	H	J	L
801 to	1,300	E	G	I	K	L
1,301 to	3,200	F	H	J	L	M
3,201 to	8,000	G	I	L	M	N
8,001 to	22,000	H	J	M	N	O
22,001 to	110,000	I	K	N	O	P
110,001 to	550,000	I	K	O	P	Q
550,001 and over		I	K	P	Q	Q

[a]Sample size code letters given in body of table are applicable when the indicated inspection levels are to be used.

corresponding to various lot sizes. For a given code letter, the required sample varies depending on the *level* of inspection used (i.e., normal, reduced, or tightened). Notice that, like 105E, there are five inspection levels from which to choose, but, unlike 105E, these levels are numbered I through V. In essence, the first two levels (I and II) correspond to the special inspection levels of 105E and the remaining levels (III, IV, and V) correspond to the general levels. Viewed in this manner, Normal Inspection, Level IV becomes the starting point for using MIL-STD-414; that is, unless otherwise specified, MIL-STD-414 should be started by using tables corresponding to Normal Inspection, Level IV.

The sample plan tables for MIL-STD-414 are used in exactly the same fashion as those in 105E. First, one selects the Master Table corresponding to the desired inspection level (normal, reduced, or tightened). Then, the particular combination of AQL and Code Letter determines the sample size and acceptance values (or percentages) for the plan. Because the MIL-STD-414 plan and its counterparts have many pages of such tables, only a few tables are included in this book for the purpose of illustrating the procedures. Complete copies of MIL-STD-414 (or its counterparts) should be obtained before these systems are used.

IMPLEMENTATION

Form 2 plans are the ones most frequently chosen for work with MIL-STD-414 because the switching rules are stated only in terms of the percent nonconforming. For this reason, we illustrate the implementation of MIL-STD-414 only in the Form 2 setting.

Suppose, for example, that a measured characteristic on a certain item has specification limits of 5.000 ±0.050 in. and that the items are shipped in lots of size 200. Suppose further that the lower specification limit is the more critical of the two, in the sense that items that exceed the upper specification limit can still be salvaged by reworking them, whereas items that fall below the lower specification limit must necessarily be scrapped. To reflect this difference in the sampling plan, different AQLs are selected at the two specification limits. Suppose, then, that an AQL of 0.80% is selected for the lower specification limit, and an AQL of 2.00% is used for the upper specification limit.

The next step is to translate these AQLs into ones that are available in the MIL-STD-414 tables. From the conversion table (Table 11.1), the upper and lower AQLs that will be used in the plan are 1.0% and 2.5%, respectively. After MIL-STD-414 is used a few times, most practitioners simply choose from the AQLs available in MIL-STD-414 at the outset, thereby eliminating the AQL conversion step. These AQLs are listed here for convenience: 0.04, 0.065, 0.10, 0.15, 0.25, 0.40, 0.65, 1.0, 1.5, 2.5, 4.0, 6.5, 10.0, and 15.0. The reader should also be aware that this AQL list is somewhat shorter in the national standard ANSI/ASQC Z1–1980.

From the lot size of 200, the Code Letter Table (see Table 11.2) indicates that Code Letter H corresponds to Inspection Level IV used to start the plan. From the Master Table for Normal and Tightened Inspection based on Variability Unknown: Standard Deviation Method, Form 2 (Table 11.3), the required sample size for the plan is $n = 20$. Table 11.4 shows data from a sample of size 20 drawn from a given lot.

The Master Table also gives the maximum percent nonconforming that corresponds to each AQL; that is, $M_L = 2.95\%$ at the AQL of 1% and $M_U = 6.17\%$ at the AQL of 2.5%. These are to be compared to the estimated percent nonconforming in the sample of Table 11.4. The sample estimates are found by first calculating the two quality ratios, $Q_L = (\bar{x} - LSL)/s$ and $Q_U = (USL - \bar{x})/s$:

$$Q_L = \frac{4.994 - 4.95}{0.0310} = 1.42 \quad \text{and} \quad Q_U = \frac{5.050 - 4.994}{0.0310} = 1.81$$

which are translated into percentage estimates by means of special conversion tables based on the values of Q_L (or Q_U) and the sample size n. Table 11.5 gives the portion of this conversion table relevant to our example. From Table 11.5 we see that the *estimated* percentages outside the two specification limits are

$$\hat{p}_L = 7.49\% \quad \text{and} \quad \hat{p}_U = 3.05\%$$

Notice that, in fact, *none* of the 20 sample measurements exceeds the USL of 5.050 in. The estimate $\hat{p}_U = 3.05\%$ results from the assumption of normality and indicates that 3.05% of the *population* of items from which the sample was drawn exceed the USL. The lot determination then follows immediately from the comparisons:

$p_L = 7.49\%$ exceeds $M_L = 2.95\%$
$p_U = 3.05\%$ does not exceed $M_U = 6.17\%$
$p_L + p_U = 10.54\%$ exceeds $\max[M_L, M_U] = 6.17\%$

TABLE 11.3 Master Table for Normal and Tightened Inspection for Plans Based on Variability Unknown: Standard Deviation Method (Double Specification Limit and Form 2—Single Specification Limit)

Acceptable Quality Levels (normal inspection)

Sample Size Code Letter	Sample Size	.04 M	.065 M	.10 M	.15 M	.25 M	.40 M	.65 M	1.00 M	1.50 M	2.50 M	4.00 M	6.50 M	10.00 M	15.00 M
B	3	→	→	→	→	→	→	→	↓	↓	7.59	18.86	26.94	33.69	40.47
C	4	→	→	→	→	→	→	→	1.53	5.50	10.92	16.45	22.86	29.45	36.90
D	5	→	→	→	→	→	→	1.33	3.32	5.83	9.80	14.39	20.19	26.56	33.99
E	7	→	→	→	→	0.422	1.06	2.14	3.55	5.35	8.40	12.20	17.35	23.29	30.50
F	10	→	→	→	0.349	0.716	1.30	2.17	3.26	4.77	7.29	10.54	15.17	20.74	27.57
G	15	0.099	0.186	0.312	0.503	0.818	1.31	2.11	3.05	4.31	6.56	9.46	13.71	18.94	25.61
H	20	0.135	0.228	0.365	0.544	0.846	1.29	2.05	2.95	4.09	6.17	8.92	12.99	18.03	24.53
I	25	0.155	0.250	0.380	0.551	0.877	1.29	2.00	2.86	3.97	5.97	8.63	12.57	17.51	23.97
J	30	0.179	0.280	0.413	0.581	0.879	1.29	1.98	2.83	3.91	5.86	8.47	12.36	17.24	23.58
K	35	0.170	0.264	0.388	0.535	0.847	1.23	1.87	2.68	3.70	5.57	8.10	11.87	16.65	22.91
L	40	0.179	0.275	0.401	0.566	0.873	1.26	1.88	2.71	3.72	5.58	8.09	11.85	16.61	22.86
M	50	0.163	0.250	0.363	0.503	0.789	1.17	1.71	2.49	3.45	5.20	7.61	11.23	15.87	22.00
N	75	0.147	0.228	0.330	0.467	0.720	1.07	1.60	2.29	3.20	4.87	7.15	10.63	15.13	21.11
O	100	0.145	0.220	0.317	0.447	0.689	1.02	1.53	2.20	3.07	4.69	6.91	10.32	14.75	20.66
P	150	0.134	0.203	0.293	0.413	0.638	0.949	1.43	2.05	2.89	4.43	6.57	9.88	14.20	20.02
Q	200	0.135	0.204	0.294	0.414	0.637	0.945	1.42	2.04	2.87	4.40	6.53	9.81	14.12	19.92
		.065	.10	.15	.25	.40	.65	1.00	1.50	2.50	4.00	6.50	10.00	15.00	

Acceptability Quality Levels (tightened inspection)

All AQL and table values are in percent defective.
↓ Use first sampling plan below arrow, that is, both sample size as well as M value. When sample size equals or exceeds lot size, every item in the lot must be inspected.

TABLE 11.4 Measurements (in inches) from a Sample of 20 Items

5.013	5.002	5.032	4.932	4.992	5.042	5.021	5.011	5.015
5.013	4.997	4.966	4.990	5.031	4.968	5.003	4.943	4.952
4.997	4.959							

TABLE 11.5 Table for Estimating the Lot Percent Defective Using Standard Deviation Method

Q_U or Q_L	Sample Size															
	3	4	5	7	10	15	20	25	30	35	40	50	75	100	150	200
1.10	9.84	13.33	13.48	13.49	13.50	13.51	13.52	13.52	13.53	13.54	13.54	13.54	13.55	13.55	13.56	13.56
1.11	8.89	13.00	13.20	13.25	13.26	13.28	13.29	13.30	13.31	13.31	13.32	13.32	13.33	13.34	13.34	13.34
1.12	7.82	12.67	12.93	13.00	13.03	13.05	13.07	13.08	13.09	13.10	13.10	13.11	13.12	13.12	13.12	13.13
1.13	6.60	12.33	12.65	12.75	12.80	12.83	12.85	12.86	12.87	12.88	12.89	12.89	12.90	12.91	12.91	12.92
1.14	5.08	12.00	12.37	12.51	12.57	12.61	12.63	12.65	12.66	12.67	12.67	12.68	12.69	12.70	12.70	12.70
1.15	0.29	11.67	12.10	12.27	12.34	12.39	12.42	12.44	12.45	12.46	12.46	12.47	12.48	12.49	12.49	12.50
1.16	0.00	11.33	11.83	12.03	12.12	12.18	12.21	12.22	12.24	12.25	12.25	12.26	12.28	12.28	12.29	12.29
1.17	0.00	11.00	11.56	11.79	11.90	11.96	12.00	12.02	12.03	12.04	12.05	12.06	12.07	12.08	12.08	12.09
1.18	0.00	10.67	11.29	11.56	11.68	11.75	11.79	11.81	11.82	11.84	11.84	11.85	11.87	11.88	11.88	11.89
1.19	0.00	10.33	11.02	11.33	11.46	11.54	11.58	11.61	11.62	11.63	11.64	11.65	11.67	11.68	11.69	11.69
1.20	0.00	10.00	10.76	11.10	11.24	11.34	11.38	11.41	11.42	11.43	11.44	11.46	11.47	11.48	11.49	11.49
1.21	0.00	9.67	10.50	10.87	11.03	11.13	11.18	11.21	11.22	11.24	11.25	11.26	11.28	11.29	11.30	11.30
1.22	0.00	9.33	10.23	10.65	10.82	10.93	10.98	11.01	11.03	11.04	11.05	11.07	11.09	11.09	11.10	11.11
1.23	0.00	9.00	9.97	10.42	10.61	10.73	10.78	10.81	10.84	10.85	10.86	10.88	10.90	10.91	10.91	10.92
1.24	0.00	8.67	9.72	10.20	10.41	10.53	10.59	10.62	10.64	10.66	10.67	10.69	10.71	10.72	10.73	10.73
1.25	0.00	8.33	9.46	9.98	10.21	10.34	10.40	10.43	10.46	10.47	10.48	10.50	10.52	10.53	10.54	10.55
1.26	0.00	8.00	9.21	9.77	10.00	10.15	10.21	10.25	10.27	10.29	10.30	10.32	10.34	10.35	10.36	10.37
1.27	0.00	7.67	8.96	9.55	9.81	9.96	10.02	10.06	10.09	10.10	10.12	10.13	10.16	10.17	10.18	10.19
1.28	0.00	7.33	8.71	9.34	9.61	9.77	9.84	9.88	9.90	9.92	9.94	9.95	9.98	9.99	10.00	10.01
1.29	0.00	7.00	8.46	9.13	9.42	9.58	9.65	9.70	9.72	9.74	9.76	9.78	9.80	9.82	9.83	9.83
1.30	0.00	6.67	8.21	8.93	9.22	9.40	9.48	9.52	9.55	9.57	9.58	9.60	9.63	9.64	9.65	9.66
1.31	0.00	6.33	7.97	8.72	9.03	9.22	9.30	9.34	9.37	9.39	9.41	9.43	9.46	9.47	9.48	9.49
1.32	0.00	6.00	7.73	8.52	8.85	9.04	9.12	9.17	9.20	9.22	9.24	9.26	9.29	9.30	9.31	9.32
1.33	0.00	5.67	7.49	8.32	8.66	8.86	8.95	9.00	9.03	9.05	9.07	9.09	9.12	9.13	9.15	9.15
1.34	0.00	5.33	7.25	8.12	8.46	8.69	8.78	8.83	8.86	8.88	8.90	8.92	8.95	8.97	8.98	8.99
1.35	0.00	5.00	7.02	7.92	8.30	8.52	8.61	8.66	8.69	8.72	8.74	8.76	8.79	8.81	8.82	8.83
1.36	0.00	4.67	6.79	7.73	8.12	8.35	8.44	8.50	8.53	8.55	8.57	8.60	8.63	8.65	8.66	8.67
1.37	0.00	4.33	6.56	7.54	7.95	8.18	8.28	8.33	8.37	8.39	8.41	8.44	8.47	8.49	8.50	8.51
1.38	0.00	4.00	6.33	7.35	7.77	8.01	8.12	8.17	8.21	8.24	8.25	8.28	8.31	8.33	8.35	8.35
1.39	0.00	3.67	6.10	7.17	7.60	7.85	7.96	8.01	8.05	8.08	8.10	8.12	8.16	8.18	8.19	8.20
1.40	0.00	3.33	5.88	6.98	7.44	7.69	7.80	7.80	7.90	7.92	7.94	7.97	8.01	8.02	8.04	8.05
1.41	0.00	3.00	5.66	6.80	7.27	7.53	7.64	7.70	7.74	7.77	7.79	7.82	7.86	7.87	7.89	7.90
1.42	0.00	2.67	5.44	6.62	7.10	7.37	7.49	7.55	7.59	7.62	7.64	7.67	7.71	7.73	7.74	7.75
1.43	0.00	2.33	5.23	6.45	6.94	7.22	7.34	7.40	7.44	7.47	7.50	7.52	7.56	7.58	7.60	7.61
1.44	0.00	2.00	5.01	7.27	6.78	7.07	7.19	7.26	7.30	7.33	7.35	7.38	7.42	7.44	7.46	7.47
1.45	0.00	1.67	4.81	6.10	6.63	6.92	7.04	7.11	7.15	7.16	7.21	7.24	7.28	7.30	7.31	7.33
1.46	0.00	1.33	4.60	5.93	6.40	6.77	6.90	6.97	7.01	7.04	7.07	7.10	7.14	7.16	7.18	7.19
1.47	0.00	1.00	4.39	5.77	6.32	6.00	6.75	6.83	6.87	6.90	6.93	6.96	7.00	7.02	7.04	7.05
1.48	0.00	.67	4.19	5.60	6.17	6.48	6.61	6.69	6.73	6.77	6.79	6.82	6.86	6.88	6.90	6.91
1.49	0.00	.33	3.99	5.44	6.02	6.34	6.48	6.55	6.60	6.63	6.65	6.69	6.73	6.75	6.77	6.78

(Continued)

TABLE 11.5   Table for Estimating the Lot Percent Defective Using Standard Deviation Method (Continued)

Q_U or Q_L	Sample Size															
	3	4	5	7	10	15	20	25	30	35	40	50	75	100	150	200
1.50	0.00	0.00	3.80	5.28	5.87	6.20	6.34	6.41	6.46	6.50	6.52	6.55	6.60	6.62	6.64	6.65
1.51	0.00	0.00	3.61	5.13	5.73	6.06	6.20	6.28	6.33	6.34	6.39	6.42	6.47	6.49	6.51	6.52
1.52	0.00	0.00	3.42	4.97	5.59	5.93	6.07	6.15	6.20	6.23	6.26	6.29	6.34	6.36	6.38	6.39
1.53	0.00	0.00	3.23	4.82	5.45	5.80	5.94	6.02	6.07	6.11	6.13	6.17	6.21	6.24	6.26	6.27
1.54	0.00	0.00	3.03	4.67	5.31	5.67	5.81	5.89	5.95	5.96	6.01	6.04	6.09	6.11	6.13	6.15
1.55	0.00	0.00	2.87	4.52	5.18	5.54	5.69	5.77	5.82	5.86	5.88	5.92	5.97	5.99	6.01	6.02
1.56	0.00	0.00	2.69	4.38	5.05	5.41	5.56	5.65	5.70	5.74	5.76	5.80	5.85	5.87	5.89	5.90
1.57	0.00	0.00	2.52	4.34	4.92	5.29	5.44	5.53	5.58	5.62	5.66	5.68	5.73	5.75	5.78	5.79
1.58	0.00	0.00	2.35	4.10	4.79	5.16	5.32	5.41	5.46	5.50	5.53	5.56	5.61	5.64	5.66	5.67
1.59	0.00	0.00	2.19	3.96	4.66	5.04	5.20	5.29	5.34	5.38	5.41	5.45	5.50	5.52	5.54	5.56
1.60	0.00	0.00	2.03	3.83	4.54	4.92	5.09	5.17	5.23	5.27	5.30	5.33	5.38	5.41	5.43	5.44
1.61	0.00	0.00	1.87	3.69	4.41	4.81	4.97	5.06	5.12	5.16	5.18	5.22	5.27	5.30	5.32	5.33
1.62	0.00	0.00	1.72	3.57	4.30	4.69	4.86	4.95	5.01	5.04	5.07	5.11	5.16	5.19	5.21	5.23
1.63	0.00	0.00	1.57	3.44	4.18	4.58	4.75	4.84	4.90	4.94	4.97	5.01	5.06	5.08	5.11	5.12
1.64	0.00	0.00	1.42	3.31	4.06	4.47	4.64	4.73	4.79	4.83	4.86	4.90	4.95	4.98	5.00	5.01
1.65	0.00	0.00	1.28	3.19	3.95	4.36	4.53	4.62	4.68	4.72	4.75	4.79	4.85	4.87	4.90	4.91
1.66	0.00	0.00	1.15	3.07	3.84	4.25	4.43	4.52	4.58	4.62	4.65	4.69	4.74	4.77	4.80	4.81
1.67	0.00	0.00	1.02	2.95	3.73	4.15	4.32	4.42	4.48	4.52	4.55	4.59	4.64	4.67	4.70	4.71
1.68	0.00	0.00	0.89	2.84	3.62	4.05	4.22	4.32	4.36	4.42	4.45	4.49	4.55	4.57	4.60	4.61
1.69	0.00	0.00	0.77	2.73	3.52	3.94	4.12	4.22	4.26	4.32	4.35	4.39	4.45	4.47	4.50	4.51
1.70	0.00	0.00	0.66	2.62	3.41	3.84	4.02	4.12	4.16	4.22	4.25	4.30	4.35	4.38	4.41	4.42
1.71	0.00	0.00	0.55	2.51	3.73	3.75	3.93	4.02	4.09	4.13	4.16	4.20	4.26	4.29	4.31	4.32
1.72	0.00	0.00	0.45	2.41	3.21	3.65	3.83	3.93	3.99	4.04	4.07	4.11	4.17	4.19	4.22	4.23
1.73	0.00	0.00	0.38	2.30	3.11	3.56	3.74	3.84	3.90	3.94	3.98	4.02	4.08	4.10	4.13	4.14
1.74	0.00	0.00	0.27	2.20	3.02	3.46	3.65	3.75	3.81	3.85	3.89	3.93	3.99	4.01	4.04	4.05
1.75	0.00	0.00	0.19	2.11	2.93	3.37	3.56	3.66	3.72	3.77	3.80	3.84	3.90	3.93	3.95	3.97
1.76	0.00	0.00	0.12	2.01	2.83	3.28	3.47	3.57	3.63	3.68	3.71	3.76	3.81	3.84	3.87	3.88
1.77	0.00	0.00	0.06	1.92	2.74	3.20	3.36	3.48	3.55	3.59	3.63	3.67	3.73	3.76	3.78	3.80
1.78	0.00	0.00	0.02	1.83	2.66	3.11	3.30	3.40	3.47	3.51	3.54	3.59	3.64	3.67	3.70	3.71
1.79	0.00	0.00	0.00	1.74	2.57	3.03	3.21	3.32	3.38	3.43	3.46	3.51	3.56	3.59	3.63	3.63
1.80	0.00	0.00	0.00	1.65	2.49	2.94	3.13	3.24	3.30	3.35	3.38	3.43	3.48	3.51	3.54	3.55
1.81	0.00	0.00	0.00	1.57	2.40	2.86	3.05	3.16	3.22	3.27	3.30	3.35	3.40	3.43	3.46	3.47
1.82	0.00	0.00	0.00	1.49	2.32	2.79	2.98	3.08	3.15	3.19	3.22	3.27	3.33	3.36	3.38	3.40
1.83	0.00	0.00	0.00	1.41	2.25	2.71	2.90	3.00	3.07	3.11	3.15	3.19	3.25	3.28	3.31	3.32
1.84	0.00	0.00	0.00	1.34	2.17	2.63	2.82	2.93	2.99	3.04	3.07	3.12	3.18	3.21	3.23	3.25
1.85	0.00	0.00	0.00	1.26	2.09	2.56	2.75	2.85	2.92	2.97	3.00	3.05	3.10	3.13	3.16	3.17
1.86	0.00	0.00	0.00	1.19	2.02	2.48	2.68	2.76	2.85	2.89	2.93	2.97	3.03	3.06	3.09	3.10
1.87	0.00	0.00	0.00	1.12	1.95	2.41	2.60	2.71	2.78	2.82	2.86	2.90	2.96	2.99	3.02	3.03
1.88	0.00	0.00	0.00	1.06	1.88	2.34	2.54	2.64	2.71	2.73	2.79	2.83	2.89	2.92	2.95	2.96
1.89	0.00	0.00	0.00	0.99	1.81	2.28	2.47	2.57	2.64	2.69	2.72	2.77	2.83	2.85	2.88	2.90

Two of these comparisons, $p_L > M_L$ and $p_L + p_U > \max[M_L, M_U]$, give signals that the lot should be rejected.

OPERATION OF THE MIL-STD-414 SAMPLING SYSTEM

1. Determine the AQL(s); for double specification limits, different AQLs may be chosen for each limit if desired.
2. If necessary, use the AQL conversion table (Table 11.1) to obtain AQLs consistent with the MIL-STD-414 plans.

(Continued)

(Continued)

---

3. Decide on the inspection mode and level; unless otherwise specified, use Normal Inspection, Level IV to start the system.
4. Use the table of Sample Size Code Letters (Table 11.2) to select the proper Code Letter.
5. Decide whether to use a Form 1 or Form 2 plan.
6. Establish the mode of handling the process variation: σ known, σ estimated by the sample standard deviation, or σ estimated by the average range method, $\overline{R}/d_2$.
7. Use the Master Table corresponding to the above choices to determine the sample size and acceptance values (or percentages). In cases where a plan does not exist for a given lot size and AQL combination, follow the arrows in the table to the nearest available plan.
8. Begin by using the plan selected in step 7 and keep a record of acceptances and rejections so that the switching rules can be applied. (Consult the MIL-STD-414 publication for a list of the switching rules.)

ANSI/ASQC Z1.9–1980 AND ISO 3951

MIL-STD-414 has commercial counterparts for use in nongovernment applications. The **American National Standards Institute** (ANSI) and the **American Society for Quality Control** (ASQC) jointly approved the ANSI/ASQC Z1.9–1980 sampling system in March 1980. This system is equivalent to the international system ISO 3951, and both systems are based on MIL-STD-414.

One of the main reasons for developing these alternative sampling schemes is to create systems that are more closely matched with the popular MIL-STD-105E system of attributes plans. In this section, we give a brief summary of the differences between MIL-STD-414 and the newer ANSI/ASQC Z1.9–1980 and ISO 3951 systems.

In ANSI/ASQC Z1.9–1980 and ISO 3951, the Inspection Levels are changed from the I, II, III, IV, V of MIL-STD-414 and are denoted S3, S4, I, II, III so that they match 105E's numbering system. Thus, as in 105E, General Inspection Level II is used to start these systems. Recall that under the older numbering of MIL-STD-414, General Inspection, Level IV is used to start the system.

Lot size ranges and Sample Size Code Letters have also been revised in ANSI/ASQC Z1.9–1980 and ISO 3951 to match those of 105E. Table 11.6 shows these revisions. For comparison, Table 11.2 gives the lot size ranges and code letters for MIL-STD-414. In this renumbering, plans associated with MIL-STD-414's Code Letters J and L were eliminated first; then all remaining plans were relettered from B up through Q (except the letter 'O' was omitted to avoid any confusion with the number zero).

TABLE 11.6   Sample Size Code Letters

			Inspection Levels		
		Special		General	
Lot Size		S3	S4	I	II	III
2 to 8	B	B	B	B	C
9 to 15	B	B	B	B	D
16 to 25	B	B	B	C	E
26 to 50	B	B	C	D	F
51 to 90	B	B	D	E	G
91 to 150	B	C	E	F	H
151 to 280	B	D	F	G	I
281 to 400	C	E	G	H	J
401 to 500	C	E	G	I	J
501 to 1,200	D	F	H	J	K
1,201 to 3,200	E	G	I	K	L
3,201 to 10,000	F	H	J	L	M
10,001 to 35,000	G	I	K	M	N
35,001 to 150,000	H	J	L	N	P
150,001 to 500,000	H	K	M	P	P
500,001 and over	H	K	N	P	P

While MIL-STD-414 trimmed off some of the 105E AQLs at both ends of the spectrum, ANSI/ASQC Z1.9–1980 and ISO 3951 trim even more. Specifically, these systems drop the AQLs of 0.04%, 0.065%, and 15.00% from the list used in MIL-STD-414. This leaves AQLs ranging from 0.10% up to 10.00%, in steps matching those in 105E. The shortened list of AQLs is: 0.10%, 0.15%, 0.25%, 0.40%, 0.65%, 1.00%, 1.50%, 2.50%, 4.00%, 6.50%, and 10.00%.

Changes in the switching rules have also been made. Here, all three systems differ slightly from one another, primarily in the rules for switching from Normal to Reduced Inspection. The reader is referred to Duncan (1986) as well as the ANSI/ASQC Z1.9–1980 and ISO 3951 publications for more discussion of these differences.

One of the most notable differences between the plans is the use of graphical procedures in ISO 3951, whereas ANSI/ASQC Z1.9–1980 retains the table format of MIL-STD-414. For example, under ISO 3951, graphs whose axes measure $(\bar{x} - \text{LSL})/(\text{USL} - \text{LSL})$ and $s/(\text{USL} - \text{LSL})$ are used to immediately determine lot acceptability, with no need for intermediate tables for converting quality indexes [for example, $Q_L = (\bar{x} - \text{LSL})/s$ and $Q_U = (\text{USL} - \bar{x})/s$] into percent nonconforming, as is done in MIL-STD-414 and ANSI/ASQC Z1.9–1980. Another distinctive difference between ISO 3951 and the other two plans is that ISO 3951 uses only the standard deviation method (and excludes the average range method) for estimating the process variation.

Just as in MIL-STD-414, the primary assumption underlying the plans and OC curves in ANSI/ASQC Z1.9–1980 and ISO 3951 is that the measured characteristics follow a *normal* distribution (see Section 5.6). Beyond the technical differences mentioned in this section, ANSI/ASQC Z1.9–1980 and ISO 3951 also use modern terminology such as 'nonconformity' in place of 'defect,' and so forth.

EXERCISES FOR CHAPTER 11

Unless otherwise stated, all sampling plans in the following exercises are assumed to be of Type B (see Section 11.5).

11.1 (a) Sketch the OC curve of the single sampling plan with $n = 10$ and $c = 0$.
   (b) On the same graph as in part (a), sketch the OC curve of the plan with $n = 15$ and $c = 2$.

11.2 Suppose you take a random sample of 25 from a lot of 1,000 items but later realize that you should have selected a sample of 40 items instead. How do you correct this error? Should you put the 25 items back and draw the required random sample of 40 from the original lot, or should you simply draw an additional random sample of 15 from the 975 remaining items?

11.3 (a) Suppose that lots of the same size, L_1 and L_2, are inspected separately using a zero acceptance plan with $n = 20$. Further suppose that type L_1 lots always contain 2% defective items, while type L_2 always have 8% defectives. If 100 lots of each kind are inspected, what is the average percent defective in the *accepted* lots?
   (b) In order to cut down on the numbers of lots inspected, suppose someone suggests that each L_1 lot be combined with one of the L_2 lots to form a single larger lot (each lot now contains 5% defectives). Applying the $n = 20$ plan to these lots, what is the average percent defective in the accepted lots?

11.4 (a) Under rectifying inspection, plot the AOQ curve of a zero acceptance plan with $n = 10$.
   (b) (*Calculus required*) At what point p does the AOQ curve achieve its maximum? What is this maximum?
   (c) Find the AOQL of this plan.
   (d) Generalize the result in parts (a) and (b) to *any* zero acceptance plan (i.e., for any sample size n).
   (e) For lots of size 250, plot the ATI curve for the $n = 10$ plan.

11.5 (*Calculus required*) For a single sampling plan with acceptance number c and sample size n, show that the point of inflection of the plan's OC curve occurs exactly at $p = c/(n - 1)$.[12]

11.6 (a) Sketch the OC curve of the double sampling plan $\{n_1 = 15, c_1 = 1, r_1 = 2\}$ and $\{n_2 = 20, c_2 = 4, r_2 = 5\}$.
   (b) Plot the ASN curve for this plan.
   (c) Under rectifying inspection and lots of size 400, sketch the ATI curve for this plan.

---

[12]See Polya (1968).

11.7 Do zero acceptance plans make up a relatively large or small percentage of the plans in the MIL-STD-105E system?

In the following exercises, assume General Inspection, Level II is used to start the MIL-STD-105E system.

11.8 Use MIL-STD-105E with an AQL of 1% and lot sizes of 4,000.
(a) Find the single sampling plan for Normal Inspection.
(b) Find the single sampling plan for Reduced Inspection.
(c) Find the single sampling plan for Tightened Inspection.
(d) Find the double sampling plan for Normal Inspection.
(e) Find the double sampling plan for Reduced Inspection.
(f) Find the double sampling plan for Tightened Inspection.

11.9 (a) Under MIL-STD-105E with AQL = 0.40% and lots of size 300, find the single sampling plan.
(b) Answer the question in part (a) for lots of size 600.

11.10 Use MIL-STD-105E, begin with Normal Inspection, and suppose that the following sequence of rejected (R) and accepted (A) lots occur:

A, A, R, A, R, A, A, R, A, A, A, A, A, A, A, A, A, A, A, R

Mark the points in this sequence where the inspection levels change, and at each such point, explain why the change occurs.

In the following exercises, assume General Inspection, Level IV is used to start the MIL-STD-414 system.

11.11 Using MIL-STD-414 with an AQL of 1% and lots of size 4,000, find the sampling plan for Normal Inspection, Form 2 with the standard deviation method.

11.12 Use MIL-STD-414 with an AQL of 2.5% and lots of size 550.
(a) Find the sampling plan for Normal Inspection, Form 2 using the standard deviation method.
(b) Under the plan in part (a), suppose the sampled items from a particular lot yield an average and standard deviation of 10.1 and 1.9, respectively. Furthermore, the specifications on these measurements are USL = 14.0 and LSL = 8.0. Should this lot be accepted or rejected?

REFERENCES FOR CHAPTER 11

Baker, R. C. 1988. "Zero Acceptance Plans: Expected Cost Increases." *Quality Progress,* January, pp. 43–46.

David, H. T. 1986. "Variations on Two Familiar Industrial Statistical Themes," *Statistical Laboratory Reprint Series,* p. 3, Iowa State University.

Duncan, A. J. 1986. *Quality Control and Industrial Statistics,* 5th ed. Homewood, IL: Irwin.

Farnum, N. R., and R. C. Suich. 1986. "Comment on: Determining Sample Sizes When Searching for Rare Events." *IEEE Transactions on Reliability* R-35:584–585.

Grant, E. L., and R. S. Leavenworth. 1988. *Statistical Quality Control,* 6th ed., chap. 13. New York: McGraw-Hill.

MIL-STD-105E. 1989. "Sampling Procedures and Tables for Inspection by Attributes," May 10. Department of Defense, Washington, DC.

Mood, A. M. 1943. "On the Dependence of Sampling Inspection Upon the Population Distributions." *Annals of Mathematical Statistics* 14:415–425.

Mudryk, W. 1988. "Quality Control Processing System for Survey Operations." *Survey Methodology* 14 (no. 2):309–316.

Mundel, A. B. 1990. "MIL-STD-105E—Sampling Procedures and Tables for Inspection by Attributes, Standards Column." *Quality Engineering* 2 (no. 3):353–363.

Polya, G. 1968. *Mathematics and Plausible Reasoning,* Vol. II, p. 80. Princeton, NJ: Princeton University Press.

Schilling, E. G. 1978. "A Lot Sensitive Sampling Plan for Compliance Testing and Acceptance Inspection." *Journal of Quality Technology* 10 (no. 2):47–51.

Schilling, E. G. 1982. *Acceptance Sampling in Quality Control.* New York: Marcel Dekker.

Schilling, E. G. 1984. "An Overview of Acceptance Sampling." *Quality Progress,* April, pp. 22–25.

Schilling, E. G. 1990. "Acceptance Control in a Modern Quality Program." *Quality Engineering* 3 (no. 2).

Schilling, E. G., and L. I. Johnson. 1980. "Tables for the Construction of Matched Single, Double, and Multiple Sampling Plans with Applications to MIL-STD-105D." *Journal of Quality Technology* 12 (no. 4):220–229.

Shewhart, W. A., and W. E. Deming. 1939. *Statistical Method from the Viewpoint of Quality Control.* The Graduate School, Department of Agriculture, Washington, DC.

Vardeman, S. B. 1986. "The Legitimate Role of Inspection in Modern SQC." *American Statistician* 40 (no. 4):325–328.

Wright, T., and H. Tsao. 1985. "Some Useful Notes on Simple Random Sampling." *Journal of Quality Technology* 17 (no. 2):67–73.

12

ADVANCED TOPICS

This chapter discusses special charting methods for the data-poor environment of short-run production processes and, at the other extreme, methods for high-volume processes. In addition, the methods of time series analysis are used to analyze processes with measurements that may be correlated.

CHAPTER OUTLINE

12.1 INTRODUCTION

Statistical methods of quality improvement are tailored to current trends in technology, production methods, fields of application, and philosophies. When control charts were introduced, Shewhart (1939) stated that they were best suited to the "continuing mass production" of his era, where data collection and chart construction were done without the aid of computers. Since that time, however, control chart methods have been refined, some entirely new quality control methods have been introduced, and there has been a shift in emphasis from 'control' to 'continuous improvement.'

From the standpoint of technology, inexpensive computing power has played a major role in altering production methods while simultaneously making statistical techniques easier to implement. Numerically controlled (NC) machines rely on built-in microprocessors to guide production steps, automatic test equipment (ATE) use microprocessors to perform measurement and testing functions, and online gages and data links automatically transfer measurements to databases or SPC software. Control chart calculations and graphical displays now require only a few keystrokes with SPC software.

One effect of these advances is that data can be overabundant in some cases and yet scarce in others. In high-volume production, the ability to instantaneously make and then download product measurements can result in huge amounts of data to analyze. Speed in handling and analyzing such data becomes a critical factor for effective quality control. Such frequently recorded data also allow the methods of statistical time series analysis to be used to model trends and autocorrelations in the data stream. These methods are described in Section 12.3

On the other hand, computerization has given rise to 'flexible manufacturing' and 'build to order' production systems. In this setting, it is possible for only a few parts of the same type to be produced at a given time before machine settings are changed for the next batch. In such a short-run or low-volume environment, even the fairly minimal data requirements of standard control charts may not be achieved, making it necessary to modify the usual control chart procedures. In this chapter, control charts that use 'deviations from nominal' are suggested for use with short-run production (see Section 12.2).

12.2 METHODS FOR SHORT-RUN PRODUCTION

Short-run or low-volume production refers to processes that generate only a small number of *items* at any one time. It also refers to *continuous processes* (e.g., chemical processes) that yield only a small number of measurements per time period. In manufacturing, computerization and an increasing trend toward 'build to order' production and quick turn around times have given rise to short-run production. Some process industries (e.g., chemical industries) are often associated with short-run data. In the service sector, long periods between successive measurements, sometimes dictated by accounting and record-keeping procedures, also impose limitations on available data. As an operational definition, by 'short-run' production, practitioners usually mean any production process that yields data less frequently than that required by standard threshold control charts.

GENERAL RECOMMENDATIONS

The following procedures have all been used to handle the scarcity of data in short-run production:

<table>
<tr>
<td>STATISTICAL
CONTROL OF
SHORT-RUN
PRODUCTION</td>
<td>
1. Use one of the more 'sensitive' control charts (e.g., CUSUM, EWMA) that have the ability to react more quickly to small process changes.
2. Control secondary or 'operating' process variables (ambient temperature, humidity, chemical concentrations, raw materials, etc.) for which abundant control chart data already exist or are easy to obtain.
3. Control the measurement system. (*Note:* This is *always* a good practice.)
4. Use a control chart procedure that is specifically designed for the short-run environment (e.g., the 'deviations from nominal' chart or an individuals chart).
5. Examine the process for repetitive operations that may be suitable for statistical control.
</td>
</tr>
</table>

USING CUSUM AND EWMA CHARTS

CUSUM and EWMA charts (see Sections 7.6 and 7.7) both have the desirable feature of being more sensitive than Shewhart charts to small process shifts. For example, for a given subgroup size n, the average run length (ARL; see Section 6.4) prior to the detection of a process shift is smaller for the CUSUM chart than for an $\bar{x}$ chart for process shifts of less than about 2 sigmas (Ewan 1963). For shifts larger than 2 sigmas, the $\bar{x}$ chart is better. Since large numbers of subgroups are not common in short-run production, it makes sense to appeal to charts like the CUSUM and EWMA for early problem detection.

To use these charts for short-run production, however, somewhat of a 'chicken or egg' problem must be solved because both charts require the user first to provide a good estimate of the process variance. Since the usual method for estimating process variation is to bring a process into statistical control, an obvious problem arises: variation cannot be meaningfully estimated until control is established, but control cannot be established until variation is estimated.

One way to solve this problem is to use historic process data to calculate an R chart (or s chart) based on the 'deviations from nominal' in each subgroup. Even though successive subgroups in a short-run process may represent different parts and different target dimensions, the *deviations* of the measured values in each subgroup from that subgroup's nominal (i.e., target) dimension ought to provide estimates of the overall process variation. Thus, *if it can be assumed that the process variation does not depend on the different part types produced,* then the deviations from nominal in the subgroups may be combined into an R (or s) chart so that control over the variation can be established. Example 12.1 illustrates this method.

EXAMPLE 12.1

A milling machine is used to grind down the ends of small metal bars to form plugs that enable the bars to connect to larger assemblies. The critical characteristic in the milling process is the resulting width of the end plug, which determines whether and how well the bar will fit into other assemblies. The machine is set up to mill plugs of various sizes, ranging from $\frac{1}{8}$ to $\frac{1}{2}$ in. The size of the plugs milled can vary from hour to hour, depending on the particular customer order being processed, so a continuous stream of data on a *single* plug size is normally unavailable. Table 12.1(a) shows subgroups of four rods selected from the most recent 20 batches milled. The assumption is made that the standard deviation of the plug width is approximately constant across all plug sizes, so the deviations from nominal method is applied. Table 12.1(b) gives these deviations and their subgroup ranges, which are then used to make the R chart of Figure 12.1. Since the R chart shows no 'out of control' conditions, the variation in the milling process is deemed to be stable and the process standard deviation is then estimated in the usual way:

$$\hat{\sigma} = \frac{\overline{R}}{d_2} = \frac{0.004612}{2.059} = 0.00224$$

TABLE 12.1 Raw Measurements (in inches) and Deviations from Their Nominal Values for Example 12.1

Part type	Nominal	x_1	x_2	x_3	x_4
P2	0.250	0.251	0.252	0.250	0.249
P3	0.375	0.372	0.378	0.379	0.375
P2	0.250	0.247	0.249	0.254	0.251
P2	0.250	0.248	0.247	0.250	0.252
P2	0.250	0.249	0.249	0.250	0.249
P1	0.125	0.125	0.127	0.125	0.126
P3	0.375	0.372	0.374	0.375	0.376
P4	0.500	0.499	0.502	0.495	0.503
P1	0.125	0.124	0.121	0.123	0.126
P1	0.125	0.126	0.126	0.130	0.122
P3	0.375	0.375	0.374	0.378	0.379
P2	0.250	0.249	0.249	0.250	0.247
P2	0.250	0.250	0.253	0.251	0.248
P2	0.250	0.249	0.250	0.249	0.249
P2	0.250	0.252	0.250	0.251	0.247
P2	0.250	0.251	0.249	0.250	0.250
P1	0.125	0.126	0.127	0.122	0.125
P1	0.125	0.123	0.123	0.123	0.128
P2	0.250	0.252	0.250	0.247	0.248
P4	0.500	0.502	0.496	0.502	0.502

(a)

Continued

(Continued)

Part type	Nominal	Deviations from nominal				Range
P2	0.250	0.001	0.002	0.000	−0.001	0.003
P3	0.375	−0.003	0.003	0.004	0.000	0.007
P2	0.250	−0.003	−0.001	0.004	0.001	0.007
P2	0.250	−0.002	−0.003	0.000	0.002	0.005
P2	0.250	−0.001	−0.001	0.000	−0.001	0.001
P1	0.125	0.000	0.002	0.000	0.001	0.002
P3	0.375	−0.003	−0.001	0.000	0.001	0.004
P4	0.500	−0.001	0.002	−0.005	0.003	0.008
P1	0.125	−0.001	−0.004	−0.002	0.001	0.005
P1	0.125	0.001	0.001	0.005	−0.003	0.008
P3	0.375	0.000	−0.001	0.003	0.004	0.005
P2	0.250	−0.001	−0.001	0.000	−0.003	0.003
P2	0.250	0.000	0.003	0.001	−0.002	0.005
P2	0.250	−0.001	0.000	−0.001	−0.001	0.001
P2	0.250	0.002	0.000	0.001	−0.003	0.005
P2	0.250	0.001	−0.001	0.000	0.000	0.002
P1	0.125	0.001	0.002	−0.003	0.000	0.005
P1	0.125	−0.002	−0.002	−0.002	0.003	0.005
P2	0.250	0.002	0.000	−0.003	−0.002	0.005
P4	0.500	0.002	−0.004	0.002	0.002	0.006

(b)

In turn, this value of can be used to set up a CUSUM or EWMA chart for monitoring the subgroup means (i.e., the means of the *deviations from nominal*). Because the measurements are deviations from a target value, the target value of the corresponding CUSUM chart and the centerline for an EWMA chart are both taken to be 0.

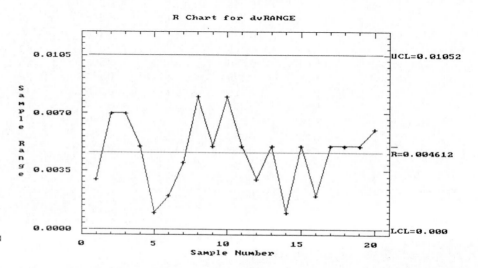

FIGURE 12.1
R chart of deviations from nominal for data of Table 12.1(b)

The procedure outlined in Example 12.1 is not restricted to CUSUM and EWMA charts. The concept of deviations from nominal values can be applied to threshold control charts as well. The next section explores such applications in more detail.

DEVIATIONS FROM NOMINAL: CONSTANT PROCESS VARIATION

The general idea underlying this deviations from nominal (hereafter abbreviated as DNOM) procedure is that, even though many small batches of differing part types may be run through a particular process, the measured differences of these parts from their respective nominal (or target) values can be plotted on the *same* control chart. To be specific, suppose that k batches are used to construct a DNOM chart, where the measured values and their deviations from the subgroup nominal dimensions are as follows:

Batch	Measurements	Nominal	Deviations
1	$x_{11}, x_{12}, \ldots, x_{1n}$	T_1	$x_{11} - T_1, x_{12} - T_1, \ldots, x_{1n} - T_1$
2	$x_{21}, x_{22}, \ldots, x_{2n}$	T_2	$x_{21} - T_2, x_{22} - T_2, \ldots, x_{2n} - T_2$
.	.	.	.
.	.	.	.
.	.	.	.
k	$x_{k1}, x_{k2}, \ldots, x_{kn}$	T_k	$x_{k1} - T_k, x_{k2} - T_k, \ldots, x_{kn} - T_k$

For simplicity, we denote the n deviations in the ith subgroup by $d_1 = x_{i1} - T_i$, $d_2 = x_{i2} - T_i, \ldots, d_n = x_{in} - T_i$. Then, for each of the k subgroups, control chart statistics such as $\bar{x}$, R, and s may be computed from the deviations and plotted. As in the previous section, CUSUM and EWMA charts may be constructed from these statistics, but the most common form of the DNOM chart consists of simply running $\bar{x}$ and R charts on the deviations. In Example 12.1, for instance, $k = 20$ subgroups of size $n = 4$ each were used to construct a DNOM R chart.

The assumption that is most critical to this DNOM procedure is that the subgroup variances are relatively constant (i.e., the process variation does not change as the part type changes). To check whether or not this assumption holds, a formal hypothesis test can be performed on the subgroup data. Such tests are usually called variance homogeneity tests. Several are available, and we refer the reader to Sachs (1982, pp. 495–500) for their description. An easier, but less formal, method is to simply plot the deviations for each subgroup against the corresponding subgroup nominal value. Visual inspection of the resulting plot is often sufficient for deciding whether or not variance homogeneity exists.

EXAMPLE 12.2   As a quick check for variance homogeneity in the data of Table 12.1, the deviations from nominal for each of the 20 subgroups are plotted against their corresponding nominal values (Figure 12.2). Since there are no pronounced

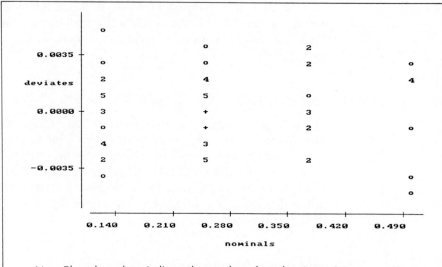

FIGURE 12.2
Checking for variance homogeneity in the data of Table 12.1(b)

Note: Plotted numbers indicate the number of overlapping points at a position; + indicates ten or more plotted points.

differences between the spreads of the plotted points among the four nominal values 0.125, 0.250, 0.375, and 0.500, we conclude that variance homogeneity is not an unreasonable assumption for these data.

DEVIATIONS FROM NOMINAL: NONCONSTANT PROCESS VARIATION

If the subgroup variances are found to change as the part type changes, then another approach is needed. One suggestion is to plot standardized subgroup statistics (see Section 6.3). Suppose that the average range $\overline{R}_j$ is known or is estimable for the jth part type. Then, the subgroup ranges R and the means $\overline{x}$ for this part type fall within the control limits

$$D_3 \overline{R}_j \leq R \leq D_4 \overline{R} \tag{12.1}$$

$$T_j - A_2 \overline{R}_j \leq \overline{x} \leq T_j + A_2 \overline{R}_j \tag{12.2}$$

for centered stable processes with a nominal (target) value of T_j. When we divide equation (12.1) through by $\overline{R}_j$, the **standardized ranges** $R/\overline{R}_j$ fall within the limits

$$D_3 \leq \frac{R}{\overline{R}_j} \leq D_4 \tag{12.3}$$

while the **standardized means** $(\overline{x} - T)/\overline{R}_j$ lie in the interval[1]

$$-A_2 \leq \frac{\overline{x} - T_j}{\overline{R}_j} \leq A_2 \tag{12.4}$$

---

[1] The term 'standardization,' as it is used here, differs slightly from that introduced in Section 6.3. The control limits in equations (12.3) and (12.4) are revised accordingly.

Using the notation introduced at the beginning of the section, we can easily compute the subgroup statistics from the deviations $d_1, d_2, \ldots, d_n$ because of the following facts. First, the range of the deviations $d_1, d_2, \ldots, d_n$ (call it R_d) always equals the range R of the original subgroup measurements. Second, the average deviation $\bar{d}$ in any subgroup must equal $\bar{x} - T_j$ for that subgroup. Thus, in terms of the deviations from nominal in a subgroup, one can plot $R_d/\bar{R}_j$ on a chart with a centerline of 1 and control limits LCL $= D_3$, UCL $= D_4$ (D_3 and D_4 are found from Appendix 1), while $\bar{d}/\bar{R}_j$ is plotted on a chart with centerline 0 and limits LCL $= -A_2$, UCL $= A_2$.

The critical part of the standardized DNOM procedure lies in obtaining good estimates of each $\bar{R}_j$ for the different part types. This can be done by initially running R charts on *each* different part type. The effort involved is necessarily greater than that required by the homogeneous variance case.

A general DNOM procedure that allows a variety of models for the process variation and measurement error variation is given in Farnum (1992). It includes the standardized DNOM approach of the preceding paragraph as a special case. With this general procedure, however, one can also set up DNOM charts for processes that have an approximately *constant coefficient of variation*—that is, for processses whose variation (as measured by the standard deviation) tends to grow as the nominal value T_i grows. In such cases, the recommended DNOM chart monitors how much $\bar{x}_i/T_i$ deviates from 1, instead of how much $\bar{x}_i - T_i$ deviates from 0, where $\bar{x}_i/T_i$ has the convenient interpretation 'percent of nominal.' The reader is referred to the paper by Farnum (1992) for the required control limit formulas.

CONTROLLING OPERATING VARIABLES

Nelson (1988) defined **fundamental variables** as those upon which a process actually depends. An **operating variable** is defined as any secondary variable that has some, but usually imperfect, correlation with one of the fundamental variables. In chemical processes, for example, fundamental variables such as hardness or thickness might have related operating variables such as reaction time, temperature, voltage, or pressure.

In short-run production, even when data on a fundamental variable exist, they may still be too scarce for running a standard control chart. Data on related operating variables, though, may be easier to obtain. When this is this case, the effectiveness of charting an operating variable depends directly on how strongly this variable is related to the fundamental variable. Assessing the strength of the relationship between two variables usually requires some kind of 'off-line' study. Figure 12.3 depicts some possible relationships between fundamental and operating variables.

ATTRIBUTES CHARTS FOR SHORT-RUN PRODUCTION

Standardized control charts (see Section 6.3), which were originally developed to handle the problem of variable subgroup sizes, can also be used as a solution to the problem of charting short-run attributes data. The approach is similar to that

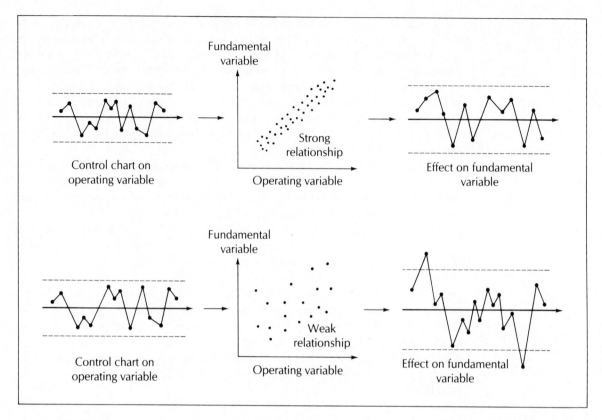

FIGURE 12.3   Effectiveness of charting operating variables depends on the relationship between operating and fundamental variables

for short-run variables data given earlier: subgroup statistics are standardized and then plotted on charts with constant control limits. Furthermore, the control limits for standardized attributes charts are always LCL = −3 and UCL = 3 (since, unlike earlier, the standard errors are used rather than the range to 'standardize' the statistics).

As before, the critical step in using standardized charts for short runs is *first* to get good estimates of the centerlines of the charts *for each different part type*. In general, if the attribute (i.e., p, np, c, or u) is denoted by θ, then the subgroup statistic to plot is given by equation (6.1) as

$$\frac{\theta - \bar{\theta}_j}{\hat{\sigma}_\theta} \tag{12.5}$$

where $\bar{\theta}_j$ is the estimate of the centerline for the jth part type. By substituting the various attribute types in equation (12.5), we summarize the standardized statistics to chart in Table 12.2.

EXAMPLE 12.3   Three different part types (P1, P2, P3) are run through the same painting station and the numbers of surface flaws are counted for a fixed number of items in each batch (i.e., for each inspection unit; see Section 9.4). For each part type, the average number of flaws per inspection unit has been running at an average

TABLE 12.2   Standardized Statistics for Attributes Control Charts

Attribute	Estimated Centerline	Subgroup Statistic
Proportion nonconforming	$\bar{p}_j$	$\dfrac{p - \bar{p}_j}{\sqrt{\bar{p}_j(1 - \bar{p}_j)/n}}$
Number nonconforming	$n\bar{p}_j$	$\dfrac{np - n\bar{p}_j}{\sqrt{n\bar{p}_j(1 - \bar{p}_j)}}$
Number of nonconformities	$\bar{c}_j$	$\dfrac{c - \bar{c}_j}{\sqrt{\bar{c}_j}}$
Number of nonconformities per unit	$\bar{u}_j$	$\dfrac{u - \bar{u}_j}{\sqrt{\bar{u}_j/n}}$

TABLE 12.3

Part Type	Number of Surface Flaws, c	Estimated Average for this Part Type, $\bar{c}_j$	Standardized Statistic, $\dfrac{c - \bar{c}_j}{\sqrt{\bar{c}_j}}$
P1	14	10	1.26
P1	7	10	−0.95
P3	7	2	3.54*
P1	10	10	0.00
P2	10	5	2.24
P2	2	5	−1.34
P3	5	2	2.12

*The upper control limit +3 is exceeded.

level of $\bar{c}_1 = 10$, $\bar{c}_2 = 5$, and $\bar{c}_3 = 2$. For the batches of parts painted, the standardized subgroup statistics are calculated as in Table 12.3. Note that only subgroup 3 indicates a possible 'out of control' condition.

A relatively recent addition to the literature on short-run attributes data is by Quesenberry (1991). In this work, Quesenberry suggests a different type of transformation (i.e., different from standardization) for charting short-run data on a binomial proportion, p. In other words, he presents what is essentially a p chart for short-run data. For the reward of handling short-run binomial data, one must pay the price of increased computational complexity. Computerization of the technique is required but straightforward, and we refer the interested reader to the paper cited above for the necessary formulas.

PROCESS ORIENTATION

Many of the suggestions for handling short-run production have a common thread: by searching for *repetitive* operations within a process, one may discover those parts of the process that are truly controllable in a statistical sense. Repetitive processes, by definition, generally involve data in amounts large enough for the effective use of standard control chart methods. Thus, the DNOM technique focuses on a single repetitive *process* involved in a short-run operation, instead of focusing on the several individual part types run through such a process. 'Operating variables' are the result of another attempt to find repetitive subprocesses. Some authors have summarized this approach to short-run production as *examining the process, not the product.* A helpful analogy might be that of an artist working on a single painting (the product). Beyond the product design, it is the artist's control over basic techniques (repetitive processes) such as brush strokes, color theory, perspective, and lighting that has a major effect on the quality of the finished work.

One repetitive process common to all production, short- or long-run, is the measurement process. Even businesses that make only a few customer-specific products at a time (e.g., satellites) repeatedly use the *same* measuring instruments to ascertain critical parameters. Control over the measurement system, which is *always* important, is even more important in the short-run setting, sometimes becoming part of the production process itself (Donmez 1989).

12.3   HIGH-VOLUME TECHNIQUES

INTRODUCTION

Computerization, accompanied by new technology, has greatly increased access to process data. Much inspection and measurement are now done automatically, with little or no human intervention. Automated inspection uses camera, electronic, X-ray, photoelectric, and other technologies. Equipment such as ATE (automatic test equipment) and CNC (computer numerically controlled) machines use these technologies to produce and analyze data from high-volume production processes.

High-volume production coupled with automated measurement systems presents a problem almost exactly opposite that of short-run production: How does one handle extremely large volumes of process data? The problem is compounded by the fact that automated measurements also make it possible to measure *several* characteristics simultaneously on each item.

This section considers some approaches to modeling high-volume data on a single quality characteristic (i.e., univariate data). Besides some general recommendations, the techniques of time series analysis and narrow-limit gaging are presented.

GENERAL RECOMMENDATIONS

With high-volume production, a good process control system should possess the following qualities (among others):

1.   The control point (where the measurements are taken) should be as close as possible to the production point (where the items are produced).

2.   The time spent analyzing process data should be minimized.

These qualities address the *performance* of the control system. The economic aspect is another matter. Although economic factors (e.g., the cost of measuring equipment, specialized sensors, data systems) certainly affect performance, the capital outlays necessary to attain performance improvements remain an individual company decision.

Moving the control point close to the production point lies at the heart of moving from the 'detection' to the 'prevention' approach to process control (see Section 2.5). High-volume production only accentuates the importance of this move because of the increased speed of production. When problems are detected, they must necessarily be caught and fixed quickly, since, in the interim, large numbers of substandard items may be produced. The greater the distance between the control and production points, the more possibly substandard items are in the pipeline between these points when problems occur. Thus, even when a problem is fixed, confirmation may not be immediate because many of the substandard items have yet to pass through the control point (Figure 12.4).

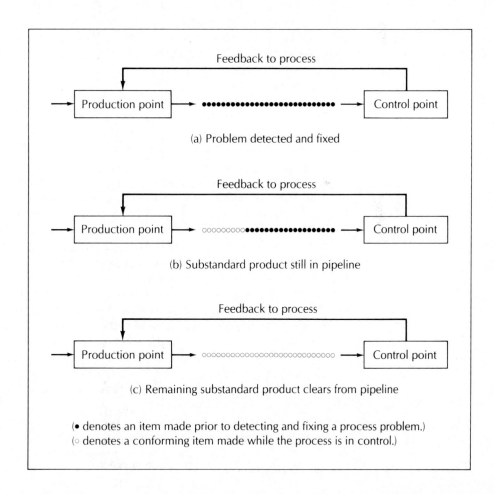

FIGURE 12.4
Accuracy and timeliness of control chart signals are a function of the distance between the control and production points

Similarly, the speed with which data are analyzed is as important as the position of the control point. Any positive effects gained from placing the control point near the production point can easily be erased by slow data analysis.

CORRELATED DATA AND TIME SERIES METHODS

Control charts were originally designed to handle streams of process data x_1, $x_2, \ldots, x_n$, in which each measurement is *statistically independent* of the others. In the language of probability, the x_i's are assumed to be independent and identically distributed random variables. By combining the x_i's into rational subgroups, one can pinpoint process problems by looking for deviations from these assumptions (as evidenced, for example, by shifts in the process mean or variance). Furthermore, because of the central limit theorem (see Section 5.4), combining data into subgroup *statistics* also helps justify use of the normal distribution and related distributions (e.g., the chi square) for calculating control limits for the charts.

For some processes, however, it is unreasonable to expect the data to be independent. Instead, $x_1, x_2, \ldots, x_n$ may be *correlated* with one another. In an electroplating process, for example, the chemical concentration in a plating bath decreases as more and more parts are plated, until the chemical in the bath is periodically replenished. If x_i represents the chemical concentration at time period i, then one would expect any sequence of measurements $x_1, x_2, \ldots, x_n$ to be interrelated. Most likely, each x_{i+1} is slightly lower than the preceding x_i, with the magnitude of the shift depending on the number of parts run through the bath between time periods i and $i + 1$. In addition, if the bath is chemically replenished on a periodic basis, say, every four time periods, we also expect to find some correlation between measurements that are four periods apart (i.e., x_i and x_{i+4} tend to be related).

Periodicity can also be found in measurements taken downstream from multiple-head machines. Many high-volume processes achieve their high-volume levels by using machines that have multiple production points (sometimes called 'heads' or 'spindles'), all making the same item. As a machine finishes making one group of parts, in many automated processes these parts tend to be collected in a certain order and then sent downstream to the next production step. Suppose, for example, that parts from one six-spindle plastic blow-molding machine are created in one step, ejected onto a conveyor system (maintaining the same order as the spindles), followed by the next six parts, and so on. The desired goal, of course, is to have all six heads producing identical parts and, if this was indeed the case, any long sequence $x_1, x_2, \ldots, x_n$ of measurements should appear to be independent of one another, thereby satisfying the usual control chart assumptions. However, if one of the spindles is making parts of a different dimension than the other spindles, then measurements six periods apart should show some degree of correlation (i.e., x_i and x_{i+6} would be correlated).

When measurements are correlated, it makes more sense to work with the *individual* values rather than collecting them into subgroups. For one reason, unlike the case when measurements are independent, the statistical properties of subgroup statistics such as $\bar{x}$, R, and s cannot easily be described when some correlation exists among the $x_1, x_2, \ldots, x_n$. Furthermore, combining correlated measurements into subgroup statistics usually *reduces* one's ability to detect correlations within the data.

Imagine, for example, the effect of averaging a subgroup of five successive chemical concentrations $x_1, x_2, \ldots, x_5$ in the electroplating bath described above. Assuming for the moment that no replenishments are made during these five periods, one would expect a general decreasing trend in the measurements, with the most recent one, x_5, giving the best indication of the *current* concentration. Similarly, since the average of the five readings likely falls somewhere between x_1 and x_5, using subgroup means gives estimates of the concentration that lag behind the true reading.

Some basic tools of time series analysis can be used to detect correlations within process data. From the terminology of time series analysis, the correlation sought is *within* a single set of readings $x_1, x_2, \ldots, x_n$, and is referred to as **autocorrelation** instead of 'correlation,' with the latter term reserved to describe the association between two or more *different* sets of data. The main diagnostic tool for detecting autocorrelation is the autocorrelation function (described below), which is capable of detecting relationships between measurements that are k periods apart (e.g., x_i and x_{i+k}) for any integer value of k. When significant autocorrelation is found for some value of k, the process can be examined to detect the cause of such autocorrelation. For example, if significant autocorrelation is found for $k = 6$ in the plastic molding example described above, one would examine the molding machine (or process) for reasons *why* a six-period autocorrelation exists. In this example, the fact that the molding machine has six spindles immediately suggests that differences between the spindles might account for the autocorrelation, so problem-solving efforts can be redirected accordingly.

THE AUTOCORRELATION FUNCTION (acf)

Measuring Autocorrelation. For any sequence of measurements $x_1, x_2, \ldots, x_n$, **autocorrelation of** (or **at**) **lag k** refers to the strength of the association between readings that are k time periods apart (k is an integer), with the measurements taken at successive *time periods* $i = 1, 2, 3, \ldots, n$. This association can be positive, negative, or 0. Positive autocorrelation at lag k means that there is a positive relationship between values that are k periods apart in the given sequence $x_1, x_2, \ldots, x_n$; that is, with positive lag k autocorrelation, the $n - k$ pairs (x_1, x_{k+1}), $(x_2, x_{k+2}), \ldots, (x_{n-k}, x_n)$ show an upward trend when plotted as pairs on a scatter plot. To illustrate, Figure 12.5 shows a plot of x_{i+2} versus x_i for data with positive lag 2 autocorrelation. Similarly, negative lag k autocorrelation is evidenced by a negative trend in the scatter plot of x_{i+k} versus x_i. If there is no autocorrelation at lag k, the scatter plot shows no discernible trend in either direction.

To create a *measure* of the lag k autocorrelation, a statistic similar (but not identical) to the familiar correlation coefficient, r, is used. First, the average $\bar{x}$ of all

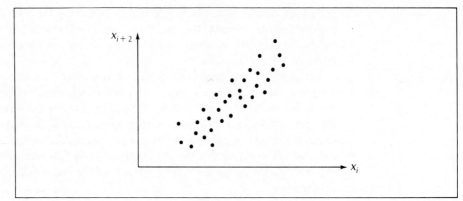

FIGURE 12.5
Positive
autocorrelation at lag
$k = 2$ appears as
positively trended
data in a plot of x_{i+2}
versus x_i

n readings is found. From this, the **autocorrelation coefficient of lag k**, denoted by r_k, is computed:

$$r_k = \frac{\sum\limits_{i=1}^{n-k} (x_i - \overline{x}) \cdot (x_{i+k} - \overline{x})}{\sum\limits_{i=1}^{n} (x_i - \overline{x})^2} \tag{12.6}$$

For any integer $k \le n - 1$, the r_k calculated from equation (12.6) enjoys the usual properties of a correlation measure:

INTERPRETING r_k

1. For any integer k, r_k must always lie between -1 and $+1$.

2. The closer r_k is to -1 or $+1$, the stronger the autocorrelation at lag k.

Instead of calculating selected r_k's, it is more revealing to calculate the first several and then plot them together versus their corresponding lags. Since the numerator in equation (12.6) contains fewer and fewer terms as k increases, it is apparent that the r_k's at the lower lags are more *statistically* reliable than the r_k's for large values of k. Thus, as a general rule, many authors recommend looking at r_k for only the first $n/4$ or so values of k (Box and Jenkins 1970, p. 33). When r_k is plotted versus k, the resulting chart is called the **autocorrelation function** (abbreviated **acf**). An example of an acf is shown in Figure 12.6. At each k, lines of length r_k are drawn to make the autocorrelation at lag k easier to see on the graph. These lines are also called 'spikes' in the acf.

Software programs that calculate the acf sometimes depict it as a vertical plot instead of the horizontal plot in Figure 12.6. For example, Figure 12.7 shows a *Minitab* plot of the acf of the same data used to create Figure 12.6.

Testing for Autocorrelation. To make use of the acf, one must know which of the r_k's are statistically different from 0 and which are not. Specifically, if ρ_k denotes the **population autocorrelation coefficient of lag k**, then we need to test the hypotheses $H_0: \rho_k = 0$ versus $H_a: \rho_k \ne 0$ for each value of k. For this purpose, a

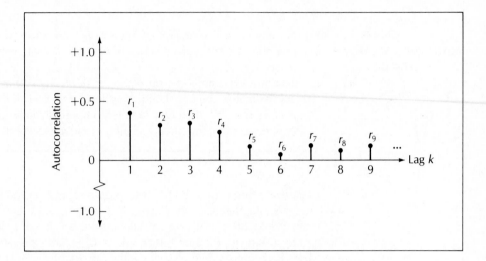

FIGURE 12.6
A typical
autocorrelation
function (acf)

```
ACF of data

          -1.0 -0.8 -0.6 -0.4 -0.2  0.0  0.2  0.4  0.6  0.8  1.0
           +----+----+----+----+----+----+----+----+----+----+
    1  0.379                            XXXXXXXXXX
    2  0.271                            XXXXXXXX
    3  0.289                            XXXXXXXX
    4  0.243                            XXXXXXX
    5  0.131                            XXXX
    6  0.059                            XX
    7  0.140                            XXXXX
    8  0.098                            XXX
    9  0.134                            XXXX
   10  0.132                            XXXX
   11  0.015                            X
   12  0.133                            XXXX
```

FIGURE 12.7
Minitab printout of an
acf

"rule of thumb" procedure based on a large-sample normal approximation can be used (Bartlett 1946). The test relies on the fact that, for large n, *each r_k is approximately normally distributed with a mean of 0 and a standard deviation of $1/\sqrt{n}$*, whenever H_o is true (i.e., whenever $\rho_k = 0$).[2] Since roughly 95% of a normal population lies within 2 standard deviations of its mean, the "rule of thumb" test simply uses a significance of 5% and rejects H_o whenever $|r_k|$ exceeds $2/\sqrt{n}$ (other significance levels can be used, if desired). The following list summarizes this test:

---

[2] Bartlett's formula for the standard deviation of r_k is quite general; the 'rule of thumb' applies his formula to the case of independent observations from a normal distribution to find $\sigma_{r_k} = 1/\sqrt{n}$.

TESTING
AUTOCORRELATION
COEFFICIENTS

1. For any lag k (k is always an integer), use equation (12.6) to compute r_k from a given series of process measurements $x_1, x_2, \ldots, x_n$.
2. To run a hypothesis test on H_o: $\rho_k = 0$, H_a: $\rho_k \neq 0$ at approximately the 5% significance level, reject H_o if $|r_k| > 2/\sqrt{n}$.
3. If H_o is rejected, then conclude (with approximately 95% confidence) that an autocorrelation of lag k is present.

To illustrate the 'rule of thumb' test, suppose that $n = 100$ measurements are used to generate the acf's in Figures 12.6 and 12.7. After the critical values $\pm 2/\sqrt{n} = \pm 2/\sqrt{100} = \pm 0.2$ are calculated for this value of n, lines are drawn on the acf at a distance of 0.2 on either side of 0 (Figure 12.8). Any r_k that exceeds these lines is then considered 'significantly' different from 0; that is, we conclude that the process contains a nonzero autocorrelation at lag k.

With its 'rule of thumb' lines affixed at $\pm 2/\sqrt{n}$, the acf provides a clear view of the various correlations contained in a given series of process data $x_1, x_2, \ldots, x_n$. In particular, the acf makes it relatively easy to check whether or not the measurements can be considered statistically independent. Since independent measurements have $\rho_k = 0$ for every value of k, the acf of data from such a process should have no 'significant' spikes; that is, the acf is entirely contained between the $\pm 2/\sqrt{n}$ lines. On the other hand, any spikes that exceed the $\pm 2/\sqrt{n}$ lines indicate the presence of autocorrelation in the data.

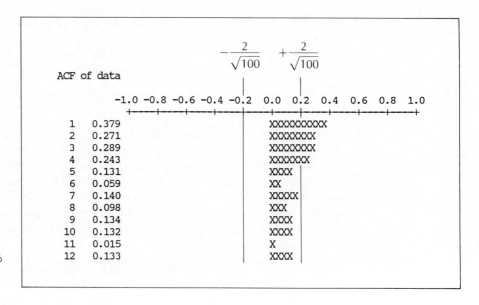

FIGURE 12.8
The acf of Figure 12.6 with $\pm 2/\sqrt{n}$ lines included ($n = 100$)

Interpreting Autocorrelation. What happens when autocorrelation is detected? Should a process whose measurements are autocorrelated be considered 'out of control'? As Alwan and Roberts (1988) have pointed out, Shewhart's original definition of statistical control did not exclude autocorrelated processes from being in 'statistical control.' Shewhart (1931) demanded only that "we can predict, at least within limits, how the phenomenon may be expected to vary in the future." To decide whether or not an autocorrelated process may be considered 'in control,' one must first investigate the reasons for the autocorrelation. Essentially, the presence of autocorrelation affords one the opportunity to better understand a process. If the autocorrelation can be explained by some system of common causes, then the process can be considered 'in control.' On the other hand, autocorrelation can also be the sign of a *reoccurring special cause,* in which case the process is not considered to be in a state of control. In the final analysis, we use the information from the acf to help make this determination.

EXAMPLE 12.4

Table 12.4 contains a sample of measurements on 25 successive parts from the cumulated output of five stamping machines, each set to produce the same metal part. Due to the high-volume nature of the stamping operation, parts from each machine are collected by slightly offsetting each machine's start and stop times so that the parts can be output to a conveyor belt system without having five parts arriving at the belt at exactly the same moment. As a by-product of this procedure, the order in which the parts are produced (i.e., which machine produces a given part) is maintained in the measured data.

To test this relatively small data set for signs of autocorrelation, the acf is computed (Figure 12.9). Since $n = 25$ for these data, the 'rule of thumb' values for testing the autocorrelation coefficients are $\pm 2/\sqrt{n} = \pm 2/\sqrt{25} = \pm 0.40$. Drawing these lines on the acf, we see that the only 'significant' spike appears at lag $k = 5$. For illustration, the acf is computed for 15 lags, but normally one should adhere to the $n/4$ rule (i.e., one should not try to interpret more than the first $n/4 = 25/4 \approx 6$ coefficients). Notice that rather large (but not quite significant) spikes also occur at lags that are multiples of 5. This is because autocorrelation at lag k often (but not always) induces some residual autocorrelation at lags $2k$, $3k$, $4k$, and so on. That only r_5 is significant and that there are five machines involved lead to the obvious question of whether or not all five machines are producing identical parts. In fact, in this example it turns out that machines 1 through 4 are 'in control' and producing parts close to the desired nominal value of 0.250 in., whereas machine five is out of adjustment

TABLE 12.4 Measurements (in inches) For 25 Successive Parts from the Output of Five Identical Metal Stamping Machines

				(read across)				
0.258	0.255	0.248	0.235	0.283	0.251	0.248	0.233	0.240
0.281	0.235	0.241	0.251	0.244	0.273	0.247	0.264	0.243
0.255	0.277	0.253	0.258	0.261	0.240	0.287		

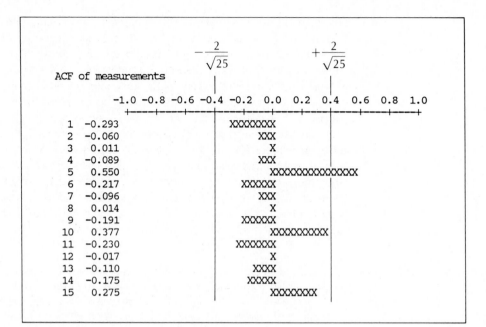

FIGURE 12.9
Autocorrelation
function of the data
in Table 12.4

and producing parts with a nominal measurement of 0.280 in. In this case, then, the presence of autocorrelation is a sign that the overall stamping process is *not* in control.

For comparison, suppose that an individuals chart is run on the data in Table 12.4. Figure 12.10 shows such a chart created by the *Minitab* software

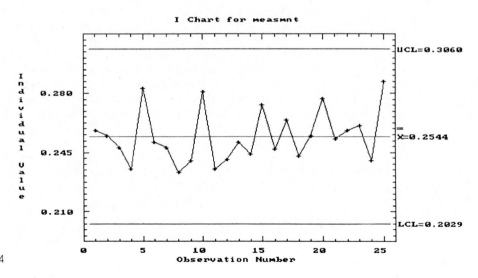

FIGURE 12.10
Individuals chart for
the data of Table 12.4

package. While it is possible that a trained observer might sense the periodic behavior in this chart, the 3-sigma control limits alone give no 'out of control' signals. The fact that machine 5 is making parts that are slightly larger than the other machines has inflated the standard deviation, thereby inflating the control limits. Thus, the fact that the process is not in control may escape notice if only an individuals chart is used.

EXAMPLE 12.5   Table 12.5 contains 85 successive readings on the concentration of a certain chemical in an electroplating solution used for making printed circuit boards. Before the boards are put into the solution, the chemical content is measured at the beginning of each shift. The concentration is recorded and, if needed, more of the chemical is added to the bath. The bath is used 7 days a week and there are two workshifts (day/night) each day.

TABLE 12.5   Chemical Concentrations (in Percent) of a Chemical Used in an Electroplating Solution ($n = 85$)

(read across)

8.80	8.53	8.33	8.44	8.53	8.49	8.09	7.87	8.34	8.21	8.23
8.56	8.14	8.20	8.44	8.32	8.76	8.39	8.36	8.30	8.40	8.46
8.52	8.66	8.40	8.52	8.05	8.11	8.45	8.62	8.43	8.39	8.76
8.14	8.11	8.44	8.33	8.22	8.31	8.45	8.27	8.25	8.16	8.47
8.75	8.51	8.45	8.32	8.35	8.36	8.70	8.56	8.16	8.21	8.25
8.30	8.20	8.28	8.18	8.28	8.47	8.12	8.27	8.30	8.28	8.46
8.46	8.40	8.44	8.38	8.39	8.46	8.54	8.35	8.24	8.35	8.29
8.69	8.49	8.73	8.49	8.54	8.50	8.76	8.59			

The acf of the data in Table 12.5 is shown in Figure 12.11 along with the $\pm 2/\sqrt{n}$ lines for testing each coefficient. Since $n = 85$ for these data, the only autocorrelation coefficient that exceeds $\pm 2/\sqrt{85} = \pm 0.217$ is $r_1 = 0.286$. Note also that $r_{10} = -0.213$ comes close to the lower cutoff of -0.217 and might bear some investigation. In fact, further inquiry reveals that the small lag 10 autocorrelation arose because there were two readings per day (one for each workshift) and weekend shifts tended to process fewer boards than weekday shifts. The lag 1 autocorrelation of 0.286 indicates that there is a small, but statistically significant, association between measurements taken one period apart. While it is possible that some special cause may be at work here, a more likely explanation is that there might be some similarity between any two readings taken only one shift apart. For example, it is reasonable to expect both readings to be high on days when relatively few boards are plated. Conversely, both readings could be low during higher-volume days. Thus, it is very possible that this process is 'in control' in the sense that it is relatively predictable and that only common causes are at work. In any event, the acf provides clues for further examining and understanding the plating process.

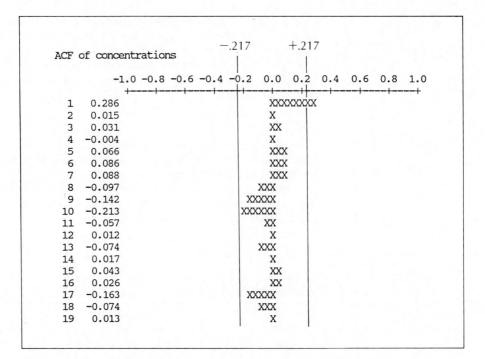

FIGURE 12.11
Autocorrelation
function of the data
in Table 12.5

ARIMA MODELS OF PROCESS DATA

The Form of ARIMA Models. As described earlier, Shewhart's definition of 'statistical control' is broader than that assumed by standard threshold control charts. Although Shewhart originally demanded only that a process be *predictable,* not that its measurements necessarily be random (i.e., uncorrelated), he nonetheless developed the theory of control charts under the assumption of uncorrelated process data, partly because it makes the statistical theory very easy to implement and partly because randomness does adequately describe many processes. Extending control chart methodology to the case of autocorrelated data is a fairly recent development.

When autocorrelated data are modeled, it is best to work with individual measurements rather than with subgroups of correlated data. Typically, then, we think of autocorrelated data as a sequence of measurements $x_1, x_2, x_3, \ldots, x_n$ on some quality characteristic X. If, as is usually the case, these measurements are taken at equally spaced time periods, then the sequence $x_1, x_2, \ldots, x_n$ can also be thought of as a time series of length n. Viewed in this manner, the techniques of time series analysis can be applied to model the autocorrelation in the data. When this is done, it is convenient to change the subscript notation from x_i to x_t (the t is intended to stand for 'time period t'), since this is the standard notation used with time series methods. One example of this approach is the EWMA chart introduced in Section 7.6. This chart is based on the well-known time series method of exponential smoothing.

The exponential smoothing model, however, is but one from a large class of time series models known as ARIMA models. ARIMA stands for **autoregressive integrated moving average,** an acronym intended to describe the types of models available in this class. The general form of an ARIMA model is given by

$$x_t = \phi_1 x_{t-1} + \phi_2 x_{t-2} + \cdots + \phi_p x_{t-p} + \epsilon_t + \theta_0 - \theta_1 \epsilon_{t-1} - \theta_2 \epsilon_{t-2}$$
$$- \cdots - \theta_q \epsilon_{t-q} \tag{12.7}$$

where $\phi_1, \phi_2, \ldots, \phi_p$ and $\theta_1, \theta_2, \ldots, \theta_q$ are constants (called 'parameters') whose values must be estimated from the process data. Equation (12.7) roughly says that the current process reading x_t may depend, to some extent, on the previous p process readings $x_{t-1}, x_{t-2}, \ldots, x_{t-p}$ as well as on the previous q random 'shocks' to the system $\epsilon_{t-1}, \epsilon_{t-2}, \ldots, \epsilon_{t-q}$. The shocks can be thought of as unexpected noise or random variation that the process has experienced at times $t - 1, t - 2, \ldots,$ $t - q$, and they are *assumed to be uncorrelated with one another.* The auto-correlations within the data $x_1, x_2, x_3, \ldots, x_t$ are captured in the parameters $\phi_1, \phi_2, \ldots, \phi_p$, and $\theta_1, \theta_2, \ldots, \theta_q$. The goal of ARIMA modeling is first to select the *form* of the model (i.e., choose p and q) that best describes the behavior of the process readings and then to estimate the parameters for that model. With these estimates in hand, equation (12.7) can then be used to *predict* the future course of the process.

The expression $\phi_1 x_{t-1} + \phi_2 x_{t-2} + \cdots + \phi_p x_{t-p}$ is called the **autoregressive** part of the model, while $-\theta_1 \epsilon_{t-1} - \theta_2 \epsilon_{t-2} - \cdots - \theta_q \epsilon_{t-q}$ is the **moving average** part. In practice, many processes follow fairly simple ARIMA models that are either purely autoregressive models (where $q = 0$ and p is small)

$$x_t = \theta_0 + \phi_1 x_{t-1} + \phi_2 x_{t-2} + \cdots + \phi_p x_{t-p} + \epsilon_t \tag{12.8}$$

or purely moving average models ($p = 0$ and q is small)

$$x_t = \epsilon_t + \theta_0 - \theta_1 \epsilon_{t-1} - \theta_2 \epsilon_{t-2} - \cdots - \theta_q \epsilon_{t-q} \tag{12.9}$$

The integer p is referred to as the *order* of the autoregressive part of the model, q is the *order* of the moving average part. Since equation (12.8) refers to a process that is purely autoregressive, it is often abbreviated as **AR(p)**, an **autoregressive process of order p.** Similarly, equation (12.9) describes a **moving average process of order q,** which is denoted **MA(q).** When both p and q are nonzero, equation (12.7) is said to be a *mixed* model of order (p, q).

Underlying most threshold control charts is the assumption that process measurements are centered around a fixed level (i.e., mean) and are independent of one another; that is, the process is assumed to follow the simple model $x_t = \theta_0 + \epsilon_t$, which is a special case of equation (12.7) (in which p and q are both taken to be 0). From this point of view, ARIMA models provide a natural extension to processes whose readings are autocorrelated with one another.

Selecting an ARIMA Model. The procedure for selecting an ARIMA model requires the use of three common time series tools: differencing, the autocorrelation function (acf), and the partial autocorrelation function (pacf). Recall that the acf has already been described. One of the roles of the acf is to check the process data to see whether they have one of the characteristic properties of an ARIMA model: an acf that either *cuts off* sharply after a certain lag or *exponentially decays*. As the lag k increases, the spikes in the acf should appear (approximately) to cut off after some lag number or, failing that, the spikes should rapidly become smaller in magnitude. For example, the acf in Figure 12.12(a) has spikes that cut off sharply after lag 2, while Figure 12.12(b) shows acf's whose spikes die out in an exponential fashion. To ascertain whether the decay in an acf is exponential or not, one can use the rule of thumb that the r_k's should have *magnitudes* that behave roughly like $r_1, r_1^2, r_1^3, r_1^4, \ldots$.

If the acf displays neither of these characteristic properties of an ARIMA model, then it may be necessary to transform the process data by differencing them. **Differencing** a series $x_t \{t = 1, 2, 3, \ldots, n\}$ amounts to replacing it by the series $x_t - x_{t-1} \{t = 2, 3, 4, \ldots, n\}$. The series $x_t - x_{t-1}$ is called the series of **first differences** of the original series x_t, and it is compactly denoted by

$$\Delta x_t = x_t - x_{t-1} \tag{12.10}$$

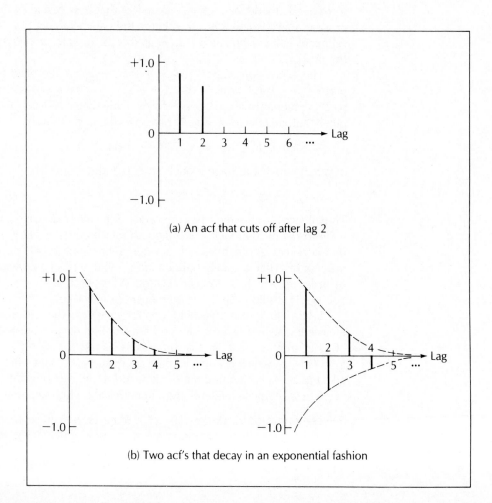

(a) An acf that cuts off after lag 2

(b) Two acf's that decay in an exponential fashion

FIGURE 12.12
Autocorrelation
functions of ARIMA
processes

Differencing is sometimes described as applying a 'high-pass filter' to the data because it tends to eliminate the low-frequency cycles in the data (such as trends) while retaining the higher frequencies (i.e., allowing the higher frequencies to 'pass' through the filter unaltered).

If first differencing is not sufficient to produce an acf that either cuts off or dies out, it may be necessary to difference once more (i.e., to difference the differenced data). This is called finding the **second differences** of the series x_t. Thinking of differencing as an *operation* applied to a time series, we call the symbol Δ the **differencing operator** and then denote repeated applications of differencing by Δ^d, where d is the number of times the series x_t is differenced. Thus, the series of second differences of x_t is denoted by $\Delta^2 x_t$, which can be written in terms of the original data by applying the Δ operator twice:

$$\Delta^2 x_t = \Delta(\Delta x_t) = \Delta(x_t - x_{t-1}) = (x_t - x_{t-1}) - (x_{t-1} - x_{t-2})$$

$$= x_t - 2x_{t-1} + x_{t-2}$$

Similarly, $\Delta^3 x_t$ refers to the series of third differences of the series x_t, and so forth. Fortunately, if differencing needs to be applied at all, then usually no more than first or second differencing is needed to produce an acf that cuts off or dies out (Box and Jenkins 1970). Many processes do not require differencing. The number of times that the data must be differenced is denoted by the letter d, which is called the *order* of differencing.

The general notation for an ARIMA model is **ARIMA(p, d, q),** where p and q are, as above, the orders of autoregressive and moving average parts of the model, respectively, and d is the order of differencing. In this general format, an AR(p) model could also be written ARIMA($p, 0, 0$). As another example, a series that is differenced once, prior to fitting a moving average model of order 1, is denoted ARIMA(0, 1, 1).

When differencing is required, an ARIMA model is fit to the *differenced* data, not the original series x_t. In order to convert this model's predictions back into predictions about the original series, the model must be 'undifferenced,' a procedure that is called 'integration' in the time series literature. (The term 'integration' is used because differencing bears a close relationship to the familiar calculus procedure of differentiation, and consequently 'undifferencing' is then the counterpart of integration.) To illustrate, suppose for a particular set of process data that it was necessary to use first differences, after which an AR(2) model would be estimated. Integrating such a model is most easily accomplished by first denoting Δx_t in a simpler fashion as w_t, writing out the AR(2) model in terms of w_t, substituting Δx_t in for w_t throughout this model, and finally solving for x_t:

Step 1: $w_t = \theta_0 + \phi_1 w_{t-1} + \phi_2 w_{t-2} + \epsilon_t$

[AR(2) model for the first differences $w_t = \Delta x_t$]

Step 2: $\Delta x_t = \theta_0 + \phi_1 \Delta x_{t-1} + \phi_2 \Delta x_{t-2} + \epsilon_t$

[Substitute Δx_t for w_t, Δx_{t-1} for w_{t-1}, and so on]

Step 3: $(x_t - x_{t-1}) = \theta_0 + \phi_1(x_{t-1} - x_{t-2}) + \phi_2(x_{t-2} - x_{t-3}) + \epsilon_t$

$x_t = \theta_0 + (1 + \phi_1)x_{t-1} + (\phi_2 - \phi_1)x_{t-2} - \phi_2 x_{t-3} + \epsilon_t$

[The final 'integrated' model for x_t]

TABLE 12.6   Characteristic Behavior of the acf and pacf of ARIMA Models

Model type	Autocorrelation function (acf)	Partial autocorrelation function (pacf)
AR (p)	Dies out	Cuts off after lag p
MA (q)	Cuts off after lag q	Dies out
Mixed (p, q)	Dies out	Dies out

The third tool needed for identifying an appropriate ARIMA model is the partial autocorrelation function (abbreviated pacf). It is similar in appearance to the acf, except the pacf plots the estimated partial autocorrelation coefficient of lag k, $\hat{\phi}_{kk}$, versus k. Each $\hat{\phi}_{kk}$ measures the autocorrelation between x_t and x_{t+k} *after adjusting for the effect of all intermediate (i.e., shorter lag) autocorrelation.* That is, it is possible that some of the autocorrelation between x_t and x_{t+k} might be induced by the lag 1, lag 2, . . . , lag $k - 1$ autocorrelation in the series, and the pacf attempts to eliminate this effect in its estimate of the lag k autocorrelation. As an example, suppose a certain series exhibits strong lag 1 autocorrelation—say, $r_1 = 0.8$. Then it is possible that some of the lag 2 autocorrelation could be *caused* by the lag 1 autocorrelation; that is, because of the lag 1 autocorrelation, x_t and x_{t+1} tend to be on the same side of the mean as are x_{t+1} and x_{t+2}, which implies that x_t and x_{t+2} may also tend to be on the same side of the mean (i.e., lag 2 autocorrelation). Calculating each $\hat{\phi}_{kk}$ and the pacf is somewhat complicated and is normally done by computer.

It is a fortunate circumstance that testing the pacf can be done with the same 'rule of thumb' test used for the acf. In other words, for any lag k, the hypothesis H_0: $\phi_{kk} = 0$ can be tested (with approximately a 5% significance level) by using rejection zones placed at $\pm 2/\sqrt{n}$. Just as with the acf, lines drawn through the pacf at distances of $\pm 2/\sqrt{n}$ from 0 allow all the lags to be tested at once.

Together, characteristic patterns in the acf and pacf allow one to determine the orders, p and q, of an ARIMA model. These patterns are summarized in Table 12.6. Determining p and q for *mixed* models is the hardest task, but two additional rules are of help in this case:

- For a mixed model, if $q \geq p$, then the **acf** should decay after lag $q - p$. (If $q < p$, the acf should decay from the beginning.)

- For a mixed model, if $p \geq q$, then the **pacf** should decay after lag $p - q$. (If $p < q$, the pacf should decay from the beginning.)

In practice, a great many processes tend to follow very simple ARIMA models, especially purely autoregressive AR(p) models or purely moving average MA(q) models where the orders of p and q are relatively small (often 2 or less). Mixed models occur much less frequently. This makes model identification even easier, since the acf/pacf patterns for AR(p) and MA(q) models are so simple and distinctive.

EXAMPLE 12.6   In a high-volume drilling operation, holes with a target diameter of 0.250 in. are drilled in metal plates. Every 15 min one of the plates is selected and the diameter of the drilled hole is measured. Table 12.7 shows a group of such measurements taken over a 2-day period ($n = 4$ measurements/hr · 8 hr/workshift · 2 workshifts $= 64$ measurements). A run chart of the data

---

TABLE 12.7   Diameters of Drilled Holes (nominal diameter = 0.250 in.)

(read across)

0.249	0.253	0.247	0.248	0.250	0.256	0.245	0.253	0.267
0.258	0.248	0.249	0.259	0.248	0.251	0.252	0.236	0.249
0.246	0.242	0.247	0.254	0.256	0.253	0.235	0.247	0.253
0.245	0.249	0.251	0.253	0.254	0.252	0.247	0.247	0.250
0.245	0.254	0.254	0.252	0.257	0.250	0.257	0.244	0.240
0.244	0.245	0.238	0.246	0.247	0.245	0.248	0.251	0.253
0.259	0.254	0.248	0.247	0.243	0.247	0.248	0.247	0.244
0.242								

---

appears in Figure 12.13. The data were collected over a period when the drill was operating normally and when there were no changes in materials, work-force, or operating procedures; that is, no obvious special causes were present and the process was thought to be 'in control.'

From the acf of these data in Figure 12.14, the 'rule of thumb' test indicates that there is a significant lag 1 autocorrelation coefficient (since $r_1 = 0.296$ exceeds $2/\sqrt{64} = 0.25$). Furthermore, the acf appears to die out fairly rapidly.

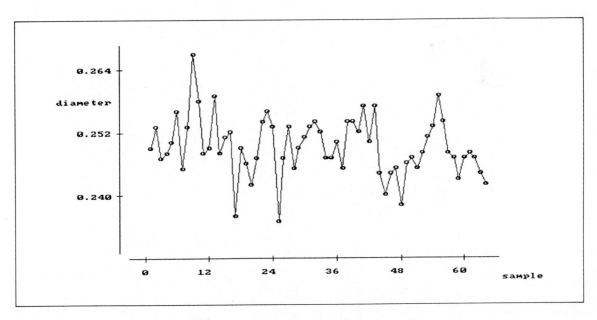

FIGURE 12.13   Run chart for the data of Table 12.7

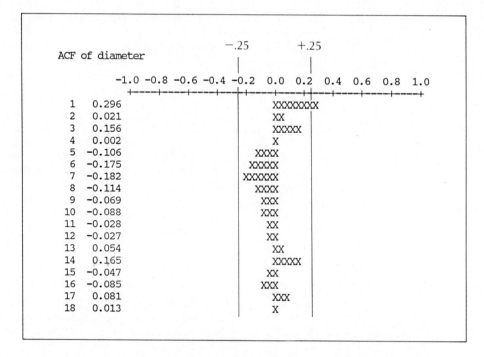

FIGURE 12.14
Autocorrelation
function of the data
in Table 12.7

From this we conclude that no differencing is needed before attempting to fit an ARIMA model to these data. From the pacf in Figure 12.15, the spike at lag 1 appears to be followed by a sharp drop off at lag 2, while the rest of the pacf

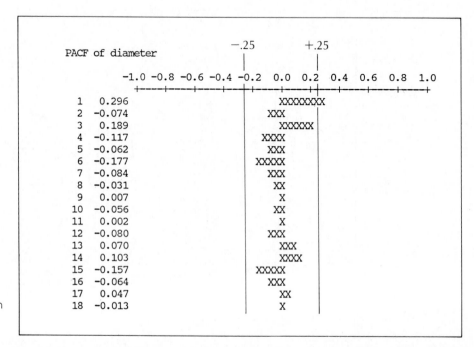

FIGURE 12.15
Partial autocorrelation
function of the data
in Table 12.7

also falls within the $\pm 2/\sqrt{n}$ lines. Tentatively, then, an AR(1) model is proposed as a possible fit to these data (since the pacf cuts off after lag 1 and the acf dies out rapidly).

To check the adequacy of this model, the *Minitab* package is used to estimate the parameters of an AR(1) model for the data of Table 12.7. The resulting printout is shown in Figure 12.16. A detailed analysis of Figure 12.16 is beyond the scope of our presentation. Interested readers may refer to time series texts for a complete discussion of ARIMA printouts such as the one in Figure 12.16. For the purpose of this example, suffice it to say that *Minitab* encounters no problems in estimating the model parameters and there are no diagnostic warnings; that is, the model provides an adequate fit to the data. Thus, when the drilling process is 'in control,' it can be modeled by the equation $x_t = 0.1767 + 0.2907x_{t-1}$.

FIGURE 12.16
Estimating the
parameters of an
AR(1) model for the
data of Table 12.7

```
Estimates at each iteration
Iteration        SSE      Parameters
    0         0.522209    0.100   0.314
    1         0.005955    0.138   0.223
    2         0.001961    0.271   0.183
    3         0.001841    0.290   0.177
    4         0.001840    0.291   0.177
    5         0.001840    0.291   0.177
Relative change in each estimate less than  0.0010


Final Estimates of Parameters
Type       Estimate     St. Dev.   t-ratio
AR   1       0.2907       0.1224      2.37
Constant  0.176717      0.000681    259.44
Mean      0.249155      0.000960


No. of obs.:   64
Residuals:     SS = 0.00184034  (backforecasts excluded)
               MS = 0.00002968  DF = 62


Modified Box-Pierce chisquare statistic
Lag               12            24           36             48
Chisquare   9.8(DF=11)   24.5(DF=23)  37.4(DF=35)   44.6(DF=47)
```

After an ARIMA model has been estimated for a process that is 'in control,' the question arises of how to detect special causes. Running a standard threshold control chart on the data, for example, is not advisable because the autocorrelation in the data will cause many of the 'out of control' rules to be violated.[3] For instance, an individuals chart of the data in Table 12.7 would reveal a run of more than eight points on one side of the centerline, among other violations.

---

[3] The rules for special causes in Section 6.4 are based on the assumption of independent readings, so they do not apply to autocorrelated data.

One suggested method for detecting special causes is to run a standard threshold control chart on the **residuals** from the estimated ARIMA model (Alwan and Roberts 1988). The residuals, or error terms, are defined to be the series $e_t = x_t - \hat{x}_t$, where $\hat{x}_t$ is the model's predicted value of x_t. In other words, after the natural behavior of the process is accounted for (i.e., the ARIMA model), special causes should create anomalous patterns in the residual series.

EXAMPLE 12.7

After the model $x_t = 0.1767 + 0.2907x_{t-1}$ is fit to the data of Table 12.7, the residual series is calculated. Since most software for fitting ARIMA models allows the predicted values and the residuals to be stored automatically, obtaining the residual series should require minimal effort in practice. From the residuals from the *Minitab* fit of the AR(1) model in Example 12.6, an individuals chart is constructed in Figure 12.17. The chart shows only one rule violation (a point above the upper control limit at subgroup 9), in contrast to the many (false) signals that arise in an individuals chart of the raw data. Even though the process was thought to be in control when the model was estimated, subgroup 9 still has to be investigated to see whether or not an identifiable special cause was indeed present.

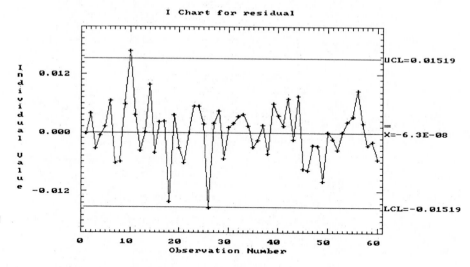

FIGURE 12.17
Individuals chart of the residuals from fitting an AR(1) model to the data of Table 12.7

NARROW-LIMIT GAGING IN HIGH-VOLUME OPERATIONS

Narrow-limit gaging (NLG), introduced in Section 7.8, is especially well suited for high-volume work. The speed with which items can be compared to the two narrow-limit gages makes data available quickly and, from these data, the determination of whether or not there has been a process shift is equally rapid, requiring only a glance at a predetermined chart to see whether too many items have fallen beyond one of the gages. Furthermore, in many cases the gaging process can itself be automated, thus speeding up data acquisition even more. Consider, for example, the production of ball bearings of a given diameter D. After suitable gage limits are

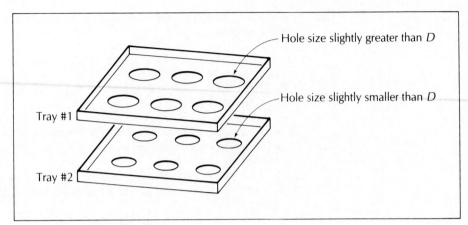

FIGURE 12.18
Narrow limit-gages for
ball bearings with
nominal diameter D

established, two trays with holes of these dimensions can be fabricated and used
to rapidly classify fairly large samples of bearings (see Figure 12.18).

Many times, automatic testing equipment (ATE) or numerically controlled (NC)
machines have the capability of measuring *every* item produced. In such cases,
data need only be accessed electronically, and the NLG test procedure can be accomplished with a few software commands.

It should be noted that NLG is applied under the same assumptions as for
standard threshold control charts—that is, when the data are assumed to be independent and identically distributed. NLG makes a further assumption that the
measurements are approximately normal. Thus, NLG cannot be applied to autocorrelated data for the same reason that threshold charts are not used; because it
would generate too many false 'out of control' signals.

EXERCISES FOR CHAPTER 12

12.1   In the short-run production of certain machined parts, four different part types are
machined in a 2-day period. The nominal dimensions of these parts are: P1 (nominal = 1.50), P2 (nominal = 0.50), P3 (nominal = 2.00), and P4 (nominal = 2.50).
Five parts of each type are selected and measured during this period, yielding the
following results:

P1	P1	P2	P3	P4	P4	P1	P1
1.50	1.51	0.39	2.08	2.57	2.38	1.63	1.54
1.56	1.57	0.47	2.04	2.50	2.50	1.42	1.57
1.56	1.52	0.48	1.96	2.55	2.45	1.50	1.42
1.41	1.37	0.34	2.10	2.47	2.40	1.52	1.67
1.53	1.62	0.43	1.92	2.54	2.54	1.65	1.62

Construct a deviations from nominal chart for these data. What is the estimated
process standard deviation?

12.2 The following 40 measurements are taken from the output of a computer-controlled grinding machine:

2.00	1.64	1.68	2.20	1.69	1.92	2.17	2.58	1.88	2.06	1.54
2.15	2.12	1.74	1.98	2.14	1.79	1.56	0.99	1.19	1.27	1.87
1.99	2.46	2.75	1.91	1.77	1.77	1.38	2.24	2.32	2.24	2.06
2.36	1.76	1.65	1.50	1.58	1.41	1.42	1.53			

(a) Find the autocorrelation function for these data. Do the measurements appear to be autocorrelated, or can they be considered to be independent of one another?

(b) Find the partial autocorrelation function for these data.

(c) From the results of parts (a) and (b), what ARIMA model best describes these data?

12.3 Show that when an ARIMA(0, 1, 1) model with *no constant term* is fit to any set of process data, the resulting values are identical to those of an EWMA chart with parameter $\lambda = 1 - \theta_1$, where θ_1 is the moving average parameter of the ARIMA(0, 1, 1) model.

12.4 The autocorrelation function of a set of 50 process measurements is shown here:

```
                                         ACF

         -1.0 -0.8 -0.6 -0.4 -0.2  0.0  0.2  0.4  0.6  0.8  1.0
         +----+----+----+----+----+----+----+----+----+----+
  1   0.288                          XXXXXXXX
  2   0.046                          XX
  3   0.023                          XX
  4   0.111                          XXXX
  5   0.057                          XX
  6  -0.151                     XXXXX
  7  -0.142                     XXXXX
  8  -0.143                     XXXXX
  9  -0.097                      XXX
 10   0.028                          XX
```

Does this acf indicate that the process data are autocorrelated? If so, which spikes in the acf appear to be the important ones?

12.5 The autocorrelation function of 36 process measurements appears as follows:

```
                                         ACF
         -1.0 -0.8 -0.6 -0.4 -0.2  0.0  0.2  0.4  0.6  0.8  1.0
         +----+----+----+----+----+----+----+----+----+----+
  1   0.093                          XXX
  2   0.125                          XXXX
  3  -0.031                          XX
  4   0.135                          XXXX
  5  -0.098                        XXX
  6  -0.090                        XXX
  7  -0.061                        XXX
  8  -0.001                          X
  9  -0.067                        XXX
 10  -0.064                        XXX
```

Is there evidence of autocorrelation in these data? If so, indicate which spikes in the acf are important.

12.6 An autocorrelation function for some process data has a significant negative spike at lag 1 and a significant positive spike at lag 2. Based on this result, describe the behavior of the process. Suppose further that the measurements are taken twice per day, once during the morning shift and once during the afternoon shift. What do the lag 1 and 2 spikes indicate about the two workshifts?

12.7 Explain why processes in which tool wear is an important factor may sometimes produce autocorrelated data. For such processes, would you expect the lag 1 auto-correlation to be positive or negative?

12.8 Sheets of metal 4 ft long travelling on a conveyor system are cut in half at a certain point in a process. The cut sheets continue, in order, on the conveyor system. Each cut sheet is automatically *weighed* as it passes over a certain point on the conveyor system. Explain how the autocorrelation function of such data could be used to signal when the cutting process is not evenly dividing the sheets of metal.

12.9 Consider the following set of process measurements (read across):

5.0	5.1	5.1	5.1	5.3	5.0	5.0	4.9	4.9	5.4	4.9
4.8	4.8	5.0	4.9	4.5	4.8	4.4	3.8	4.3	4.1	4.8
4.5	4.7	4.7	5.5	5.5	4.7	4.6	4.8	4.7	4.8	4.5
4.5	5.1	5.2	5.0	5.5	5.7	5.3	4.4	4.9	5.0	5.1
4.7	4.5	4.9	4.7	5.3	5.4					

The acf and pacf of these data are:

```
                               ACF

        -1.0 -0.8 -0.6 -0.4 -0.2  0.0  0.2  0.4  0.6  0.8  1.0
         +----+----+----+----+----+----+----+----+----+----+
    1    0.508                         XXXXXXXXXXXXX
    2    0.209                         XXXXX
    3    0.129                         XXXX
    4    0.122                         XXXX
    5    0.031                         XX
    6   -0.189                   XXXXX
    7   -0.204                   XXXXX
    8   -0.167                   XXXXX
    9   -0.069                     XXX
   10    0.033                         XX


                               PACF

        -1.0 -0.8 -0.6 -0.4 -0.2  0.0  0.2  0.4  0.6  0.8  1.0
         +----+----+----+----+----+----+----+----+----+----+
    1    0.508                         XXXXXXXXXXXXX
    2   -0.066                      XXX
    3    0.066                         XXX
    4    0.054                         XX
    5   -0.079                      XXX
    6   -0.245                 XXXXXX
    7   -0.005                        X
    8   -0.054                       XX
    9    0.079                         XXX
   10    0.128                         XXXX
```

(a) From the acf and pacf, what ARIMA model(s) would best describe these data?

(b) Use a computer package with ARIMA capabilities to fit the model(s) in part (a) to the data.

(c) Construct an individuals chart of the residuals from the model(s) fit in part (b). Are there any points beyond the control limits?

(d) Construct the acf of the residuals from the models in part (b). Does this acf indicate that there is any autocorrelation in the residuals?

REFERENCES FOR CHAPTER 12

Alwan, L. C., and H. V. Roberts. 1988. "Time Series Modeling for Statistical Process Control," *Journal of Business and Economic Statistics* 6 (no. 1):87–95.

Bartlett, M. S. 1946. "On the Theoretical Specification of Sampling Properties of Autocorrelated Time Series." *Journal of the Royal Statistical Society,* B, 8:27–41.

Box, G. E. P., and G. M. Jenkins. 1970. *Time Series Analysis: Forecasting and Control.* New York: Holden Day.

Donmez, M. A. 1989. "A Real-Time Control System for a CNC Machine Tool Based on Deterministic Metrology." In *Statistical Process Control in Automated Manufacturing,* edited by J. B. Keats and N. F. Hubele, pp. 271–290. New York: Marcel Dekker.

Ewan, W. D. 1963. "When and How to Use cu-sum Charts." *Technometrics.* 8:1–22.

Farnum, N. R. 1992. "Control Charts for Short Runs: Nonconstant Process and Measurement Error," *Journal of Quality Technology,* 24 (no. 3):138–144.

Nelson, L. S. 1988. "Control Charts: Rational Subgroups and Effective Applications." *Journal of Quality Technology,* 20 (no. 1):3–75.

Quesenberry, C. P. 1991. "SPC Q Charts for a Binomial Parameter p: Short and Long Runs." *Journal of Quality Technology* 23 (no. 3):239–246.

Sachs, L. 1982. *Applied Statistics: A Handbook of Techniques.* New York: Springer-Verlag.

Shewhart, W. A. 1939. *Statistical Method from the Viewpoint of Quality Control,* p. 46. The Graduate School, Department of Agriculture, Washington, DC.

Shewhart, W. A. 1931. *Economic Control of Quality of Manufactured Product,* p. 6. New York: Van Nostrand.

13

EXPERIMENTAL DESIGN

This chapter gives an introduction to statistical experimental design, including one- and two-way analysis of variance (ANOVA), the analysis of means (ANOM), factorial designs, main effects, interactions, 2^k designs, and the Taguchi methodology.

CHAPTER OUTLINE

13.1	EXPERIMENTAL DESIGN

Experimental design is a statistical method for evaluating the effects of different treatments on a response variable. In the field of agronomy, where these designs were introduced, different fertilizer blends (treatments) were applied to crops in an effort to maximize crop yield (response).[1] The essential idea underlying a designed experiment is that some methods of collecting data on treatment and response variables are more efficient and powerful than others. In particular, the intuitive method of analyzing one treatment at a time, while holding the others fixed, turns out to be one of the less effective designs.

In quality control, these methods are often called **design of experiments** (abbreviated **DOE**). In addition, the designs used in quality control are generally from a special class called **factorial designs.** These designs examine various **factors** thought to affect a process by considering a range of possible values for each factor. Each combination of values of the factors then forms a 'treatment,' and the goal is to find the particular combination(s) of factor levels that have the most desirable effect on the process. If we think of each factor as a dial or knob, a factorial experiment attempts to find the right amount to turn each dial in order to achieve the optimum improvement in a process.

Though DOE has long been used in the sciences, many industries have not embraced these methods since they were introduced in manufacturing in the 1940s. In recent years, however, DOE has gained great popularity, primarily because of its notable successes in Japan.

This section introduces the concepts, terminology, and calculations of experimental designs, especially factorial designs. After introducing basic analysis of variance calculations, we describe the analysis of 2^k **factorial designs.** References for more advanced designs and related methods are given throughout. In addition, the special contributions of Taguchi's approach to experimental design are reserved for a separate section (Section 13.3), as is the related method of analysis of means (ANOM; see Section 13.2).

ONE FACTOR AT A TIME EXPERIMENTS

If we continue with the dial analogy, then from the practitioner's viewpoint, a most desirable experimental outcome is to find out how much (and in what direction) to turn the various dials that control a process. Given that information, it would also be nice if turning each dial has a fixed effect on the process, regardless of the settings on the other dials.

With these goals in mind, one approach to experimentation is to simply turn one dial at a time, while keeping the others fixed, and to record the effect on the process. Then, after the dial is returned to its original position, a second dial is

---

[1] R. A. Fisher introduced the concept of statistically designed experiments in the 1920s.

moved while keeping the others fixed, and so forth. After all the dials are tested in this manner, each is then turned to its optimum setting. By doing this, the experimenter hopes that the individual effects of each dial will *add* to the effects of the others, creating a larger total effect on the process. Figure 13.1 illustrates what the experimenter hopes to achieve by this method.

An alternative strategy is to test the first dial, while holding the others fixed, and then to *leave* this dial at its 'best' setting. Next, a second dial is tested, while holding the others fixed (including dial 1, which is now at its *new* setting). Dial 2 is also left in its optimum position and held fixed as the remaining dials are tested. In this manner, the experimenter hopes that all the dials are optimally positioned at the end of the testing process. Figure 13.2 shows the experimenter's view of this procedure.

For obvious reasons, both of the methods are said to use the **one factor at a time** approach to optimizing a process. Intuitive as both approaches may be, most of the time they do *not* result in finding the optimal settings for each factor. The following example illustrates how this can happen. As R. A. Fisher showed in the 1930s, the best approach is to turn various *combinations* of the dials simultaneously, recording their combined effect on the response variable. By carefully choosing the combinations so that each setting is equally represented and contrasted

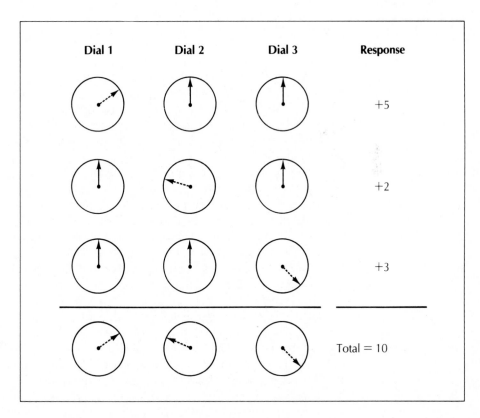

FIGURE 13.1
One factor at a time approach; dials are returned to the starting positions at each test

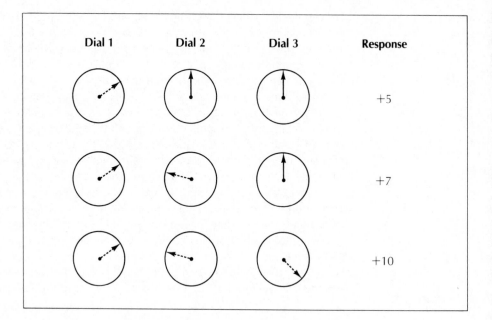

FIGURE 13.2
One at a time
approach; dials are
left at optimum
setting at each trial

with the other settings, one can obtain much more information about a process than is possible with the one factor at a time method.

As an example, consider a hypothetical chemical operation in which parts on a conveyor belt pass through a coating station where a chemical is sprayed on the moving parts. Let factor A be the 'speed of the conveyor belt' and factor B be the 'amount of chemical sprayed.' Suppose that one wants to set these two factors to achieve the most uniform coating on the parts while also speeding the throughput. With the one factor at a time approach, increasing the conveyor speed (which increases the throughput) while not changing the spray volume might have the effect shown in Figure 13.3(a); that is, the increased speed causes the spray to miss some areas, resulting in spotty coverage in the final coating. Similarly, keeping the belt speed fixed and increasing the spray volume might have the effect shown in Figure 13.3(b). Here, the increased spray volume puts too much chemical on each part, resulting in an uneven bumpy coverage. From the graphs in Figure 13.3(a) and (b), the one at a time experimenter might conclude that changing either factor is harmful and, therefore, that nothing can be done to increase throughput.

However, the experimenter may consider adding an extra experimental run, one in which *both* the belt speed and the spray volume are increased [see Figure 13.3(c)]. Here, increased belt speed is compensated by an increased spray volume, resulting in even better (more uniform) coverage than before and with an increased throughput. This information would not have been available with the one at a time approach, since the experimenter would have returned each factor to its best setting (its original level) before proceeding to examine the other factor. By simultaneously testing the *combination* of factor levels, both the experimenter's goals are achieved in this example.

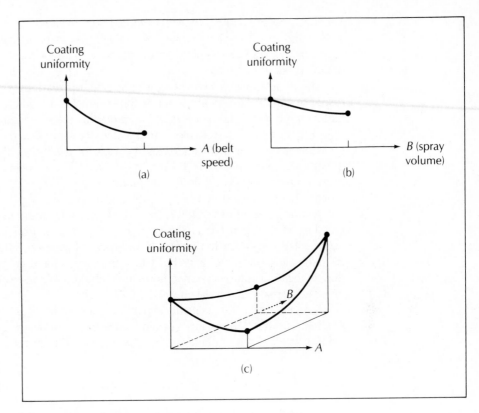

FIGURE 13.3
One at a time
experiments (a) and
(b) versus a factorial
design (c)

ANALYSIS OF VARIANCE—ONE FACTOR

HOW ANALYSIS OF VARIANCE WORKS.   In order to sort out which factors have a real effect on a process and which do not, statistical hypothesis tests must be conducted. Data collection is dictated by an appropriate experimental design, followed by a statistical evaluation of the data. The most popular technique for carrying out the evaluation is the **analysis of variance** (abbreviated **ANOVA**). Another method, the **analysis of means** (abbreviated **ANOM**), is discussed in Section 13.2.

The ANOVA technique is quite general and can be used to study the effects of one factor or several factors. This section introduces **one-factor** or **one-way analysis of variance.** Two-way designs are considered in the next section.

The one-way ANOVA technique is used to detect differences between the means of two or more populations by statistically examining differences between samples from these populations. To be specific, the hypotheses tested by an ANOVA are

$$H_o: \mu_1 = \mu_2 = \mu_3 = \cdots = \mu_c$$

versus

$$H_a: \text{not all the } \mu_j\text{'s are equal} \tag{13.1}$$

There can be any number ($c \geq 2$) of population means $\mu_j \{j = 1, 2, 3, \ldots, c\}$ of interest, and the samples drawn from these populations can be of any size $n_1, n_2, n_3, \ldots, n_c$. The sample sizes do not have to be identical, although it is often desirable that they be so.[2]

The populations can be thought of as a set of **treatments** whose effects on a **response** variable are sought. Usually, the treatments are different levels of a single **factor** of interest. Each mean μ_j is interpreted as the average response caused by the jth treatment. For example, if four brands of some raw material are tested to see whether any of them increases the yield of a process, then the brands are 'treatments' and process yield is the 'response.' The treatments can be considered as four different levels of the factor 'brand.' Each $\mu_j \{j = 1, 2, 3, 4\}$ is the average process yield obtained by using the jth brand.

To understand how ANOVA tests work, consider the two hypotheses in expression (13.1). When $H_o: \mu_1 = \mu_2 = \mu_3 = \cdots = \mu_c$ is *true*, the data from the c treatments look similar to those in Figure 13.4. All the sample mean responses are approximately the same, and the data from each sample are packed around the sample mean in approximately a normal distribution.[3] Conversely, Figure 13.5 illustrates what one expects when H_o does *not* hold.

To construct a test statistic capable of discerning these two situations, the ANOVA approach compares certain *variation measures*. Suppose, for the moment, that H_o is not true and that the data obtained from the treatments look like

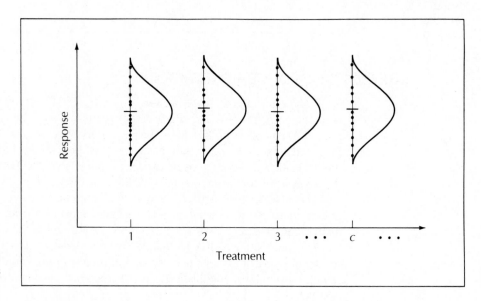

FIGURE 13.4
Typical data from c treatments when H_o: $\mu_1 = \mu_2 = \cdots = \mu_c$ is true

---

[2] With equal sample sizes, ANOVA tests are more robust against departures from the assumption of normality.

[3] ANOVA tests are based on the assumptions that each treatment population is normal and that the variance is the **same** for each population.

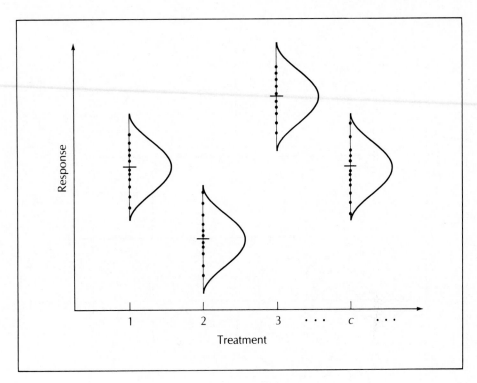

FIGURE 13.5
Typical data from c
treatments when H_O:
$\mu_1 = \mu_2 = \cdots = \mu_c$ is
not true

those in Figure 13.6. Assuming that the variation in responses for each treatment is the same as for any other (an ANOVA assumption), one can **pool** the sample variances $s_1^2, s_2^2, s_3^2, \ldots, s_c^2$ to obtain an overall estimate of the **within-treatments** variation, σ^2. Similarly, computing the variation among the sample means $\bar{x}_1, \bar{x}_2, \bar{x}_3, \ldots, \bar{x}_c$ gives an estimate of the **between-treatments** variation. The ANOVA method then compares these two variation estimates. If they are relatively the same, one concludes that H_o cannot be rejected; that is, there does not seem to be a significant difference between the treatment means $\mu_1, \mu_2, \mu_3, \ldots, \mu_c$. If the two variation estimates differ markedly, it is concluded that there *are* differences between the μ_j's. Figure 13.7 demonstrates how this procedure works.

WITHIN- AND BETWEEN-TREATMENTS VARIATION.   These concepts can be formalized as follows. Denoting the sample sizes from each treatment as $n_1, n_2, n_3, \ldots, n_c$ and letting x_{ij} denote the measurement of the ith item from the jth treatment, we calculate each treatment mean by

$$\text{sample mean of } j\text{th treatment} = \bar{x}_j = \frac{1}{n_j} \sum_{i=1}^{n_j} x_{ij} \qquad \{j = 1, 2, 3, \ldots, c\} \quad (13.2)$$

Next, a **grand mean** $\bar{\bar{x}}$ is computed:

$$\text{grand mean} = \bar{\bar{x}} = \frac{1}{n_T} \sum_{j=1}^{c} \sum_{i=1}^{n_j} x_{ij} \qquad (13.3)$$

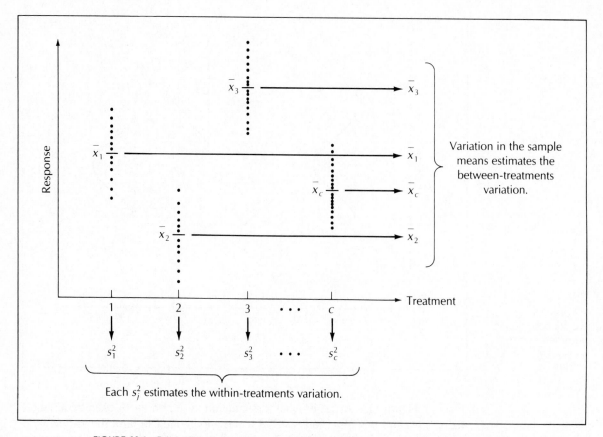

FIGURE 13.6 Estimating between- and within-treatments variation in one-way ANOVA

where $n_T = n_1 + n_2 + n_3 + \cdots + n_c$. In simpler terms, the grand mean is just the average of all n_T sample values.

The **between-treatments variation** (denoted **SSB**) is then estimated by

$$SSB = n_1(\overline{x}_1 - \overline{\overline{x}})^2 + n_2(\overline{x}_2 - \overline{\overline{x}})^2 + n_3(\overline{x}_3 - \overline{\overline{x}})^2 + \cdots + n_c(\overline{x}_c - \overline{\overline{x}})^2 \quad (13.4)$$

The notation SSB (read 'sum of squares between treatments') is meant to be a reminder that the expression in equation (13.4) is a sum of squares. Similarly, the **within-treatments variation,** which is also thought of as the *error* variation, is denoted by **SSE** and calculated by pooling the sample variations:

$$SSE = \sum_{i=1}^{n_1} (x_{i1} - \overline{x}_1)^2 + \sum_{i=1}^{n_2} (x_{i2} - \overline{x}_2)^2 + \cdots + \sum_{i=1}^{n_c} (x_{ic} - \overline{x}_c)^2 \quad (13.5)$$

It is also possible to derive an estimate of the **total variation (SST)** of all the n_T readings:

$$SST = \sum_{j=1}^{c} \sum_{i=1}^{n_j} (x_{ij} - \overline{\overline{x}})^2 \quad (13.6)$$

It can be shown mathematically that SSB, SSE, and SST are always related by the expression

$$SST = SSB + SSE \quad (13.7)$$

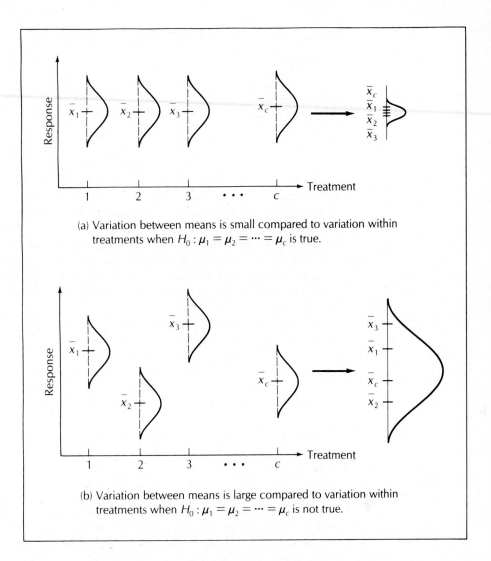

(a) Variation between means is small compared to variation within treatments when $H_0 : \mu_1 = \mu_2 = \cdots = \mu_c$ is true.

(b) Variation between means is large compared to variation within treatments when $H_0 : \mu_1 = \mu_2 = \cdots = \mu_c$ is not true.

FIGURE 13.7
How the ANOVA
method works

Equation (13.7) is called the *decomposition* of the total variation into components SSB (due to the differences between the $\bar{x}_j$'s) and SSE (the natural error variation around each population mean). Roughly speaking, the larger SSB, the more evidence one has for rejecting H_o and concluding that there is a difference between the treatments.

THE ANOVA TABLE.   A formal hypothesis test is conducted by organizing the above results into an **ANOVA table** as shown in Table 13.1. Computational formulas are available for efficiently calculating SSB, SSE, and SST, although one commonly relies on computer software to perform this task (as we do in this text). The reader is referred to other texts for these shortcut formulas and for more detailed analyses of the ANOVA assumptions and tests (Dixon and Massey 1983, Snedecor and Cochran 1989).

TABLE 13.1 ANOVA Table for the General One-Way Design

Source of variation	d.f.	SS	MS	F
Between treatments	$c - 1$	SSB	MSB	MSB/MSE
Within treatments (error)	$n_T - c$	SSE	MSE	
Total variation	$n_T - 1$	SST		

Note: $n_T = n_1 + n_2 + n_3 + \cdots + n_c$

Each sum of squares in equations (13.4) and (13.5) must be *averaged* by dividing by a factor related to the number of terms in the associated summation. This makes intuitive sense; otherwise, one could inflate or deflate SSB and SSE merely by changing the number of treatments or by changing the sample sizes n_j. These divisors not only take into account the number of terms summed, but also adjust for any restrictions that the ANOVA method imposes on these sums. In essence, the more restrictions, the fewer *effective* terms there are in the sum of squares. The statistical name for such a divisor is the **degrees of freedom** associated with the particular sum of squares.

For example, SSB contains c terms in its summation, but there are effectively only $c - 1$ terms. Similarly, SSE effectively has $n_T - c$ terms in its summation. In general, it can be shown mathematically that the number of terms in a sum of squares can be *reduced* by the number of restrictions that a statistical procedure places on the data. The reader may consult more theoretical texts for this development (Brownlee 1965). The following example demonstrates how terms are 'lost' in a sum of squares.[4]

Recall that the sample variance (see Section 3.3),

$$s^2 = \frac{1}{n - 1} \sum_{i=1}^{n} (x_i - \bar{x})^2$$

has a divisor (i.e., degrees of freedom) of $n - 1$, not n. The reason is that $\bar{x}$, not μ, is used to calculate the squared deviations in the formula. Using the data to calculate $\bar{x}$ and then using $\bar{x}$ along with the very same data in the formula for s^2 effectively reduce the number of terms in the summation. This can be seen most easily for the case $n = 2$:

$$\sum_{i=1}^{2} (x_i - \bar{x})^2 = (x_1 - \bar{x})^2 + (x_2 - \bar{x})^2$$

$$= \left(x_1 - \frac{x_1 + x_2}{2} \right)^2 + \left(x_2 - \frac{x_1 + x_2}{2} \right)^2$$

$$= \left(\frac{x_1 - x_2}{2} \right)^2 + \left(\frac{x_2 - x_1}{2} \right)^2 = \frac{1}{2}(x_1 - x_2)^2$$

That is, the two terms $(x_1 - \bar{x})^2$ and $(x_2 - \bar{x})^2$ in the sum can effectively be replaced by just *one* term, $\frac{1}{2}(x_1 - x_2)^2$.

---

[4]This example was originally described in Mosteller and Rourke (1973).

The sums of squares, averaged by their degrees of freedom, are called **mean squares.** They estimate the between- and within-treatments variation illustrated in Figure 13.6. Their ratio, MSB/MSE, measures the extent to which the data support or refute H_o. Under the ANOVA assumptions, that treatment responses have normal distributions with the same variances, MSB/MSE can be shown to follow an F distribution with degrees of freedom $(c - 1, n_T - c)$. To conduct a test of the hypotheses in equation (13.1), the following procedure is used:

ONE-WAY ANALYSIS
OF VARIANCE TEST

1. From the number of treatments, c, and the sample sizes, $n_1, n_2, \ldots, n_c$, find F_α, the upper $(1 - \alpha) \cdot 100\%$ point for an F distribution with $(c - 1, n_T - c)$ degrees of freedom (see Appendices 6, 7, and 8).
2. If $F = $ MSB/MSE $> F_\alpha$, conclude with $(1 - \alpha) \cdot 100\%$ confidence that there is a difference in the treatment means. Otherwise, conclude that the data do not indicate that a difference between the treatment means exists.

EXAMPLE 13.1

Numerous factors contribute to the smooth running of an electric motor (Anand 1991, pp. 361–369). In particular, it is desirable to keep motor noise and vibration to a minimum. To study the effect that the brand of bearing has on motor vibration, five different bearing brands are examined. Each type of bearing is installed in six individual motors and, while all 30 motors are running, the amount of vibration in each is measured (in microns). The data for this study are given in Table 13.2. With vibration considered as the response and each brand as a different *treatment,* a one-way ANOVA test is conducted. A *Minitab*

TABLE 13.2   Vibration (in microns) in Five Groups of Electric Motors, with Each Group Using a Different Bearing Brand

Brand 1	Brand 2	Brand 3	Brand 4	Brand 5
13.1	16.3	13.7	15.7	13.5
15.0	15.7	13.9	13.7	13.4
14.0	17.2	12.4	14.4	13.2
14.4	14.9	13.8	16.0	12.7
14.0	14.4	14.9	13.9	13.4
11.6	17.2	13.3	14.7	12.3

printout of this ANOVA test is shown in Figure 13.8. Since $n_T = n_1 + n_2 + \cdots + n_5 = 30$, the degrees of freedom for the F test are $(c - 1, n_T - c) = (4, 25)$. Using a significance level of $\alpha = 0.01$, Appendix 8 gives a critical value of $F_{0.01}(4, 25) = 4.18$. Since the ratio $F = $ MSB/MSE $= 7.714/0.914 = 8.44$

```
ANALYSIS OF VARIANCE
SOURCE      DF        SS        MS        F        p
FACTOR       4      30.855    7.714     8.44    0.000
ERROR       25      22.838    0.914
TOTAL       29      53.694

                                        INDIVIDUAL 95 PCT CI'S FOR MEAN
                                        BASED ON POOLED STDEV
LEVEL       N       MEAN      STDEV   ---------+---------+---------+-----
C1          6      13.683     1.194       (-----*-----)
C2          6      15.950     1.167                        (-----*-----)
C3          6      13.667     0.816     (-----*-----)
C4          6      14.733     0.940            (-----*-----)
C5          6      13.083     0.479   (-----*-----)

POOLED STDEV =     0.956                ---------+---------+---------+-----
                                           13.5      15.0      16.5
```

FIGURE 13.8
Analysis of variance
for the data of
Table 13.2

exceeds $F_{0.01}$, it can be concluded that there is a significant difference in the average vibration caused by the different brands of bearings.[5] In particular, it appears that bearings of brands 1, 3, and 5 are the best for reducing vibration.

ANALYSIS OF VARIANCE—TWO FACTORS

TWO-WAY ANOVA—NO REPEATED MEASURES. The ANOVA decomposition in equation (13.7) in one-factor analyses can be extended to analyses of more than one factor. This section considers the extension to the two-factor case. As a general notation, the factors are labeled I and II, where factor I is examined at c different levels and factor II at k levels.

The layout of a two-way classification is shown in Figure 13.9. Technically, this is called a two-way **crossed** design, since every level of factor I is paired (or 'crossed') with every level of factor II. There are ck cells in the two-way table, each containing one or more measurements on the particular response variable. Figure 13.9 shows the case of a single observation per cell. The case of multiple observations per cell is discussed later.

As in the one-way classification, sums of squares are used to detect differences in the levels of a factor. As a reference point, the grand mean $\bar{\bar{x}}$ of all the readings is found:

$$\bar{\bar{x}} = \frac{1}{ck} \sum_{i=1}^{k} \sum_{j=1}^{c} x_{ij} \tag{13.8}$$

---

[5] The test could also have been conducted by comparing the p value in the printout ($p = 0.000$) to the significance level $\alpha = 0.01$. When $p < \alpha$, H_O is rejected.

Factor I

	1	2	3	$\cdots$	c	
1	x_{11}	x_{12}	x_{13}	$\cdots$	x_{1c}	$\bar{B}_1$
2	x_{21}	x_{22}	x_{23}	$\cdots$	x_{2c}	$\bar{B}_2$
3	x_{31}	x_{32}	x_{33}	$\cdots$	x_{3c}	$\bar{B}_3$
$\vdots$	$\vdots$	$\vdots$	$\vdots$	$\vdots$	$\vdots$	$\vdots$
k	x_{k1}	x_{k2}	x_{k3}	$\cdots$	x_{kc}	$\bar{B}_k$
	$\bar{A}_1$	$\bar{A}_2$	$\bar{A}_3$	$\cdots$	$\bar{A}_c$	$\bar{\bar{x}}$

Factor I means: $\bar{A}_j$ is the mean of the readings in column j.
Factor II means: $\bar{B}_i$ is the mean of the readings in row i.
Grand mean: $\bar{\bar{x}}$ is the mean of all kc readings.

FIGURE 13.9
General layout of a
two-way analysis of
variance with one
observation per cell

Then the sum of squares for factor I is

$$SS_I = k \sum_{j=1}^{c} (\bar{A}_j - \bar{\bar{x}})^2 \tag{13.9}$$

which has an associated **$c - 1$ degrees of freedom.** The sum of squares for factor II is computed from

$$SS_{II} = c \sum_{i=1}^{k} (\bar{B}_i - \bar{\bar{x}})^2 \tag{13.10}$$

an expression that has **$k - 1$ degrees of freedom.** The calculation of the total sum of squares, SST, is similar to that used earlier. SST is just the sum of squares of all ck readings from the grand mean $\bar{\bar{x}}$:

$$SST = \sum_{i=1}^{k} \sum_{j=1}^{c} (x_{ij} - \bar{\bar{x}})^2 \tag{13.11}$$

A final sum of squares, the sum of squares for error, SSE, can be found from the fundamental ANOVA decomposition:

$$SST = SS_I + SS_{II} + SSE \tag{13.12}$$

The degrees of freedom associated with SSE are the product of those for the two factors—that is, $(c - 1) \cdot (k - 1)$.

TABLE 13.3   ANOVA Table for the General Two-Way Crossed Design With One Observation Per Cell

Source of variation	d.f.	SS	MS	F
Factor I	$c - 1$	SS_I	MS_I	MS_I/MSE
Factor II	$k - 1$	SS_{II}	MS_{II}	MS_{II}/MSE
Error	$(c - 1)(k - 1)$	SSE	MSE	
Total variation	$ck - 1$	SST		

After the sums of squares are averaged by their corresponding degrees of freedom, the results are collected in an ANOVA table as shown in Table 13.3. The mean squares for the two factors are compared to the mean square for error exactly as in the one-factor case, and F tests are run for each factor.

TWO-WAY ANALYSIS
OF VARIANCE TEST

1. From the c levels of factor I and k levels of factor II, find F_α, the upper $(1 - \alpha) \cdot 100\%$ points for F distributions with $[c - 1, (c - 1)(k - 1)]$ and $[k - 1, (c - 1)(k - 1)]$ degrees of freedom (see Appendices 6, 7, and 8).

2. If $F = MS_I/MSE > F_\alpha[c - 1, (c - 1)(k - 1)]$, conclude with $(1 - \alpha) \cdot 100\%$ confidence that there is a significant difference in the levels of factor I. If $F = MS_{II}/MSE > F_\alpha[k - 1, (c - 1)(k - 1)]$, conclude that significant differences exist between the levels of factor II.

EXAMPLE 13.2

In the study of electric motor vibration in Example 13.1, suppose that, for each brand of bearing, six different motor casings are tested; that is, suppose that the six readings in column 1 of Table 13.2 are the result of testing brand 1 bearings on motors with six different types of casing (steel casing, aluminum casing, plastic casing, etc.). Similarly, brand 2 bearings are used in an additional six motors, and so forth. In this type of experiment, 'brand' is factor I and 'casing' is factor II. The numbers of levels of these factors are $c = 5$ and $k = 6$, respectively. With *Minitab*, a two-way ANOVA is performed (Figure 13.10). For a significance level $\alpha = 0.01$, the critical F values for testing the two factors are

$$F_{0.01}[c - 1, (c - 1)(k - 1)] = F_{0.01}(4, 20) = 4.43$$

$$F_{0.01}[k - 1, (c - 1)(k - 1)] = F_{0.01}(5, 20) = 4.10$$

Only factor I is significant at $\alpha = 0.01$, since $MS_I/MSE = 7.14 > F_{0.01}(4, 20)$. Thus, if the experiment is run in this fashion, one can conclude that some bearing brands have significantly different effects on motor vibration, while the different casing materials do not differ in their effect on vibration.

```
              ANALYSIS OF VARIANCE of vibration

              SOURCE         DF        SS        MS
              FactorI         4      30.86      7.71
              FactorII        5       1.31      0.26
              ERROR          20      21.53      1.08
              TOTAL          29      53.69


           Factor I :  Brand of Bearing Used
           Factor II:  Casing Material
```

FIGURE 13.10
Two-way ANOVA
table for the data of
Table 13.2

TWO-WAY ANOVA—REPEATED MEASURES AND INTERACTION. If possible, two-factor designs should be augmented by running *additional* tests in each cell. By doing this, it becomes possible to test for the presence of **interactions** between the factors. Factors are said to **interact** when the effect of one factor is influenced by the particular level(s) of another factor. For instance, in Example 13.2, if a 'brand–casing' interaction was present, it would mean that the different brands' ability to reduce motor vibration depends on which casings are used. That is, the difference between the effects of brand 1 and brand 2 on vibration might be large for motors with plastic casings, but small for motors with steel casings.

The general two-factor design with *n repeated* measures per cell is shown in Figure 13.11. The sums of squares for the two factors, the error variation, and the total variation are given by

$$SS_I = nk \sum_{j=1}^{c} (\overline{A}_j - \overline{\overline{x}})^2 \tag{13.13}$$

$$SS_{II} = nc \sum_{i=1}^{k} (\overline{B}_i - \overline{\overline{x}})^2 \tag{13.14}$$

$$SSE = \sum_{j=1}^{c} \sum_{i=1}^{k} \sum_{r=1}^{n} (x_{ijr} - \overline{\overline{x}}_{ij})^2 \tag{13.15}$$

$$SST = \sum_{j=1}^{c} \sum_{i=1}^{k} \sum_{r=1}^{n} (x_{ijr} - \overline{\overline{x}})^2 \tag{13.16}$$

where $\overline{x}_{ij}$ denotes the mean of the n readings in cell ij (i.e., the cell for factor I at level j and factor II at level i). The fundamental decomposition of SST for the two-way repeated measures design is

$$SST = SS_I + SS_{II} + SS_{Int} + SSE \tag{13.17}$$

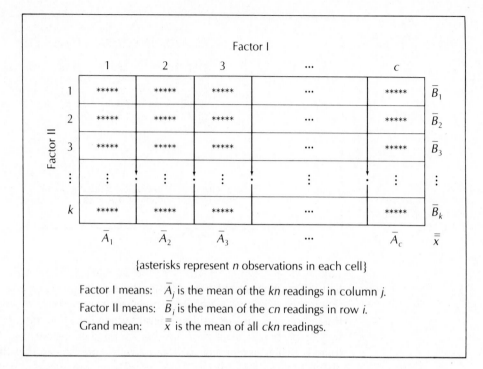

FIGURE 13.11
General layout of a
two-way analysis of
variance with *n*
observations per cell

where SS_{Int} denotes the sum of squares due to interaction between the factors. From this equation, SS_{Int} can be calculated by subtracting SS_I, SS_{II}, and SSE from SST. These results are summarized in the usual ANOVA table format in Table 13.4.

As noted in Section 10.7, before the *F* tests are conducted, a decision must be made concerning the exact hypotheses under consideration. If the levels of each factor are the *only* ones of interest in the experiment, then the design is said to be a **fixed-effects model.** If the levels of the factors are considered to be only *samples* from the possible levels of each factor, then the design is a **random-effects model.** In this chapter, we consider only fixed-effects models. A more complete discussion of the differences between the two models can be found in Duncan (1986). Operationally, the differences between random- and fixed-effects models appear in the construction of the *F* tests. For the fixed-effects case, *F* ratios are found for each factor, and for the interaction term, by dividing their mean squares by the mean square error, MSE. In a random-effects model, only factor mean squares are divided by the mean square for *interaction,* MS_{Int}, to form the *F* ratios. Table 13.4 shows general ANOVA tables for both fixed- and random-effects models.

EXAMPLE 13.3   Section 10.7 presented a repeatability study for separating out the effects of operator variation, product variation, instrument variation, and part–operator variation. The design for such a study, depicted in Figure 10.14, is exactly that of the two-way repeated measures design shown in Figure 13.11. In the repeatability context, factor I is 'operator' and factor II is 'part,' where each part is measured *n* times by each operator. In Section 10.7, operators and parts were considered

TABLE 13.4   ANOVA Table for a Two-Way Crossed Design With n Observations Per Cell

Source of Variation	Fixed-Effects Model d.f.	SS	MS	F
Factor I	$c - 1$	SS_I	MS_I	MS_I/MSE
Factor II	$k - 1$	SS_{II}	MS_{II}	MS_{II}/MSE
Interaction	$(c - 1)(k - 1)$	SS_{Int}	MS_{Int}	MS_{Int}/MSE
Error	$ck(n - 1)$	SSE	MSE	
Total variation	$ckn - 1$	SST		

The differences between the ANOVA tables of the fixed- and random-effects models occur in their F ratio calculations.

Source of Variation	Random-Effects Model d.f.	SS	MS	F
Factor I	$c - 1$	SS_I	MS_I	MS_I/MS_{Int}
Factor II	$k - 1$	SS_{II}	MS_{II}	MS_{II}/MS_{Int}
Interaction	$(c - 1)(k - 1)$	SS_{Int}	MS_{Int}	MS_{Int}/MSE
Error	$ck(n - 1)$	SSE	MSE	
Total variation	$ckn - 1$	SST		

to be *samples* of the possible operators and parts that could occur in production; that is, a *random-effects* model was assumed. As mentioned, the F ratios for testing each factor and the interaction term are slightly different from those for the fixed-effects model. These F ratios are shown in the ANOVA table of Figure 10.15. Note that, for the random-effects model, factor mean squares are divided by the *interaction mean square*:

$$F \text{ ratio for factor I} = \frac{MS_I}{MS_{Int}}$$

$$= \frac{0.214207}{0.003523} = 60.80$$

$$F \text{ ratio for factor II} = \frac{MS_{II}}{MS_{Int}}$$

$$= \frac{0.005249}{0.003523} = 1.49$$

whereas the F ratio for the interaction term uses a denominator of MSE:

$$F \text{ ratio for interaction} = \frac{MS_{Int}}{MSE}$$

$$= \frac{0.003523}{0.000505} = 6.98$$

Although the interaction between parts and operators appears to be statistically significant, Example 10.9 showed that the magnitude of this effect is small compared to the product variation.

EXAMPLE 13.4 One-factor and multifactor ANOVA can easily be performed as a regression analysis. For example, the general one-way analysis with c treatments can be written as a regression analysis of the response variable, now denoted by Y, on $c - 1$ *indicator variables:*

$$X_1 = \begin{cases} 1 & \text{if data are from treatment 1} \\ 0 & \text{if data are not from treatment 1} \end{cases}$$

$$X_2 = \begin{cases} 1 & \text{if data are from treatment 2} \\ 0 & \text{if data are not from treatment 2} \end{cases}$$

$$\vdots$$

$$X_{c-1} = \begin{cases} 1 & \text{if data are from treatment } c - 1 \\ 0 & \text{if data are not from treatment } c - 1 \end{cases}$$

Only $c - 1$ indicator variables are needed, since data from treatment c correspond to the case $X_1 = X_2 = \cdots = X_{c-1} = 0$. Coded in this manner, the data from Example 13.1 take the form shown in Table 13.5. A printout of a regression of Y on $X_1, X_2, \ldots, X_{c-1}$ is given in Figure 13.12. Notice that the ANOVA table in Figure 13.12 exactly matches the ANOVA table in Figure 13.8; that is, the variation due to the regression model is identical to the between-treatments variation.

In addition, the regression coefficients can always be interpreted relative to the c treatment means. For example, data from treatment 5 are coded $X_1 = X_2 = \cdots = X_4 = 0$, which can be substituted into the regression model to obtain $\mu_5 = E(Y|X_1 = 0, X_2 = 0, \ldots, X_4 = 0) = \beta_0 + \beta_1(0) + \beta_2(0) + \cdots + \beta_4(0) = \beta_0$ or, more simply, $\mu_5 = \beta_0$.[6] Similarly, for treatment 1, $\mu_1 = E(Y|X_1 = 1, X_2 = 0, \ldots, X_4 = 0) = \beta_0 + \beta_1(1) + \beta_2(0) + \cdots + \beta_4(0) = \beta_0 + \beta_1$, or $\beta_1 = \mu_1 - \beta_0 = \mu_1 - \mu_5$. In general, for the coding scheme used above, one always has $\beta_0 = \mu_c$, $\beta_1 = \mu_1 - \mu_c$, $\beta_2 = \mu_2 - \mu_c$, $\ldots$, $\beta_{c-1} = \mu_{c-1} - \mu_c$. This means that the coefficients $\hat{\beta}_1, \hat{\beta}_2, \ldots, \hat{\beta}_{c-1}$ are estimates of $\mu_1 - \mu_c, \mu_2 - \mu_c, \ldots, \mu_{c-1} - \mu_c$. In fact, it can be shown that $\hat{\beta}_0 = \bar{x}_c, \hat{\beta}_1 = \bar{x}_1 - \bar{x}_c, \hat{\beta}_2 = \bar{x}_2 - \bar{x}_c, \ldots, \hat{\beta}_{c-1} = \bar{x}_{c-1} - \bar{x}_c$. As an example, compare the sample means of Figure 13.8 to the regression coefficients of Figure 13.12.

If the c treatment means are equal [i.e., if the hypothesis H_0 of equation (13.1) is true], then the $\hat{\beta}_i$'s should be close to 0. Otherwise, if at least one of the $\hat{\beta}_i$'s is statistically significant, then it can be concluded that there is a difference in the treatment means.

EXAMPLE 13.5 The coding scheme in Example 13.4 can also be used for multifactor ANOVA (with or without repeated measures per cell). For example, in two-way ANOVA with c levels of factor I and k levels of factor II, one simply uses $c - 1$ indicator variables $X_1, X_2, X_3, \ldots, X_{c-1}$ for factor 1, and $k - 1$ indicator variables $Z_1, Z_2, \ldots, Z_{k-1}$ for factor 2. To model the interaction between the two factors, all cross-products of the form $X_i Z_j$ are also included in the regression model. For a more in-depth presentation of the regression approach to ANOVA designs, the reader is referred to the text by Mendenhall and McClave (1981).

---

[6] Recall that regression models express the **conditional** mean $E(Y|X_1, x_2, \ldots, X_c)$ in terms of the independent variables.

TABLE 13.5   Data of Table 13.2 Coded for a Regression Analysis

i	Y_i	$X1_i$	$X2_i$	$X3_i$	$X4_i$
1	13.1	1	0	0	0
2	15.0	1	0	0	0
3	14.0	1	0	0	0
4	14.4	1	0	0	0
5	14.0	1	0	0	0
6	11.6	1	0	0	0
7	16.3	0	1	0	0
8	15.7	0	1	0	0
9	17.2	0	1	0	0
10	14.9	0	1	0	0
11	14.4	0	1	0	0
12	17.2	0	1	0	0
13	13.7	0	0	1	0
14	13.9	0	0	1	0
15	12.4	0	0	1	0
16	13.8	0	0	1	0
17	14.9	0	0	1	0
18	13.3	0	0	1	0
19	15.7	0	0	0	1
20	13.7	0	0	0	1
21	14.4	0	0	0	1
22	16.0	0	0	0	1
23	13.9	0	0	0	1
24	14.7	0	0	0	1
25	13.5	0	0	0	0
26	13.4	0	0	0	0
27	13.2	0	0	0	0
28	12.7	0	0	0	0
29	13.4	0	0	0	0
30	12.3	0	0	0	0

2^k FACTORIAL DESIGNS

FACTORIAL DESIGNS.   The experimental designs most frequently used by quality practitioners come from the class of **factorial designs.** Recall the analogy of dials whose settings control a process characteristic. It is convenient to think of the myriad factors affecting a process as dials whose optimal settings are sought. For a given factor, the various settings considered in an experiment are called its **factor levels.** Factor levels can be either different *values* of a measurable characteristic (e.g., different temperature settings) or different *categories* of a qualitative characteristic (e.g., different brands of a raw material).

To find the minimum number of experimental runs required for a complete factorial experiment, one simply multiplies the numbers of factor levels. If factor 1 has c_1 levels, factor 2 has c_2 levels, ..., and factor k has c_k levels, then the experiment is called a "c_1 by c_2 by...by c_k" factorial design and denoted $c_1 \times c_2 \times c_3 \times \cdots \times c_k$. This notation is convenient because it indicates how many

```
The regression equation is
vibrate = 13.1 + 0.600 X1 + 2.87 X2 + 0.583 X3 + 1.65 X4

Predictor        Coef       Stdev     t-ratio       p
Constant      13.0833      0.3902       33.53     0.000
X1             0.6000      0.5518        1.09     0.287
X2             2.8667      0.5518        5.19     0.000
X3             0.5833      0.5518        1.06     0.301
X4             1.6500      0.5518        2.99     0.006

s = 0.9558      R-sq = 57.5%      R-sq(adj) = 50.7%

Analysis of Variance

SOURCE       DF         SS          MS        F        p
Regression    4    30.8553      7.7138     8.44    0.000
Error        25    22.8383      0.9135
Total        29    53.6937

SOURCE       DF      SEQ SS
X1            1      2.1870
X2            1     20.2672
X3            1      0.2336
X4            1      8.1675

Unusual Observations
Obs.      X1    vibrate      Fit Stdev.Fit  Residual   St.Resid
  6     1.00    11.600    13.683    0.390    -2.083      -2.39R

R denotes an obs. with a large st. resid.
```

FIGURE 13.12
Regression analysis of
data in Table 13.5

factors there are and the number of factor levels in each factor, and the multiplication signs remind us that the minimum number of experimental runs required will be the product of the c_i's. Thus, a $3 \times 2 \times 4$ factorial design involves 3 factors with 3, 2, and 4 levels each, and at least $3 \cdot 2 \cdot 4 = 24$ runs will be needed to conduct the study of these factors.

When one is starting a study, it is desirable to test as large a number of different factors as possible. However, because of the multiplication formula for the number of experimental runs, the size of the study can easily exceed the available resources if there are too many levels of each factor. One way of limiting the number of runs is to use only two levels for each factor. Thus, if k two-level factors are studied, then at least $2 \times 2 \times \cdots \times 2 = 2^k$ experimental runs will be needed. Designs of this type are called **2^k factorial designs.**

CODING SCHEMES FOR FACTOR LEVELS.   With exactly two levels of each factor, the ensuing analysis is particularly simple, making these designs well suited to the production environment. By taking advantage of the special structures in a 2^k design, we can evaluate factor effects and interactions even *without* the aid of a computer.

Various schemes are used to denote the two levels of each factor, the most useful employing 'plus' and 'minus' signs.[7] In this scheme, for measurable characteristics one generally uses the $+$ sign to denote the *larger* of the two measured values, with $-$ denoting the smaller value. For example, if the factor 'temperature' is studied at the two levels 60°F and 80°F, then we normally use $-$ to denote the factor level of 60°F and $+$ for the 80°F level. For qualitative characteristics, either factor level may receive the $+$ sign, with the other level receiving the $-$ sign.

With the factors of a 2^k design denoted by $F_1, F_2, F_3, \ldots, F_k$, the **design matrix** of the experiment is constructed in the manner shown in Table 13.6. Starting with factor F_1, a string of 2^k alternating $-$ and $+$ signs is put in the column under F_1. Next, for F_2, the $-$ and $+$ signs are entered in alternating blocks of *two* each. Factor F_3 uses alternating blocks of *four* $-$ and $+$ signs each, and, in general, factor F_i has alternating blocks of 2^{i-1} $-$ and $+$ signs. The last factor, F_k, always contains a string of 2^{k-1} $-$ signs followed by 2^{k-1} $+$ signs. Each of the 2^k rows in the matrix dictates the settings of a particular experimental run in the study.

To conduct the study, the 2^k experimental runs in the design matrix should be run in *random* order if possible. By **randomizing** the runs, one minimizes the effects of possible biases. Suppose, for example, that the runs are conducted in **standard order** (i.e., the order listed in the design matrix), half of them in the morning and the other half in the afternoon on the day of the study. This means that factor k is always at its low level in the morning and at its high level in the afternoon. Therefore, if some factor *not* included in the study happens to influence the process differently in the morning than in the afternoon, its effect would be attributed (by the design matrix) to F_k, even though F_k may itself not even influence the process. Randomization helps prevent the effects of extraneous variables from being confused with the effects of the factors used in the study.

TABLE 13.6 Design Matrix for a 2^k Factorial Design

| Run | Factors | | | | | | Response |
	F_1	F_2	F_3	F_4	$\cdots$	F_k	y
1	$-$	$-$	$-$	$-$		$-$	y_1
2	$+$	$-$	$-$	$-$		$-$	y_2
3	$-$	$+$	$-$	$-$		$-$	y_3
4	$+$	$+$	$-$	$-$		$-$	y_4
5	$-$	$-$	$+$	$-$		$-$	y_5
6	$+$	$-$	$+$	$-$		$-$	y_6
7	$-$	$+$	$+$	$-$		$-$	y_7
8	$+$	$+$	$+$	$-$		$-$	y_8
$\vdots$	$\vdots$	$\vdots$	$\vdots$	$\vdots$		$\vdots$	$\vdots$
2^k							y_{2^k}

---

[7] Another notation uses letters $a, b, c, \ldots$ to denote 'high' levels of factors $A, B, C, \ldots$, and **no** letter to denote their low levels. Thus, the combination '*bdf*' denotes a run with factors A and C set at their low levels and factors $B, D,$ and F set at their high levels.

MAIN EFFECTS AND INTERACTION EFFECTS.   The goal of the 2^k study is to determine which factors have significant effects on a process characteristic, and whether there are any interactions between the factors. The **main effect** of factor F_i is defined to be the average response in the experimental runs where F_i has its + setting, minus the average response in all runs where F_i is set at $-$. From the way the design matrix is structured, the runs where F_i is + will include all combinations of levels of the *other* factors, which is also the case for the runs where F_i is $-$. In this sense, the average 'effect' of changing F_i from its low setting to its high setting is estimated. The **interaction** between two factors, F_i and F_j, is also measured by subtracting two averages: the average effect of F_i for F_j set at its *low* level is subtracted from the average effect of F_i when F_j is at its *high* level. Whenever a significant interaction exists between two (or more) factors, the individual factors that make up that interaction cannot be simply interpreted by adding their separate main effects. From the dial analogy again, if two factors A and B interact, then the effect of turning dial A will be *different* for each particular setting of dial B.

It is also possible to measure three-factor, four-factor, on up to k-factor interactions, with the definitions becoming more tedious as k increases. Fortunately, though, the + and $-$ notation is designed to make the computational work easy and mechanical. The first step is to create columns for the various interactions. These columns are appended to the right side of the design matrix. To illustrate, Table 13.7 shows a 2^4 design, with some of its interaction columns completed. To avoid subscripts, the factors are simply denoted as A, B, C, and D. To fill in the appropriate + and $-$ signs in these extra columns, one simply multiplies the corresponding signs in the factor columns that make up the particular interaction.[8] Thus, the column for the AB interaction is simply the 'product' of the signs in columns A and B. The second and final step is to calculate the main effects and interactions by affixing each column's signs to the response column, calculating the resulting sums (called **contrasts**), and dividing each sum by 2^{k-1}. Thus, in Table 13.7, the AB interaction is found by putting the signs in the AB column on the readings in the response column and then summing and dividing the result by 8. The next example illustrates this procedure.

EXAMPLE 13.6   In our earlier discussion of one-factor-at-a-time experiments, spray volume and belt speed were two factors thought to affect surface uniformity in a chemical coating process. Suppose that the brand of chemical is also under consideration. Table 13.8 shows the result of a 2^3 factorial experiment for assessing the effects of these three factors. Spray volume (V) is tested at two levels, high (+) and low ($-$). Similarly, belt speed (S) is studied at two particular settings, a high one (+) and a low one ($-$). Finally, two brands of chemical are used, brand I (+) and brand II ($-$). For convenience, Table 13.8 also shows the interaction columns for this design. The contrast for spray volume is found by affixing the

---

[8]The familiar rules "plus times plus is plus," "minus times minus is plus," and "plus times minus is minus" are used to find the signs for the interaction columns.

signs in column V to the response column: $-40 + 25 - 30 + 50 - 45 + 25 - 30 + 52 = 7$. The main effect for spray volume is then $7/2^{3-1} = 1.75$. Similarly, the interaction contrast for factors V and S is $+40 - 25 - 30 + 50 + 45 - 25 - 30 + 52 = 77$, so the VS interaction effect is $77/2^{3-1} = 19.25$. The remaining effects are found in the same fashion and are given in Table 13.9.

TABLE 13.7   Creating the Interaction Columns for a 2^4 Design

Run	Factors				Interactions										Response
	A	B	C	D	AB	AC	AD	BC	BD	CD	ABC	ABD	BCD	$ABCD$	y
1	−	−	−	−	+						−			+	y_1
2	+	−	−	−	−						−			−	y_2
3	−	+	−	−	−						+			−	y_3
4	+	+	−	−	+						−			+	y_4
5	−	−	+	−	−						+			−	y_5
6	+	−	+	−	+						−			+	y_6
7	−	+	+	−	+						−			+	y_7
8	+	+	+	−	−						+			−	y_8
9	−	−	−	+	−						−			−	y_9
10	+	−	−	+	+						+			+	y_{10}
11	−	+	−	+	+						+			+	y_{11}
12	+	+	−	+	−						−			−	y_{12}
13	−	−	+	+	+						+			+	y_{13}
14	+	−	+	+	−						−			−	y_{14}
15	−	+	+	+	−						−			−	y_{15}
16	+	+	+	+	+						+			+	y_{16}

Notes: (1) To find the signs in an interaction column, multiply the corresponding signs in the individual factor columns. (2) For purposes of illustration, the signs in only three of the interaction columns (AB, ABC, and $ABCD$) are filled in.

TABLE 13.8   Design Matrix, Interaction Columns, and Response Column for the 2^3 Factorial Study in Example 13.6

Spray Volume, V	Belt Speed, S	Brand of Chemical, B	Interactions				Surface Uniformity, y
			VS	VB	SB	VSB	
−	−	−	+	+	+	−	40
+	−	−	−	−	+	+	25
−	+	−	−	+	−	+	30
+	+	−	+	−	−	−	50
−	−	+	+	−	−	+	45
+	−	+	−	+	−	−	25
−	+	+	−	−	+	−	30
+	+	+	+	+	+	+	52

Note: Larger y values correspond to more uniform surfaces.

TABLE 13.9   Contrasts and Effects for the Data of Example 13.6

Factor/Interaction	Contrast	Divisor	Effect
V	7	4	1.75
S	27	4	6.75
B	7	4	1.75
VS	77	4	19.25
VB	−3	4	−0.75
SB	−3	4	−0.75
VSB	7	4	1.75

ANOVA FOR 2^k DESIGNS.   In Example 13.6, it is possible to obtain estimates for every factor and every interaction, but there is *no* estimate of experimental error against which to evaluate these effects. This is always the case in a 2^k design (Caulcutt 1983, p. 177). There are two solutions to this problem. One method is to simply **replicate** the 2^k experiment n times, where $n \geq 2$; that is, each experimental condition is run n times, for a total of $n2^k$ runs. In this way, each combination of factor levels has its own set of n measurements from which an estimate of the error variation can be obtained. A second solution is to *assume* that various higher-order interaction terms are unimportant or negligible, so that their estimated effects can be combined into an estimate of the error variation. Both of these options are now discussed in greater detail.

A replicated 2^k factorial design is one in which all 2^k runs have been repeated n times, where $n \geq 2$. As a result, there are n response columns instead of 1. To estimate main effects and interactions, the n response columns are first summed, and then the sequences of + and − signs from the extended design matrix are applied to the column of sums, just as was done for the unreplicated design. The resulting **contrasts** are converted into main effects and interactions by dividing each by $n2^{k-1}$. Similarly, each main effect and interaction contributes its own sum of squares to the total treatment sum of squares:

$$\text{factor or interaction sum of squares} = \frac{(\text{contrast})^2}{n2^k} \qquad (13.18)$$

Table 13.10 summarizes these steps.

Because of the balanced nature of factorial designs, the sums of squares for each factor and interaction can be added to form the total treatment sum of squares. As always, the total variation, SST, is found by summing the squared deviations of all $n2^k$ measurements from their grand mean, and the usual ANOVA decomposition holds:

$$\text{SST} = \text{SS(treatments)} + \text{SS(error)} \qquad (13.19)$$

Therefore, if any of the higher-order interaction terms (especially the three-factor and above interactions) are thought to be negligible, then their contribution to the total variation can be shifted from SS(treatment) to SS(error). In fact, because the

TABLE 13.10   Main Effects, Interactions, and Sums of Squares for n Replications of a 2^k Factorial Design

Factors $F_1, F_2, F_3, \ldots, F_k$				Interactions $F_1 F_2, \ldots, F_1 F_2 \ldots F_k$			Response columns		Total
−	−	− ...	−	+	...	·	$y_{11}, y_{21}, \ldots, y_{n1}$	→	Y_1
+	−	− ...	−	−	...	·	$y_{12}, y_{22}, \ldots, y_{n2}$	→	Y_2
−	+	− ...	−	−	...	·	$y_{13}, y_{23}, \ldots, y_{n3}$	→	Y_3
+	+	− ...	−	+	...	·	$y_{14}, y_{24}, \ldots, y_{n4}$	→	Y_4
·	·	· ...	·	·	...	·	·	·	·
·	·	· ...	·	·	...	·	·	·	·
·	·	· ...	·	·	...	·	$y_{12^k}, y_{22^k}, \ldots, y_{n2^k}$	→	Y_{2^k}

$$\text{Factor or interaction } \textbf{contrast} = \left(\begin{array}{c}\text{factor or interaction}\\\text{column}\end{array}\right) \cdot \left(\begin{array}{c}\text{column of}\\\text{response totals}\end{array}\right)$$

$$\text{Factor or interaction } \textbf{effect} = \frac{\text{factor or interaction contrast}}{n2^{k-1}}$$

$$\text{Factor or interaction } \textbf{SS} = \frac{(\text{factor or interaction contrast})^2}{n2^k}$$

accumulated evidence of many studies suggests that higher-order interactions do not occur with great frequency, the method of relegating their sums of squares to SSE is often applied, especially when $k \geq 4$. The advantage of pooling the high-order interactions is that it then becomes possible to *avoid* replicating the experiment, since an estimate of SSE is available (i.e., SSE is the sum of squares of the higher-order interactions). Figure 13.13 summarizes the method of pooling interaction terms.

EXAMPLE 13.7

Main effects and interactions for spray volume, belt speed, and brand of chemical were computed in Example 13.6. However, with only one set of experimental runs in the 2^3 design, no SSE can be calculated and, hence, no F ratios can be formed to determine which effects are significant and which are not. Suppose, then, that a second set of runs is obtained (Table 13.11). With $n = 2$ measurements for each experimental condition, the error variation can be calculated by pooling the sample variances:

$$\text{SSE} = \frac{(n-1)s_1^2 + (n-1)s_2^2 + \cdots + (n-1)s_8^2}{(n-1) + (n-1) + \cdots + (n-1)} = \frac{s_1^2 + s_2^2 + \cdots + s_8^2}{8}$$

$$= 9.00$$

In general, the degrees of freedom $n - 1$ are used to weight each s_i^2 in the calculation, and the divisor is always 2^k. The degrees of freedom associated with SSE are $(n - 1)2^k$, which can be divided into SSE to obtain the mean square error, MSE. F ratios for each effect are formed by dividing each effect's mean square by the MSE. Since effect mean squares always have 1 d.f. in a

Factor/interaction effect	Factor/interaction sum of squares	d.f.
Main effects:		
F_1	$SS(F_1)$	1
F_2	$SS(F_2)$	1
⋮	⋮	⋮
F_k	$SS(F_k)$	1
Two-factor interactions:		
F_1F_2	$SS(F_1F_2)$	1
F_1F_3	$SS(F_1F_3)$	1
⋮	⋮	⋮
$F_{k-1}F_k$	$SS(F_{k-1}F_k)$	1
Three-factor interactions:		
$F_1F_2F_3$	$SS(F_1F_2F_3)$	1
⋮	⋮	⋮
$F_{k-2}F_{k-1}F_k$	$SS(F_{k-2}F_{k-1}F_k)$	1
⋮	⋮	⋮
k-factor interaction:		
$F_1F_2F_3\cdots F_k$	$SS(F_1F_2F_3\cdots F_k)$	1
	↓ Total SS	↓ Total d.f.

(Three-factor interactions and above) → MSE

FIGURE 13.13
Pooling high-order interactions

2^k design, the effect mean square is just the sum of squares for the effect. Each F ratio then has $[1, (n - 1)2^k]$ degrees of freedom. Table 13.11 summarizes these calculations. At a significance level of $\alpha = 0.05$, only the the main effects for belt speed (S) and the volume–speed interaction (VS) have significant F ratios. Brand of chemical is not significant, so it should not matter which brand is selected for production. Because of the significant VS interaction, the belt speed (S) does not have the simple interpretation that one might wish; that is, the main effect of increasing belt speed appears to be 6.00, but the *actual* effect depends on the spray volume setting. These results indicate that belt speed and spray volume are factors that should be considered *together* when choosing levels of either factor.

TABLE 13.11 Evaluating Main Effects and Interactions for Example 13.7

	Replication 1	Replication 2	Total
	40	36	76
	25	28	53
	30	32	62
	50	48	98
	45	43	88
	25	30	55
	30	29	59
	52	49	101

Factor/interaction	Effect	SS	d.f.	MS	F ratio
V	2.75	30.25	1	30.25	3.361
S	6.00	144.00	1	144.00	16.000
B	1.75	12.25	1	12.25	1.361
VS	16.75	1,122.25	1	1,122.25	124.691
VB	−0.50	1.00	1	1.00	0.111
SB	−1.75	12.25	1	12.25	1.361
VSB	2.00	16.00	1	16.00	1.778
Error		72.00	8	9.00	

13.2   ANALYSIS OF MEANS

The **analysis of means (ANOM)** is another method for testing for differences between k population or treatment means. The ANOM technique was invented by Ott in 1957–1958 (Ott 1957, 1958) for use with variables data and was subsequently extended to include attributes data (Ott 1960). ANOM differs from ANOVA in that the k sample means are compared directly to the grand mean of the data (or to a target mean, if one is given) *without* indirectly comparing two variation estimates. The ANOM procedure is graphically based and can be considered an extension of the Shewhart chart method.

Most of the time, the ANOM and ANOVA methods reach the same conclusions when applied to the same data sets, but there are some differences. ANOM is somewhat more sensitive than ANOVA for detecting when *one* mean differs significantly from the others, whereas ANOVA is more sensitive to detecting when *groups* of means differ (Ott 1967). Like ANOVA, the ANOM method also assumes that the k populations (treatments) have the *same* variance. Unlike ANOVA, ANOM can be applied only to **fixed-effects** models and *not* to **random-effects models** (Ramig 1983). Among the advantages of ANOM listed by Ott (1967), the following two are important:

ADVANTAGES OF
THE ANOM
PROCEDURE

> 1. ANOM is a *graphical* method that compares the grand mean to the individual treatment means; it is similar to the familiar Shewhart chart in its construction.
> 2. ANOM pinpoints which individual treatments have significant effects, whereas with ANOVA, detection of significant between-treatments variation must be followed by additional comparisons of sample means to determine which treatments are the important ones.

ANOM PROCEDURE—STANDARD GIVEN

As in ANOVA, ANOM tests the hypothesis $H_o; \mu_1 = \mu_2 = \mu_3 = \cdots = \mu_k = \mu$ against the alternative that at least one of the means differs from the rest.[9] The mechanics of ANOM are most easily illustrated in the case where both the common mean and variance, μ and σ^2, are known or can be specified. This is sometimes called the **standard given** case. In our discussion, it is also assumed that each of the k samples is of the same size, n.

The goal of the ANOM procedure is to find **decision lines** at a fixed distance on either side of μ, such that sample means falling outside the decision lines are the ones that are 'significantly different from the other means.' Figure 13.14 shows a typical ANOM chart for comparing k treatment means.

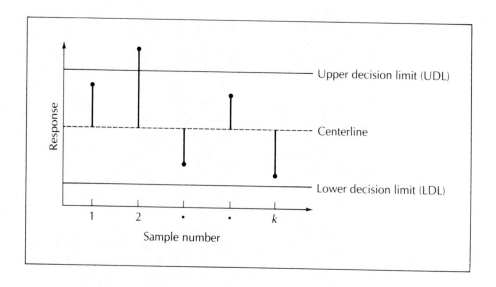

FIGURE 13.14
The ANOM chart

---

[9] Instead of using c to denote the number of treatments (as was done in Section 13.1), we use k in this discussion in order to adhere to the traditional ANOM notation.

To find the decision lines, a risk level α must be chosen. If H_o is true, then α is the probability of a false positive; that is, α is the probability that at least one of the k means will fall outside the decision lines when, in fact, the k means $\mu_1, \mu_2, \ldots, \mu_k$ are equal. Letting p denote the probability that any *one* mean falls inside the decision lines, we have

$$\alpha = 1 - p^k \tag{13.20}$$

which, when solved for p, gives

$$p = (1 - \alpha)^{1/k} \tag{13.21}$$

We assume that all treatment populations can be described by a *normal* distribution with mean μ and variance σ^2, so the percentile of the standard normal distribution corresponding to an upper tail area of $(1 - p)/2$ can be found (Figure 13.15). With this z value denoted by *uppercase Z_α*, the decision lines of the ANOM chart are given by

$$\mu \pm Z_\alpha \left(\frac{\sigma}{\sqrt{n}} \right) \tag{13.22}$$

Tables of Z_α can be constructed for values of α commonly used in hypothesis testing, but such tables are not necessary if one has access to a table of the normal distribution. For example, suppose that $k = 4$ samples of size $n = 25$ are drawn from four populations, each thought to have a common mean $\mu = 50$ and a variance $\sigma^2 = 16$. Using a risk level of $\alpha = 0.05$, one solves equation (13.21) for p to find $p = (1 - \alpha)^{1/k} = (1 - 0.05)^{1/4} = 0.98726$. Then, $Z_{0.05}$ is the standard z value that gives an upper tail area of $(1 - p)/2 = 0.0064$. In this case, $Z_{0.05} \approx 2.49$.

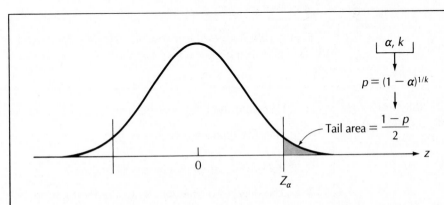

FIGURE 13.15
Calculation of Z_α for an ANOM chart (standard given)

Note: In ANOM charts, α refers to the overall risk level; it is related, but not equal, to the tail area to the right of Z_α.

Thus, decision lines for the $k = 4$ means are set at $\mu \pm Z_{0.05}\sigma/\sqrt{n} = 50 \pm$ $2.49(4)/\sqrt{25} = 50 \pm 1.992$. To construct the ANOM chart, upper and lower limit lines of $50 + 1.992$ and $50 - 1.992$ are drawn parallel to a horizontal centerline of $\mu = 50$. If any of the four sample means falls outside the limit lines, it would be concluded that there are significant differences between the means.

ANOM PROCEDURE—NO STANDARD GIVEN

There are many cases in which the common mean and variance of the k treatment populations are not known. Such is the case, for example, when studying the yields of k different machines, where neither the average yield nor the variation in yields is known in advance. Another example is testing five brands of bearings for their effects on motor vibration (see Example 13.1). Such situations are described as **no standard given** cases.

Calculating decision lines now becomes much harder. The goal of placing upper and lower limits around the mean remains the same, but this unknown population mean must be *estimated* by the grand mean $\overline{\overline{x}}$ of the data. Because the k sample means are used to find $\overline{\overline{x}}$, the deviations of the sample means from $\overline{\overline{x}}$ can no longer be considered statistically independent.[10] Adding to the difficulty is the fact that the variance σ^2 must also be estimated from the samples.

Special tables are a necessity for the no standards given case. These tables have been constructed by Nelson (1983) and they appear in Appendix 10. For a given α ($\alpha = 0.10, 0.05, 0.01, 0.001$), the critical values listed in the tables are denoted by:

$$\text{critical value} = h_{\alpha,k,\nu}$$

where $\alpha = $ the risk level to be used for the test, $k = $ the number of treatments, and $\nu = $ the number of degrees of freedom associated with the estimate of σ^2. Decision lines are then placed around $\overline{\overline{x}}$ at

$$\overline{\overline{x}} \pm s \cdot h_{\alpha,k,\nu} \sqrt{\frac{k-1}{kn}} \tag{13.23}$$

Ramig (1983) gives the following step-by-step procedure for constructing the ANOM chart in the case where each sample is of size n:

CONSTRUCTING AN ANOM CHART—NO STANDARD GIVEN	1. Calculate the treatment means: $\overline{x}_1, \overline{x}_2, \overline{x}_3, .., \overline{x}_k$. 2. Find the grand mean: $\overline{\overline{x}} = \dfrac{1}{k} \sum\limits_{j=1}^{k} \overline{x}_j$. 3. Pool the k sample variances to find $s^2 = \dfrac{1}{k} \sum\limits_{j=1}^{k} s_j^2$. 4. The degrees of freedom associated with s^2 are $\nu = (n-1)k$.

---

[10]The deviations from the grand mean now have a correlation of $-1/(k-1)$ with one another.

5. Find $h_{\alpha,k,\nu}$ from Appendix 10 and form the upper and lower decision limits:

$$\text{upper decision limit} = \bar{\bar{x}} + s \cdot h_{\alpha,k,\nu} \sqrt{\frac{k-1}{kn}}$$

$$\text{lower decision limit} = \bar{\bar{x}} - s \cdot h_{\alpha,k,\nu} \sqrt{\frac{k-1}{kn}}$$

6. Plot $\bar{\bar{x}}$ and the sample means on the same chart. Any sample mean that falls outside a decision limit is an indication that there is a difference between the treatment means.

The ANOM can be extended to many types of experimental designs, including the factorial designs of Section 13.1. In these applications, it should be kept in mind that the critical values used for main effects are slightly different from those for interactions. For further reading on ANOM applied to experimental design, the reader is referred to Schilling (1973).

EXAMPLE 13.8

In Example 13.1, five brands of bearings used in electric motors were studied to determine whether or not there was a significant difference in their effects on motor vibration. The data for the study consisted of five samples of six measurements each (see Table 13.2). The ANOVA table in Figure 13.8 indicated that there were indeed significant differences between the brands. To perform an ANOM test on the same data, the grand mean of the five samples,

$$\bar{\bar{x}} = \frac{13.683 + 15.950 + 13.667 + 14.733 + 13.083}{5} = 14.223$$

and their pooled variance,

$$s^2 = \frac{(1.194)^2 + 1.167)^2 + (0.816)^2 + (0.940)^2 + (0.479)^2}{5} = 0.913$$

are found. The degrees of freedom associated with s^2 are $\nu = (n-1)k = (6-1) \cdot 5 = 25$. Using an α of 0.05, we find the critical ANOM value is $h_{0.05,5,25} \approx 2.74$ (from Appendix 10), so decision lines are set at:

$$\text{upper limit} = \bar{\bar{x}} + s \cdot h_{\alpha,k,\nu} \sqrt{\frac{k-1}{kn}}$$

$$= 14.223 + (0.956)(2.74) \sqrt{\frac{5-1}{5 \cdot 6}} = 15.179$$

$$\text{lower limit} = \bar{\bar{x}} - s \cdot h_{\alpha,k,\nu} \sqrt{\frac{k-1}{kn}}$$

$$= 14.223 - (0.956)(2.74) \sqrt{\frac{5-1}{5 \cdot 6}} = 13.267$$

Figure 13.16 shows the ANOM chart with the decision lines. The chart clearly shows that two means (treatments 2 and 5) are outside the decision lines, so it can be concluded that the five treatment means are not the same. In particular, bearing brand 5 not only differs from the others, but also causes *lower* motor vibration. Thus, brand 5 becomes the logical choice to use in production.

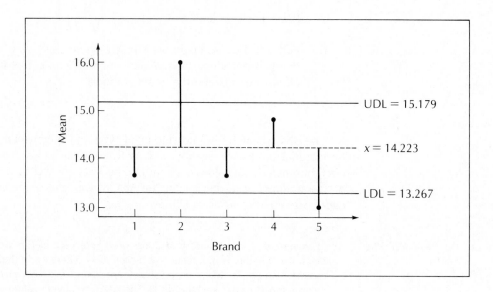

FIGURE 13.16
ANOM chart for the
data of Table 13.2

Attributes data can also be analyzed with the ANOM method. For such applications, it is assumed that sample sizes are large enough to invoke the normal approximation to the underlying binomial or Poisson distributions. Therefore, for large n, attributes ANOM charts use critical values $h_{\alpha,k}$ from the $\nu = \infty$ row in the ANOM tables in Appendix 10. In lieu of using tables, we note that the Z_α calculations in Section 13.2 provide excellent approximations to the values in the $\nu = \infty$ row of Appendix 10. ANOM procedures for attributes data are summarized by Ramig (1983) as follows:

ANOM PROCEDURE
FOR THE
PROPORTION OF
NONCONFORMING
ITEMS (p)

1. Obtain k samples of size n each.
2. Find the sample proportions nonconforming: $\hat{p}_1, \hat{p}_2, \ldots, \hat{p}_k$.
3. Find the grand proportion: $\hat{p} = \dfrac{1}{k} \sum\limits_{j=1}^{k} \hat{p}_j$.
4. Find the standard deviation: $s = \sqrt{\hat{p}(1 - \hat{p}/n)}$.
5. Find $h_{\alpha,k\infty}$ in the $\nu = \infty$ row of the ANOM tables in Appendix 10, or use Z_α (as calculated in Section 13.2).
6. Put decision lines around $\hat{p}$ at $\hat{p} \pm h_{\alpha,k\infty} s \sqrt{(k - 1)/k}$.

ANOM
PROCEDURE FOR
THE NUMBER OF
NONCONFORMITIES
(c)

1. Obtain a sample of k inspection units.
2. Count the number of nonconformities: $c_1, c_2, c_3, \ldots, c_k$.
3. Find the grand average: $\bar{c} = \dfrac{1}{k} \sum_{j=1}^{k} \bar{c}_j$.
4. Find $h_{\alpha, k\infty}$ in the $\nu = \infty$ row of the ANOM tables in Appendix 10, or use Z_α (as calculated in Section 13.2).
5. Put decision lines around $\bar{c}$ at $\bar{c} \pm h_{\alpha, k\infty} \sqrt{\bar{c} \cdot (k-1)/k}$.

13.3 ROBUST DESIGN

An important aspect of quality is a product's ability to perform its tasks in a variety of circumstances. Because variation in the operating environment is normally beyond the control of designers, it is desirable that products be *robust* enough to successfully handle such variation. Box, Bisgaard, and Fung (1988) give the example of designing cake mixes, where the product must be robust against the variability that is likely to be encountered in customers' ovens, baking times, and cooking styles. **Robust design** refers to the use of DOE techniques for finding product parameter settings that make products resilient to variations in working environments.

Much of the credit for introducing the *statistical* study of robust design goes to Dr. Genichi Taguchi, originator of the loss function definition of quality (see Section 1.2). Although Taguchi's statistical methods have been used in Japan for a few decades, his techniques were only introduced in the United States in around 1980 (Gunter 1987). The loss function approach makes variance reduction a primary goal for quality improvement. Unlike both the 'within specifications' and 'zero defects' goals, when the loss function definition is used, efforts to improve quality are not discontinued for processes that already operate within specifications.

Taguchi's contribution goes further than simply emphasizing continuous improvement. He provides a *mechanism* for finding parameter settings that both reduce product variation and increase product robustness. His statistical approach to these problems is based on the method of factorial designs (see Section 13.1), with alterations that are uniquely his own. In particular, Taguchi's approach to DOE uses the loss function concept to study both the product parameters and key environmental variables.

Taguchi suggests a threefold approach to designing quality into a product or process:

STAGES OF
ROBUST DESIGN

1. **System design** The use of current technology, processes, materials, and engineering methods to define and build a new 'system.' The system may be a new product or process or an improved version of an existing one (Ross 1988, p. 168).

(continued)

(*continued*)

> 2. **Parameter design** From the initial system design, key product or process parameters are identified for study. DOE methods are used to find optimal parameter settings.
> 3. **Tolerance design** The scientific determination of parameter specification limits.

The three stages are performed in the order stated. Most important, tolerance design is applied only after the parameter design phase. The reason is that quality improvements from parameter design can often be made at little or no extra cost, while tolerance design incurs the costs of having to improve other processes, buy better materials, accept more scrap and rework, and so on.

This section presents only the fundamentals of the 'Taguchi' methods. The references given throughout the section provide many examples and embellishments of these methods. It should also be noted that concerns about the validity of these methods have been raised in the statistical and quality communities. The last section of the chapter summarizes these concerns.

THE LOSS FUNCTION

Taguchi's loss function concept was introduced in Section 1.2. For a given product or process characteristic, Y, the simplest symmetric loss function that measures the 'loss to society' when Y deviates from its target dimension T is given by

$$l(y) = k(y - T)^2 \tag{13.24}$$

where the lowercase y represents a *measured* value of the characteristic Y (see Appendix 12 for a derivation of this loss function).

Viewing the measurements of Y as having some probability distribution with mean μ and variance σ^2, we can use equation (13.24) to calculate the expected loss:[11]

$$\begin{aligned}
\text{expected loss} &= kE(Y - T)^2 = kE[(Y - \mu) + (\mu - T)]^2 \\
&= kE[(Y - \mu)^2 + 2(Y - \mu)(\mu - T) + (\mu - T)^2] \\
&= k[\sigma^2 + (\mu - T)^2] \tag{13.25}
\end{aligned}$$

That is, the average loss is $k[\sigma^2 + (\mu - T)^2]$, highlighting the fact that some of the loss is caused by variation and some is caused by **bias** (the amount by which the process is off target).

From equation (13.25), it is clear that reductions in loss require that both the variation and bias be reduced. We also use the average loss to justify the use of signal-to-noise ratios in the parameter design phase.

---

[11] The middle term $E[2(Y - \mu)(\mu - T)] = 2(\mu - T)E(Y - \mu) = 0$, since $E(Y) = \mu$.

SIGNAL-TO-NOISE RATIOS

Traditionally, DOE methods have evaluated the effects of factors on a quality characteristic Y by means of single-run or replicated experiments. In the single-run case, the measured responses at each factor–level combination in the design matrix are combined to estimate main effects and interactions. With replicated runs, error variation can also be estimated; the analysis involves the averages of the replicated measurements at each factor–level combination.

When the loss function concept is used, however, equation (13.25) suggests that it may be advantageous to combine *both* average response and variation into a single measure, which Taguchi calls a **signal-to-noise ratio.** The terminology 'signal to noise' comes from the field of engineering, where the average response $\bar{y}$ is the 'signal' and the variation σ^2 is the 'noise.' Other authors refer to signal-to-noise ratios as **performance statistics** (Kackar 1985).

Depending on the design goal, many different signal-to-noise ratios are possible. Three frequently occurring goals are the following:

PARAMETER
DESIGN GOALS

1. *Nominal is best.*   The characteristic Y has a target (nominal) value T, and the loss increases the more Y differs from T.
2. *Lower is better.*   The characteristic Y is nonnegative with a target value of 0. Losses increase as Y becomes larger.
3. *Higher is better.*   The characteristic is nonnegative, with a target value of infinity. Losses increase the closer Y is to 0.

Examples of characteristics where these goals apply are the length of a 4-in. steel rod (nominal is best; target = 4 in.), the breaking strength of a material (higher is better), and the amount of impurities in a product (lower is better). In each case, Taguchi suggests using a different type of signal-to-noise ratio.

For the 'nominal is best' case, the recommended approach is to first find an *adjustment factor* that can be set at a level that will eliminate the bias. Adjustment factors are design parameters that are within the control of a designer to change. Sometimes, adjustment factors can be found that control the mean *without* affecting the variance. With such an adjustment used to eliminate the bias term, or at least to substantially reduce it, the expected loss from equation (13.25) becomes $k\sigma^2$, and the goal becomes to reduce the variation. For each of the n replicated measurements at a given factor–level combination, Taguchi recommends the signal-to-noise ratio

$$\text{SN} = -10 \log_{10}(s^2) \tag{13.26}$$

where s^2 is the sample standard deviation of the n measurements at a given factor–level combination. The goal of minimizing s^2 is equivalent to that of *maximizing* $-10 \log_{10}(s^2)$, and Taguchi chose the latter expression in order to keep all quality goals uniform (i.e., he expresses every goal as a maximization).

At other times, adjustment factors can be found that affect only the mean μ *along with* the variance σ^2, but in a such a manner that the coefficient of variation σ/μ remains relatively constant. Letting CV denote the coefficient of variation, we can write the expected loss as

$$k \cdot [\sigma^2 + (\mu - T)^2] = k \cdot [(CV)^2 \mu^2 + (\mu - T)^2]$$

Then, as the adjustment factors are changed to bring the mean on target, the CV remains unchanged, and the expected loss after adjustment becomes $k(CV)^2 T^2$ (Phadke 1989, p 100). From this expression, it is clear that the *other* parameters can be studied with the goal of minimizing the CV. For this purpose, Taguchi recommends the signal-to-noise ratio

$$SN = -10 \log_{10}\left(\frac{\bar{y}^2}{s^2}\right) \tag{13.27}$$

where $\bar{y}$ and s^2 are the sample mean and variance, respectively, of the n readings at a given factor–level combination.

For the 'lower is better' criterion, the target is $T = 0$, so the average loss is given by $E(Y^2)$, which suggests minimizing the average of the squared measurements $(1/n)\sum_{i=1}^{n} y_i^2$. In keeping with the goal of finding statistics to *maximize*, Taguchi recommends the signal-to-noise ratio

$$SN = -10 \log_{10}\left(\frac{1}{n}\sum_{i=1}^{n} y_i^2\right) \tag{13.28}$$

For the 'higher is better' case, by noting that maximizing Y is the same as minimizing $1/Y$, we can apply the 'lower is better' solution to $1/Y$; that is, the signal-to-noise ratio for a 'higher is better' characteristic is

$$SN = -10 \log_{10}\left(\frac{1}{n}\sum_{i=1}^{n} \frac{1}{y_i^2}\right) \tag{13.29}$$

These three signal-to-noise ratios are the most commonly used, although Taguchi has developed more than 70 such measures, each taking into account different engineering requirements (Barker 1986, p. 35). Table 13.12 summarizes the four signal-to-noise ratios mentioned above.

TABLE 13.12   Signal-to-Noise Ratios for Three Types of Criteria

Nominal is best:

Variance independent of mean: $SN = -10 \log_{10}(s^2)$

Variance dependent on mean: $SN = -10 \log_{10}\left(\dfrac{\bar{y}^2}{s^2}\right)$

Lower is better:

$SN = -10 \log_{10}\left(\dfrac{1}{n}\sum_{i=1}^{n} y_i^2\right)$

Higher is better:

$SN = -10 \log_{10}\left(\dfrac{1}{n}\sum_{i=1}^{n} \dfrac{1}{y_i^2}\right)$

ORTHOGONAL ARRAYS

L_N ARRAYS.   To study the various factors that affect a product or process, Taguchi uses a small collection of specially constructed design matrices. These matrices are also called **orthogonal arrays** because of an 'orthogonality' property between any two columns in the array. Thinking of the columns as sequences of $+$ and $-$ signs, we say two columns are orthogonal when the component-by-component product of the columns results in a third column whose entries sum to 0. In particular, the 2^k designs have design matrices that are orthogonal (see Section 13.1).

When factors are studied at two levels, Taguchi uses '1' and '2' as the array entries to denote these levels, instead of the $-$ and $+$ notation of Section 13.1. The reason for changing from the $+/-$ scheme to a numbering scheme is that Taguchi also allows factors to be studied at *more* than two levels, if desired. Thus, a factor with three levels would use an orthogonal array whose column entries consist of a certain sequence of 1's, 2's, and 3's.

The particular arrays used by Taguchi are usually *subarrays* of larger matrices. By constructing orthogonal arrays in this manner, one can study more factors with fewer experimental runs. For example, to study seven two-level factors, a 2^7 experiment could be performed. However, it would require $2^7 = 128$ separate runs, which, in many cases, might exhaust the resources available for the study. Instead, Taguchi suggests a particular subset of the *rows* of the 2^7 design matrix:

$$\begin{bmatrix} 1 & 1 & 1 & 1 & 1 & 1 & 1 \\ 1 & 1 & 1 & 2 & 2 & 2 & 2 \\ 1 & 2 & 2 & 1 & 1 & 2 & 2 \\ 1 & 2 & 2 & 2 & 2 & 1 & 1 \\ 2 & 1 & 2 & 1 & 2 & 1 & 2 \\ 2 & 1 & 2 & 2 & 1 & 2 & 1 \\ 2 & 2 & 1 & 1 & 2 & 2 & 1 \\ 2 & 2 & 1 & 2 & 1 & 1 & 2 \end{bmatrix}$$

This matrix, called the L_8 array, has seven orthogonal columns and requires only eight experimental runs, in contrast to the 128 runs for the full 2^7 experiment. Furthermore, the L_8 design can be used to study *less* than seven factors if desired (some of the columns are simply not assigned to factors), so this one matrix serves in a variety of applications. In this manner, Taguchi provides a very small catalogue of orthogonal arrays that can handle a large number of engineering experiments. For example, Taguchi's list of two-level arrays includes the L_4, L_8, L_{12}, L_{16}, L_{32}, and L_{64}.[12] The list of three-level arrays includes the L_9, L_{18}, L_{27}, and L_{81}. In general, the L_N array has N experimental runs and is capable of handling up to $N - 1$ factors.

LINEAR GRAPHS.   Intuitively, something must be lost when the eight runs of an L_8 are used instead of the 128 runs of the full 2^7 design. What is lost is (1) the ability to estimate some of the interactions between factors and (2) the ability to clearly

---

[12]Higher-order arrays are available, but this list covers a large number of cases likely to arise.

estimate some of the main effects. As an example, consider columns 2, 4, and 6 of the L_8 matrix (the $+/-$ scheme is used to facilitate the calculation of interaction columns):

$$
\begin{array}{ccc}
A & B & C \\
\downarrow & \downarrow & \downarrow \\
\end{array}
\begin{bmatrix}
- & - & - \\
- & + & + \\
+ & - & + \\
+ & + & - \\
- & - & - \\
- & + & + \\
+ & - & + \\
+ & + & -
\end{bmatrix}
\rightarrow
\begin{array}{cc}
\{AB & = -C\} \\
\downarrow & \downarrow \\
+ & + \\
- & - \\
- & - \\
+ & + \\
+ & + \\
- & - \\
- & - \\
+ & +
\end{array}
$$

When the signs in columns A and B are multiplied to form the AB interaction column, the resulting column is exactly the negative of column C; that is, the main effect for factor C and the AB interaction (after multiplying by -1) are estimated by the *same* column in the L_8 array. There is no way to disassociate these effects, so an AB interaction could be mistaken for the main effect of C, and vice versa. When this happens, the two effects are said to be **aliases** of each other. Because of their reduced number of experimental runs, the orthogonal arrays used by Taguchi contain a large amount of aliasing.

However, not all of the interactions are confounded with main effects in the L_N arrays. In order to keep track of what effects each column in an L_N array estimates, Taguchi provides a set of **linear graphs** with each array in his catalogue. Linear graphs are collections of nodes, representing columns in the L_N, that are connected by lines that indicate the interaction between two columns. For example, Figure 13.17 shows the linear graphs associated with the L_8 array. The graphs are used for assigning interaction terms to columns in order to reduce the

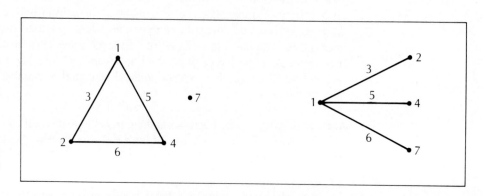

FIGURE 13.17
Linear graphs for the
L_8 array

aliasing problem.[13] Thus, if an experimenter believes factors A and B and their interaction are worthy of study, then a linear graph is consulted to determine where to assign the AB interaction. Any node in the graph can be assigned any factor, so suppose A is assigned to column 1. Since interactions are represented by the lines between nodes, B could be assigned to column 2 or 4. If B was assigned to column 2, then the linear graph indicates that the AB interaction is associated with column 3. Otherwise, if column 4 was selected for B, then the AB interaction would be found in column 5.

By using linear graphs, one can avoid certain columns of an L_N when assigning the remaining factors in the study. Suppose, for example, that factors A, B, and C are to be studied along with the AB interaction. With the linear graph, A and B can be assigned to columns 1 and 2, which means that column 3 (the AB interaction) should be avoided in the assignment of factor C. Factor C can be assigned to any column other than column 3 in this study.

CHOOSING AN L_N ARRAY.   In order to select the right L_N array for a study, the following method is used. First, note that the total degrees of freedom associated with an L_N are always $N - 1$. Then, the L_N array appropriate for the study is the smallest one whose degrees of freedom just exceed the total degrees of freedom associated with all the factors and interactions. The selection of an orthogonal array proceeds as follows:

CHOOSING AN
ORTHOGONAL
ARRAY

1. Each factor with k levels has $k - 1$ degrees of freedom.
2. The degrees of freedom for a two-factor interaction term is the product of the degrees of freedom for the individual factors.
3. The individual degrees of freedom for all factors and interactions to be studied are summed to form the total degrees of freedom, d.f.$_{total}$, required by the experiment.
4. The smallest L_N array for which $N - 1 \geq$ d.f.$_{total}$ *and* which can accommodate the factors and interactions is selected.

Note that in step 4, it is possible to select combinations of factors and interactions for which $N - 1 \geq$ d.f.$_{total}$, yet the L_N array fails to handle all the terms. This usually happens when a large number of interaction terms are included in the study. In such cases, the next larger L_N array is used for the study. Even though the larger array will contain more columns than needed for the study, no problem is encountered, since it is allowable to leave columns unassigned in any L_N.

---

[13]The problem is never entirely eliminated. Linear graphs only provide a method to avoid confounding a two-factor interaction with a main effect. Higher-order interactions may still be aliased with the main effect.

EXAMPLE 13.9

Crossfield and Dale (1991) apply L_N arrays and effects plots (see Example 13.10 for an illustration of effects plotting) to the improvement of a manufacturing process, the machining of shafts for turbochargers. Three factors pertaining to shaft machining are selected for study: the presence or absence of a thread relief undercut on the shaft (U), the presence or absence of a sand-blasted thrust collar on the shaft (S), and the effect of repeated stretchings of the shaft (R). Two response characteristics are measured: the shaft stretch and the impeller torque. The interactions between the three factors are also to be included in the study.

With the above abbreviations for the factors, the total degrees of freedom associated with this study are as follows:

Factor	Levels	Degrees of Freedom
U	With, without	1
S	With, without	1
R	Once, ten times	1
US		1
UR		1
SR		1
	Total	6

The L_8 array, with 7 degrees of freedom, is the smallest array for which $N - 1$ exceeds the degrees of freedom for factors and interactions. Before the L_8 can be used, however, its linear graph must be consulted to ascertain that all three factors can be assigned to columns without being aliased with one of the interactions. In fact, such an assignment is possible, as shown by the L_8 linear graph in Figure 13.18.

FIGURE 13.18
Column assignments in an L_8 array for Example 13.9. (Note: the second linear graph could not accommodate all main effects and interactions and would therefore not be used.)

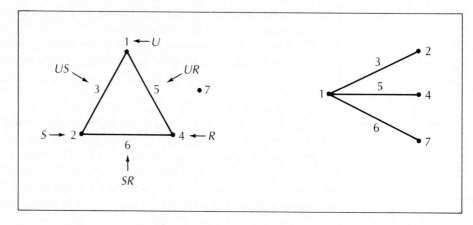

EFFECTS PLOTS.   For two-level arrays (i.e., all factors are studied at two levels), main effects and interactions can be estimated in the same fashion as in the 2^k designs of Section 13.1. However, Taguchi's approach also recommends plotting all main effects and interactions to show their relative magnitudes and directions. In particular, to plot the main effect of factor A, two points are plotted: the average of all responses corresponding to a '1' in column A, and the average of all responses associated with a '2' in column A. To show the effect of moving A from level 1 to level 2, a straight line is drawn between the two plotted points. Similarly, the interaction between two factors A and B is depicted by plotting the two average B responses when A is at level 1 and the two average B responses for A at level 2. Figure 13.19 shows two typical plots of the main effects of two factors, A and B, and their interaction term. When factors are studied at more than two levels, the same graphical technique can be used.

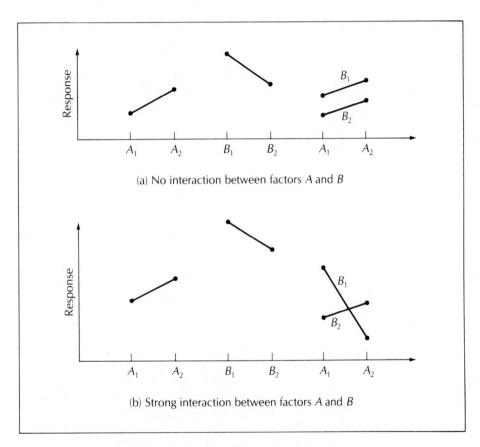

FIGURE 13.19
Plotting main effects
and two-factor
interactions

(a) No interaction between factors A and B

(b) Strong interaction between factors A and B

EXAMPLE 13.10

Main effects and two-factor interactions can be graphed for any design, not just the special L_N designs of Taguchi. For example, a full 2^3 design was used to study surface uniformity in the chemical coating of Example 13.6. Using the responses listed in Table 13.8, we know the average effect of spray volume (factor V) at its low level is $(40 + 30 + 45 + 30)/4 = 36.25$. Using the numbering notation, this value is denoted V_1. The average effect of V at its high volume is $(25 + 50 + 25 + 52)/4 = 38$. We call this value V_2. In a similar manner, the effects of factors S and B can be summarized by $S_1 = 33.75$, $S_2 = 40.5$, $B_1 = 36.25$, and $B_2 = 38$. Notice that the main effects calculated in Example 13.6 are the *differences* of these values—for example, the main effect for $V = V_2 - V_1 = 38 - 36.26 = 1.75$. The two-factor interactions require four averages: $V_1 S_1$, $V_1 S_2$, $V_2 S_1$, and $V_2 S_2$. $V_i S_j$ is the average response when *both* V is at level i and S is at level j. Thus, $V_1 S_1 = (40 + 45)/2 = 42.5$, $V_1 S_2 = (30 + 30)/2 = 30$, $V_2 S_1 = (25 + 25)/2 = 25$, and $V_2 S_2 = (50 + 52)/2 = 51$. The VS interaction calculated in Example 13.6 is the average *difference* of $V_2 S_2 - V_1 S_2$ and $V_2 S_1 - V_1 S_1$—that is, $\frac{1}{2}[(51 - 30) - (25 - 42.5)] = 19.25$. A plot of all the main effects and two-factor interactions for Table 13.8 is shown in Figure 13.20. The ease with which the effects can be compared is an advantage of this technique.

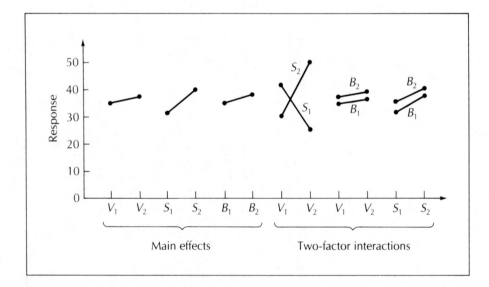

FIGURE 13.20
Main effects and interactions for Example 13.10 (data from Table 13.8)

ENVIRONMENTAL NOISE

A primary goal of robust design is to find parameter settings that give products or processes the ability to resist the adverse effects of changes in the operating environment. As an example, Box and coworkers (1988) describe the efforts of a Japanese manufacturer to analyze the effects of environmental conditions such as temperature, humidity, dust, and size and shape of paper on the performance of a new photocopy machine.

One way that designers reduce the effects of factors external to the product (i.e., its operating environment) is by including a list of recommended operating conditions with the product. Thus, some battery chargers on cordless appliances come with the warning that charging the appliance is best done when the ambient temperature is within a specified range. Of course, some environmental factors cannot be so controlled and, more important, it is simply not desirable to include long lists of restrictions on how or when to operate a product, since a product's ability to tolerate change is a measure of its quality.

Making products and processes robust requires an *active* study of environmental factors. (Waiting for customer returns or repairs is not an active approach.) Taguchi incorporates the study of environmental factors *along with* the parameter design phase. Specifically, he recommends constructing *two* orthogonal arrays, one for the design parameters and one for the 'noise' factors.[14] The procedure for studying the noise variables is the same as that for studying the design parameters: decide which noise factors are to be considered, and then choose an appropriate orthogonal array for these factors. The design parameters are assigned to a matrix that Taguchi calls the **inner array,** while noise factors are put in an **outer array.**

Figure 13.21 shows the data-collection scheme for studying design parameters and noise parameters. Each combination of noise factors in the outer array determines a different set of environmental conditions for each experimental run in the inner array. Thus, each row of the design matrix yields a set of m measurements, where m is the number of runs in the outer array. Each set of m measurements is combined into an appropriate signal-to-noise ratio and, since the design parameters are chosen so as to maximize the signal-to-noise ratio, settings are found that should make the product robust against the particular noise factors in the study. The following example shows the steps in using Taguchi's method of robust design.

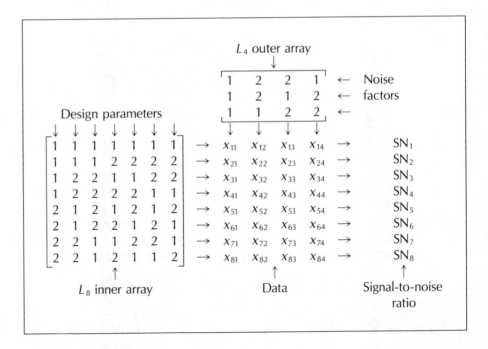

FIGURE 13.21
Experimental layout with inner and outer arrays

---

[14]Taguchi refers to environmental variables beyond the control of the designer as noise variables.

EXAMPLE 13.11   Three raw materials are used in varying amounts to form a certain plastic product. Depending on the quantities of each material used, the product will vary in strength, with the goal being to make the strongest product possible. It is also known that ambient temperature and humidity can affect the formation, and therefore the strength, of the resulting product. With the three materials denoted as A, B, and C, a study is conducted to select levels of these factors that will make the product's manufacture robust against changes in temperature and humidity. The design engineers also consider the AB and BC interactions important enough to include in the study. The engineers decide on reasonable 'low' and 'high' levels of each material and on 'low' and 'high' values of temperature and humidity to use in the tests.

After the total degrees of freedom are determined (three for factors plus two for interactions), an L_8 array is chosen as the inner array. By consulting the triangular linear graph for the L_8, the experimenters choose the following column assignments: $A \rightarrow 1$, $B \rightarrow 2$, $AB \rightarrow 3$, $C \rightarrow 4$, and $BC \rightarrow 6$ (the remaining columns are not assigned). Similarly, an L_4 is chosen for the outer array, with temperature assigned to the first column and humidity to the second (the third column is not assigned to any effect). Figure 13.22 shows the layout of this study, along with the data collected from running the designated combinations of design factors and noise variables. Interactions and main effects are calculated from the eight signal-to-noise ratios shown in Figure 13.22. These effects are plotted in Figure 13.23. Although the main effects for A, B, and C indicate that each of their 'high' levels is best (in terms of maximizing the SN ratio), the presence of a strong AB interaction means that B's effect cannot be interpreted

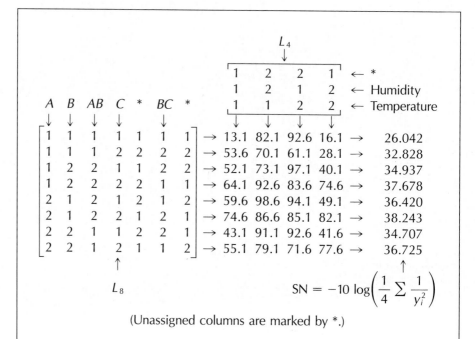

FIGURE 13.22
Experimental layout
and data for
Example 13.11

separately from that of A. Therefore, based on A, B, and the AB interaction, the best settings for this process are A_2, B_1, and C_2 (*not A_2, B_2, and C_2*). Notice in Figure 13.22 that the row corresponding to A_2, B_1, C_2 gives the largest SN ratio, even though other rows in the data matrix contain higher y_i readings in some places.

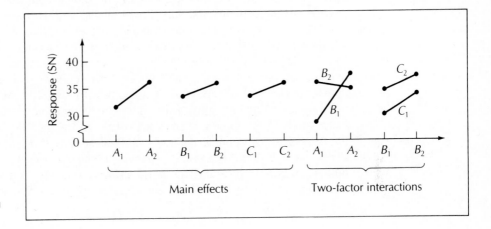

FIGURE 13.23
Main effects and
interactions for data
of Example 13.11

    In Example 13.11, effects plots are used to select optimal factor levels without statistical tests being done first to determine which effects are indeed significant. Normally, the examination of effects plots should be accompanied by an analysis of variance study. The calculations are similar to those in Section 13.1, and we refer the reader to other texts for these developments (Ross 1988). In addition, it is sometimes the case that the 'optimal' levels indicated by the effects plots correspond to a set of experimental conditions that are not included in the original design. For this reason, and because the L_N designs are so sparse, Taguchi recommends that a follow-up or **confirmatory** experiment be run with factors set at their 'optimal' levels (as determined from the effects plots). Some practitioners have also argued that when an effect is not significant, it should not matter *which* of its levels is selected; that is, adjusting the significant factors is necessary, but the unnecessary adjustment of insignificant factors should not hurt anything. Unfortunately, this argument does not work in all cases, so it is recommended that ANOVA tests accompany any robust design.

COMMENTS ON THE TAGUCHI METHOD

There has been a good deal of controversy surrounding Taguchi's methodology since its introduction to the United States. The concern centers around Taguchi's use of statistical methods, not his philosophy of 'loss to society' or robust design.

One concern involves the very sparse and highly aliased L_N arrays used by Taguchi. Although Taguchi expressly chose sparse arrays in accordance with his experimental philosophy of "digging wide, not deep," some of these arrays are, nonetheless, not necessarily the best choices. In some cases, there are other arrays of the same size that allow the same number of factors to be estimated, but without aliasing any main effects with two-factor interactions (Ryan 1988).

Another problem is that the signal-to-noise ratios may not always work in the desired manner. In some cases, it can also be shown that they do not make efficient use of the experimental data (Box, Bisgaard, and Fung 1988).

Apart from its statistical shortcomings, Taguchi's philosophy has had a strong influence on the use of DOE methods worldwide. Perhaps the most important reason for this success is that he provides a "cookbook" approach to DOE: there are only a small number of L_N arrays from which to choose, there is a simple procedure for deciding which array to select and how to assign the factors to columns of the array, and simple graphs of the main effects and interactions allow 'optimal' factor levels to be chosen. Even though some of the statistical components may not themselves be optimal, Taguchi's methods are based on standard factorial designs, and this alone represents a great improvement over the one at a time approach. Although the statistical issues are not completely settled, the end result has been a greatly enhanced use and appreciation of DOE techniques in general.

EXERCISES FOR CHAPTER 13

13.1   If five process variables are thought to influence process yield, and if each factor is studied at three levels, how many experimental runs are necessary to run a complete factorial design?

13.2   Five raw materials are tested for their effect on a process yield. Samples of size 10 are used for each of the materials. Complete the following ANOVA table for this experiment:

Source of variation	d.f.	SS	MS	F
Brands	_____	_____	_____	15.32
Error	_____	_____	0.64	
Total variation	_____	_____		

13.3   A fixed-effects model is used to analyze two factors, each of which has five levels. Three replicated measurements are available for each combination of factor levels. Complete the following ANOVA table for this experiment:

Source of variation	d.f.	SS	MS	F
Factor I	_____	20	_____	_____
Factor II	_____	_____	_____	8.1
Interaction	_____	_____	_____	_____
Error	_____	_____	2	
Total variation	_____	200		

13.4 In Exercise 13.3, suppose that the analysis was instead run as a random-effects experiment. Complete the following ANOVA table for this experiment:

Source of variation	d.f.	SS	MS	F
Factor I	_____	20	_____	_____
Factor II	_____	_____	_____	_____
Interaction	_____	_____	_____	1.5
Error	_____	_____	2	
Total variation	_____	200		

13.5 Decide whether or not the following pairs of columns can be considered orthogonal.
(a) $[-, +, -, -, -, +, +, -, +, -]$ and $[+, -, +, +, +, +, +, -, +, -]$
(b) $[-, -, -, -, -, +, +, +, +, +]$ and $[+, +, +, +, +, -, -, -, -, -]$
(c) $[-, -, -, -, -, +, +, +, +, +]$ and $[+, -, +, -, +, -, +, -, +, -]$
(d) $[-, -, -, -, -, +, +, +, +, +]$ and $[-, +, -, +, -, +, -, +, -, +]$

13.6 For a given column of length 2^k (k is a positive integer), how many distinct columns of length 2^k can be formed, each of which is orthogonal to the given column? (Assume that all columns are restricted to containing an equal number of + and − signs.)

13.7 "Screening designs" are experimental designs with many variables and relatively few experimental runs that are used to screen out unnecessary factors in the early stages of experimentation. Suppose that ten factors are to be used in a study. Explain and compare the merits of each approach.
(a) Use two levels of each of the ten factors in one experiment.
(b) Use a more in-depth study of the first five factors, with three levels of each, followed by a similar study of the remaining five factors (again with three levels each).

13.8 A fast method for finding the signs in the interaction columns of a 2^k design involves counting the number of − signs in each row of the corresponding factor columns. State this rule precisely.

13.9 In an effort to reduce variation in copper plating thickness on printed circuit boards, Delott and Gupta (1990) used a half-fraction of a 2^3 factorial design to study the effects of the factors anode height (up or down), circuit board orientation (in or out), and anode placement (spread or tight). The response variable of interest was the variance in plating thickness, and the experimental results were as follows:

Anode height	Board orientation	Anode placement	Variance
−	−	−	11.63
−	+	+	3.57
+	−	+	5.57
+	+	−	7.36

Assuming that all interactions are insignificant, estimate the main effects of each of these factors. Which factor appears to have the greatest effect on plating variation?

13.10   Strength is an important characteristic of woven textiles. In a study of a high-speed weaving process, four factors thought to influence fabric strength are selected for study: side-to-side differences in strength (nozzle side of fabric versus opposite side), yarn type (air spun versus ring spun), pick density (the number of strands per unit inch; 35 strands versus 50 strands), and air pressure (30 versus 45 psi).[15] A half-fraction of the full 2^4 experiment is used to study these factors.

Side-to-side	Yarn type	Pick density	Air pressure	Strength
−	−	−	−	24.50
+	−	−	+	22.05
−	+	−	+	24.52
+	+	−	−	25.00
−	−	+	+	25.68
+	−	+	−	24.51
−	+	+	−	24.68
+	+	+	+	24.23

Although this design confounds the main effects with three-factor interactions, the effect of three-factor interactions is assumed to be negligible. Estimate the main effects and two-factor interactions from these data.

13.11   The yield of a chemical reaction (in pounds) is measured at each of four different temperatures, T_1, T_2, T_3, and T_4. The reaction is run five times at each temperature:

T_1	T_2	T_3	T_4
9.91	4.99	10.31	11.04
10.64	11.63	9.95	13.44
3.54	10.43	12.22	14.30
10.91	9.65	7.69	15.43
10.96	12.41	13.55	15.39

Perform an analysis of variance test to see whether there is a difference in yield across the four temperatures. (Use $\alpha = 0.05$.)

13.12   For the experiment described in Exercise 13.11, run the ANOM procedure to see whether there is a difference between the yields at the four temperatures. (Use $\alpha = 0.05$.)

13.13   For each of the following quality characteristics, identify the most appropriate quality objective in terms of 'larger is better,' 'nominal is best,' or 'smaller is better.'
(a) Average fuel consumption of an automobile
(b) The strength of a weld between metal parts
(c) The yield of a process that makes silicon wafers for computer circuits
(d) The width of a car door
(e) The proportion of nonconforming items from a process
(f) The amount of galvanized plating on a sheet of metal

---

[15]See Johnson, Clapp, and Baqai, (1989).

13.14 In a chemical company, an experimental design is used to study the effect of four process variables on the first stage of a chemical reaction. The measured response is the percent of a critical chemical that is converted during the first stage of the reaction. The process variables and their levels for this study are:

	Variable	Low level	High level
A	pressure (psi)	14.0	20.0
B	steam ratio	7.5	11.5
C	throughput rate	0.52	0.66
D	temperature (Fahrenheit)	1,150	1,200

One experimental run is made for each of the 16 treatment combinations and the percent conversion is recorded:

A	B	C	D	Percent conversion
14	7.5	.52	1,150	27.22
20	7.5	.52	1,150	25.19
14	11.5	.52	1,150	23.23
20	11.5	.52	1,150	18.93
14	7.5	.66	1,150	25.32
20	7.5	.66	1,150	22.61
14	11.5	.66	1,150	26.80
20	11.5	.66	1,150	20.20
14	7.5	.52	1,200	44.53
20	7.5	.52	1,200	42.44
14	11.5	.52	1,200	43.78
20	11.5	.52	1,200	37.66
14	7.5	.66	1,200	42.16
20	7.5	.66	1,200	38.97
14	11.5	.66	1,200	48.85
20	11.5	.66	1,200	42.05

(a) What kind of experimental design is this?
(b) Calculate and plot the main effects and two-factor interactions.
(c) From the graphs in part (b), which variables appear to be important in maximizing the percent conversion?

REFERENCES FOR CHAPTER 13

Anand, K. N. 1991. "Increasing Market Share Through Improved Product and Process Design: An Experimental Approach." *Quality Engineering* 3 (no. 3).

Barker, T. B. 1986. "Quality Engineering by Design: Taguchi's Philosophy." *Quality Progress* 19 (no. 12):32–42.

Box, G., S. Bisgaard, and C. Fung. 1988. "An Explanation and Critique of Taguchi's Contributions to Quality Engineering." *Quality and Reliability Engineering International* 4:123–131.

Box, G. E. P., R. N. Kackar, V. J. Nair, M. Phadke, A. C. Shoemaker, and C. F. Jeff Wu. 1988. "Quality Practices in Japan." *Quality Progress,* March, pp. 37–41.

Brownlee, K. A. 1965. *Statistical Theory and Methodology in Science and Engineering.* New York: John Wiley.

Caulcutt, R. (1983). *Statistics in Research and Development,* London: Chapman and Hall.

Crossfield, R. T., and B. G. Dale. 1991. "Applying Taguchi Methods to the Design Improvement Process of Turbochargers." *Quality Engineering* 3 (no. 4):501–516.

Delott, C., and P. Gupta. 1990. "Characterization of Copper Plating Process for Ceramic Substrates." *Quality Engineering* 2 (no. 3):269–284.

Dixon, W. J., and F. J. Massey. 1983. *Introduction to Statistical Analysis,* 4th ed. New York: McGraw-Hill.

Duncan, A. J. 1986. *Quality Control and Industrial Statistics,* 5th ed. Homewood, IL: Irwin.

Gunter, B. 1987. "A Perspective on the Taguchi Methods." *Quality Progress* 20 (no. 6):44–51.

Johnson, R., T. Clapp, and N. Baqai. 1989. "Understanding the Effect of Confounding in Design of Experiments: A Case Study in High-Speed Weaving." *Quality Engineering* 1 (no. 4):501–508.

Kackar, R. N. 1985. "Off-Line Quality Control, Parameter Design, and the Taguchi Method." *Journal of Quality Technology* 17 (no. 4):176–188.

Mendenhall, W., and J. T. McClave. 1981. *A Second Course in Business Statistics: Regression Analysis.* San Francisco: Dellen.

Mosteller, F., and R. E. K. Rourke. 1973. *Sturdy Statistics: Nonparametrics and Order Statistics.* Reading, MA: Addison Wesley.

Nelson, L. S. 1983. "Exact Critical Values for Use with the Analysis of Means." *Journal of Quality Technology* 15 (no. 1):40–44.

Ott, E. R. 1957. "A Graphical Analysis of Means." *Proceedings of the All-Day Conference on Quality Control,* ASQC Metropolitan Section and Rutgers University.

Ott, E. R. 1958. "Analysis of Means." Technical Report No. 1, prepared for Army, Navy, and Air Force under contract No. 404(11) (Task NR 042-021) with the Office of Naval Research.

Ott, E. R. 1960. "Analysis of Means Applied to Percent Defective Data." Technical Report No. 2, prepared for Army, Navy, and Air Force under contract No. 404(11) (Task 042-021) with the Office of Naval Research.

Ott, E. R. 1967. "Analysis of Means—A Graphical Procedure." *Industrial Quality Control* 24 (no. 2):101–109.

Phadke, M. S. 1989. *Quality Engineering Using Robust Design.* New York: Prentice Hall.

Ramig, P. F. 1983. "Applications of the Analysis of Means." *Journal of Quality Technology* 15 (no. 1):19–25.

Ross, P. J. 1988. *Taguchi Techniques for Quality Engineering.* New York: McGraw-Hill.

Ryan, T. P. 1988. "Taguchi's Approach to Experimental Design: Some Concerns." *Quality Progress,* May, pp. 34–36.

Sachs, L. 1982. *Applied Statistics: A Handbook of Techniques.* New York: Springer Verlag.

Schilling, E. G. 1973. "A Systematic Approach to the Analysis of Means." *Journal of Quality Technology* 5:93–108, 147–159.

Snedecor, G. W., and W. G. Cochran. 1989. *Statistical Methods,* 8th ed. Ames: Iowa State University Press.

14

IMPLEMENTATION

Significant quality improvements require an organized approach to implementation. Different approaches have been suggested and a knowledge of each is essential. This chapter addresses issues of training, computerization, customer and supplier relationships, organizing for quality, total quality management, and problem-solving guidelines.

CHAPTER OUTLINE

14.1   INTRODUCTION

Possessing a tool is one thing, but using it effectively is quite another. This chapter discusses methods for implementing statistical tools for quality improvement. Successful approaches, pitfalls, and strategies are examined.

From the time Shewhart invented the first control chart until the modern resurgence of experimental design, there have been numerous applications of statistical techniques to production processes. The recommendations and discoveries resulting from these efforts have consistently identified two primary components for success: a commitment to improving quality and a plan for effecting this goal. Conversely, the lack of either component is usually a formula for failure.

Taken to its logical end, the commitment to quality improvement implies the cooperation of the entire organization, not just selected departments within the organization. Feigenbaum's total quality control[1] movement and its modern incarnations are typical of the level of commitment needed. Focusing an entire company on a particular goal requires careful planning (Section 14.2), especially in the areas of training (Section 14.3), problem solving (Section 14.4), and customer and supplier relationships (Sections 14.7 and 14.8). A carefully constructed implementation plan outlines the particular steps and intermediate goals for applying statistical methods.

The points made in this chapter are gleaned from the published experience of many quality practitioners and companies. They can be summarized briefly as follows:

FACTORS IN
SUCCESSFUL
QUALITY
PROGRAMS

- The commitment to quality improvement must begin from top management, not as a ground swell from below.
- A well-constructed implementation plan is required.
- Training in statistical methods is required of all employees, with different emphases depending on one's function.
- Simple statistical methods (Pareto charts, cause-and-effect diagrams, location diagrams, checksheets) are very effective and should not be bypassed in an attempt to use more 'sophisticated' methods.
- Pilot projects are an excellent vehicle for translating theory into practice, but they should not take the place of an implementation plan.
- The commitment to quality should be companywide; most important, it should not reside solely within the quality department.
- Computerization and data gathering definitely help, but they are not the driving factors in success.

(continued)

---

[1]Several acronyms have been used to describe this goal: **TQC** (total quality control), **TQM** (total quality management), **TQA** (total quality assurance), and **CWQC** (companywide quality control).

(continued)

- In some format (quality circles or organized groups), the team approach to problem solving is best.
- Management's attitude can make or break a quality program.
- Quality improvement and reduced cycle time are strategic competitive goals.
- Without attention to quality, the advantages of advanced technology quickly diminish.
- Quality improvement drives down quality costs. However, simply reducing quality costs does not necessarily improve quality.
- JIT ('just in time') complements and requires TQM ('total quality management') and SPC ('statistical process control').

14.2 TOTAL QUALITY MANAGEMENT (TQM)

Feigenbaum's original concept of TQC has evolved into the modern TQM program. Although common sense dictates that more attention to quality is better than less attention, TQM also arose in order to combat the practice of relegating quality activities to a single department or group of people within a company. Prior to TQM, quality was considered the responsibility of quality control inspectors and quality assurance departments. Under such a system, the quality department often took the role of 'police' looking for 'quality violators,' which created an adversarial relationship between the production and quality functions. Furthermore, training in statistical methods was given only to those in quality departments, leading others in the organization to misunderstand and mistrust the use of statistical methods.

In theory, TQM spreads the responsibility for assuring and improving quality throughout the organization. In practice, a concerted effort is required to keep this goal on track, partly because of the natural tendency for the statistical portions of TQM to slip back into the hands of the quality department. Initially, quality practitioners tend to be more experienced in the statistical tools than those whose primary function is sales, purchasing, or production. This makes it very easy for a company to give a member of the quality department the responsibility of creating and monitoring control chart activities, which defeats both the goal of TQM and the immediate usefulness of the control chart to the workers involved. Successfully implementing a TQM program requires that the role of the quality department be carefully redefined (Aquino 1987).

The scope of a TQM program is vast. Effectively involving every arm of an organization in the quality movement is understandably a larger task than just running control charts on production processes. Some have described TQM as nothing short of changing the corporate 'culture' (*Business Week* 1991, p. 11). For this reason, an implementation plan is important, and many have been tried.

Pavsidis (1984) summarized early experiences of TQM efforts in the United States, noting that, as of 1984, no U.S. company had stated in writing that it had achieved a TQM environment. This is not the case today, especially among the winners of the Malcolm Baldrige Award, which "focuses on the applicant's total quality system" (*Malcomb Baldrige National Quality Award*, 1992 Guidelines).[2] According to Pavsidis's review, several components are common to successful TQM programs:

COMMON ELEMENTS IN SUCCESSFUL TQM PROGRAMS	1. Specific company goals and objectives must be established. 2. Sufficient resources must be devoted to the TQM effort. 3. The team approach and pilot projects should be used. 4. A TQM resource person should facilitate or lead the activities. 5. Successful programs can take 3–7 years to develop.

To a large extent, TQM programs must be custom fit (*Business Week* 1991, p. 57). As a result, TQM programs usually start with rather generic lists of requirements (such as those above) and proceed by translating them into requirements specific to the given company. Most published TQM "success stories" follow this format by describing the steps a particular company took to effect such a translation. It is unfortunate that more TQM failures have not been documented, since it is equally important to know where the pitfalls lie. The remainder of this section gives some alternative lists of TQM components and ends with a recommended plan for starting a TQM program.

Kenworthy (1986) highlights four TQM components: top management commitment, top-down training, pilot projects, and follow-up training, all held together by a comprehensive implementation plan. Snee (1986) divides TQM into three basic elements: philosophy, policies and procedures, and tools. Philosophy includes things like management commitment, an emphasis on continuous improvement, a companywide focus, and a customer focus. Policies and procedures consist of training, resources, rewards, and organization. Tools include computers, technology, SPC, DOE, and measurement systems.

Mozer (1984) describes the six-step approach taken by a Japan-based Hewlett Packard plant in winning the Deming Prize:

ONE COMPANY'S APPROACH TO TQM	1. Having a commitment to continuous quality improvement 2. Using data to analyze and solve problems 3. Identifying exactly who is responsible for taking action 4. Actively obtaining feedback from customers 5. Using the Deming cycle ("plan–do–check–act") 6. Using statistics as a tool

---

[2]This award was first given in 1988.

Under TQM, each functional unit in an organization is responsible for improving quality, but Brache and Rummler (1988) warn that departments should not operate alone but in conjunction with the overall TQM effort. They give several examples of how one department's efforts to optimize quality can have a negative impact on another department's efforts. This could happen, for example, if the marketing department decides to improve customer relations by promising delivery times and custom features that turn out to be difficult for the manufacturing department to achieve. Coordination between departments is as important as quality within departments.

Combining the above recommendations, we can outline a general plan for starting and continuing a TQM program:

STEPS IN
ACHIEVING TQM

1. Begin by understanding exactly what TQM is. Study the experience of others. Create a general statement of the company's mission, level of commitment, goals, and objectives regarding TQM.
2. Translate this list into company-specific requirements, goals, and objectives for each department.
3. Establish the organizational structure necessary to support the program. Delineate responsibilities for TQM, training schedules, resources, project teams, reporting and documentation procedures, cross-functional communication, and vendor and supplier participation.
4. Conduct regular reviews and evaluations of progress. Use these results to make improvements in the previous steps.

Concerning step 1, with the advent of the Malcolm Baldrige Award (see Section 14.6), a convenient way of studying the experience of others is to obtain published reports of award winners. Since one condition imposed on awardees is their promise to disseminate information on their quality system, these data are readily available. Step 3 requires an investigation of the types and level of training needed (see Section 14.3), a consideration of how to organize the collection and analysis of data (Section 14.5), a method for communicating results and needs across departments, and procedures for involving suppliers and customers in the program (Sections 14.7 and 14.8). The reader is referred to Damon (1990) for a good example of translating general TQM goals into company-specific ones, complete with time schedules, stated goals, training topics, strategic objectives, and organizational structure.

14.3   TRAINING

Training makes up a large part of most TQM programs, both in time and in money. With several topics and several hours multiplied by several people, effective training is certainly a subject worthy of attention. Since training is so essential to TQM, there is no shortage of literature and advice on the topic. This section attempts to answer the questions of how much training should be done, who gets trained, and what methods work best.

In a TQM program, it stands to reason that *everyone* should be trained in quality improvement. Most recommend that training be done in a **top-down** fashion; that is, CEOs and senior management are trained first, followed by middle managers and line workers. The rationale is that, knowing immediate supervisors are knowledgeable and supportive of training, each lower level in the company is much more likely to give training the serious attention it deserves (Hooper 1989, p. 25). The only drawback to this approach is that time can exact a toll on a company's enthusiasm if training takes too long to filter through the organization. Noting this, Roth (1989, p. 64) stresses the importance of proper timing and suggests that training begin at both the top and bottom simultaneously. Making the same observation, Feninger (1988) advises that quality improvement methods be put into practice quickly.

Training topics in quality improvement include quality philosophy, statistical concepts, descriptive statistics and graphical methods, control charts, acceptance sampling, problem-solving methods, experimental design, and reliability. Different levels in the company usually get different mixes of these topics. In the United States it is common to find statistical fundamentals and control charts covered in two- to five-day seminars. In contrast, Japanese training in basic QC methods often averages around 188 hr spread over months (Box et al. 1988, p. 38). Although it is tempting to offer more classes rather than fewer, theory and experience indicate that students should put knowledge into practice immediately for best results; that is, there should be sufficient time between successive classes for students to apply what they have learned.

Concerning the *approach* to training, there are many suggestions. Most important, experience has shown that quality training should be reality-based and results-oriented (Roth 1989, p. 82; Hooper 1989, p. 26; Pennucci 1985, p. 80; Mandel 1989, p. 54). General examples that do not have immediate relevance to an employee's job are not as easily applied as job-related examples. In order of effectiveness, job-related examples using company data are preferable to case studies, which in turn are preferable to general examples and theory. Case studies, even those from a relevant job or industry, tend to be fairly structured and do not actively involve students in learning the techniques (Hooper 1989, p. 26). Of course, there is a downside to using company data and real problems. First, there are no guarantees that a problem will be pushed through to a successful conclusion. Second, real problems tend to take time, and their completion may not coincide with the length of the training classes. For these reasons, students should choose class problems that have a reasonable chance of being solved. Examining and measuring process variation on a key operation, for example, would be a better choice than trying to optimize a process yield. Harder problems can be tackled as problem-solving experience is gained.

Homework (Mandel 1989, p. 54) and sometimes even exams (Pennucci 1985, p. 81) are good methods for ensuring mastery of the details of computational procedures, but the focus should be on problem solving, not statistics per se (Hooper 1989, p. 25). Describing the Japanese experience with statistical methods,

Ishikawa (1985) recommends that simple graphical and numerical methods be emphasized first.

Training works well when classes are homogeneous in their abilities and job functions. The greater the disparity in ability or experience, the greater the difficulty in using real examples that are understandable by all. Homogeneity can be achieved by training within departments, across the same managerial levels, within the same operations, and so on. Conversely, nonhomogeneity plays a role in forming project teams after initial training is completed. Teams that consist of internal 'customers' and 'suppliers' (see Section 2.4) are often more able to articulate and solve problems than teams whose members have identical backgrounds.

14.4   STEPS IN PROBLEM SOLVING

After organizing a TQM program, individuals and teams must eventually begin to tackle quality problems. Initially, problems are easily articulated: reduce scrap, improve yields, reduce nonconformities, and so on. However, *solving* these problems requires more careful descriptions of problems and the underlying processes, with an eye toward *decomposing* the initial problems into simpler, more readily attackable ones. It is this second step in the problem-solving sequence that can cause difficulties for those who are first using statistical methods. Specifically, the most asked questions at this stage are: What tools do I use? When do I use them? Where do I use them?

To this end, it is useful to have a structured problem-solving sequence to follow. An excellent SPC problem-solving technique has been outlined by Butler and Bryce (1986), who give a step-by-step method for clearly defining and understanding a problem, as in Table 14.1. Notice that control charts and DOE methods occur toward the end of the list, whereas the simpler descriptive methods (Pareto charts, cause-and-effect diagrams, location charts, etc.) come first. This is in line with the recommendations of many others, including Ishikawa, who, in describing the Japanese experience with statistical methods, states that the 1950s were a period of overemphasis on 'sophisticated' methods (control charts, etc.) when "simpler methods would have sufficed" (Ishikawa 1985).

Deming and Juran have also attempted to structure the problem-solving process. Deming suggests the Deming cycle ('Plan, Do, Act, Evaluate'; Gitlow et al. 1989, p. 19 and Deming 1982), while Juran uses his 'breakthrough sequence' (see Section 2.3 and Juran and Gryna 1980).

Other quality-specific problem-solving methods have been offered by Pyzdec (1985) and Skrabec (1986). Pyzdec's scope is slightly different from the others, in that he focuses on process control. He recommends creating process flow diagrams, but then addresses determining the adequacy of the measurement system, identifying key process variables, choosing an appropriate control chart, eliminating special causes, and estimating process capability.

One reference that is invaluable to any problem solver is *How To Solve It* by George Polya (1945). To a large extent, Polya's four-step procedure is at the heart of all problem-solving techniques, including those outlined above. Polya's book

TABLE 14.1   A Step-By-Step Method for Solving Quality Problems

Steps	Tools
1. Select a problem	Brainstorming
2. Investigate and thoroughly understand the current practice	Process flow chart
3. Decompose the problem into simpler parts and determine the type of data to collect	Cause-and-effect diagram
4. Collect the data	Check sheets; measurement systems
5. Analyze the data	Paretos, histograms, run charts, scatter plots, location diagrams, descriptive statistics
6. If the analysis suggests a solution, make the necessary fix, return to step 4, collect more data to see if improvement is attained If no solution is suggested, go to step 7	Management decision
7. Investigate further and return to step 2 (if necessary) or 4	Experimental design
8. When sufficient improvements have been made, establish regular monitoring of process	Control charts

embellishes upon these steps and provides numerous examples of their application.[3] The following is a list of these steps, along with a brief translation of its application to quality improvement problems:

POLYA'S
PROBLEM-SOLVING
STEPS

1.   *Understand the problem.* For quality improvement problems, this is best done by understanding the *process*, which can be accomplished by creating a process flow diagram.

2.   *Devise a plan.* For quality problems, this usually means using cause-and-effect diagrams and Pareto charts to decompose a problem or to identify an auxiliary problem to solve. It also means choosing the statistical method of attack.

3.   *Carry out the plan.* Collect the data and identify a possible solution.

4.   *Evaluate the solution.* Implement the proposed solution. Check to see whether it has solved or reduced the initial quality problem.

---

[3]Polya's applications are primarily in mathematics, but his methods can be applied to any problem.

| 14.5 | COMPUTERIZATION |

Both the repetitive environment of high-volume production and the continuously changing environment of custom-built systems are well served by computers. In terms of quality, computers run manufacturing and test equipment (NC machines, ATE, robots), collect data (on-line automated gaging and measurement), store data (quality databases, daily logs of operations), disseminate data (local area networks), and summarize data (SPC and DOE software).

An obvious advantage of computer-controlled production is the elimination of human error in highly repetitive operations. Automated equipment tends to be faster, less subject to errors caused by fatigue (e.g., wear), and generally capable of higher precision than human workers. Each one of these virtues translates into a reduction in part-to-part variation, one of the prime goals of continuous improvement.

Because of on-line gages that transfer measurements electronically to databases, the reliability and speed of data acquisition are improved. Better data surely foster better decisions, but speed may be the more important of the two factors. Much less time away from production is needed when data are gathered by computer. In addition, the increased timeliness with which data become available for analysis enhances its usefulness for process control.

Computers are an essential tool for problem solving generally and for finding special causes, in particular. The key to effectively using control charts is traceability, and quality databases can help in tracking down special causes. To be effective, the database should be designed to support traceability. Some common factors to include are the following:

COMMON FIELDS
IN A QUALITY
DATABASE

Date	Raw material batch number
Time	Raw material supplier
Operator	Customer ID number
Operation	Customer returns/problems/dates
Part type/name	Shipping date
Part ID number	Labor/material cost
Quality characteristic	Disposition of nonconforming product

Each company must personalize its own list of items for the database. One recommendation for doing this is to focus on four types of product attributes (*SPC in Automated Manufacturing*, p. 203):

TYPES OF PRODUCT
ATTRIBUTES

1. Product-related (e.g., which dimension on which part type)
2. Chronology-related (e.g., the date a part was made)
3. Equipment-related (e.g., the machine used to make the part)
4. Workforce-related (e.g., which operator made the part)

The expansion of each category produces a customized list of items for the database.

One popular method for obtaining and recording data on complex products is to use a document (sometimes called a 'traveller') that stays with a product as it passes through various production steps. The traveller specifies what actions are to be taken at which production steps, what materials are needed, the specification limits, and other relevant information. Travellers usually have sign-off and date boxes for recording completed process steps. When such documents are used, attention must be paid to the very first step, properly filling out the traveller. Catching errors at this stage is crucial, since faulty documents become recipes for making defective or nonconforming products.

With computers, creating and analyzing control charts are not the laborious tasks they once were. Software for creating control charts is abundant. As with data collection, automating the data *analysis* step increases the chance that the data will provide useful process information. To increase access to process data, several on-site computers can be connected in a **local area network** (LAN). LANs make it possible for workers in different parts of the company to trace through various operations that might affect quality in their jobs ("Software Automates Pharmaceutical Data Analysis" 1991).

Since computer hardware and software decisions can involve substantial monetary commitments, a question arises as to the most cost-effective approach to computerization. In this quest, it is natural to ask how successful SPC users—for example, the Japanese—have handled this question. Interestingly, it appears that the Japanese have not stressed SPC software, DOE software, or computer graphics for analyzing process data. Much of their analysis has been done 'by hand' (Box et al. 1988, p 39). The conclusion one must draw is that, although extensive computerization must certainly help a TQM or SPC program, it is not necessarily required for success. However, as Box points out, computerization could help successful programs be even more successful.

14.6   THE MALCOLM BALDRIGE NATIONAL QUALITY AWARD

The Malcolm Baldrige National Quality Improvement Act was signed into being on August 20, 1987, after about five years of work by various individuals and professional organizations. The award, inspired by the desire to increase U.S. quality and productivity, was intended to serve the same purpose in the United States that the yearly Deming Prize has served in Japan since 1951. The first Baldrige Awards were given in 1988 (DeCario and Sterett 1990).

The National Institute of Standards and Technology (NIST; see Section 10.3) is charged with the responsibility for the Baldrige Award, although the administration of the award activities is handled by another organization. In 1991, NIST selected ASQC as the sole administrator of the award. Prior to 1991, administration was handled by ASQC along with the American Productivity and Quality Center.

The award administrator evaluates applications on three levels: by a board of judges, senior examiners, and examiners.

The Baldrige Award, which itself undergoes improvements each year, focuses on a company's total quality system by dividing it into several categories. Each category and subcategory are assigned a point value (Figure 14.1), with a possible total of 1,000. There are three award categories: manufacturing companies, service companies, and small businesses. No more than two awards per category may be given in any one year.

Since the Baldrige Award was patterned after the Deming Prize, comparisons between the two are inevitable. Bush and Dooley (1989) highlight several differences. Most notably, the Deming Prize guidelines consist of a single page of categories, whereas the Baldrige Award guidelines contain more than 20 pages of detailed explanations of categories and subcategories. The Deming Prize administrators justify the short list of categories on the grounds that every company is different and should be given some latitude in responding to the categories. Although both awards are based on achieving minimum standards, the Baldrige Award further specifies that there is a limit of two awards per category in any one year. The Deming Prize, on the other hand, may be awarded to any number of companies or individuals that meet the minimum requirements. Perhaps the greatest difference is that the Baldrige Award places much more emphasis on customer satisfaction (300 out of the 1,000 points) than does the Deming Prize.

The Baldrige Award influences quality in a number of ways. As many applicants have stated, self-evaluation and improvement are the most important part of the award process. Filling out the application is just the last step in this process. Furthermore, companies that apply for the award are visited by examiners and supplied with written evaluations of their quality systems (Stratton 1990, p. 30). Given the nominal application fee, this unbiased evaluation is a service that would otherwise be fairly expensive for most companies. Finally, the Baldrige Award has been suggested as a template for those initiating TQM systems, since it attempts to subdivide the various quality categories necessary in any TQM program.

FIGURE 14.1
Examination
categories in the
Malcolm Baldrige
Award (1993)

Examination Categories/Items	Point Values
1.0 Leadership	**95**
1.1 Senior Executive Leadership. 45	
1.2 Management for Quality 25	
1.3 Public Responsibility. 25	
2.0 Information and Analysis	**75**
2.1 Scope and Management of Quality and Performance Data and Information 15	
2.2 Competitive Comparisons and Benchmarks 20	
2.3 Analysis and Uses of Company-Level Data 40	

3.0 Strategic Quality Planning		**60**
3.1	Strategic Quality and Company Performance Planning Process	35
3.2	Quality and Performance Plans	25
4.0 Human Resource Development and Management		**150**
4.1	Human Resource Management	20
4.2	Employee Involvement	40
4.3	Employee Education and Training	40
4.4	Employee Performance and Recognition	25
4.5	Employee Well-Being and Morale	25
5.0 Management of Process Quality		**140**
5.1	Design and Introduction of Quality Products and Services	40
5.2	Process Management—Product and Service Production and Delivery Processes	35
5.3	Process Management—Business Processes and Support Services	30
5.4	Supplier Quality	20
5.5	Quality Assessment	15
6.0 Quality and Operational Results		**180**
6.1	Product and Service Quality Results	70
6.2	Company Operational Results	50
6.3	Business Process and Support Service Results	25
6.4	Supplier Quality Results	35
7.0 Customer Focus and Satisfaction		**300**
7.1	Future Requirements and Expectations of Customers	35
7.2	Customer Relationship Management	65
7.3	Commitment to Customers	15
7.4	Customer Satisfaction Determination	30
7.5	Customer Satisfaction Results	85
7.6	Customer Satisfaction Comparison	70
TOTAL POINTS		**1000**

14.7   VENDOR PARTICIPATION

Vendors are a company's external suppliers of raw materials, parts, subassemblies, and services. Because the quality of a company's products is linked to the quality of the vendor's product, vendors must rightly be considered a part of a company's production and quality systems.

Vendors, by definition, are external to the company, and this poses special problems in regard to quality. Almost all the current recommendations for interacting with vendors can be simplified to one requirement: communication. In particular, a company must seek to establish *active* communication with vendors, instead of communicating only through invoices and purchase orders. Some examples of active communications are given by Schrock (1991):

<table>
<tr>
<td>COMMUNICATING
WITH VENDORS</td>
<td>

1.   Be clear about which quality characteristics or tests are important.
2.   Make specification limits realistic; do not choose them arbitrarily. Give reasons for your choices to the vendor.
3.   To avoid confusion, put product requirements in writing.
4.   Periodically visit the vendor's facility (source inspection).
5.   Have the vendor visit your facility.

</td>
</tr>
</table>

Beyond that, various methods have been used to include vendors in a company's quality system:

<table>
<tr>
<td>THE VENDOR'S
ROLE IN THE
QUALITY SYSTEM</td>
<td>

1.   Sponsor an SPC/TQM awareness seminar for vendors.
2.   Ask vendors to supply statistical evidence of process control efforts on key quality characteristics.
3.   Report inspection and test results back to vendors.

</td>
</tr>
</table>

Assessing and improving vendor performance are accomplished in the same manner as for any process, with statistical methods. Because they are external to the company, directly assessing performance with control charts is not possible. At best, vendors can be asked to submit control chart data along with each shipment. The traditional approach is to apply acceptance sampling to incoming products (see Chapter 11). This offers more immediate control over which shipments are acceptable, but, except for gross problems, it offers little in the way of process improvement at the vendor's facility. However, documenting inspection and test results can still be of value in monitoring a vendor's long-term performance. For example, various performance statistics (delivery times, rejection rates at incoming inspection) can be tracked. Control charts and Pareto charts are often used for this purpose. Performance monitoring allows for feedback to help the vendor improve quality.

14.8   CUSTOMER PARTICIPATION AND QFD

Modern quality programs use customer input earlier in the production and design stages than ever before. Traditionally, customer involvement was limited to direct inputs such as customer complaints and/or returns, warranty data, marketing focus groups, surveys, and indirect measures such as changes in sales volume. Unfortunately, most of these inputs occur *after* production is complete, which greatly diminishes the effectiveness of using such data for process or product improvement.

Since the 1970s, one of the most significant improvements in using customer data has been in the way customer input is translated into specific product characteristics and specifications. One tool for accomplishing this is **quality function deployment (QFD).** The QFD technique was developed by the Japanese in the late 1960s and early 1970s at Bridgestone Tire Corporation and Mitsubishi Heavy Industries, Ltd. It was given its name in a 1978 book on the subject by professors Yoji Akao and Shigeru Mizuno (*Business Week* 1991, p. 21).

At the heart of QFD is the desire to transmit customer input throughout the design and production stages, starting with design and cascading specific requirements through the various departments. The usual sequence would be first to translate customer requirements into engineering requirements. This is followed by a translation of engineering requirements into specific part dimensions, which then leads to specific process and production requirements. The discussion in Section 2.5 describes the first step: converting customer requirements into design requirements.

The primary tool for effecting each translation is the QFD matrix, which simply lists the input requirements on the left side and the resulting output requirements along the top (e.g., recall Figure 2.8). Two matrices are linked by making the output requirements of one the inputs to the second. Figure 14.2 illustrates a typical sequence, beginning with customer requirements and progressing through production requirements.

Sullivan (1986, p. 41) gives an eight-step procedure for creating a QFD matrix. Without going into this level of detail, we summarize the basic steps as follows. Among the first tasks is the collection and ranking of customer requirements in order of relative importance. Since some customer concerns are more important than others, ranking them provides useful information for choosing the exact set of engineering requirements that will be changed. The next steps involve showing the

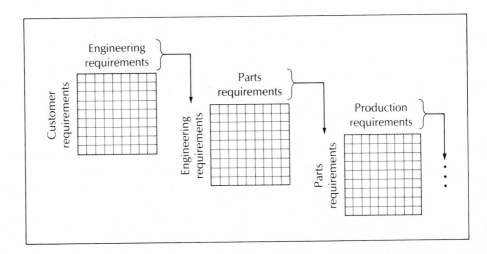

FIGURE 14.2
A sequence of QFD matrices

customer input list to design engineers and having them create a list of engineering attributes that are related to the customer inputs. Within the matrix, at the intersection of each customer input row with each engineering requirement column, an indication is made of the relationship between the input and output. For example, if one customer requirement is that a car be 'quieter' when the windows are rolled up, then this requirement would be positively related to the engineering attribute 'increased window seal thickness' but negatively related to the attribute 'thinner window glass.' Many QFD matrices also include a triangular grid affixed to the top of the matrix showing some of the relationships between the output requirements. Thus, 'increased seal thickness' might be negatively related to another engineering attribute such as 'window closing resistance.' The final steps involve the selection of the particular set of output requirements that will be changed or 'deployed' to the next level.

QFD matrices are often quite large. As an example, Hauser and Clausing (1988) report that matrices with as many as 30 or 100 customer requirements are not uncommon. The end result, however, is an integrated set of design and production plans that not only meets end user requirements but also enhances communication within the company. The latter benefit necessarily has a positive effect on increasing quality and decreasing production ('cycle') time.

REFERENCES FOR CHAPTER 14

Aquino, M. A. 1987. "Tomorrow's Total Quality Manager." *Quality Progress,* November, pp. 48–50.

Box, G. E. P., R. N. Kackar, V. N. Nair, M. Phadke, A. C. Shoemaker, and C. F. Jeff Wu. 1988. "Quality Practices in Japan." *Quality Progress,* March, pp. 37–41.

Brache, A. P., and G. A. Rummler. 1988. "The Three Levels of Quality." *Quality Progress,* October, pp. 46–51.

Bush, D., and K. Dooley. 1989. "The Deming Prize and Baldrige Award: How They Compare." *Quality Progress,* January, pp. 28–30.

Business Week. 1991. Special Issue: The Quality Imperative, "Dueling Pioneers." January, p. 17.

Butler, C., and G. Rex Bryce. 1986. "Implementing SPC with Signetics Production Personnel." *Quality Progress,* April, pp. 42–50.

Damon, G. A. 1990. "Total Quality Management (TQM) at Pearl Harbor Naval Shipyard (NSY)." *Quality Engineering* (no. 4):421–437.

DeCario, N. J., and W. Kent Sterett. 1990. "History of the Malcolm Baldrige National Quality Award." *Quality Progress,* March, pp. 21–27.

Deming, W. E. 1982. *Quality, Productivity, and Competitive Position.* Cambridge, MA: MIT Press.

Feninger, J. E. 1988. "Quality Training." *Quality Progress,* March.

Gitlow, H., S. Gitlow, A. Oppenheim, and R. Oppenheim. 1989. *Tools and Methods for the Improvement of Quality.* Homewood, IL: Irwin.

Hauser, J. R., and D. Clausing. 1988. "The House of Quality." *Harvard Business Review,* May–June, pp. 63–73.

Hooper, J. H. 1989. "Making Statistical Training Effective." *Quality Progress,* February, pp. 24–27.

Ishikawa, K. 1985. *What Is Total Quality Control?: The Japanese Way.* New York: Prentice Hall.

Juran, J. M., and F. M. Gryna. 1957. *Quality Control Handbook.* New York: McGraw-Hill.

Juran, J. M., and F. M. Gryna. 1980. *Quality Planning and Analysis,* New York: McGraw-Hill.

Keats, J. B., and N. F. Hubele (editors). *Statistical Process Control in Automated Manufacturing.* New York: Marcel Dekker.

Kenworthy, H. W. 1986. "Total Quality Concept: A Proven Path to Success." *Quality Progress,* July, pp. 21–24.

Mandel, B. J. 1989. "Teaching Statistics at the Adult Employee Level." *Quality Progress,* October, pp. 53–55.

Mozer, C. 1984. "TQC: Science, Not Witchcraft." *Quality Progress,* September, pp. 30–33.

Pavsidis, C. 1984. "Total Quality Control: An Overview of Current Efforts." *Quality Progress,* September, pp. 28–29.

Pennucci, N. J. 1985. "A Model Training Program." *Quality Progress,* November, pp. 80–81.

Polya, G. 1945. *How To Solve It.* Princeton, NJ: Princeton University Press.

Pyzdec, T. 1985. "A Ten-Step Plan for Process Control Studies." *Quality Progress,* July, pp. 18–22.

Roth, W. R. 1989. "Get Training Out of the Classroom." *Quality Progress,* May, pp. 62–64.

Schrock, E. M. 1991. "Improving Vendor-Consumer Relationships." *Quality Engineering* (no. 3):425–432.

Skrabec, Q. R. 1986. "Process Diagnostics: Seven Steps for Problem-Solving." *Quality Progress,* November, pp. 40–44.

Snee, R. D. 1986. "In Pursuit of Total Quality." *Quality Progress,* August, pp. 25–28.

"Software Automates Pharmaceutical Data Analysis." 1991. *Process Industry Quarterly,* 3rd quarter.

Stratton, B. 1990. "Making the Malcolm Baldrige Award Better." *Quality Progress,* March, pp. 30–32.

Sullivan, L. P. 1986. "Quality Function Deployment." *Quality Progress,* June, pp. 39–50.

APPENDICES

APPENDIX OUTLINE

APPENDIX 1: CONTROL CHART CONSTANTS*

Sample Size (n)	Process Variation				Process Average				Process Standard Deviation		
	D_3	D_4	B_3	B_4	A_2	A_3	A_6	A_7	d_2	c_4	d_3
2	0	3.267	0	3.267	1.880	2.659	1.880	1.880	1.128	0.7979	0.853
3	0	2.574	0	2.568	1.023	1.954	1.187	1.067	1.693	0.8862	0.888
4	0	2.282	0	2.266	0.729	1.628	0.796	0.796	2.059	0.9213	0.880
5	0	2.114	0	2.089	0.577	1.427	0.691	0.660	2.326	0.9400	0.864
6	0	2.004	0.030	1.970	0.483	1.287	0.549	0.580	2.534	0.9515	0.848
7	0.076	1.924	0.118	1.882	0.419	1.182	0.509	0.521	2.704	0.9594	0.833
8	0.136	1.864	0.185	1.815	0.373	1.099	0.434	0.477	2.847	0.9650	0.820
9	0.184	1.816	0.239	1.761	0.337	1.032	0.412	0.444	2.970	0.9693	0.808
10	0.223	1.777	0.284	1.716	0.308	0.975	0.365	0.419	3.078	0.9727	0.797
11	0.256	1.744	0.321	1.679	0.285	0.927	0.350	0.399	3.173	0.9754	0.787
12	0.283	1.717	0.354	1.646	0.266	0.886	0.317	0.382	3.258	0.9776	0.778
13	0.307	1.693	0.382	1.618	0.249	0.850	0.306	0.368	3.336	0.9794	0.770
14	0.328	1.672	0.406	1.594	0.235	0.817	0.282	0.356	3.407	0.9810	0.763
15	0.347	1.653	0.428	1.572	0.223	0.789	0.274	0.346	3.472	0.9823	0.756
16	0.363	1.637	0.448	1.552	0.212	0.763	0.257	0.337	3.532	0.9835	0.750
17	0.378	1.622	0.466	1.534	0.203	0.739	0.250	0.329	3.588	0.9845	0.744
18	0.391	1.608	0.482	1.518	0.194	0.718	0.237	0.322	3.640	0.9854	0.739
19	0.403	1.597	0.497	1.503	0.187	0.698	0.231	0.315	3.689	0.9862	0.734
20	0.415	1.585	0.510	1.490	0.180	0.680	0.218	0.308	3.735	0.9869	0.729
21	0.425	1.575	0.523	1.477	0.173	0.663	0.215	0.303	3.778	0.9876	0.724
22	0.434	1.566	0.534	1.466	0.167	0.647	0.204	0.298	3.819	0.9882	0.720
23	0.443	1.557	0.545	1.455	0.162	0.633	0.202	0.292	3.858	0.9887	0.716
24	0.451	1.548	0.555	1.445	0.157	0.619	0.192	0.288	3.895	0.9892	0.712
25	0.459	1.541	0.565	1.435	0.153	0.606	0.191	0.284	3.931	0.9896	0.708

*Values in this table were generated using MathCAD version 3.1 software.

2

APPENDIX 2: STANDARD NORMAL DISTRIBUTION (RIGHT-TAIL PROBABILITIES)

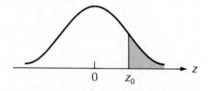

z_0	$P(z > z_0)$	z_0	$P(z > z_0)$	z_0	$P(z > z_0)$	z_0	$P(z > z_0)$	z_0	$P(z > z_0)$	z_0	$P(z > z_0)$
0.00	0.500000	0.31	0.378281	0.62	0.267629	0.93	0.176185	1.24	0.107488	1.55	0.0605708
0.01	0.496011	0.32	0.374484	0.63	0.264347	0.94	0.173609	1.25	0.105650	1.56	0.0593799
0.02	0.492022	0.33	0.370700	0.64	0.261086	0.95	0.171056	1.26	0.103835	1.57	0.0582076
0.03	0.488033	0.34	0.366928	0.65	0.257846	0.96	0.168528	1.27	0.102042	1.58	0.0570534
0.04	0.484047	0.35	0.363169	0.66	0.254627	0.97	0.166023	1.28	0.100273	1.59	0.0559174
0.05	0.480061	0.36	0.359424	0.67	0.251429	0.98	0.163543	1.29	0.098525	1.60	0.0547993
0.06	0.476078	0.37	0.355691	0.68	0.248252	0.99	0.161087	1.30	0.096801	1.61	0.0536989
0.07	0.472097	0.38	0.351973	0.69	0.245097	1.00	0.158655	1.31	0.095098	1.62	0.0526161
0.08	0.468119	0.39	0.348268	0.70	0.241964	1.01	0.156248	1.32	0.093418	1.63	0.0515507
0.09	0.464144	0.40	0.344578	0.71	0.238852	1.02	0.153864	1.33	0.091759	1.64	0.0505026
0.10	0.460172	0.41	0.340903	0.72	0.235762	1.03	0.151505	1.34	0.090123	1.65	0.0494714
0.11	0.456205	0.42	0.337243	0.73	0.232695	1.04	0.149170	1.35	0.088508	1.66	0.0484572
0.12	0.452242	0.43	0.333598	0.74	0.229650	1.05	0.146859	1.36	0.086915	1.67	0.0474597
0.13	0.448283	0.44	0.329969	0.75	0.226627	1.06	0.144572	1.37	0.085343	1.68	0.0464786
0.14	0.444330	0.45	0.326355	0.76	0.223627	1.07	0.142310	1.38	0.083793	1.69	0.0455139
0.15	0.440382	0.46	0.322758	0.77	0.220650	1.08	0.140071	1.39	0.082264	1.70	0.0445654
0.16	0.436441	0.47	0.319178	0.78	0.217695	1.09	0.137857	1.40	0.080757	1.71	0.0436329
0.17	0.432505	0.48	0.315614	0.79	0.214764	1.10	0.135666	1.41	0.079270	1.72	0.0427162
0.18	0.428576	0.49	0.312067	0.80	0.211855	1.11	0.133500	1.42	0.077804	1.73	0.0418151
0.19	0.424655	0.50	0.308538	0.81	0.208970	1.12	0.131357	1.43	0.076359	1.74	0.0409295
0.20	0.420740	0.51	0.305026	0.82	0.206108	1.13	0.129238	1.44	0.074934	1.75	0.0400591
0.21	0.416834	0.52	0.301532	0.83	0.203269	1.14	0.127143	1.45	0.073529	1.76	0.0392039
0.22	0.412936	0.53	0.298056	0.84	0.200454	1.15	0.125072	1.46	0.072145	1.77	0.0383635
0.23	0.409046	0.54	0.294599	0.85	0.197662	1.16	0.123024	1.47	0.070781	1.78	0.0375379
0.24	0.405165	0.55	0.291160	0.86	0.194894	1.17	0.121001	1.48	0.069437	1.79	0.0367269
0.25	0.401294	0.56	0.287740	0.87	0.192150	1.18	0.119000	1.49	0.068112	1.80	0.0359303
0.26	0.397432	0.57	0.284339	0.88	0.189430	1.19	0.117023	1.50	0.0668072	1.81	0.0351478
0.27	0.393580	0.58	0.280957	0.89	0.186733	1.20	0.115070	1.51	0.0655217	1.82	0.0343794
0.28	0.389739	0.59	0.277595	0.90	0.184060	1.21	0.113140	1.52	0.0642555	1.83	0.0336249
0.29	0.385908	0.60	0.274253	0.91	0.181411	1.22	0.111233	1.53	0.0630084	1.84	0.0328841
0.30	0.382089	0.61	0.270931	0.92	0.178786	1.23	0.109349	1.54	0.0617802	1.85	0.0321567

z_0	$P(z > z_0)$	z_0	$P(z > z_0)$	z_0	$P(z > z_0)$	z_0	$P(z > z_0)$	z_0	$P(z > z_0)$	z_0	$P(z > z_0)$
1.86	0.0314427	2.09	0.0183088	2.32	0.0101704	2.55	0.00538617	2.78	0.00271803	3.10	0.00096760
1.87	0.0307419	2.10	0.0178643	2.33	0.0099031	2.56	0.00523365	2.79	0.00263548	3.20	0.00068714
1.88	0.0300540	2.11	0.0174291	2.34	0.0096418	2.57	0.00508493	2.80	0.00255519	3.30	0.00048342
1.89	0.0293789	2.12	0.0170029	2.35	0.0093867	2.58	0.00494003	2.81	0.00247711	3.40	0.00033693
1.90	0.0287165	2.13	0.0165858	2.36	0.0091375	2.59	0.00479883	2.82	0.00240123	3.50	0.00023263
1.91	0.0280665	2.14	0.0161773	2.37	0.0088940	2.60	0.00466120	2.83	0.00232744	3.60	0.00015911
1.92	0.0274289	2.15	0.0157775	2.38	0.0086563	2.61	0.00452715	2.84	0.00225574	3.70	0.00010780
1.93	0.0268034	2.16	0.0153863	2.39	0.0084242	2.62	0.00439650	2.85	0.00218600	3.80	0.00007235
1.94	0.0261898	2.17	0.0150034	2.40	0.00819755	2.63	0.00426930	2.86	0.00211829	3.90	0.00004810
1.95	0.0255880	2.18	0.0146286	2.41	0.00797623	2.64	0.00414532	2.87	0.00205243	4.00	0.00003167
1.96	0.0249978	2.19	0.0142621	2.42	0.00776023	2.65	0.00402462	2.88	0.00198847	4.10	0.00002066
1.97	0.0244191	2.20	0.0139034	2.43	0.00754941	2.66	0.00390708	2.89	0.00192630	4.20	0.00001355
1.98	0.0238517	2.21	0.0135525	2.44	0.00734365	2.67	0.00379258	2.90	0.00186586	4.30	0.00000854
1.99	0.0232954	2.22	0.0132093	2.45	0.00714284	2.68	0.00368118	2.91	0.00180721	4.40	0.00000541
2.00	0.0227501	2.23	0.0128736	2.46	0.00694686	2.69	0.00357264	2.92	0.00175023	4.50	0.00000340
2.01	0.0222155	2.24	0.0125454	2.47	0.00675565	2.70	0.00346702	2.93	0.00169486	4.60	0.00000211
2.02	0.0216916	2.25	0.0122244	2.48	0.00656915	2.71	0.00336421	2.94	0.00164115	4.70	0.00000130
2.03	0.0211782	2.26	0.0119106	2.49	0.00638717	2.72	0.00326413	2.95	0.00158894	4.80	0.00000079
2.04	0.0206751	2.27	0.0116038	2.50	0.00620967	2.73	0.00316679	2.96	0.00153828	4.90	0.00000048
2.05	0.0201821	2.28	0.0113038	2.51	0.00603658	2.74	0.00307202	2.97	0.00148904	5.00	0.00000029
2.06	0.0196992	2.29	0.0110106	2.52	0.00586778	2.75	0.00297982	2.98	0.00144130		
2.07	0.0192261	2.30	0.0107241	2.53	0.00570315	2.76	0.00289011	2.99	0.00139493		
2.08	0.0187627	2.31	0.0104440	2.54	0.00554264	2.77	0.00280285	3.00	0.00134999		

Selected Percentage Points of the Standard Normal							
α	0.10	0.05	0.025	0.01	0.001	0.0001	0.00001	0.000001
z_α	1.282	1.645	1.960	2.326	3.090	3.719	4.265	4.753

APPENDIX 3: CUMULATIVE BINOMIAL DISTRIBUTION

n = 5

x	.01	.05	.10	.20	.30	.40	p .50	.60	.70	.80	.90	.95	.99	x
0	0.9510	0.7738	0.5905	0.3277	0.1681	0.0778	0.0313	0.0102	0.0024	0.0003	0.0000	0.0000	0.0000	0
1	0.9990	0.9774	0.9185	0.7373	0.5282	0.3370	0.1875	0.0870	0.0308	0.0067	0.0005	0.0000	0.0000	1
2	1.0000	0.9988	0.9914	0.9421	0.8369	0.6826	0.5000	0.3174	0.1631	0.0579	0.0086	0.0012	0.0000	2
3	1.0000	1.0000	0.9995	0.9933	0.9692	0.9130	0.8125	0.6630	0.4718	0.2627	0.0815	0.0226	0.0010	3
4	1.0000	1.0000	1.0000	0.9997	0.9976	0.9898	0.9688	0.9222	0.8319	0.6723	0.4095	0.2262	0.0490	4

n = 10

x	.01	.05	.10	.20	.30	.40	p .50	.60	.70	.80	.90	.95	.99	x
0	0.9044	0.5987	0.3487	0.1074	0.0282	0.0060	0.0010	0.0000	0.0000	0.0000	0.0000	0.0000	0.0000	0
1	0.9957	0.9139	0.7361	0.3758	0.1493	0.0464	0.0107	0.0017	0.0001	0.0000	0.0000	0.0000	0.0000	1
2	0.9999	0.9885	0.9298	0.6778	0.3828	0.1673	0.0547	0.0123	0.0016	0.0001	0.0000	0.0000	0.0000	2
3	1.0000	0.9990	0.9872	0.8791	0.6496	0.3823	0.1719	0.0548	0.0106	0.0009	0.0000	0.0000	0.0000	3
4	1.0000	0.9999	0.9984	0.9672	0.8497	0.6331	0.3770	0.1662	0.0473	0.0064	0.0001	0.0000	0.0000	4
5	1.0000	1.0000	0.9999	0.9936	0.9527	0.8338	0.6230	0.3669	0.1503	0.0328	0.0016	0.0001	0.0000	5
6	1.0000	1.0000	1.0000	0.9991	0.9894	0.9452	0.8281	0.6177	0.3504	0.1209	0.0128	0.0010	0.0000	6
7	1.0000	1.0000	1.0000	0.9999	0.9984	0.9877	0.9453	0.8327	0.6172	0.3222	0.0702	0.0115	0.0001	7
8	1.0000	1.0000	1.0000	1.0000	0.9999	0.9983	0.9893	0.9536	0.8507	0.6242	0.2639	0.0861	0.0043	8
9	1.0000	1.0000	1.0000	1.0000	1.0000	0.9999	0.9990	0.9940	0.9718	0.8926	0.6513	0.4013	0.0956	9

n = 15

x	.01	.05	.10	.20	.30	.40	p .50	.60	.70	.80	.90	.95	.99	x
0	0.8601	0.4633	0.2059	0.0352	0.0047	0.0005	0.0000	0.0000	0.0000	0.0000	0.0000	0.0000	0.0000	0
1	0.9904	0.8290	0.5490	0.1671	0.0353	0.0052	0.0005	0.0000	0.0000	0.0000	0.0000	0.0000	0.0000	1
2	0.9996	0.9638	0.8159	0.3980	0.1268	0.0271	0.0037	0.0003	0.0000	0.0000	0.0000	0.0000	0.0000	2
3	1.0000	0.9945	0.9444	0.6482	0.2969	0.0905	0.0176	0.0019	0.0001	0.0000	0.0000	0.0000	0.0000	3
4	1.0000	0.9994	0.9873	0.8358	0.5155	0.2173	0.0592	0.0093	0.0007	0.0000	0.0000	0.0000	0.0000	4
5	1.0000	0.9999	0.9978	0.9389	0.7216	0.4032	0.1509	0.0338	0.0037	0.0001	0.0000	0.0000	0.0000	5
6	1.0000	1.0000	0.9997	0.9819	0.8689	0.6098	0.3036	0.0950	0.0152	0.0008	0.0000	0.0000	0.0000	6
7	1.0000	1.0000	1.0000	0.9958	0.9500	0.7869	0.5000	0.2131	0.0500	0.0042	0.0000	0.0000	0.0000	7
8	1.0000	1.0000	1.0000	0.9992	0.9848	0.9050	0.6964	0.3902	0.1311	0.0181	0.0003	0.0000	0.0000	8
9	1.0000	1.0000	1.0000	0.9999	0.9963	0.9662	0.8491	0.5968	0.2784	0.0611	0.0022	0.0001	0.0000	9
10	1.0000	1.0000	1.0000	1.0000	0.9993	0.9907	0.9408	0.7827	0.4845	0.1642	0.0127	0.0006	0.0000	10
11	1.0000	1.0000	1.0000	1.0000	0.9999	0.9981	0.9824	0.9095	0.7031	0.3518	0.0556	0.0055	0.0000	11
12	1.0000	1.0000	1.0000	1.0000	1.0000	0.9997	0.9963	0.9729	0.8732	0.6020	0.1841	0.0362	0.0004	12
13	1.0000	1.0000	1.0000	1.0000	1.0000	1.0000	0.9995	0.9948	0.9647	0.8329	0.4510	0.1710	0.0096	13
14	1.0000	1.0000	1.0000	1.0000	1.0000	1.0000	1.0000	0.9995	0.9953	0.9648	0.7941	0.5367	0.1399	14

n = 20

x	.01	.05	.10	.20	.30	.40	p .50	.60	.70	.80	.90	.95	.99	x
0	0.8179	0.3585	0.1216	0.0115	0.0008	0.0000	0.0000	0.0000	0.0000	0.0000	0.0000	0.0000	0.0000	0
1	0.9831	0.7358	0.3917	0.0692	0.0076	0.0005	0.0000	0.0000	0.0000	0.0000	0.0000	0.0000	0.0000	1
2	0.9990	0.9245	0.6769	0.2061	0.0355	0.0036	0.0002	0.0000	0.0000	0.0000	0.0000	0.0000	0.0000	2
3	1.0000	0.9841	0.8670	0.4114	0.1071	0.0160	0.0013	0.0000	0.0000	0.0000	0.0000	0.0000	0.0000	3
4	1.0000	0.9974	0.9568	0.6296	0.2375	0.0510	0.0059	0.0003	0.0000	0.0000	0.0000	0.0000	0.0000	4
5	1.0000	0.9997	0.9887	0.8042	0.4164	0.1256	0.0207	0.0016	0.0000	0.0000	0.0000	0.0000	0.0000	5
6	1.0000	1.0000	0.9976	0.9133	0.6080	0.2500	0.0577	0.0065	0.0003	0.0000	0.0000	0.0000	0.0000	6
7	1.0000	1.0000	0.9996	0.9679	0.7723	0.4159	0.1316	0.0210	0.0013	0.0000	0.0000	0.0000	0.0000	7
8	1.0000	1.0000	0.9999	0.9900	0.8867	0.5956	0.2517	0.0565	0.0051	0.0001	0.0000	0.0000	0.0000	8
9	1.0000	1.0000	1.0000	0.9974	0.9520	0.7553	0.4119	0.1275	0.0171	0.0006	0.0000	0.0000	0.0000	9
10	1.0000	1.0000	1.0000	0.9994	0.9829	0.8725	0.5881	0.2447	0.0480	0.0026	0.0000	0.0000	0.0000	10
11	1.0000	1.0000	1.0000	0.9999	0.9949	0.9435	0.7483	0.4044	0.1133	0.0100	0.0001	0.0000	0.0000	11
12	1.0000	1.0000	1.0000	1.0000	0.9987	0.9790	0.8684	0.5841	0.2277	0.0321	0.0004	0.0000	0.0000	12
13	1.0000	1.0000	1.0000	1.0000	0.9997	0.9935	0.9423	0.7500	0.3920	0.0867	0.0024	0.0000	0.0000	13
14	1.0000	1.0000	1.0000	1.0000	1.0000	0.9984	0.9793	0.8744	0.5836	0.1958	0.0113	0.0003	0.0000	14
15	1.0000	1.0000	1.0000	1.0000	1.0000	0.9997	0.9941	0.9490	0.7625	0.3704	0.0432	0.0026	0.0000	15
16	1.0000	1.0000	1.0000	1.0000	1.0000	1.0000	0.9987	0.9840	0.8929	0.5886	0.1330	0.0159	0.0000	16
17	1.0000	1.0000	1.0000	1.0000	1.0000	1.0000	0.9998	0.9964	0.9645	0.7939	0.3231	0.0755	0.0010	17
18	1.0000	1.0000	1.0000	1.0000	1.0000	1.0000	1.0000	0.9995	0.9924	0.9308	0.6083	0.2642	0.0169	18
19	1.0000	1.0000	1.0000	1.0000	1.0000	1.0000	1.0000	1.0000	0.9992	0.9885	0.8784	0.6415	0.1821	19

n = 25

x	.01	.05	.10	.20	.30	.40	p .50	.60	.70	.80	.90	.95	.99	x
0	0.7778	0.2774	0.0718	0.0038	0.0001	0.0000	0.0000	0.0000	0.0000	0.0000	0.0000	0.0000	0.0000	0
1	0.9742	0.6424	0.2712	0.0274	0.0016	0.0001	0.0000	0.0000	0.0000	0.0000	0.0000	0.0000	0.0000	1
2	0.9980	0.8729	0.5371	0.0982	0.0090	0.0004	0.0000	0.0000	0.0000	0.0000	0.0000	0.0000	0.0000	2
3	0.9999	0.9659	0.7636	0.2340	0.0332	0.0024	0.0001	0.0000	0.0000	0.0000	0.0000	0.0000	0.0000	3
4	1.0000	0.9928	0.9020	0.4207	0.0905	0.0095	0.0005	0.0000	0.0000	0.0000	0.0000	0.0000	0.0000	4
5	1.0000	0.9988	0.9666	0.6167	0.1935	0.0294	0.0020	0.0001	0.0000	0.0000	0.0000	0.0000	0.0000	5
6	1.0000	0.9998	0.9905	0.7800	0.3407	0.0736	0.0073	0.0003	0.0000	0.0000	0.0000	0.0000	0.0000	6
7	1.0000	1.0000	0.9977	0.8909	0.5118	0.1536	0.0216	0.0012	0.0000	0.0000	0.0000	0.0000	0.0000	7
8	1.0000	1.0000	0.9995	0.9532	0.6769	0.2735	0.0539	0.0043	0.0001	0.0000	0.0000	0.0000	0.0000	8
9	1.0000	1.0000	0.9999	0.9827	0.8106	0.4246	0.1148	0.0132	0.0005	0.0000	0.0000	0.0000	0.0000	9
10	1.0000	1.0000	1.0000	0.9944	0.9022	0.5858	0.2122	0.0344	0.0018	0.0000	0.0000	0.0000	0.0000	10
11	1.0000	1.0000	1.0000	0.9985	0.9558	0.7323	0.3450	0.0778	0.0060	0.0001	0.0000	0.0000	0.0000	11
12	1.0000	1.0000	1.0000	0.9996	0.9825	0.8462	0.5000	0.1538	0.0175	0.0004	0.0000	0.0000	0.0000	12
13	1.0000	1.0000	1.0000	0.9999	0.9940	0.9222	0.6550	0.2677	0.0442	0.0015	0.0000	0.0000	0.0000	13
14	1.0000	1.0000	1.0000	1.0000	0.9982	0.9656	0.7878	0.4142	0.0978	0.0056	0.0000	0.0000	0.0000	14
15	1.0000	1.0000	1.0000	1.0000	0.9995	0.9868	0.8852	0.5754	0.1894	0.0173	0.0001	0.0000	0.0000	15
16	1.0000	1.0000	1.0000	1.0000	0.9999	0.9957	0.9461	0.7265	0.3231	0.0468	0.0005	0.0000	0.0000	16
17	1.0000	1.0000	1.0000	1.0000	1.0000	0.9988	0.9784	0.8464	0.4882	0.1091	0.0023	0.0000	0.0000	17
18	1.0000	1.0000	1.0000	1.0000	1.0000	0.9997	0.9927	0.9264	0.6593	0.2200	0.0095	0.0002	0.0000	18
19	1.0000	1.0000	1.0000	1.0000	1.0000	0.9999	0.9980	0.9706	0.8065	0.3833	0.0334	0.0012	0.0000	19
20	1.0000	1.0000	1.0000	1.0000	1.0000	1.0000	0.9995	0.9905	0.9095	0.5793	0.0980	0.0072	0.0000	20
21	1.0000	1.0000	1.0000	1.0000	1.0000	1.0000	0.9999	0.9976	0.9668	0.7660	0.2364	0.0341	0.0001	21
22	1.0000	1.0000	1.0000	1.0000	1.0000	1.0000	1.0000	0.9996	0.9910	0.9018	0.4629	0.1271	0.0020	22
23	1.0000	1.0000	1.0000	1.0000	1.0000	1.0000	1.0000	0.9999	0.9984	0.9726	0.7288	0.3576	0.0258	23
24	1.0000	1.0000	1.0000	1.0000	1.0000	1.0000	1.0000	1.0000	0.9999	0.9962	0.9282	0.7226	0.2222	24

APPENDIX 4: CUMULATIVE POISSON DISTRIBUTION*

k	.10	.20	.30	.40	.50	1.0	1.5	2.0	2.5	3.0	3.5	4.0	4.5	5.0	5.5	6.0	6.5	7.0	7.5	8.0	8.5	9.0	9.5	10
0	.905	.819	.741	.670	.607	.368	.223	.135	.082	.050	.030	.018	.011	.007	.004	.002	.002	.001	.001	.000	.000	.000	.000	.000
1	.995	.982	.963	.938	.910	.736	.558	.406	.287	.199	.136	.092	.061	.040	.027	.017	.011	.007	.005	.003	.002	.001	.001	.000
2	1.000	.999	.996	.992	.986	.920	.809	.677	.544	.423	.321	.238	.174	.125	.088	.062	.043	.030	.020	.014	.009	.006	.004	.003
3		1.000	1.000	.999	.998	.981	.934	.857	.758	.647	.537	.433	.342	.265	.202	.151	.112	.082	.059	.042	.030	.021	.015	.010
4				1.000	1.000	.996	.981	.947	.891	.815	.725	.629	.532	.440	.358	.285	.224	.173	.132	.100	.074	.055	.040	.029
5						.999	.996	.983	.958	.916	.858	.785	.703	.616	.529	.446	.369	.301	.241	.191	.150	.116	.089	.067
6						1.000	.999	.995	.986	.966	.935	.889	.831	.762	.686	.606	.527	.450	.378	.313	.256	.207	.165	.130
7							1.000	.999	.996	.988	.973	.949	.913	.867	.809	.744	.673	.599	.525	.453	.386	.324	.269	.220
8								1.000	.999	.996	.990	.979	.960	.932	.894	.847	.792	.729	.662	.593	.523	.456	.392	.333
9									1.000	.999	.997	.992	.983	.968	.946	.916	.877	.830	.776	.717	.653	.587	.522	.458
10										1.000	.999	.997	.993	.986	.975	.957	.933	.901	.862	.816	.763	.706	.645	.583
11											1.000	.999	.998	.995	.989	.980	.966	.947	.921	.888	.849	.803	.752	.697
12												1.000	.999	.998	.996	.991	.984	.973	.957	.936	.909	.876	.836	.792
13													1.000	.999	.998	.996	.993	.987	.978	.966	.949	.926	.898	.864
14														1.000	.999	.999	.997	.994	.990	.983	.973	.959	.940	.917
15															1.000	.999	.999	.998	.995	.992	.986	.978	.967	.951
16																1.000	.999	.999	.998	.996	.993	.989	.982	.973
17																	1.000	1.000	.999	.998	.997	.995	.991	.986
18																			1.000	.999	.999	.998	.996	.993
19																				1.000	.999	.999	.998	.997
20																					1.000	1.000	.999	.998
21																							1.000	.999
22																								1.000
23																								
24																								
25																								

*Tabled entries are $P(x \leq k)$, where x is a Poisson random variable with parameter λ.

APPENDIX 5: PERCENTAGE POINTS OF STUDENT'S t

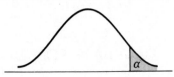

d.f.	$t_{0.100}$	$t_{0.050}$	$t_{0.025}$	$t_{0.010}$	$t_{0.005}$
1	3.078	6.314	12.706	31.820	63.655
2	1.886	2.920	4.303	6.965	9.925
3	1.638	2.353	3.182	4.541	5.841
4	1.533	2.132	2.776	3.747	4.601
5	1.476	2.015	2.571	3.365	4.032
6	1.440	1.943	2.447	3.143	3.707
7	1.415	1.895	2.365	2.998	3.499
8	1.397	1.860	2.306	2.896	3.355
9	1.383	1.833	2.262	2.821	3.250
10	1.372	1.812	2.228	2.764	3.169
11	1.363	1.796	2.201	2.718	3.106
12	1.356	1.782	2.179	2.681	3.055
13	1.350	1.771	2.160	2.650	3.012
14	1.345	1.761	2.145	2.624	2.977
15	1.341	1.753	2.131	2.602	2.947
16	1.337	1.746	2.120	2.583	2.921
17	1.333	1.740	2.110	2.567	2.898
18	1.330	1.734	2.101	2.552	2.878
19	1.328	1.729	2.093	2.539	2.861
20	1.325	1.725	2.086	2.528	2.845
21	1.323	1.721	2.080	2.518	2.831
22	1.321	1.717	2.074	2.508	2.819
23	1.319	1.714	2.069	2.500	2.807
24	1.318	1.711	2.064	2.492	2.797
25	1.316	1.708	2.060	2.485	2.787
26	1.315	1.706	2.056	2.479	2.779
27	1.314	1.703	2.052	2.473	2.771
28	1.313	1.701	2.048	2.467	2.763
29	1.311	1.699	2.045	2.462	2.756
30	1.310	1.697	2.042	2.457	2.750
40	1.303	1.684	2.021	2.423	2.704
60	1.296	1.671	2.000	2.390	2.660
120	1.289	1.658	1.980	2.358	2.617
∞	1.282	1.645	1.960	2.326	2.576

Source: Computed using *Minitab* software.

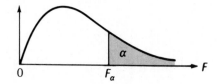

APPENDIX 6: PERCENTAGE POINTS OF THE F DISTRIBUTION ($\alpha = 0.10$)

ν_1 ν_2	Numerator Degrees of Freedom								
	1	2	3	4	5	6	7	8	9
1	39.86	49.50	53.59	55.83	57.24	58.20	58.91	59.44	59.86
2	8.53	9.00	9.16	9.24	9.29	9.33	9.35	9.37	9.38
3	5.54	5.46	5.39	5.34	5.31	5.28	5.27	5.25	5.24
4	4.54	4.32	4.19	4.11	4.05	4.01	3.98	3.95	3.94
5	4.06	3.78	3.62	3.52	3.45	3.40	3.37	3.34	3.32
6	3.78	3.46	3.29	3.18	3.11	3.05	3.01	2.98	2.96
7	3.59	3.26	3.07	2.96	2.88	2.83	2.78	2.75	2.72
8	3.46	3.11	2.92	2.81	2.73	2.67	2.62	2.59	2.56
9	3.36	3.01	2.81	2.69	2.61	2.55	2.51	2.47	2.44
10	3.29	2.92	2.73	2.61	2.52	2.46	2.41	2.38	2.35
11	3.23	2.86	2.66	2.54	2.45	2.39	2.34	2.30	2.27
12	3.18	2.81	2.61	2.48	2.39	2.33	2.28	2.24	2.21
13	3.14	2.76	2.56	2.43	2.35	2.28	2.23	2.20	2.16
14	3.10	2.73	2.52	2.39	2.31	2.24	2.19	2.15	2.12
15	3.07	2.70	2.49	2.36	2.27	2.21	2.16	2.12	2.09
16	3.05	2.67	2.46	2.33	2.24	2.18	2.13	2.09	2.06
17	3.03	2.64	2.44	2.31	2.22	2.15	2.10	2.06	2.03
18	3.01	2.62	2.42	2.29	2.20	2.13	2.08	2.04	2.00
19	2.99	2.61	2.40	2.27	2.18	2.11	2.06	2.02	1.98
20	2.97	2.59	2.38	2.25	2.16	2.09	2.04	2.00	1.96
21	2.96	2.57	2.36	2.23	2.14	2.08	2.02	1.98	1.95
22	2.95	2.56	2.35	2.22	2.13	2.06	2.01	1.97	1.93
23	2.94	2.55	2.34	2.21	2.11	2.05	1.99	1.95	1.92
24	2.93	2.54	2.33	2.19	2.10	2.04	1.98	1.94	1.91
25	2.92	2.53	2.32	2.18	2.09	2.02	1.97	1.93	1.89
26	2.91	2.52	2.31	2.17	2.08	2.01	1.96	1.92	1.88
27	2.90	2.51	2.30	2.17	2.07	2.00	1.95	1.91	1.87
28	2.89	2.50	2.29	2.16	2.06	2.00	1.94	1.90	1.87
29	2.89	2.50	2.28	2.15	2.06	1.99	1.93	1.89	1.86
30	2.88	2.49	2.28	2.14	2.05	1.98	1.93	1.88	1.85
40	2.84	2.44	2.23	2.09	2.00	1.93	1.87	1.83	1.79
60	2.79	2.39	2.18	2.04	1.95	1.87	1.82	1.77	1.74
120	2.75	2.35	2.13	1.99	1.90	1.82	1.77	1.72	1.68
∞	2.71	2.30	2.08	1.94	1.85	1.77	1.72	1.67	1.63

Denominator Degrees of Freedom

Source: Adapted from M. Merrington and C.M. Thompson, 1943, "Tables of Percentage Points of the Invested Beta (*F*)-Distribution," *Biometrika* 33:73–88. Used by permission of the *Biometrika* Trustees.

ν_1	Numerator Degrees of Freedom									
ν_2	10	12	15	20	24	30	40	60	120	∞
1	60.19	60.71	61.22	61.74	62.00	62.26	62.53	62.79	63.06	63.33
2	9.39	9.41	9.42	9.44	9.45	9.46	9.47	9.47	9.48	9.49
3	5.23	5.22	5.20	5.18	5.18	5.17	5.16	5.15	5.14	5.13
4	3.92	3.90	3.87	3.84	3.83	3.82	3.80	3.79	3.78	3.76
5	3.30	3.27	3.24	3.21	3.19	3.17	3.16	3.14	3.12	3.10
6	2.94	2.90	2.87	2.84	2.82	2.80	2.78	2.76	2.74	2.72
7	2.70	2.67	2.63	2.59	2.58	2.56	2.54	2.51	2.49	2.47
8	2.54	2.50	2.46	2.42	2.40	2.38	2.36	2.34	2.32	2.29
9	2.42	2.38	2.34	2.30	2.28	2.25	2.23	2.21	2.18	2.16
10	2.32	2.28	2.24	2.20	2.18	2.16	2.13	2.11	2.08	2.06
11	2.25	2.21	2.17	2.12	2.10	2.08	2.05	2.03	2.00	1.97
12	2.19	2.15	2.10	2.06	2.04	2.01	1.99	1.96	1.93	1.90
13	2.14	2.10	2.05	2.01	1.98	1.96	1.93	1.90	1.88	1.85
14	2.10	2.05	2.01	1.96	1.94	1.91	1.89	1.86	1.83	1.80
15	2.06	2.02	1.97	1.92	1.90	1.87	1.85	1.82	1.79	1.76
16	2.03	1.99	1.94	1.89	1.87	1.84	1.81	1.78	1.75	1.72
17	2.00	1.96	1.91	1.86	1.84	1.81	1.78	1.75	1.72	1.69
18	1.98	1.93	1.89	1.84	1.81	1.78	1.75	1.72	1.69	1.66
19	1.96	1.91	1.86	1.81	1.79	1.76	1.73	1.70	1.67	1.63
20	1.94	1.89	1.84	1.79	1.77	1.74	1.71	1.68	1.64	1.61
21	1.92	1.87	1.83	1.78	1.75	1.72	1.69	1.66	1.62	1.59
22	1.90	1.86	1.81	1.76	1.73	1.70	1.67	1.64	1.60	1.57
23	1.89	1.84	1.80	1.74	1.72	1.69	1.66	1.62	1.59	1.55
24	1.88	1.83	1.78	1.73	1.70	1.67	1.64	1.61	1.57	1.53
25	1.87	1.82	1.77	1.72	1.69	1.66	1.63	1.59	1.56	1.52
26	1.86	1.81	1.76	1.71	1.68	1.65	1.61	1.58	1.54	1.50
27	1.85	1.80	1.75	1.70	1.67	1.64	1.60	1.57	1.53	1.49
28	1.84	1.79	1.74	1.69	1.66	1.63	1.59	1.56	1.52	1.48
29	1.83	1.78	1.73	1.68	1.65	1.62	1.58	1.55	1.51	1.47
30	1.82	1.77	1.72	1.67	1.64	1.61	1.57	1.54	1.50	1.46
40	1.76	1.71	1.66	1.61	1.57	1.54	1.51	1.47	1.42	1.38
60	1.71	1.66	1.60	1.54	1.51	1.48	1.44	1.40	1.35	1.29
120	1.65	1.60	1.55	1.48	1.45	1.41	1.37	1.32	1.26	1.19
∞	1.60	1.55	1.49	1.42	1.38	1.34	1.30	1.24	1.17	1.00

Denominator Degrees of Freedom

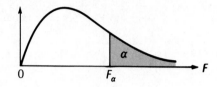

7

APPENDIX 7: PERCENTAGE POINTS OF THE F DISTRIBUTION ($\alpha = 0.05$)

ν_2 \ ν_1	Numerator Degrees of Freedom								
	1	2	3	4	5	6	7	8	9
1	161.4	199.5	215.7	224.6	230.2	234.0	236.8	238.9	240.5
2	18.51	19.00	19.16	19.25	19.30	19.33	19.35	19.37	19.38
3	10.13	9.55	9.28	9.12	9.01	8.94	8.89	8.85	8.81
4	7.71	6.94	6.59	6.39	6.26	6.16	6.09	6.04	6.00
5	6.61	5.79	5.41	5.19	5.05	4.95	4.88	4.82	4.77
6	5.99	5.14	4.76	4.53	4.39	4.28	4.21	4.15	4.10
7	5.59	4.74	4.35	4.12	3.97	3.87	3.79	3.73	3.68
8	5.32	4.46	4.07	3.84	3.69	3.58	3.50	3.44	3.39
9	5.12	4.26	3.86	3.63	3.48	3.37	3.29	3.23	3.18
10	4.96	4.10	3.71	3.48	3.33	3.22	3.14	3.07	3.02
11	4.84	3.98	3.59	3.36	3.20	3.09	3.01	2.95	2.90
12	4.75	3.89	3.49	3.26	3.11	3.00	2.91	2.85	2.80
13	4.67	3.81	3.41	3.18	3.03	2.92	2.83	2.77	2.71
14	4.60	3.74	3.34	3.11	2.96	2.85	2.76	2.70	2.65
15	4.54	3.68	3.29	3.06	2.90	2.79	2.71	2.64	2.59
16	4.49	3.63	3.24	3.01	2.85	2.74	2.66	2.59	2.54
17	4.45	3.59	3.20	2.96	2.81	2.70	2.61	2.55	2.49
18	4.41	3.55	3.16	2.93	2.77	2.66	2.58	2.51	2.46
19	4.38	3.52	3.13	2.90	2.74	2.63	2.54	2.48	2.42
20	4.35	3.49	3.10	2.87	2.71	2.60	2.51	2.45	2.39
21	4.32	3.47	3.07	2.84	2.68	2.57	2.49	2.42	2.37
22	4.30	3.44	3.05	2.82	2.66	2.55	2.46	2.40	2.34
23	4.28	3.42	3.03	2.80	2.64	2.53	2.44	2.37	2.32
24	4.26	3.40	3.01	2.78	2.62	2.51	2.42	2.36	2.30
25	4.24	3.39	2.99	2.76	2.60	2.49	2.40	2.34	2.28
26	4.23	3.37	2.98	2.74	2.59	2.47	2.39	2.32	2.27
27	4.21	3.35	2.96	2.73	2.57	2.46	2.37	2.31	2.25
28	4.20	3.34	2.95	2.71	2.56	2.45	2.36	2.29	2.24
29	4.18	3.33	2.93	2.70	2.55	2.43	2.35	2.28	2.22
30	4.17	3.32	2.92	2.69	2.53	2.42	2.33	2.27	2.21
40	4.08	3.23	2.84	2.61	2.45	2.34	2.25	2.18	2.12
60	4.00	3.15	2.76	2.53	2.37	2.25	2.17	2.10	2.04
120	3.92	3.07	2.68	2.45	2.29	2.17	2.09	2.02	1.96
∞	3.84	3.00	2.60	2.37	2.21	2.10	2.01	1.94	1.88

Denominator Degrees of Freedom

ν_1 ν_2	Numerator Degrees of Freedom									
	10	12	15	20	24	30	40	60	120	∞
1	241.9	243.9	245.9	248.0	249.1	250.1	251.1	252.2	253.3	254.3
2	19.40	19.41	19.43	19.45	19.45	19.46	19.47	19.48	19.49	19.50
3	8.79	8.74	8.70	8.66	8.64	8.62	8.59	8.57	8.55	8.53
4	5.96	5.91	5.86	5.80	5.77	5.75	5.72	5.69	5.66	5.63
5	4.74	4.68	4.62	4.56	4.53	4.50	4.46	4.43	4.40	4.36
6	4.06	4.00	3.94	3.87	3.84	3.81	3.77	3.74	3.70	3.67
7	3.64	3.57	3.51	3.44	3.41	3.38	3.34	3.30	3.27	3.23
8	3.35	3.28	3.22	3.15	3.12	3.08	3.04	3.01	2.97	2.93
9	3.14	3.07	3.01	2.94	2.90	2.86	2.83	2.79	2.75	2.71
10	2.98	2.91	2.85	2.77	2.74	2.70	2.66	2.62	2.58	2.54
11	2.85	2.79	2.72	2.65	2.61	2.57	2.53	2.49	2.45	2.40
12	2.75	2.69	2.62	2.54	2.51	2.47	2.43	2.38	2.34	2.30
13	2.67	2.60	2.53	2.46	2.42	2.38	2.34	2.30	2.25	2.21
14	2.60	2.53	2.46	2.39	2.35	2.31	2.27	2.22	2.18	2.13
15	2.54	2.48	2.40	2.33	2.29	2.25	2.20	2.16	2.11	2.07
16	2.49	2.42	2.35	2.28	2.24	2.19	2.15	2.11	2.06	2.01
17	2.45	2.38	2.31	2.23	2.19	2.15	2.10	2.06	2.01	1.96
18	2.41	2.34	2.27	2.19	2.15	2.11	2.06	2.02	1.97	1.92
19	2.38	2.31	2.23	2.16	2.11	2.07	2.03	1.98	1.93	1.88
20	2.35	2.28	2.20	2.12	2.08	2.04	1.99	1.95	1.90	1.84
21	2.32	2.25	2.18	2.10	2.05	2.01	1.96	1.92	1.87	1.81
22	2.30	2.23	2.15	2.07	2.03	1.98	1.94	1.89	1.84	1.78
23	2.27	2.20	2.13	2.05	2.01	1.96	1.91	1.86	1.81	1.76
24	2.25	2.18	2.11	2.03	1.98	1.94	1.89	1.84	1.79	1.73
25	2.24	2.16	2.09	2.01	1.96	1.92	1.87	1.82	1.77	1.71
26	2.22	2.15	2.07	1.99	1.95	1.90	1.85	1.80	1.75	1.69
27	2.20	2.13	2.06	1.97	1.93	1.88	1.84	1.79	1.73	1.67
28	2.19	2.12	2.04	1.96	1.91	1.87	1.82	1.77	1.71	1.65
29	2.18	2.10	2.03	1.94	1.90	1.85	1.81	1.75	1.70	1.64
30	2.16	2.09	2.01	1.93	1.89	1.84	1.79	1.74	1.68	1.62
40	2.08	2.00	1.92	1.84	1.79	1.74	1.69	1.64	1.58	1.51
60	1.99	1.92	1.84	1.75	1.70	1.65	1.59	1.53	1.47	1.39
120	1.91	1.83	1.75	1.66	1.61	1.55	1.50	1.43	1.35	1.25
∞	1.83	1.75	1.67	1.57	1.52	1.46	1.39	1.32	1.22	1.00

Denominator Degrees of Freedom

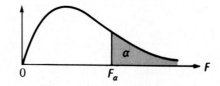

8

APPENDIX 8: PERCENTAGE POINTS OF THE F DISTRIBUTION ($\alpha = 0.01$)

ν_1				Numerator Degrees of Freedom					
ν_2	1	2	3	4	5	6	7	8	9
1	4,052	4,999.5	5,403	5,625	5,764	5,859	5,928	5,982	6,022
2	98.50	99.00	99.17	99.25	99.30	99.33	99.36	99.37	99.39
3	34.12	30.82	29.46	28.71	28.24	27.91	27.67	27.49	27.35
4	21.20	18.00	16.69	15.98	15.52	15.21	14.98	14.80	14.66
5	16.26	13.27	12.06	11.39	10.97	10.67	10.46	10.29	10.16
6	13.75	10.92	9.78	9.15	8.75	8.47	8.26	8.10	7.98
7	12.25	9.55	8.45	7.85	7.46	7.19	6.99	6.84	6.72
8	11.26	8.65	7.59	7.01	6.63	6.37	6.18	6.03	5.91
9	10.56	8.02	6.99	6.42	6.06	5.80	5.61	5.47	5.35
10	10.04	7.56	6.55	5.99	5.64	5.39	5.20	5.06	4.94
11	9.65	7.21	6.22	5.67	5.32	5.07	4.89	4.74	4.63
12	9.33	6.93	5.95	5.41	5.06	4.82	4.64	4.50	4.39
13	9.07	6.70	5.74	5.21	4.86	4.62	4.44	4.30	4.19
14	8.86	6.51	5.56	5.04	4.69	4.46	4.28	4.14	4.03
15	8.68	6.36	5.42	4.89	4.56	4.32	4.14	4.00	3.89
16	8.53	6.23	5.29	4.77	4.44	4.20	4.03	3.89	3.78
17	8.40	6.11	5.18	4.67	4.34	4.10	3.93	3.79	3.68
18	8.29	6.01	5.09	4.58	4.25	4.01	3.84	3.71	3.60
19	8.18	5.93	5.01	4.50	4.17	3.94	3.77	3.63	3.52
20	8.10	5.85	4.94	4.43	4.10	3.87	3.70	3.56	3.46
21	8.02	5.78	4.87	4.37	4.04	3.81	3.64	3.51	3.40
22	7.95	5.72	4.82	4.31	3.99	3.76	3.59	3.45	3.35
23	7.88	5.66	4.76	4.26	3.94	3.71	3.54	3.41	3.30
24	7.82	5.61	4.72	4.22	3.90	3.67	3.50	3.36	3.26
25	7.77	5.57	4.68	4.18	3.85	3.63	3.46	3.32	3.22
26	7.72	5.53	4.64	4.14	3.82	3.59	3.42	3.29	3.18
27	7.68	5.49	4.60	4.11	3.78	3.56	3.39	3.26	3.15
28	7.64	5.45	4.57	4.07	3.75	3.53	3.36	3.23	3.12
29	7.60	5.42	4.54	4.04	3.73	3.50	3.33	3.20	3.09
30	7.56	5.39	4.51	4.02	3.70	3.47	3.30	3.17	3.07
40	7.31	5.18	4.31	3.83	3.51	3.29	3.12	2.99	2.89
60	7.08	4.98	4.13	3.65	3.34	3.12	2.95	2.82	2.72
120	6.85	4.79	3.95	3.48	3.17	2.96	2.79	2.66	2.56
∞	6.63	4.61	3.78	3.32	3.02	2.80	2.64	2.51	2.41

Denominator Degrees of Freedom

ν_1 ν_2	10	12	15	20	24	30	40	60	120	∞
				Numerator Degrees of Freedom						
1	6,056	6,106	6,157	6,209	6,235	6,261	6,287	6,313	6,339	6,366
2	99.40	99.42	99.43	99.45	99.46	99.47	99.47	99.48	99.49	99.50
3	27.23	27.05	26.87	26.69	26.60	26.50	26.41	26.32	26.22	26.13
4	14.55	14.37	14.20	14.02	13.93	13.84	13.75	13.65	13.56	13.46
5	10.05	9.89	9.72	9.55	9.47	9.38	9.29	9.20	9.11	9.02
6	7.87	7.72	7.56	7.40	7.31	7.23	7.14	7.06	6.97	6.88
7	6.62	6.47	6.31	6.16	6.07	5.99	5.91	5.82	5.74	5.65
8	5.81	5.67	5.52	5.36	5.28	5.20	5.12	5.03	4.95	4.86
9	5.26	5.11	4.96	4.81	4.73	4.65	4.57	4.48	4.40	4.31
10	4.85	4.71	4.56	4.41	4.33	4.25	4.17	4.08	4.00	3.91
11	4.54	4.40	4.25	4.10	4.02	3.94	3.86	3.78	3.69	3.60
12	4.30	4.16	4.01	3.86	3.78	3.70	3.62	3.54	3.45	3.36
13	4.10	3.96	3.82	3.66	3.59	3.51	3.43	3.34	3.25	3.17
14	3.94	3.80	3.66	3.51	3.43	3.35	3.27	3.18	3.09	3.00
15	3.80	3.67	3.52	3.37	3.29	3.21	3.13	3.05	2.96	2.87
16	3.69	3.55	3.41	3.26	3.18	3.10	3.02	2.93	2.84	2.75
17	3.59	3.46	3.31	3.16	3.08	3.00	2.92	2.83	2.75	2.65
18	3.51	3.37	3.23	3.08	3.00	2.92	2.84	2.75	2.66	2.57
19	3.43	3.30	3.15	3.00	2.92	2.84	2.76	2.67	2.58	2.49
20	3.37	3.23	3.09	2.94	2.86	2.78	2.69	2.61	2.52	2.42
21	3.31	3.17	3.03	2.88	2.80	2.72	2.64	2.55	2.46	2.36
22	3.26	3.12	2.98	2.83	2.75	2.67	2.58	2.50	2.40	2.31
23	3.21	3.07	2.93	2.78	2.70	2.62	2.54	2.45	2.35	2.26
24	3.17	3.03	2.89	2.74	2.66	2.58	2.49	2.40	2.31	2.21
25	3.13	2.99	2.85	2.70	2.62	2.54	2.45	2.36	2.27	2.17
26	3.09	2.96	2.81	2.66	2.58	2.50	2.42	2.33	2.23	2.13
27	3.06	2.93	2.78	2.63	2.55	2.47	2.38	2.29	2.20	2.10
28	3.03	2.90	2.75	2.60	2.52	2.44	2.35	2.26	2.17	2.06
29	3.00	2.87	2.73	2.57	2.49	2.41	2.33	2.23	2.14	2.03
30	2.98	2.84	2.70	2.55	2.47	2.39	2.30	2.21	2.11	2.01
40	2.80	2.66	2.52	2.37	2.29	2.20	2.11	2.02	1.92	1.80
60	2.63	2.50	2.35	2.20	2.12	2.03	1.94	1.84	1.73	1.60
120	2.47	2.34	2.19	2.03	1.95	1.86	1.76	1.66	1.53	1.38
∞	2.32	2.18	2.04	1.88	1.79	1.70	1.59	1.47	1.32	1.00

APPENDIX 9: PERCENTAGE POINTS OF VON NEUMANN'S RATIO (MEAN SQUARE SUCCESSIVE DIFFERENCE TEST)

$$M = \frac{SS_{\Delta y}}{SS_{yy}} = \frac{\sum_{2}^{n} (\Delta y)^2}{\sum_{1}^{n} (y - \bar{y})^2}$$

Lower critical value: M_{1-p} in table

Upper critical value: $M_p = 4 - M_{1-p}$

One-tailed test: $p = \alpha$ Two-tailed test: $p = \alpha/2$

n	p 0.10	0.05	0.01	n	p 0.10	0.05	0.01	n	p 0.10	0.05	0.01
10	1.251	1.062	0.752	30	1.543	1.418	1.195	100	1.745	1.674	1.542
11	1.280	1.096	0.792	32	1.557	1.436	1.218	110	1.757	1.689	1.563
12	1.306	1.128	0.828	34	1.569	1.451	1.239	120	1.767	1.702	1.581
13	1.329	1.156	0.862	36	1.581	1.466	1.259	130	1.776	1.714	1.597
14	1.351	1.182	0.893	38	1.592	1.480	1.277	140	1.784	1.724	1.611
15	1.370	1.205	0.922	40	1.602	1.492	1.293	150	1.792	1.733	1.624
16	1.388	1.227	0.949	42	1.611	1.504	1.309	160	1.798	1.741	1.636
17	1.405	1.247	0.974	44	1.620	1.515	1.324	170	1.804	1.749	1.647
18	1.420	1.266	0.998	46	1.628	1.525	1.338	180	1.810	1.756	1.656
19	1.434	1.283	1.020	48	1.635	1.534	1.351	190	1.815	1.763	1.665
20	1.447	1.300	1.041	50	1.642	1.544	1.363	200	1.819	1.768	1.674
21	1.460	1.315	1.060	55	1.659	1.564	1.391	250	1.838	1.793	1.708
22	1.471	1.329	1.078	60	1.673	1.582	1.415	300	1.852	1.811	1.733
23	1.482	1.342	1.096	65	1.685	1.598	1.437	350	1.863	1.825	1.752
24	1.492	1.355	1.112	70	1.697	1.612	1.457	400	1.872	1.836	1.768
25	1.502	1.367	1.128	75	1.707	1.625	1.474	450	1.879	1.845	1.781
26	1.511	1.378	1.143	80	1.716	1.636	1.490	500	1.886	1.853	1.793
27	1.520	1.389	1.157	85	1.724	1.647	1.505	600	1.895	1.866	1.811
28	1.528	1.399	1.170	90	1.732	1.657	1.518	800	1.909	1.884	1.836
29	1.535	1.409	1.183	95	1.739	1.666	1.531	1,000	1.919	1.896	1.853

Source: Reproduced from L.S. Nelson, 1980, "The Mean Square Successive Difference Test," *Journal of Quality Technology* 12(3):175. Used by permission of the Editor, *Journal of Quality Technology*.

APPENDIX 10: EXACT CRITICAL VALUES FOR USE WITH THE ANALYSIS OF MEANS

TABLE 1 Exact Critical Values $h_{0.10}$ for the Analysis of Means

Significance Level = 0.10
Number of Means, k

df	3	4	5	6	7	8	9	10	11	12	13	14	15	16	17	18	19	20
3	3.16																	
4	2.81	3.10																
5	2.63	2.88	3.05															
6	2.52	2.74	2.91	3.03														
7	2.44	2.65	2.81	2.92	3.02													
8	2.39	2.59	2.73	2.85	2.94	3.02												
9	2.34	2.54	2.68	2.79	2.88	2.95	3.01											
10	2.31	2.50	2.64	2.74	2.83	2.90	2.96	3.02										
11	2.29	2.47	2.60	2.70	2.79	2.86	2.92	2.97	3.02									
12	2.27	2.45	2.57	2.67	2.75	2.82	2.88	2.93	2.98	3.02								
13	2.25	2.43	2.55	2.65	2.73	2.79	2.85	2.90	2.95	2.99	3.03							
14	2.23	2.41	2.53	2.63	2.70	2.77	2.83	2.88	2.92	2.96	3.00	3.03						
15	2.22	2.39	2.51	2.61	2.68	2.75	2.80	2.85	2.90	2.94	2.97	3.01	3.04					
16	2.21	2.38	2.50	2.59	2.67	2.73	2.79	2.83	2.88	2.92	2.95	2.99	3.02	3.05				
17	2.20	2.37	2.49	2.58	2.65	2.72	2.77	2.82	2.86	2.90	2.93	2.97	3.00	3.03	3.05			
18	2.19	2.36	2.47	2.56	2.64	2.70	2.75	2.80	2.84	2.88	2.92	2.95	2.98	3.01	3.03	3.06		
19	2.18	2.35	2.46	2.55	2.63	2.69	2.74	2.79	2.83	2.87	2.90	2.94	2.96	2.99	3.02	3.04	3.06	
20	2.18	2.34	2.45	2.54	2.62	2.68	2.73	2.78	2.82	2.86	2.89	2.92	2.95	2.98	3.00	3.03	3.05	3.07
24	2.15	2.32	2.43	2.51	2.58	2.64	2.69	2.74	2.78	2.82	2.85	2.88	2.91	2.93	2.96	2.98	3.00	3.02
30	2.13	2.29	2.40	2.48	2.55	2.61	2.66	2.70	2.74	2.77	2.81	2.84	2.86	2.89	2.91	2.93	2.96	2.98
40	2.11	2.27	2.37	2.45	2.52	2.57	2.62	2.66	2.70	2.73	2.77	2.79	2.82	2.85	2.87	2.89	2.91	2.93
60	2.09	2.24	2.34	2.42	2.49	2.54	2.59	2.63	2.66	2.70	2.73	2.75	2.78	2.80	2.82	2.84	2.86	2.88
120	2.07	2.22	2.32	2.39	2.45	2.51	2.55	2.59	2.62	2.66	2.69	2.71	2.74	2.76	2.78	2.80	2.82	2.84
INF	2.05	2.19	2.29	2.36	2.42	2.47	2.52	2.55	2.59	2.62	2.65	2.67	2.69	2.72	2.74	2.76	2.77	2.79

Source: Reproduced from Nelson, L. S. 1983. "Exact Critical Values for Use with the Analysis of Means," *Journal of Quality Technology,* Vol. 15, no. 1, 41–44.

TABLE 2 Exact Critical Values $h_{0.05}$ for the Analysis of Means

Significance Level = 0.05
Number of Means, k

df	3	4	5	6	7	8	9	10	11	12	13	14	15	16	17	18	19	20
3	4.18																	
4	3.56	3.89																
5	3.25	3.53	3.72															
6	3.07	3.31	3.49	3.62														
7	2.94	3.17	3.33	3.45	3.56													
8	2.86	3.07	3.21	3.33	3.43	3.51												
9	2.79	2.99	3.13	3.24	3.33	3.41	3.48											
10	2.74	2.93	3.07	3.17	3.26	3.33	3.40	3.45										
11	2.70	2.88	3.01	3.12	3.20	3.27	3.33	3.39	3.44									
12	2.67	2.85	2.97	3.07	3.15	3.22	3.28	3.33	3.38	3.42								
13	2.64	2.81	2.94	3.03	3.11	3.18	3.24	3.29	3.34	3.38	3.42							
14	2.62	2.79	2.91	3.00	3.08	3.14	3.20	3.25	3.30	3.34	3.37	3.41						
15	2.60	2.76	2.88	2.97	3.05	3.11	3.17	3.22	3.26	3.30	3.34	3.37	3.40					
16	2.58	2.74	2.86	3.95	3.02	3.09	3.14	3.19	3.23	3.27	3.31	3.34	3.37	3.40				
17	2.57	2.73	2.84	2.93	3.00	3.06	3.12	3.16	3.21	3.25	3.28	3.31	3.34	3.37	3.40			
18	2.55	2.71	2.82	2.91	2.98	3.04	3.10	3.14	3.18	3.22	3.26	3.29	3.32	3.35	3.37	3.40		
19	2.54	2.70	2.81	2.89	2.96	3.02	3.08	3.12	3.16	3.20	3.24	3.27	3.30	3.32	3.35	3.37	3.40	
20	2.53	2.68	2.79	2.88	2.95	3.01	3.06	3.11	3.15	3.18	3.22	3.25	3.28	3.30	3.33	3.35	3.37	3.40
24	2.50	2.65	2.75	2.83	2.90	2.96	3.01	3.05	3.09	3.13	3.16	3.19	3.22	3.24	3.27	3.29	3.31	3.33
30	2.47	2.61	2.71	2.79	2.85	2.91	2.96	3.00	3.04	3.07	3.10	3.13	3.16	3.18	3.20	3.22	3.25	3.27
40	2.43	2.57	2.67	2.75	2.81	2.86	2.91	2.95	2.98	3.01	3.04	3.07	3.10	3.12	3.14	3.16	3.18	3.20
60	2.40	2.54	2.63	2.70	2.76	2.81	2.86	2.90	2.93	2.96	2.99	3.02	3.04	3.06	3.08	3.10	3.12	3.14
120	2.37	2.50	2.59	2.66	2.72	2.77	2.81	2.84	2.88	2.91	2.93	2.96	2.98	3.00	3.02	3.04	3.06	3.08
INF	2.34	2.47	2.56	2.62	2.68	2.72	2.76	2.80	2.83	2.86	2.88	2.90	2.93	2.95	2.97	2.98	3.00	3.02

TABLE 3 Exact Critical Values $h_{0.01}$ for the Analysis of Means

Significance Level = 0.01
Number of means, k

df	3	4	5	6	7	8	9	10	11	12	13	14	15	16	17	18	19	20
3	7.51																	
4	5.74	6.21																
5	4.93	5.29	5.55															
6	4.48	4.77	4.98	5.16														
7	4.18	4.44	4.63	4.78	4.90													
8	3.98	4.21	4.38	4.52	4.63	4.72												
9	3.84	4.05	4.20	4.33	4.43	4.51	4.59											
10	3.73	3.92	4.07	4.18	4.28	4.36	4.43	4.49										
11	3.64	3.82	3.96	4.07	4.16	4.23	4.30	4.36	4.41									
12	3.57	3.74	3.87	3.98	4.06	4.13	4.20	4.25	4.31	4.35								
13	3.51	3.68	3.80	3.90	3.98	4.05	4.11	4.17	4.22	4.26	4.30							
14	3.46	3.63	3.74	3.84	3.92	3.98	4.04	4.09	4.14	4.18	4.22	4.26						
15	3.42	3.58	3.69	3.79	3.86	3.92	3.98	4.03	4.08	4.12	4.16	4.19	4.22					

df	3	4	5	6	7	8	9	10	11	12	13	14	15	16	17	18	19	20
16	3.38	3.54	3.65	3.74	3.81	3.87	3.93	3.98	4.02	4.06	4.10	4.14	4.17	4.20				
17	3.35	3.50	3.61	3.70	3.77	3.83	3.89	3.93	3.98	4.02	4.05	4.09	4.12	4.14	4.17			
18	3.33	3.47	3.58	3.66	3.73	3.79	3.85	3.89	3.94	3.97	4.01	404	4.07	4.10	4.12	4.15		
19	3.30	3.45	3.55	3.63	3.70	3.76	3.81	3.86	3.90	3.94	3.97	4.00	4.03	4.06	4.08	4.11	4.13	
20	3.28	3.42	3.53	3.61	3.67	3.73	3.78	3.83	3.87	3.90	3.94	3.97	4.00	4.02	4.05	4.07	4.09	4.12
24	3.21	3.35	3.45	3.52	3.58	3.64	3.69	3.73	3.77	3.80	3.83	3.86	3.89	3.91	3.94	3.96	3.98	4.00
30	3.15	3.28	3.37	3.44	3.50	3.55	3.59	3.63	3.67	3.70	3.73	3.76	3.78	3.81	3.83	3.85	3.87	3.89
40	3.09	3.21	3.29	3.36	3.42	3.46	3.50	3.54	3.58	3.60	3.63	3.66	3.68	3.70	3.72	3.74	3.76	3.78
60	3.03	3.14	3.22	3.29	3.34	3.38	3.42	3.46	3.49	3.51	3.54	3.56	3.59	3.61	3.63	3.64	3.66	3.68
120	2.97	3.07	3.15	3.21	3.26	3.30	3.34	3.37	3.40	3.42	3.45	3.47	3.49	3.51	3.53	3.55	3.56	3.58
INF	2.91	3.01	3.08	3.14	3.18	3.22	3.26	3.29	3.32	3.34	3.36	3.38	3.40	3.42	3.44	3.45	3.47	3.48

TABLE 4 Exact Critical Values $h_{0.001}$ for the Analysis of Means

Significance Level = 0.001
Number of Means, k

df	3	4	5	6	7	8	9	10	11	12	13	14	15	16	17	18	19	20
3	16.4																	
4	10.6	11.4																
5	8.25	8.79	9.19															
6	7.04	7.45	7.76	8.00														
7	6.31	6.65	6.89	7.09	7.25													
8	5.83	6.12	6.32	6.49	6.63	6.75												
9	5.49	5.74	5.92	6.07	6.20	6.30	6.40											
10	5.24	5.46	5.63	5.76	5.87	5.97	6.05	6.13										
11	5.05	5.25	5.40	5.52	5.63	5.71	5.79	5.86	5.92									
12	4.89	5.08	5.22	5.33	5.43	5.51	5.58	5.65	5.71	5.76								
13	4.77	4.95	5.08	5.18	5.27	5.35	5.42	5.48	5.53	5.58	5.63							
14	4.66	4.83	4.96	5.06	5.14	5.21	5.28	5.33	5.38	5.43	5.48	5.51						
15	4.57	4.74	4.86	4.95	5.03	5.10	5.16	5.21	5.26	5.31	5.35	5.39	5.42					
16	4.50	4.66	4.77	4.86	4.94	5.00	5.06	5.11	5.16	5.20	5.24	5.28	5.31	5.34				
17	4.44	4.59	4.70	4.78	4.86	4.92	4.98	5.03	5.07	5.11	5.15	5.18	5.22	5.25	5.28			
18	4.38	4.53	4.63	4.72	4.79	4.85	4.90	4.95	4.99	5.03	5.07	5.10	5.14	5.16	5.19	5.22		
19	4.33	4.47	4.58	4.66	4.73	4.79	4.84	4.88	4.93	4.96	5.00	5.03	5.06	5.09	5.12	5.14	5.17	
20	4.29	4.42	4.53	4.61	4.67	4.73	4.78	4.83	4.87	4.90	4.94	4.97	5.00	5.03	5.05	5.08	5.10	5.12
24	4.16	4.28	4.37	4.45	4.51	4.56	4.61	4.65	4.69	4.72	4.75	4.78	4.81	4.83	4.86	4.88	4.90	4.92
30	4.03	4.14	4.23	4.30	4.35	4.40	4.44	4.48	4.51	4.54	4.57	4.60	4.62	4.64	4.67	4.69	4.71	4.72
40	3.91	4.01	4.09	4.15	4.20	4.25	4.29	4.32	4.35	4.38	4.40	4.43	4.45	4.47	4.49	4.50	4.52	4.54
60	3.80	3.89	3.96	4.02	4.06	4.10	4.14	4.17	4.19	4.22	4.24	4.27	4.29	4.30	4.32	4.33	4.35	4.37
120	3.69	3.77	3.84	3.89	3.93	3.96	4.00	4.03	4.05	4.07	4.09	4.11	4.13	4.15	4.16	4.17	4.19	4.21
INF	3.58	3.66	3.72	3.76	3.80	3.84	3.87	3.89	3.91	3.93	3.95	3.97	3.99	4.00	4.02	4.03	4.04	4.06

TABLE I Sample Size Code Letters

Lot or batch size			Special inspection levels				General inspection levels		
			S-1	S-2	S-3	S-4	I	II	III
2	to	8	A	A	A	A	A	A	B
9	to	15	A	A	A	A	A	B	C
16	to	25	A	A	B	B	B	C	D
26	to	50	A	B	B	C	C	D	E
51	to	90	B	B	C	C	C	E	F
91	to	150	B	B	C	D	D	F	G
151	to	280	B	C	D	E	E	G	H
281	to	500	B	C	D	E	F	H	J
501	to	1200	C	C	E	F	G	J	K
1201	to	3200	C	D	E	G	H	K	L
3201	to	10000	C	D	F	G	J	L	M
10001	to	35000	C	D	F	H	K	M	N
35001	to	150000	D	E	G	J	L	N	P
150001	to	500000	D	E	G	J	M	P	Q
500001	and	over	D	E	H	K	N	Q	R

480

TABLE II-A Single Sampling Plans for Normal Inspection (Master Table)

Sample size code letter	Sample size	Acceptable Quality Levels (normal inspection)																								
		0.010	0.015	0.025	0.040	0.065	0.10	0.15	0.25	0.40	0.65	1.0	1.5	2.5	4.0	6.5	10	15	25	40	65	100	150	250	400	650	1000
		Ac Re	Ac Re	Ac Re	Ac Re	Ac Re	Ac Re	Ac Re	Ac Re	Ac Re	Ac Re	Ac Re	Ac Re	Ac Re	Ac Re	Ac Re	Ac Re	Ac Re	Ac Re	Ac Re	Ac Re	Ac Re	Ac Re	Ac Re	Ac Re	Ac Re	Ac Re

(Table body consists of acceptance/rejection number pairs arranged diagonally with directional arrows indicating "use first sampling plan above/below arrow." Sample size code letters A–R correspond to sample sizes 2, 3, 5, 8, 13, 20, 32, 50, 80, 125, 200, 315, 500, 800, 1250, 2000.)

⬇ = Use first sampling plan below arrow. If sample size equals, or exceeds, lot or batch size, do 100 percent inspection.

⬆ = Use first sampling plan above arrow.

Ac = Acceptance number.

Re = Rejection number.

TABLE II-B   Single Sampling Plans for Tightened Inspection (Master Table)

Acceptable Quality Levels (tightened inspection). Each cell shows **Ac Re** (Ac = Acceptance number, Re = Rejection number). ↓ = Use first sampling plan below arrow. ↑ = Use first sampling plan above arrow.

Sample size code letter	Sample size	0.010	0.015	0.025	0.040	0.065	0.10	0.15	0.25	0.40	0.65	1.0	1.5	2.5	4.0	6.5	10	15	25	40	65	100	150	250	400	650	1000
A	2	↓	↓	↓	↓	↓	↓	↓	↓	↓	↓	↓	↓	↓	↓	↓	↓	↓	0 1	1 2	2 3	3 4	5 6	8 9	12 13	18 19	27 28
B	3	↓	↓	↓	↓	↓	↓	↓	↓	↓	↓	↓	↓	↓	↓	↓	↓	0 1	1 2	2 3	3 4	5 6	8 9	12 13	18 19	27 28	41 42
C	5	↓	↓	↓	↓	↓	↓	↓	↓	↓	↓	↓	↓	↓	↓	↓	0 1	1 2	2 3	3 4	5 6	8 9	12 13	18 19	27 28	41 42	↑
D	8	↓	↓	↓	↓	↓	↓	↓	↓	↓	↓	↓	↓	↓	↓	0 1	1 2	2 3	3 4	5 6	8 9	12 13	18 19	27 28	41 42	↑	↑
E	13	↓	↓	↓	↓	↓	↓	↓	↓	↓	↓	↓	↓	↓	0 1	1 2	2 3	3 4	5 6	8 9	12 13	18 19	27 28	41 42	↑	↑	↑
F	20	↓	↓	↓	↓	↓	↓	↓	↓	↓	↓	↓	↓	0 1	1 2	2 3	3 4	5 6	8 9	12 13	18 19	27 28	41 42	↑	↑	↑	↑
G	32	↓	↓	↓	↓	↓	↓	↓	↓	↓	↓	↓	0 1	1 2	2 3	3 4	5 6	8 9	12 13	18 19	27 28	41 42	↑	↑	↑	↑	↑
H	50	↓	↓	↓	↓	↓	↓	↓	↓	↓	↓	0 1	1 2	2 3	3 4	5 6	8 9	12 13	18 19	27 28	41 42	↑	↑	↑	↑	↑	↑
J	80	↓	↓	↓	↓	↓	↓	↓	↓	↓	0 1	1 2	2 3	3 4	5 6	8 9	12 13	18 19	27 28	41 42	↑	↑	↑	↑	↑	↑	↑
K	125	↓	↓	↓	↓	↓	↓	↓	↓	0 1	1 2	2 3	3 4	5 6	8 9	12 13	18 19	27 28	41 42	↑	↑	↑	↑	↑	↑	↑	↑
L	200	↓	↓	↓	↓	↓	↓	↓	0 1	1 2	2 3	3 4	5 6	8 9	12 13	18 19	27 28	41 42	↑	↑	↑	↑	↑	↑	↑	↑	↑
M	315	↓	↓	↓	↓	↓	↓	0 1	1 2	2 3	3 4	5 6	8 9	12 13	18 19	27 28	41 42	↑	↑	↑	↑	↑	↑	↑	↑	↑	↑
N	500	↓	↓	↓	↓	↓	0 1	1 2	2 3	3 4	5 6	8 9	12 13	18 19	27 28	41 42	↑	↑	↑	↑	↑	↑	↑	↑	↑	↑	↑
P	800	↓	↓	↓	↓	0 1	1 2	2 3	3 4	5 6	8 9	12 13	18 19	27 28	41 42	↑	↑	↑	↑	↑	↑	↑	↑	↑	↑	↑	↑
Q	1250	↓	↓	↓	0 1	1 2	2 3	3 4	5 6	8 9	12 13	18 19	27 28	41 42	↑	↑	↑	↑	↑	↑	↑	↑	↑	↑	↑	↑	↑
R	2000	↓	↓	0 1	1 2	2 3	3 4	5 6	8 9	12 13	18 19	27 28	41 42	↑	↑	↑	↑	↑	↑	↑	↑	↑	↑	↑	↑	↑	↑
S	3150	↓	0 1	1 2	2 3	3 4	5 6	8 9	12 13	18 19	27 28	41 42	↑	↑	↑	↑	↑	↑	↑	↑	↑	↑	↑	↑	↑	↑	↑

↓ = Use first sampling plan below arrow. If sample size equals or exceeds lot or batch size, do 100 percent inspection.
↑ = Use first sampling plan above arrow.
Ac = Acceptance number.
Re = Rejection number.

TABLE II-C Single Sampling Plans for Reduced Inspection (Master Table)

Sample size code letter	Sample size	Acceptable Quality Levels (reduced inspection)†																																																			
		0.010		0.015		0.025		0.040		0.065		0.10		0.15		0.25		0.40		0.65		1.0		1.5		2.5		4.0		6.5		10		15		25		40		65		100		150		250		400		650		1000	
		Ac	Re	Ac	Re	Ac	Re	Ac	Re	Ac	Re	Ac	Re	Ac	Re	Ac	Re	Ac	Re	Ac	Re	Ac	Re	Ac	Re	Ac	Re	Ac	Re	Ac	Re	Ac	Re	Ac	Re	Ac	Re	Ac	Re	Ac	Re	Ac	Re	Ac	Re	Ac	Re	Ac	Re	Ac	Re	Ac	Re
A	2																											0	1			1	2			2	3	3	4	5	6	7	8	10	11	14	15	21	22	30	31
B	2																													0	2			1	3	2	3	3	5	5	6	7	10	10	11	14	15	21	22	30	31
C	2																															0	2	1	3	1	4	3	6	5	8	7	10	10	13	14	17	21	24	⇑	
D	3																													0	2	1	3	1	4	2	5	3	6	5	8	7	10	10	13	13	14	14	17	21	24	⇑	
E	5																						0	2	1	3	1	4	2	5	3	5	3	6	5	8	7	10	8	10	10	13	13	14	17		21		⇑	
F	8																				0	2	0	1	1	3	1	4	2	5	3	6	5	6	5	8	7	10	10	13										
G	13																			1	3	1	4	2	5	3	6	5	8	5	8	7	10	7	10	10	13												
H	20																	2	5	3	5	3	6	5	8	6	8	7	10	8	10	10	13															
J	32																	3	6	5	6	5	8	7	10	7	10	10	13																			
K	50														5	8	7	10	7	10	10	13																										
L	80													7	10	7	10	10	13																														
M	125											7	10	10	13	10	13																																
N	200									5	8																																						
P	315							5	6																																								
Q	500					5	6																																										
R	800			7	10	10	13																																										

⇩ = Use first sampling plan below arrow. If sample size equals or exceeds lot or batch size, do 100 percent inspection.
⇧ = Use first sampling plan above arrow.
Ac = Acceptance number.
Re = Rejection number.
† = If the acceptance number has been exceeded, but the rejection number has not been reached; accept the lot, but reinstate normal inspection.

TABLE III-A Double Sampling Plans for Normal Inspection (Master Table)

Each cell shows the acceptance number (Ac) and rejection number (Re) as "Ac Re". Symbols: ↓ = Use first sampling plan below arrow. ↑ = Use first sampling plan above arrow. • = Use corresponding single sampling plan (or alternatively, use double sampling plan below, where available).

Sample size code letter	Sample	Sample size	Cumulative sample size	0.010	0.015	0.025	0.040	0.065	0.10	0.15	0.25	0.40	0.65	1.0	1.5	2.5	4.0	6.5	10	15	25	40	65	100	150	250	400	650	1000
																				Acceptable Quality Levels (normal inspection)									
A				↓	↓	↓	↓	↓	↓	↓	↓	↓	↓	↓	↓	↓	↓	↓	↓	↓	↓	↓	↓	↓	↓	↓	↓	↓	↓
B	First	2	2	↓	↓	↓	↓	↓	↓	↓	↓	↓	↓	↓	↓	↓	↓	↓	•	0 2	0 3	1 4	2 5	3 7	5 9	7 11	11 16	17 22	25 31
	Second	2	4	↓	↓	↓	↓	↓	↓	↓	↓	↓	↓	↓	↓	↓	↓	↓	•	1 2	3 4	4 5	6 7	8 9	12 13	18 19	26 27	37 38	56 57
C	First	3	3	↓	↓	↓	↓	↓	↓	↓	↓	↓	↓	↓	↓	↓	↓	•	0 2	0 3	1 4	2 5	3 7	5 9	7 11	11 16	17 22	25 31	↑
	Second	3	6	↓	↓	↓	↓	↓	↓	↓	↓	↓	↓	↓	↓	↓	↓	•	1 2	3 4	4 5	6 7	8 9	12 13	18 19	26 27	37 38	56 57	↑
D	First	5	5	↓	↓	↓	↓	↓	↓	↓	↓	↓	↓	↓	↓	↓	•	0 2	0 3	1 4	2 5	3 7	5 9	7 11	11 16	17 22	25 31	↑	↑
	Second	5	10	↓	↓	↓	↓	↓	↓	↓	↓	↓	↓	↓	↓	↓	•	1 2	3 4	4 5	6 7	8 9	12 13	18 19	26 27	37 38	56 57	↑	↑
E	First	8	8	↓	↓	↓	↓	↓	↓	↓	↓	↓	↓	↓	↓	•	0 2	0 3	1 4	2 5	3 7	5 9	7 11	11 16	17 22	25 31	↑	↑	↑
	Second	8	16	↓	↓	↓	↓	↓	↓	↓	↓	↓	↓	↓	↓	•	1 2	3 4	4 5	6 7	8 9	12 13	18 19	26 27	37 38	56 57	↑	↑	↑
F	First	13	13	↓	↓	↓	↓	↓	↓	↓	↓	↓	↓	↓	•	0 2	0 3	1 4	2 5	3 7	5 9	7 11	11 16	17 22	25 31	↑	↑	↑	↑
	Second	13	26	↓	↓	↓	↓	↓	↓	↓	↓	↓	↓	↓	•	1 2	3 4	4 5	6 7	8 9	12 13	18 19	26 27	37 38	56 57	↑	↑	↑	↑
G	First	20	20	↓	↓	↓	↓	↓	↓	↓	↓	↓	↓	•	0 2	0 3	1 4	2 5	3 7	5 9	7 11	11 16	17 22	25 31	↑	↑	↑	↑	↑
	Second	20	40	↓	↓	↓	↓	↓	↓	↓	↓	↓	↓	•	1 2	3 4	4 5	6 7	8 9	12 13	18 19	26 27	37 38	56 57	↑	↑	↑	↑	↑
H	First	32	32	↓	↓	↓	↓	↓	↓	↓	↓	↓	•	0 2	0 3	1 4	2 5	3 7	5 9	7 11	11 16	17 22	25 31	↑	↑	↑	↑	↑	↑
	Second	32	64	↓	↓	↓	↓	↓	↓	↓	↓	↓	•	1 2	3 4	4 5	6 7	8 9	12 13	18 19	26 27	37 38	56 57	↑	↑	↑	↑	↑	↑
J	First	50	50	↓	↓	↓	↓	↓	↓	↓	↓	•	0 2	0 3	1 4	2 5	3 7	5 9	7 11	11 16	17 22	25 31	↑	↑	↑	↑	↑	↑	↑
	Second	50	100	↓	↓	↓	↓	↓	↓	↓	↓	•	1 2	3 4	4 5	6 7	8 9	12 13	18 19	26 27	37 38	56 57	↑	↑	↑	↑	↑	↑	↑
K	First	80	80	↓	↓	↓	↓	↓	↓	↓	•	0 2	0 3	1 4	2 5	3 7	5 9	7 11	11 16	17 22	25 31	↑	↑	↑	↑	↑	↑	↑	↑
	Second	80	160	↓	↓	↓	↓	↓	↓	↓	•	1 2	3 4	4 5	6 7	8 9	12 13	18 19	26 27	37 38	56 57	↑	↑	↑	↑	↑	↑	↑	↑
L	First	125	125	↓	↓	↓	↓	↓	↓	•	0 2	0 3	1 4	2 5	3 7	5 9	7 11	11 16	17 22	25 31	↑	↑	↑	↑	↑	↑	↑	↑	↑
	Second	125	250	↓	↓	↓	↓	↓	↓	•	1 2	3 4	4 5	6 7	8 9	12 13	18 19	26 27	37 38	56 57	↑	↑	↑	↑	↑	↑	↑	↑	↑
M	First	200	200	↓	↓	↓	↓	↓	•	0 2	0 3	1 4	2 5	3 7	5 9	7 11	11 16	17 22	25 31	↑	↑	↑	↑	↑	↑	↑	↑	↑	↑
	Second	200	400	↓	↓	↓	↓	↓	•	1 2	3 4	4 5	6 7	8 9	12 13	18 19	26 27	37 38	56 57	↑	↑	↑	↑	↑	↑	↑	↑	↑	↑
N	First	315	315	↓	↓	↓	↓	•	0 2	0 3	1 4	2 5	3 7	5 9	7 11	11 16	17 22	25 31	↑	↑	↑	↑	↑	↑	↑	↑	↑	↑	↑
	Second	315	630	↓	↓	↓	↓	•	1 2	3 4	4 5	6 7	8 9	12 13	18 19	26 27	37 38	56 57	↑	↑	↑	↑	↑	↑	↑	↑	↑	↑	↑
P	First	500	500	↓	↓	↓	•	0 2	0 3	1 4	2 5	3 7	5 9	7 11	11 16	17 22	25 31	↑	↑	↑	↑	↑	↑	↑	↑	↑	↑	↑	↑
	Second	500	1000	↓	↓	↓	•	1 2	3 4	4 5	6 7	8 9	12 13	18 19	26 27	37 38	56 57	↑	↑	↑	↑	↑	↑	↑	↑	↑	↑	↑	↑
Q	First	800	800	↓	↓	•	0 2	0 3	1 4	2 5	3 7	5 9	7 11	11 16	17 22	25 31	↑	↑	↑	↑	↑	↑	↑	↑	↑	↑	↑	↑	↑
	Second	800	1600	↓	↓	•	1 2	3 4	4 5	6 7	8 9	12 13	18 19	26 27	37 38	56 57	↑	↑	↑	↑	↑	↑	↑	↑	↑	↑	↑	↑	↑
R	First	1250	1250	↓	•	0 2	0 3	1 4	2 5	3 7	5 9	7 11	11 16	17 22	25 31	↑	↑	↑	↑	↑	↑	↑	↑	↑	↑	↑	↑	↑	↑
	Second	1250	2500	↓	•	1 2	3 4	4 5	6 7	8 9	12 13	18 19	26 27	37 38	56 57	↑	↑	↑	↑	↑	↑	↑	↑	↑	↑	↑	↑	↑	↑

⇩ = Use first sampling plan below arrow. If sample size equals or exceeds lot or batch size, do 100 percent inspection.
⇧ = Use first sampling plan above arrow.
Ac = Acceptance number.
Re = Rejection number.
• = Use corresponding single sampling plan (or alternatively, use double sampling plan below, where available).

484

TABLE III-B   Double Sampling Plans for Tightened Inspection (Master Table)

↓ = Use first sampling plan below arrow. If sample size equals or exceeds lot or batch size, do 100 percent inspection.

↑ = Use first sampling plan above arrow.

Ac = Acceptance number.

Re = Rejection number.

• = Use corresponding single sampling plan (or, alternatively, use double sampling plan below, where available).

TABLE III-C Double Sampling Plans for Reduced Inspection (Master Table)

Acceptable Quality Levels (reduced inspection)†

Sample size code letter	Sample	Sample size	Cumulative sample size	0.010	0.015	0.025	0.040	0.065	0.10	0.15	0.25	0.40	0.65	1.0	1.5	2.5	4.0	6.5	10	15	25	40	65	100	150	250	400	650	1000
A																												•
B																											•	•
C																										•	•	•
D	First	2	2	↓	↓	↓	↓	↓	↓	↓	↓	↓	↓	↓	↓	↓	↓	0 2	0 3	0 4	•	•	•	3 8	5 10	7 12	11 17	•	•
	Second	2	4															0 2	0 4	0 5	•	•	•	8 12	12 16	18 22	26 30		
E	First	3	3	↓	↓	↓	↓	↓	↓	↓	↓	↓	↓	↓	↓	↓	0 2	0 3	0 4	•	•	2 7	3 8	5 10	7 12	11 17	•	•	•
	Second	3	6														0 2	0 4	0 5	•	•	6 9	8 12	12 16	18 22	26 30			
F	First	5	5	↓	↓	↓	↓	↓	↓	↓	↓	↓	↓	↓	↓	0 2	0 3	0 4	•	1 5	2 7	3 8	5 10	7 12	11 17	•	•	•	•
	Second	5	10													0 2	0 4	0 5	•	4 7	6 9	8 12	12 16	18 22	26 30				
G	First	8	8	↓	↓	↓	↓	↓	↓	↓	↓	↓	↓	↓	0 2	0 3	0 4	0 4	1 5	2 7	3 8	5 10	7 12	11 17	•	•	•	•	•
	Second	8	16												0 2	0 4	0 5	1 6	4 7	6 9	8 12	12 16	18 22	26 30					
H	First	13	13	↓	↓	↓	↓	↓	↓	↓	↓	↓	↓	0 2	0 3	0 4	0 4	1 5	2 7	3 8	5 10	7 12	11 17	•	•	•	•	•	•
	Second	13	26											0 2	0 4	0 5	1 6	4 7	6 9	8 12	12 16	18 22	26 30						
J	First	20	20	↓	↓	↓	↓	↓	↓	↓	↓	↓	0 2	0 3	0 4	0 4	1 5	2 7	3 8	5 10	7 12	11 17	↑	↑	↑	↑	↑	↑	↑
	Second	20	40										0 2	0 4	0 5	1 6	4 7	6 9	8 12	12 16	18 22	26 30							
K	First	32	32	↓	↓	↓	↓	↓	↓	↓	↓	0 2	0 3	0 4	0 4	1 5	2 7	3 8	5 10	7 12	11 17	↑	↑	↑	↑	↑	↑	↑	↑
	Second	32	64									0 2	0 4	0 5	1 6	4 7	6 9	8 12	12 16	18 22	26 30								
L	First	50	50	↓	↓	↓	↓	↓	↓	↓	0 2	0 3	0 4	0 4	1 5	2 7	3 8	5 10	7 12	11 17	↑	↑	↑	↑	↑	↑	↑	↑	↑
	Second	50	100								0 2	0 4	0 5	1 6	4 7	6 9	8 12	12 16	18 22	26 30									
M	First	80	80	↓	↓	↓	↓	↓	↓	0 2	0 3	0 4	0 4	1 5	2 7	3 8	5 10	7 12	11 17	↑	↑	↑	↑	↑	↑	↑	↑	↑	↑
	Second	80	160							0 2	0 4	0 5	1 6	4 7	6 9	8 12	12 16	18 22	26 30										
N	First	125	125	↓	↓	↓	↓	↓	0 2	0 3	0 4	0 4	1 5	2 7	3 8	5 10	7 12	11 17	↑	↑	↑	↑	↑	↑	↑	↑	↑	↑	↑
	Second	125	250						0 2	0 4	0 5	1 6	4 7	6 9	8 12	12 16	18 22	26 30											
P	First	200	200	↓	↓	↓	↓	0 2	0 3	0 4	0 4	1 5	2 7	3 8	5 10	7 12	11 17	↑	↑	↑	↑	↑	↑	↑	↑	↑	↑	↑	↑
	Second	200	400					0 2	0 4	0 5	1 6	4 7	6 9	8 12	12 16	18 22	26 30												
Q	First	315	315	↓	↓	↓	0 2	0 3	0 4	0 4	1 5	2 7	3 8	5 10	7 12	11 17	↑	↑	↑	↑	↑	↑	↑	↑	↑	↑	↑	↑	↑
	Second	315	630				0 2	0 4	0 5	1 6	4 7	6 9	8 12	12 16	18 22	26 30													
R	First	500	500	↓	↓	0 2	0 3	0 4	0 4	1 5	2 7	3 8	5 10	7 12	11 17	↑	↑	↑	↑	↑	↑	↑	↑	↑	↑	↑	↑	↑	↑
	Second	500	1000			0 2	0 4	0 5	1 6	4 7	6 9	8 12	12 16	18 22	26 30														

⇩ = Use first sampling plan below arrow. If sample size equals or exceeds lot or batch size, do 100 percent inspection.

⇧ = Use first sampling plan above arrow.

Ac = Acceptance number.

Re = Rejection number.

• = Use corresponding single sampling plan (or alternatively, use double sampling plan below, when available).

† = If, after the second sample, the acceptance number has been exceeded, but the rejection number has not been reached, accept the lot, but reinstate normal inspection.

TABLE IV-A   Multiple Sampling Plans for Normal Inspection (Master Table)

Acceptable Quality Levels (normal inspection)

Sample size code letter	Sample	Sample size	Cumulative sample size	0.010	0.015	0.025	0.040	0.065	0.10	0.15	0.25	0.40	0.65	1.0	1.5	2.5	4.0	6.5	10	15	25	40	65	100	150	250	400	650	1000
A																												
B																												
C																												
D	First	2	2																										
	Second	2	4																										
	Third	2	6																										
	Fourth	2	8																										
	Fifth	2	10																										
	Sixth	2	12																										
	Seventh	2	14																										
E	First	3	3																										
	Second	3	6																										
	Third	3	9																										
	Fourth	3	12																										
	Fifth	3	15																										
	Sixth	3	18																										
	Seventh	3	21																										
F	First	5	5																										
	Second	5	10																										
	Third	5	15																										
	Fourth	5	20																										
	Fifth	5	25																										
	Sixth	5	30																										
	Seventh	5	35																										
G	First	8	8																										
	Second	8	16																										
	Third	8	24																										
	Fourth	8	32																										
	Fifth	8	40																										
	Sixth	8	48																										
	Seventh	8	56																										
H	First	13	13																										
	Second	13	26																										
	Third	13	39																										
	Fourth	13	52																										
	Fifth	13	65																										
	Sixth	13	78																										
	Seventh	13	91																										
J	First	20	20																										
	Second	20	40																										
	Third	20	60																										
	Fourth	20	80																										
	Fifth	20	100																										
	Sixth	20	120																										
	Seventh	20	140																										

⇩ = Use first sampling plan below arrow (refer to continuation of table on following page, when necessary). If sample size equals or exceeds lot or batch size, do 100 percent inspection.

⇧ = Use first sampling plan above arrow.

Ac = Acceptance number.

Re = Rejection number.

• = Use corresponding single sampling plan (or alternatively, use double sampling plan below, where available).

++ = Use corresponding double sampling plan (or alternatively, use multiple sampling plan below, where available).

" = Acceptance not permitted at this sample size.

487

TABLE V-A   Average Outgoing Quality Limit Factors for Normal Inspection (Single Sampling)*

Code Letter	Sample Size	0.010	0.015	0.025	0.040	0.065	0.10	0.15	0.25	0.40	0.65	1.0	1.5	2.5	4.0	6.5	10	15	25	40	65	100	150	250	400	650	1000
A	2															18											
B	3														12				42	69	97	160	220	330	470	730	1100
C	5													7.4			17	28	46	65	110	150	220	310	490	720	1100
D	8												4.6			11	17	27	39	63	90	130	190	290	430	660	
E	13											2.8			6.5	11	15	24	40	56	82	120	180	270	410		
F	20										1.8			4.2	6.9	9.7	16	24	34	50	72	110	170	250			
G	32									1.2			2.6	4.3	6.1	9.9	14	22	33	47	73						
H	50								0.74			1.7	2.7	3.9	6.3	9.0	13	21	29	46							
J	80							0.46			1.1	1.7	2.4	4.0	5.6	8.2	12	19	29								
K	125						0.29			0.67	1.1	1.6	2.5	3.6	5.2	7.5	12	18									
L	200					0.18			0.42	0.69	0.97	1.6	2.2	3.3	4.7	7.3											
M	315				0.12			0.27	0.44	0.62	1.00	1.4	2.1	3.0	4.7												
N	500			0.074			0.17	0.27	0.39	0.63	0.90	1.3	1.9	2.9													
P	800		0.046			0.11	0.17	0.24	0.40	0.56	0.82	1.2	1.8														
Q	1250	0.029			0.067	0.11	0.16	0.25	0.36	0.52	0.75	1.2															
R	2000			0.042	0.069	0.097	0.16	0.22	0.33	0.47	0.73																

Acceptable Quality Level

*Note: For the exact AOQL, the above volume must be multiplied by $\left(1 - \dfrac{\text{Sample size}}{\text{Lot or Batch size}}\right)$

TABLE VIII Limit Numbers for Reduced Inspection

Number of sample units from last 10 lots or batches	Acceptable Quality Level																									
	0.010	0.015	0.025	0.040	0.065	0.10	0.15	0.25	0.40	0.65	1.0	1.5	2.5	4.0	6.5	10	15	25	40	65	100	150	250	400	650	1000
20 - 29	•	•	•	•	•	•	•	•	•	•	•	•	•	•	•	0	0	2	4	8	14	22	40	68	115	181
30 - 49	•	•	•	•	•	•	•	•	•	•	•	•	•	•	0	0	1	3	7	13	22	36	63	105	178	277
50 - 79	•	•	•	•	•	•	•	•	•	•	•	•	•	0	0	2	3	7	14	25	40	63	110	181	301	
80 - 129	•	•	•	•	•	•	•	•	•	•	•	•	0	0	2	4	7	14	24	42	68	105	181	297		
130 - 199	•	•	•	•	•	•	•	•	•	•	•	0	0	2	4	7	13	25	42	72	115	177	301	490		
200 - 319	•	•	•	•	•	•	•	•	•	•	0	0	2	4	8	14	22	40	68	115	181	277	471			
320 - 499	•	•	•	•	•	•	•	•	•	0	0	1	4	8	14	24	39	68	113	169						
500 - 799	•	•	•	•	•	•	•	•	0	0	2	3	7	14	25	40	63	110	181							
800 - 1249	•	•	•	•	•	•	•	0	0	2	4	7	14	24	42	68	105	181								
1250 - 1999	•	•	•	•	•	•	0	0	2	4	7	13	24	40	69	110	169									
2000 - 3149	•	•	•	•	•	0	0	2	4	8	14	22	40	68	115	181										
3150 - 4999	•	•	•	•	0	0	1	4	8	14	24	38	67	111	186											
5000 - 7999	•	•	•	0	0	2	3	7	14	25	40	63	110	181												
8000 - 12499	•	•	0	0	2	4	7	14	24	42	68	105	181													
12500 - 19999	•	0	0	2	4	7	13	24	40	69	110	169														
20000 - 31499	0	0	2	4	8	14	22	40	68	115	181															
31500 & Over	0	1	4	8	14	24	38	67	111	186																

• Denotes that the number of sample units from the last tea lots or batches is not sufficient for reduced inspection for this AQL. In this instance more than tea lots or batches may be used for the calculation, provided that the lots or batches used are the most recent ones in sequence, that they have all been on normal inspection, and that none has been rejected while on original inspection.

TABLE IX   Average Sample Size Curves for Double and Multiple Sampling (Normal and Tightened
Inspection)

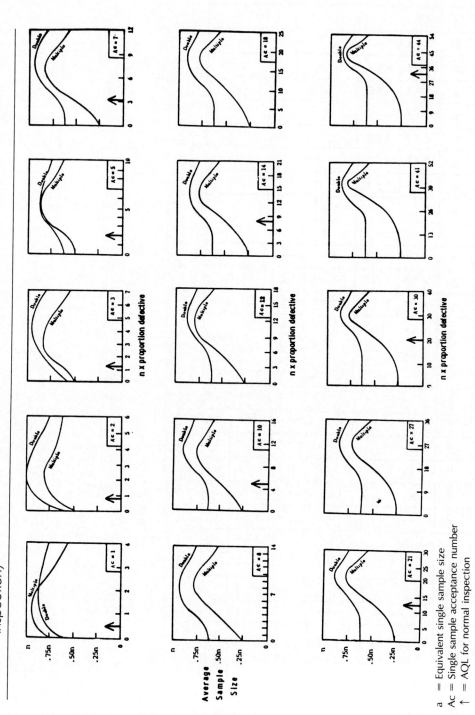

a = Equivalent single sample size
Ac = Single sample acceptance number
↑ = AQL for normal inspection

The basic property of a loss function is the assessment of a penalty for making products that deviate from a target (nominal) value T. Furthermore, this penalty increases directly with the distance between the measured characteristic, y, and the target, T. Finally, the loss is zero when $y = T$.

Suppose that $l(y)$ is a symmetric loss function (i.e., symmetric around the target value, T). That is, for any positive distance δ, $l(T + \delta) = l(T - \delta)$. Expanding $l(y)$ in a Taylor series around T, we find

$$l(y) = l(T) + l'(T)(y - T) + \frac{1}{2!}l''(T)(y - T)^2 + \frac{1}{3!}l'''(y - T)^3 + \cdots$$

Substituting the symmetry requirement $l(T + \delta) = l(T - \delta)$ into this expansion yields the equality

$$l(T + \delta) = l(T) + l'(T)(\delta) + \frac{1}{2!}l''(T)(\delta)^2 + \frac{1}{3!}l'''(\delta)^3 + \cdots$$

$$= l(T) + l'(T)(-\delta) + \frac{1}{2!}l''(T)(-\delta)^2 + \frac{1}{3!}l'''(-\delta)^3 + \cdots = l(T - \delta),$$

which reduces to

$$2l'(T)(\delta) + \frac{2}{3!}l'''(\delta)^3 + \frac{2}{5!}l'''''(\delta)^5 + \cdots = 0$$

Since this last equality must hold for any value of $\delta > 0$, and since the only way a power series can be identically zero over $\delta > 0$ is for all its coefficients to be zero, we have $l'(t) = 0$, $l'''(\delta)^3 = 0$, $l'''''(\delta)^5 = 0$, etc. Thus, any symmetric loss function must look like

$$l(y) = l(T) + \frac{1}{2!}l''(T)(y - T)^2 + \frac{1}{3!}l''''(y - T)^4 + \cdots$$

Furthermore, since the loss at $y = T$ is zero, we also have $l(T) = 0$, and the loss function becomes

$$l(y) = \frac{1}{2!}l''(T) \cdot (y - T)^2 + \frac{1}{4!}l''''(T) \cdot (y - T)^4 + \cdots$$

When all terms past the first one are dropped, the *simplest* symmetric loss function is one that looks like

$$l(y) = \frac{1}{2!}l''(T) \cdot (y - T)^2$$

which can be more compactly written $l(y) = k \cdot (y - T)^2$.

Name Index

Subject Index